Protocol Test Systems VIII

IFIP – The International Federation for Information Processing

IFIP was founded in 1960 under the auspices of UNESCO, following the First World Computer Congress held in Paris the previous year. An umbrella organization for societies working in information processing, IFIP's aim is two-fold: to support information processing within its member countries and to encourage technology transfer to developing nations. As its mission statement clearly states,

> IFIP's mission is to be the leading, truly international, apolitical organization which encourages and assists in the development, exploitation and application of information technology for the benefit of all people.

IFIP is a non-profitmaking organization, run almost solely by 2500 volunteers. It operates through a number of technical committees, which organize events and publications. IFIP's events range from an international congress to local seminars, but the most important are:

- the IFIP World Computer Congress, held every second year;
- open conferences;
- working conferences.

The flagship event is the IFIP World Computer Congress, at which both invited and contributed papers are presented. Contributed papers are rigorously refereed and the rejection rate is high.

As with the Congress, participation in the open conferences is open to all and papers may be invited or submitted. Again, submitted papers are stringently refereed.

The working conferences are structured differently. They are usually run by a working group and attendance is small and by invitation only. Their purpose is to create an atmosphere conducive to innovation and development. Refereeing is less rigorous and papers are subjected to extensive group discussion.

Publications arising from IFIP events vary. The papers presented at the IFIP World Computer Congress and at open conferences are published as conference proceedings, while the results of the working conferences are often published as collections of selected and edited papers.

Any national society whose primary activity is in information may apply to become a full member of IFIP, although full membership is restricted to one society per country. Full members are entitled to vote at the annual General Assembly, National societies preferring a less committed involvement may apply for associate or corresponding membership. Associate members enjoy the same benefits as full members, but without voting rights. Corresponding members are not represented in IFIP bodies. Affiliated membership is open to non-national societies, and individual and honorary membership schemes are also offered.

PREFACE

IWPTS'95 (International Workshop on Protocol Test Systems) is being held this year at INT (Institut National des Télécommunications), Evry, France, from 4 to 6 September, 1995. IWPTS'95 is the eighth of a series of annual meetings sponsored by the IFIP Working Group WG6.1 dedicated to "Architecture and Protocols for Computer Networks". The seven previous workshops were held in Vancouver (Canada, 1988), Berlin (Germany, 1989), Mclean (USA, 1990), Leidschendam (The Netherlands, 1991), Montréal (Canada, 1992), Pau (France, 1993) and Tokyo (Japan, 1994).

The workshop is a meeting place where both research and industry, theory and practice come together. By bringing both researchers and practitioners together, IWPTS opens up the communication between these groups. This helps keep the research vital and improves the state of the practitioner's art.

Forty-eight papers have been submitted to IWPTS'95 and all of them have been reviewed by the members of the Program Committee and additional reviewers. The completed reviewers list is included in this Proceedings. Based on these reviews, the Program Committee selected 26 for oral presentation and 4 to be presented as posters. Two specially invited papers complete the Workshop Program, which is composed of ten sessions : Testing Methods (Session 1), Test Environments (Session 2), Theoretical Framework (Session 3), Algorithms and Languages (Session 4), Test Generation 1 (Session 5), Testability (Session 6), Test Generation 2 (Session 7), Industrial Applications (Session 8), Distributed Testing and performance (Session 9) and Test Management (Session 10).

IWPTS'95 includes a special session organized by the project COST 247, a European Research Action on verification and validation methods for formal descriptions. The special session is focused on the work developed on conformance testing by some of the participants to this project.

IWPTS'95 has received financial support from the European Commission to help researchers and students from Central and Eastern European countries to participate in the workshop. These stipends cover registration fees and/or subsistence and/or travel expenses.

IWPTS'95 could not take place without the effort of a great many individuals and organizations. The editors wish to thank all of them. In particular, we would like to thank our colleagues of INT: Michel Andrieu, Yves Dumont, Marie-Laure Feral, Hacene Fouchal, Maria Guilbert, Barbara Huc, Toma Macavei, Michel Marty, Luiz Paula-Lima, Louis Rambaud, Serge Robinson, Jocelyne Vallet.

Evry, September 1995

Ana Cavalli
Stan Budkowski

PROGRAM COMMITTEE

Bernd Baumgarten, GMD-Darmstadt, Germany

Gregor von Bochmann, Université de Montreal, Canada (confirme)

Ed Brinskma, University of Twente, The Netherlands

Richard Castanet, Universite de Bordeaux, France

Samuel Chanson, University of HongKong, HongKong (confirme)

B. Chin, ETRI, Korea (confirme)

Anton Dahbura, Motorola, USA

Rachida Dssouli, Université de Montreal, Canada

Jean-Philippe Favreau, NIST, USA

Roland Groz, CNET, France

Teruo Higashino, University of Osaka, Japan (confirmé)

Dieter Hogrefe, University of Bern, Switzerland

Sung-Un Kim, Korea Telecom, Korea

Jan Kroon, PTT Research, The Netherlands

Gang Luo, University of Ottawa

Jan de Meer, GMD-Fokus, Germany (confirme)

Raymond E. Miller, University of Maryland, USA

Jose Manas, Technical University of Madrid, Spain

Tadanori Mizuno, University Shizuoka, Japan

Alexandre Petrenko, IECS, Latvia (confirme)

Marc Phalippou, CNET, France (confirme)

Omar Rafiq, Université de Pau, France (confirme)

Pierre de Saqui-Sannes, ENSICA, France (confirme)

Behcet Sarikaya, The University of Aizu, Japan

Nori Shiratori, Tohoku University, Japan

Katie Tarnay, KFKI, Hongary

Jan Tretmans, ERCIM, The Netherlands (confirme)

Hasan Ural, University of Ottawa, Canada

Son T. Vuong, University of British Columbia, Canada (confirme)

Jianping Wu, Tsinghua University, China

LIST OF REFEREES

B. Algayres
R. Anido
B. Baumgarten
O. Bellal
B. Bista B.
G.v. Bochmann
L. Boullier
E. Brinksma
R. Castanet
S.T. Chanson
O. Charles
B. Chin
M. Clatin
G. Csopaki
A. Dahbura
P. De Saqui-Sannes
K. Drira
R. Dssouli
H. Eertink
A. Ezust
J-P. Favreau
D. de Frutos
A. Giessler
K. Go
G. Grabowski
R. Groz
S. Guyot
L. Heerink
O. Henniger
M. Higuchi
G. Huecas
K. Kasama
A. Khoumsi
I. Khriss
G. Leduc
G. Luo
P. Maigron
J.A. Manas
J. de Meer
J. Miskolczi
M. Mori
A. Nakata
N. Okazaki
L.A. de Paula Lima
A. Peeters
A. Petrenko
M. Phalippou
O. Rafiq
A. Rennoch
N. Risser
R. Roth
H. Rudin
I. Sanz
B. Sarikaya
F. Sato
R. Scheurer
G. Schoemakers
I. Schubert
K. Tarnay
M.J.A. Tesselaar
J. Tretmans
A. Ulrich
H. Ural
M.U. Uyar
M. Van Essen
G. Vermeer
S. Vuong
H. Wiland
K. Yasumoto
N. Yevtushenko
S. Zhang
J. Zhu

PART ONE

Testing Methods

INVITED LECTURE

1
Testing Through the Ages

A. T. Dahbura
Motorola Cambridge Research Center
One Kendall Square, Building 200
Cambridge, MA 02139
atd@mcrc.mot.com

Abstract

This paper examines the testing challenges humans have faced through recorded history and discusses why systems and testing methods fail. The paper places in perspective the testing issues which are likely to be encountered over the next several years relative to the gains made thus far.

Keywords

Testing, failures, risks, conformance testing, requirements validation, verification.

"*What we anticipate seldom occurs; what we least expect generally happens.*"

- Henrietta Temple.

1 Introduction

Throughout history, humans have created increasingly complex systems that put themselves and their environment at risk. In fact, human-made catastrophes appear to have increased in frequency and magnitude with industrialization (Perrow, 1984). Early designers were able to more readily learn from their mistakes, leading to wondrous advances in architecture, transportation, healthcare, microelectronics, and so on. Unfortunately, the growing complexity of present-day systems and the critical application areas in which they are used has made it more difficult to produce the technologically-advanced marvels that society relies upon and even expects without also introducing the possibility of catastrophic failures.

F.G. Juenger referred to failures as "resistances", or obstacles which must be encountered during the design and implementation of any major technological achievement. He observed that these resistances are never overcome, but are merely subdued, "watching in ambush, forever ready to burst into destruction" (Juenger, 1949). Failures include not only those which affect the intended users and bystanders, but also the producers of the system, where a flaw in a product leads to massive recalls, lost profit, loss of reputation, and so on.

Testing, the process of checking that a system possesses a set of desired properties and/or behaviors, has become an integral part of the invention, production, and operation of systems in order to reduce the risk of catastrophic failures. In many cases the testing methods used are implicit in the design process. Typically, design and test have been treated as discrete components of a system's life-cycle; only recently have engineers started viewing the two as being intrinsically and inexorably related.

In this paper, we consider different types of complex systems built through history and examine why they, and their design and test methodologies, have succeeded or failed. A characterization scheme is given for systems during their life-cycle and also for the different classes of tests which must be used for different purposes at each stage of the life-cycle. The major challenges for researchers in the field of testing are described.

2 "Normal" Accidents

Charles Perrow has written one of the most insightful treatises on failures and their causes (Perrow, 1984). He refers to what he terms the "interactive complexity" of a system, and argues that accidents are caused by the way failures interact and the way systems are tied together; two or more failures interact in some unexpected way. Perrow refers to this characteristic of the system, which is not due directly to a part or an operator, as a "normal accident" or "system accident": given the system characteristics, multiple and unexpected interactions of failures are inevitable.

Normal accidents are an inherent property of the system and do not have to be expected or occur frequently. Furthermore, normal accidents are usually incomprehensible by those involved for some critical period of time.

Normal accidents are most prevalent when the system is tightly coupled, that is, processes occur in quick succession. A disturbance propagates quickly and irretrievably, and operator actions or safety systems may even make it worse. Often, portions of the disturbance are masked by other portions.

Perrow characterizes the components of normal accidents as DEPOSE: Design, Equipment, Procedures, Operators (and/or organizations), Supplies/material, and Environment. Usually, several of these factors are contributing causes to an incident becoming an accident.

Complex interactions, those of unfamiliar, unplanned, or unexpected sequences and not visible nor immediately comprehensible, are more common causes of normal accidents than so-called linear interactions, which are those which are expected in a familiar sequence.

For example, on May 25, 1979, an American Airlines DC-10 crashed upon take-off from O'Hare International Airport in Chicago, killing 273 passengers and crew. It was later determined that one of its three engines tore off due to an engine pylon failure, probably because of poor maintenance practices. In spite of this, the DC-10 is designed to fly with only two of its engines operational. Unfortunately, as the engine broke away it severed control cables in the wing, forcing the leading edge slats on one wing to retract. Although the plane is designed to be able to fly in this mode, hydraulic lines in the wing were also severed, preventing appropriate warning indicators in the cockpit to activate; since the pilots were totally unaware of this unexpected state of the aircraft, they were unable to take measures to prevent the tragedy (Perrow, 1984). In this example almost all of the DEPOSE factors (with the possible exception of the environment) played a role in the bizarre sequence of events that led to the disaster.

3 The Phases of a System's Life-Cycle

A system can be thought of as undergoing the following phases from concept to implementation to operation (see Figure 1):

- *requirements phase*: the definition of the functions the system performs and the properties it possesses. Usually, these are described at a very high-level; for example, "the boat must float", "the protocol must not contain deadlocks", "two trains should never collide", etc.
- *specification phases*: these are the various stages of system design, typically starting with a high-level specification, to successively lower-level designs which lead to one which is directly implementable. For instance, in VLSI design, an early specification consists of a block diagram of the architecture. Later phases include RTL-, gate-, and transistor-level specifications, which are then used to produce masks and ultimately, wafers.
- *prototyping phase*: the design is implemented via a model or via simulation, primarily for the purpose of checking that it meets its requirements or conforms to some level of its specification.
- *production phase*: the design is implemented (manufactured) for the purpose of delivery to the end-user. In many cases the implementation is replicated multiple times (mass production).
- *operational phase*: the period during which the system is carrying out the functions it was designed and built to perform.

The requirements phase defines the "what"; the specification phase defines the "how". Ideally, the phases of a systems life-cycle should be discrete; however, it is often the case that some of the phases are merged together or bypassed entirely. This is especially true when modifications are made over time to a system when it is already in its operational phase, and is a major cause of failures, as will be illustrated in the following sections.

4 Classes of Testing

There are several different kinds of testing which are used at the different phases of a system's life-cycle (see Figure 1) and depending on whether a given phase of the system is to be checked with the requirements or with one of the specification levels as the reference.

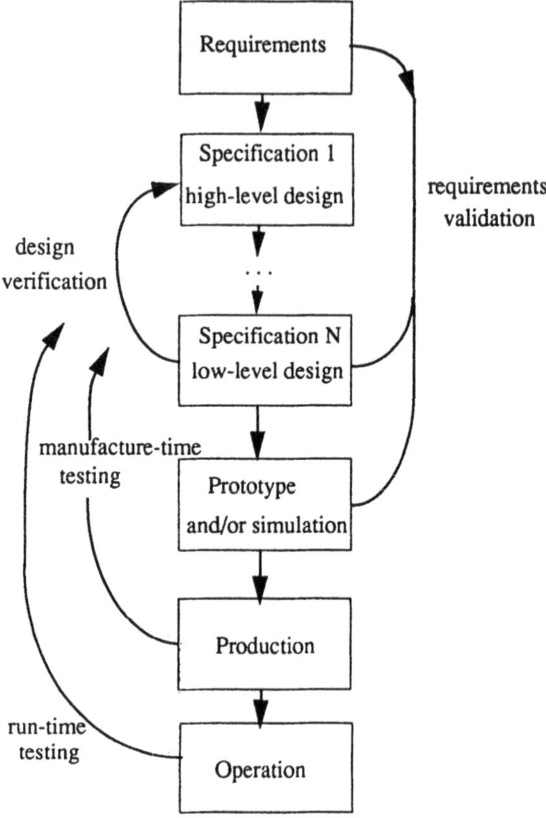

FIGURE 1. The phases of a system's life-cycle and the different classes of testing.

4.1 Checking that the requirements are met

Requirements validation is the process of checking that a system possesses the desired properties and performs its intended function (for instance, is the airplane capable of flying- see Figure 2).

4.1.1 Requirements validation from a specification

Requirements validation can be performed on a paper design which is close to the actual implementation of the system, or on any later phase. Within this classification scheme, what has traditionally been called *protocol verification* is a special case of requirements validation since, typically, the goal is to detect undesirable properties of the protocol design such as deadlocks and livelocks.

FIGURE 2. An 1876 aircraft design by W.J. Lewis of New York (Moolman, 1980).

4.1.2 Requirements validation from a prototype

The functions of a system can also be validated from its prototype, especially in the case of mechanical systems where the system must interact with its environment in order to validate it with confidence.

Prototypes are often built and validated when some degree of invention or innovation is involved in the system. There are few documented accounts of failed prototyping experiences, perhaps since it is more natural to boast about the successes. However, throughout the course of history, perhaps no experience in design validation from the prototype phase is as celebrated and popularized as the early experiments in aviation. Many of these validation exercises ended in failure, and in many cases the inventors themselves were the ones to test their own contraptions, sometimes with tragic results.

Some clever inventors of the era learned very quickly how to turn failure into success. Hiram S. Maxim, an American who later became a naturalized Briton, was one of the many participants in the frenzied race of the late 1800's to build a flying machine. Maxim's work was unique at the time because he wanted to build aircraft that could support large payloads. He opted for steam power and tubular steel construction. He spent seven years performing simulations in a wind tunnel, and on July 31, 1984 he conducted the first actual test of the machine. The test rig consisted of a rail system which was designed to prevent the machine from lifting more than .75m above the ground; however, the rig broke during the test run, leading to the outcome shown in Figure 3. In spite of the accident, there was evidence that the machine had 'flown', and the inventor claimed that "it was the first time that a powered flying machine had actually lifted itself, and its crew, into the air". In fact, Maxim was later knighted for his accomplishment (Andrews, 1977).

FIGURE 3. H.S. Maxim after the trial of his flying machine (Andrews, 1977).

4.1.3 Requirements validation from a manufactured and/or operational system

Sometimes, the checks performed on a system during the production and operational phases of its life-cycle are functional in flavor, especially in electronic systems. A common form of run-time requirements validation is a form of fault detection called *sanity checking* (Kraft and Toy, 1981). For example, a sanity check which could be used in banking systems is to ensure that an account never has a negative balance. Such checks are a powerful means for detecting software errors.

4.2 Checking that the specifications are met

Design verification is the process of checking that the different stages and/or levels of design or implementation meet a given higher level of specification.

4.2.1 Design verification from a lower-level specification

In many design processes with multiple phases of specification, it is common to check that a lower-level paper design conforms to a higher-level design. In the VLSI design area, this is usually referred to as *hardware verification.* For example, a great deal of effort is spent checking that the gate-level design of a circuit conforms to its functional-level design.

4.2.2 Design verification from a prototype

Once a design is implemented, it is likely to lose a vast degree of observability and controllability of its internal states and signals. In the protocol arena, the problem of verifying an implementation of a system to its specification is known as *protocol conformance testing (*Dahbura, Sabnani, and Uyar, 1990*)*. Note that in conformance testing, the actual function of the system is largely irrelevant; the goal is to ensure that the externally observable behavior of the implementation conforms to its high-level specification such as a finite-state machine description. If the finite-state machine contains errors from the requirments to the specification they may not be detected; however, the observer of the test process may notice that a behavior is abnormal, such as a telephone system which handles calls improperly.

4.2.3 Design verification from a manufactured system

In the *manufacture-time testing* of mass-produced devices such as integrated circuits, the amount of testing which can be economically performed is likely to be limited. Therefore, it is generally assumed that conformance testing is more thorough and is aimed at detecting design faults while manufacture-time testing is for detecting defects in a given instantiation of the implementation. Also, conformance testing is usually performed based on a higher-level specification than manufacture-time testing. For instance, the specification used for conformance testing could be a finite-state machine, while the traditional specification used for manufacture-time testing is at the gate level.

4.2.4 Design verification from an operational system

Testing during the run time of an operational system checks that the instantiation of the system continues to conform to its specification during its operation. Such checks can be performed on- or off-line. If an abnormality is detected, often there are safety systems built in to the system to diagnose, confine, and mask the fault and to reconfigure, recover, and repair the system (Siewiorek and Swarz, 1982). Usually the operational demands placed on a system during its operation limit the rigor of the tests.

4.3 Faulty Requirements

Even an ideal design and implementation of a system cannot be expected to perform properly if they are based on a flawed set of requirements. A requirement can be flawed if:

- it does not anticipate an input (such as a lightning strike) or a sequence of inputs;
- it contains conflicting requirements which are resolved in an imperfect manner in the system design and implementation.

A flawed set of requirements is the Achilles' heel of a system, since even perfect testing throughout the system's life-cycle cannot prevent a tragedy, as will be seen below.

FIGURE 4. The Dale Dyke dam immediately after the disaster of 1864 (Smith, 1972).

5 Case Studies in Failures

Given the phases of a system's life-cycle and the types of tests for each, let us examine case studies of some well-documented failures of complex systems to better understand the reasons for failure and the measures, if any, which could have been taken to prevent them.

5.1 The Bradfield Disaster of 1864

On March 11, 1864, the outlet valves of the Dale Dyke dam, which had been built in 1858 near the town of Sheffield, were closed for the first time to raise the water level in the Bradfield reservoir. That evening, a crack developed and the dam gave way (see Figure 4), setting loose an estimated 200 million gallons of water. Over 250 people were killed. Later analysis determined that the likely cause of the failure was due to the settlement of the heavy clay, shale, and rubble bank around the outlet pipes, causing a gradual erosion and weakening of the interior of the dam. The so-called Bradfield disaster raised the awareness in Great Britain about the considerable social responsibility the dam-builder has (Smith, 1972).

The water level at the time of the breach was below the maximum level provided for in the requirements. The high-level specification may have underestimated the amount and quality of material needed for such a dam, and poor design of the outlet pipes structure could have contributed to the problem; proper design validation could have detected this, although the understanding of dam construction was still limited at that time. On the other hand, it is possible that shoddy materials which did not meet the specification were called for by the detailed design; this could have been questioned by design verification. In the event that the materials that were used were inferior to those called for in the specification, the best hope of detection in this instance was at manufacture time, since the structure was sealed after that and run-time testing (inspection) would have been nearly impossible.

5.2 Chunnel ghost trains

Earlier this year, it was reported that trains in the "Chunnel", the tunnel beneath the English Channel, were forced to make emergency stops approximately five times per week due to spurious emergency signals caused by the salt-water mist raised by the trains which short-circuited sensors in the track, thereby mimicking the presence of a train. The effect of salt water on electronic equipment is well-known yet was overlooked in this instance (Wodehouse, 1995).

In this case, it is likely that the requirements did not take into account the salt-water mist as an input to the system (unexpected input). If they did, the next likely scenario is that the sensor which was used did not meet the requirements for salt-water tolerance; the fault could have been introduced in the specification process, at prototype-time, manufacture-time, or at run-time. Appropriate requirements validation, design verification, conformance testing or manufacture-time testing could have detected the problem.

5.3 New York City subway crash

On June 5, 1995, a subway train in New York City crashed into the rear of another train on the Williamsburg Bridge, killing the motorman and injuring 54 people. The cause, as reported by the New York Times, was that the distance between signals, which was set in 1918, is shorter than the stopping distance of today's longer heavier, and faster trains. The trains were upgraded without a corresponding modification in the control system (Stalzer, 1995).

There are two ways to view this system. If the trains are viewed as external inputs to the control system then the system was given new, unexpected inputs (the newer trains), for which it could not adequately respond. On the other hand, if the trains are to be viewed as being part of the system then part of the system requirements changed over time: more passengers per hour must be transported, although the requirement that there be no collisions remained throughout. As the operational system was transformed over time, the requirements validation process broke down and two conflicting system properties (fast trains, too short stopping distances) were introduced.

5.4 The cruise ship Royal Majesty

On June 10, 1995, the cruise ship Royal Majesty became grounded off the coast of the island of Nantucket, Massachusetts. The National Transportation Safety Board found that a cracked housing of an antenna for the ship's Global Positioning System (GPS) caused the antenna to fail to receive satellite information regarding the ship's position. An alarm went off and the autopilot system reverted to dead reckoning (which is less accurate) for its navigation; the alarm was so faint and poorly positioned that no one noticed, and the ship went off course. The accident could have been averted by someone simply looking out the window (Arnold, 1995).

Although one run-time testing mechanism successfully detected the failure of the antenna, the design and/or implementation failed to meet the requirement that the ship's crew be adequately

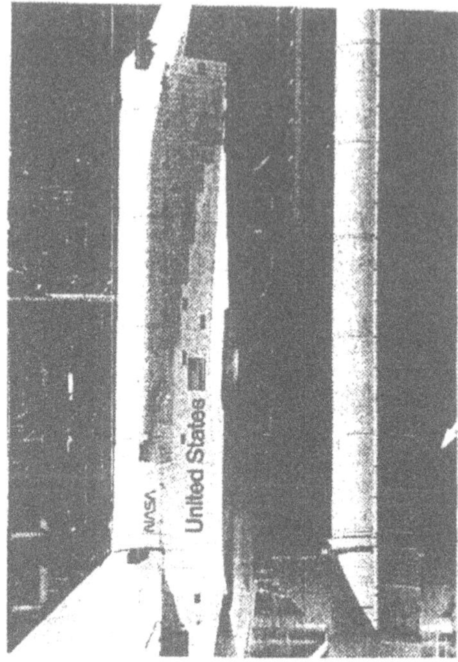

FIGURE 5. The Space Shuttle Challenger upon its fateful lift-off (United States, 1986).

warned in the event of a failure. Another run-time test, a sanity check to be performed by the crew to make sure that the ship was not too close to land, was not performed. The inadequate warning system could have been detected by appropriate requirements validation and design verification, including conformance testing to check that the alarm was audible and visible as designed.

5.5 Space Shuttle Challenger disaster

The explosion aboard the Space Shuttle Challenger is one of the most notable technological failures in modern times. On January 18, 1986, flight 51-L of the Challenger began after several launch delays. A mere 73 seconds into flight, a chain reaction of explosions enveloped the Challenger in flames and it broke into several large pieces which plummeted into the Atlantic Ocean. Seven astronauts perished. A Presidential Commission investigating the accident concluded that the cause was the "failure of the pressure seal in the aft field joint of the right Solid Rocket Motor, due to a design unacceptably sensitive to temperature, the effects of reusability, and the reaction of the joint to dynamic loading." (See Figure 5.) Also, the Commission concluded that the decision to launch was flawed since those making the key decisions were unaware of the history of problems concerning the joints; apparently engineers at the company which produced the joints opposed the design but were overruled by management (United States, 1986).

In this unfortunate situation, the specification of the pressure seal (commonly known as an O-ring) did not meet the requirements from the start. In this case, it is likely that requirements validation detected the problem but organizational issues prevented appropriate action. Run-time tests were unable to detect the failure in time to avert disaster.

5.6 The Crash of the DC-10

In the example of a catastrophic failure described in Section 2 of an American Airlines DC-10 on take-off from O'Hare International Airport, one of the early causes of the chain reaction of events was a poor run-time off-line test (in this case, inspection of the engine pylon during maintenance). However, the reaction of the system to the event of the failed pylon contributed in a major way to the crash. The requirements for the DC-10 had provided for the loss of an engine and for the loss of control of the leading edge slats on one wing, but had not explicitly required that the warning system continue to be operational in the event of simultaneous failures. This scenario was not tested (requirements validation) during any of the design or implementation phases of the aircraft.

6 Testing challenges for the future

Thus far, we have discussed the reasons why systems fail and the different classes of tests which can reduce the risk of failure if used appropriately. In some sense, the science of testing is in a race with the science of system design: as system designs introduce new challenges, the techniques for testing them must be able to keep pace.

6.1 The system complexity battle

The most notable challenge faced by testing is the ever-growing system complexity, from microprocessors to nuclear power plants. Designs are accelerated by more powerful description techniques which are able to compress the specification but, unfortunately, merely obscure the internal state of the system, making testing more difficult. In the case of software systems, designers herald specification languages which require fewer lines of code since (assuming that programmers write a constant number of lines of code per day independently of the language) the belief is that the system will be implemented in less time. Furthermore, because commonly used types of software testing claim to be effective in proportion to the size of the code, there is a common, but false, belief that more compressed programs are also more easily testable!

Complex systems which strain the limits of testing bring about the need for new and automated means of modeling the relevant portions of the system to be tested so that the unavoidable trade-offs can be more rationally made. For instance, a telephone could be modeled at the level of its electronic components (with an exponential number of states), or it could simply be modeled in terms of its function: the ability to make and receive a call. Depending on the circumstances, each view has its own set of benefits and risks which need to be exposed and understood.

Finally, it would be fruitful to investigate the impact of different fault models on the effectiveness of testing. In integrated circuit testing, the single stuck-at fault model has been used for years to generate test vectors (Siewiorek and Swarz, 1982). Although VLSI technology has changed radically over time and single stuck-at faults are not the prevalent fault in digital microcircuits, the stuck-at fault model is still used because:

- the number of single stuck-at faults to be tested for and consequently the number of generated test vectors are proportional to the number of gates in the circuit, which is marginally acceptable, and
- it is generally believed (although there is surprisingly little supporting evidence!) that a test which can detect a high percentage of stuck-at faults will also tend to detect a high percentage of actual faults in the manufactured circuits.

Discovering both enumerable and relevant fault models for different system domains could help to make the test generation process and the test sequence length more manageable.

6.2 Design and test; design for testability

The notion that design and test are inseparable tasks in system development is increasing in popularity but has not yet matured. In particular, the concept of *design for testability (DFT)* has taken hold in the VLSI area but is only nascent in other areas such as software. Researchers have started to develop ideas for DFT in software-based systems, such as improving the controllability and observability and eliminating unneeded state space. Automated tools for assisting implementers in improving these properties in their designs would be extremely useful.

7 A System Success Story: the Great Pyramid

Perhaps the most successful system story of all is also one of the oldest: that of the pyramids built by the Egyptians in ancient times.

August Mencken describes the process of designing and building the Great Pyramid of Gizeh (see Figure 6) and other Egyptian monuments around 2500 B.C. (Mencken, 1963). He points out that probable reasons for their incredible longevity include:

- *simple design and conservative structure*: it is likely that the Egyptians knew that the structures could be made much larger and more complex, although they would be less stable;
- *overengineered design*, perhaps to make up for limited analytical skills and tools;
- *careful measurement*: the Egyptians were almost fanatical about mathematically precise measurements and relationships among them and the heavenly bodies;
- *finite construction window*: pyramids were tombs for the Pharaohs, so their construction span was planned to be about 20 years, the expected duration of the Pharaoh's reign;
- *evolutionary design*: the designs changed very little over time;
- *sufficient manpower* assigned to the task: historians speculate that one of the purposes of the pyramids was to keep large segments of the population employed;
- *no major modifications to the implementation*: once the Pyramids were built, they were left, for the most part, unaltered.

FIGURE 6. The Great Pyramid at Gizeh (Mencken, 1963).

- *luck*: although some parts have been damaged by severe earthquakes and others have settled, there has been no major environmental disaster in the area.

While not all of the pyramids in Egypt have survived, the design principles used are equally applicable today.

8 Conclusions

F.G. Juenger said that "to think in terms of causes, effects, and purposes means to think one-sidedly. To see things in their whole context cannot be learned, no more than one can learn rhythm...Correlations and contexts are noticed only by those minds which think in universal and reverent terms..." (Juenger, 1949).

In this paper, we have argued that, while "normal accidents" have complex causes and are ultimately inevitable, the effective use of the several different classes of tests described earlier can go far in reducing the number of catastrophic system failures which occur. Testing has multiple facets during a system's life-cycle. It is the professional and even the social responsibility of the system designers to keep the role of testing in its proper context and to use the power of testing to its fullest to effectively reduce risk.

9 References

Andrews, A. (1977). *Back to the drawing board: the evolution of flying machines*. David and Charles, Ltd., London.

Arnold, D. (1995). U.S. points to faulty alarm in groundings; ships warned, in *The Boston Globe*, August 11, 1995.

Dahbura, A.T., Sabnani, K.K., and Uyar, M.U. (1990). Formal methods for generating conformance test sequences, in *Proceedings of the IEEE*, vol. 78, no. 8, pp. 1317-1326.

Juenger, F.G. (1949). *The failure of technology: perfection without purpose*. H. Regnery Co., Hinsdale, IL.

Kraft, G.D. and Toy, W.N. (1981). *Microprogrammed control and reliable design of computers*. Prentice-Hall, Inc., New York.

Mencken, A. (1963). *Designing and building the Great Pyramid*. Privately printed, 1963.

Moolman, V. (1980). *The road to Kitty Hawk*. Time-Life Books, Alexandria, VA.

Perrow, C. (1984). *Normal accidents: living with high-risk technology*. Basic Books, New York.

Siewiorek, D., and Swarz, B. (1982). *The Theory and Practice of Reliable System Design*. Digital Press, New Bedford, MA.

Smith, N.A.F. (1972). *A history of dams*. Citadel Press, Secaucus, NJ.

Stalzer, M. (1995) Re: the New York City subway crash, in *The Risks Digest* (P.G. Neumann, ed.), vol. 17, issue 19, June 19, 1995.

United States- Presidential Commission on the Space Shuttle Challenger Accident (1986). *Report to the President*, vol. 1. The Commission, Washington, D.C.

Wodehouse, J. (1995) Chunnel has ghost trains, in *The Risks Digest* (P.G. Neumann, ed.), vol. 17, issue 3, April 4, 1995.

10 Biography

Anton T. Dahbura received the BSEE, MSEE, and Ph.D. in Electrical Engineering and Computer Science from the Johns Hopkins University in 1981, 1982, and 1983, respectively.

In 1983 he joined the Computing Systems Research Laboratory at AT&T Bell Laboratories, Murray Hill, NJ, as a Member of Technical Staff. During his tenure at Bell Labs, he conducted research on fault detection and diagnosis algorithms for multiprocessor systems, conformance test sequence generation methodologies for communication protocols, algorithms for memory reconfiguration, and yield enhancement techniques for laser-programmable logic arrays.

The testing method for communications protocols which Dr. Dahbura co-pioneered at AT&T Bell Laboratories, called POSTMAN, has been used extensively by AT&T and others for certification of protocols and product testing, and has become part of a CCITT standard.

In 1990, Dr. Dahbura became the Research Director of the Motorola Cambridge Research Center in Cambridge, Massachusetts. His current responsibilities include management of research activities in the area of parallel and distributed computing systems. His current research interests include dependable and mobile computing, combinatorial optimization algorithms, and applications of the NII.

In 1993, Dr. Dahbura was awarded the IEEE Browder J. Thompson Memorial Prize Award for outstanding paper in any IEEE publication by an author under 30 years of age.

Dr. Dahbura is a Senior Member of the IEEE Computer Society and is a member of the ACM (SIGACT). He is also a member of IFIP Working Group 10.4 (Dependable Computing and Fault Tolerance). In 1988 he was an Invited Lecturer in the Department of Computer Science at Princeton University. Currently, he is a Research Affiliate of MIT's Laboratory for Computer Science.

Since 1991 he has served as an Editor for the IEEE Transactions on Computers. Since 1989 he has served as an Associate Editor for the Journal of Circuits, Systems, and Computers. He has served on the Program Committee of the International Symposium on Fault Tolerant Computing (FTCS) in 1988, 1989, and 1993 and served as Program Co-Chairman of FTCS-24 in 1994. He is the author of over 50 technical papers and holds three U.S. patents.

2

An Executable Protocol Test Sequence Generation Method for EFSM-specified Protocols *

Chung-Ming Huang, Yuan-Chuen Lin and Ming-Yuhe Jang
Laboratory of Computer-Aided Protocol Engineering (LOCAPE)
Institute of Information Engineering
National Cheng Kung University
Tainan, Taiwan 70101
R.O.C.
Telephone: 011-886-6-2757575 ext 62523
Fax: 011-886-6-2747076
E-mail: huangcm@locust.iie.ncku.edu.tw

Abstract

In this paper, we propose a method which can generate executable Extended Finite State Machine (EFSM)-based test sequences for data flow protocol test. In EFSM-specified protocols, the switch sequences can decide the executability of DO-paths. Based on the proposed Transition Executability Analysis (TEA) technique, executable switching sequences, executable DO-paths, and executable back paths, which are from the tail states of DO-paths to the initial state, can be derived. Then the complete executable data flow test sequence for an EFSM-specified protocol can be generated by concatenating the associated executable switching sequences, DO-paths, and back paths.

Keywords

Protocol Testing, Extended Finite State Machines (EFSMs), Data Flow Testing, (Executable) Test Sequence Generation.

1 Introduction

Protocol test sequence generation methods for Extended Finite State Machine (EFSM)-specified protocols are different from those for FSM-specified protocols, because of the existence of context variables and predicates [Chanson 93, Koh 94, Miller 92, Sarikaya 87, Ural 91]. One of the key issues in EFSM-based protocol test sequence generation is the executability problem. In [Chanson 93], Chanson and Zhu applied Constraint Satisfaction Problem (CSP) solving techniques, which essentially use transition self-loop analysis, to solve the test sequence executability problem. The principles of Chanson and Zhu's

*The research is supported by the National Science Council of the Republic of China under the grant NSC 84-2213-E-006-035.

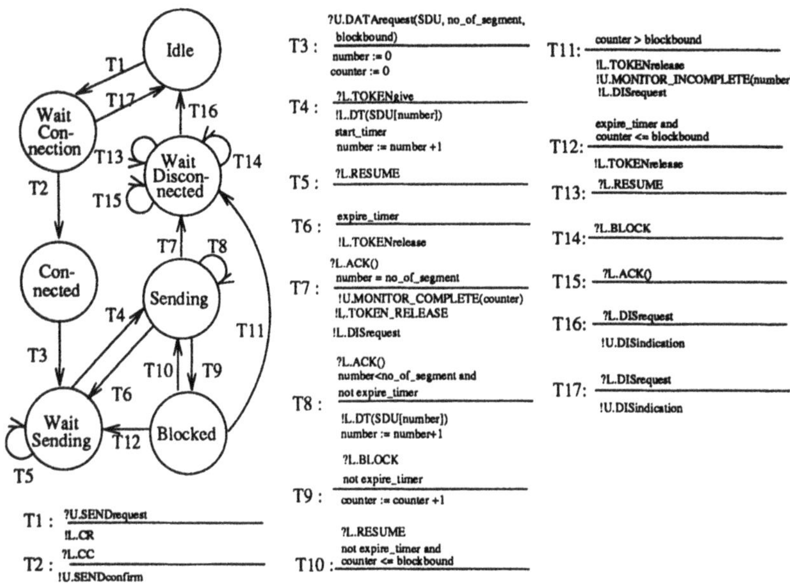

Figure 1: An EFSM-specified protocol.

method is as follows: Find all paths[1] that connect variable's definitions and uses (or outputs) at the first phase. If some paths are unexecutable, then add some self-loops to make these paths executable. However, not all unexecutable transitions can find self-loops to resolve the executability problem. An example can be found in the protocol depicted in Figure 1[2]. In Figure 1, the transition sequence (T_4, T_9, T_{11}) is an unexecutable sequence, in which variable *number* has a definition use in T_4 and has an output use in T_{11}. To have T_{11} be executable, transitions T_9 and T_{10} need to be executed twice. Thus, the final executable transition sequence from T_4 to T_{11} is (T_4, T_9, T_{10}, T_9, T_{10}, T_9, T_{11}). However, transitions T_9 and T_{10} are not self-loops of state *Blocked.*

In this paper, we propose a new executable test sequence generation method, which covers the data flow test of EFSM-specified protocols, using the Transition Executability Analysis (TEA) technique. Our method concatenates executable transitions during the derivation process for a test sequence. Thus, each executable test sequence, which contains a definition use of variable V and the corresponding output use of variable V, can be derived. Based on the initial status of an EFSM, in which the EFSM is in its initial state and all of the context variables are in their initial values, our method has three steps: (1) Explore an executable (switch) sequence TS_s whose tail transition T_d contains a definition use of a variable V; (2) derive each executable sequence TS_o rooted from the tail state of T_d, such that the tail transition T_o of TS_o contains an output use of V and there is no re-definition use of V in TS_o; and (3) explore an executable sequence TS_b rooted from the tail state of T_o, such that TS_b's tail state is the initial state of the EFSM, for each TS_o.

[1]For convenience, a path denotes a transition sequence in the following presentation.

[2]For simplicity, the initial value of each variable is 0, variables *SDU*, *no-of-segment* and *blockbound* are assigned as 2 respectively in transition T_3. The upper bounds of *number* and *counter* are 2 and 3 respectively.

In this way, an executable test sequence can be derived by concatenating the associated TS_s, TS_o and TS_b. The complete data flow test sequence for an EFSM-specified protocol is generated by concatenating these executable test sequences.

The rest of this paper is organized as follows: In Section 2, we introduce the formal EFSM protocol specification model, and some concepts and definitions for the data flow protocol test. In Section 3, our executable test sequence generation method is presented in detail. In Section 4, we have conclusion remarks.

2 Background

In this section, we introduce the formal EFSM model that is used in our method at first. Then, some concepts and related definitions for the data flow protocol test are given.

2.1 The Formal EFSM Model

An EFSM is formally represented as a seven-tuple $< \sum, S, s_o, V, P, A, F >$, where (1) $\sum$ is the set of messages that can be sent or received, (2) S is the set of states, (3) s_o is the initial state, (4) V is the set of context variables, (5) P is the set of predicates that operate on context variables, (6) A is the set of actions that operate on context variables, (7) F is the set of state transition functions, in which each state transition function can be formally represented as follows: $S \times \sum \times P(V) \rightarrow S \times \sum \times A(V)$.

For convenience, each state transition is represents as $S_1 \xrightarrow{T} S_2$, where S_1 (S_2) is called the head (tail) state of transition T, T is called the incoming (outgoing) transition of state S_2 (S_1). Each transition contains two parts, i.e., the condition part and the action part. The condition part can contain an input event and/or a predicate. An input event is represented as "?ID.mess", in which message *mess* is input from entity *ID*. The predicate is a boolean expression that operates on context variables and parameters of the input messages. The action part can contain output events and a number of statements that operate on context variables. An output event is represented as "!ID.mess", in which message *mess* is output to entity *ID*. A transition can be fired when the condition part is satisfied. When a transition T is fired, the corresponding action part is executed, and the EFSM's state is switched from T's head state to T's tail state. An EFSM-specified network management (monitor) protocol is depicted in Figure 1, in which a circle represents a state and an arrow represents a transition.

Using our method, each EFSM is assumed to be a normalized EFSM. That is, there is no conditional statements, e.g., **if-then-else**, **case**, **while** loop, **for** loop, etc, in the action part of a transition. For the details of Normal Form Specifications (NFSs) for EFSMs, please refer the paper in [Sarikaya 87].

2.2 Some Concepts for the Data Flow Test

The data flow test focuses on how variables are bound to values, and how these variables are to be used. In the process of protocol testing, an Implementation Under Test (IUT) is always regarded as a black box. In order to observe how the variables are to be used, only those paths, which contain variable's definition uses and those output interactions that are influenced by these definitions, i.e., all-definition-output-paths (all-do-paths) are tested. For convenience, the following definitions are used in the protocol test sequence generation that follows the all-do-paths criterion.

Definition 1: (state configuration)
A *state* (context variables) *configuration* $S_i(V_{i_1}, V_{i_2}, ..., V_{i_n})$, which is abbreviated as S_i,

Set / Transition	A-use	I-use	P-use	O-use
T1				
T2				
T3	number counter	SDU no_of_segment blockbound		
T4	number			SDU number
T5				
T6				
T7			number no_of_segment	counter
T8	number		number no_of_segment	SDU number
T9	counter			
T10			counter blockbound	
T11			counter blockbound	number
T12			counter blockbound	
T13				
T14				
T15				
T16				
T17				

(a)

Define Element (variable, transition_id)	Executable Switching Sequence (ESS)
(number, 3)	(T1.T2.T3)
(counter, 3)	(T1.T2.T3)
(SDU, 3)	(T1.T2.T3)
(no_of_segment, 3)	(T1.T2.T3)
(blockbound, 3)	(T1.T2.T3)
(number, 4)	(T1.T2.T3,T4)
(number, 4)	(T1.T2.T3,T4,T6,T4)
(number, 4)	(T1.T2.T3.T4.T9.T12.T4)
(number, 4)	(T1.T2.T3,T4,T9,T10,T9,T12,T4)
(number, 8)	(T1.T2.T3.T4.T8)
(number, 8)	(T1.T2.T3.T4.T9.T10.T8)
(number, 8)	(T1.T2.T3.T4.T9.T10.T9.T10.T8)
(counter, 9)	(T1.T2.T3,T4,T9)
(counter, 9)	(T1.T2.T3,T4,T8,T9)
(counter, 9)	(T1.T2.T3,T4,T9,T10,T9)
(counter, 9)	(T1.T2.T3,T4,T8,T9,T10,T9)
(counter, 9)	(T1.T2.T3,T4,T9,T10,T9,T10,T9)
(counter, 9)	(T1.T2.T3,T4,T8,T9,T10,T9,T10,T9)

(b)

Table 1: (a) The sets of A-use, I-use, P-use and O-use in each transition; (b) the ESSs for each A-use and I-use.

of an EFSM is a snapshot of the EFSM such that the EFSM is at state S_i, and the values of variables V_1, V_2 ,......, and V_n of the EFSM are V_{i_1}, V_{i_2},......,V_{i_n} respectively.

Definition 2: (reachable state configuration)
A state configuration $S_i(V_{i_1}, V_{i_2},......,V_{i_n})$ is a reachable state configuration if there is an executable transition path $(t_1,t_2,...,t_m)$, such that transition t_1's head state configuration is the initial state configuration and $S_i(V_{i_1}, V_{i_2},......,V_{i_n})$ is transition t_m's tail state configuration.

Definition 3: (A-use, I-use, P-use, O-use)
The use of a variable x is an assignment-use (*A-use*) in a transition if x appears at the left hand side of an assignment statement of the transition. When a variable x appears in the input portion of a transition, the use of x is an input-use (*I-use*) in the transition. When a variable x appears in the predicate expression of a transition, the use of x is a predicate-use (*P-use*) in the transition. When a variable x appears in an output statement of a transition, the use of x is an output-use (*O-use*) in the transition.

Both *A-use* and *I-use* are called definition-use (*D-use*). Table 1-(a) shows the uses of all variables and the associated transition identifiers in the protocol depicted in Figure 1.

Definition 4: (global use, local use)
The use of a variable x in a transition is a *global use* when x's *D-use* occurs in some other transitions, otherwise the use of x is a *local use*.

Definition 5: (global def, local def)
The definition of a variable x in a transition is a *global def* if there is a global use of x in some other transitions, otherwise the definition of x is a *local def*.

Definition 6: (def-clear path)
A path $(t_1,...,t_n)$ is a *def-clear* path with respect to a variable x if transitions $t_2,...,t_{n-1}$ do not contain any *D-use* of x.

Definition 7: (DO-path)
A path $(t_1,...,t_n)$ is a *DO-path* with respect to a variable x if t_1 has a *global def* of x and

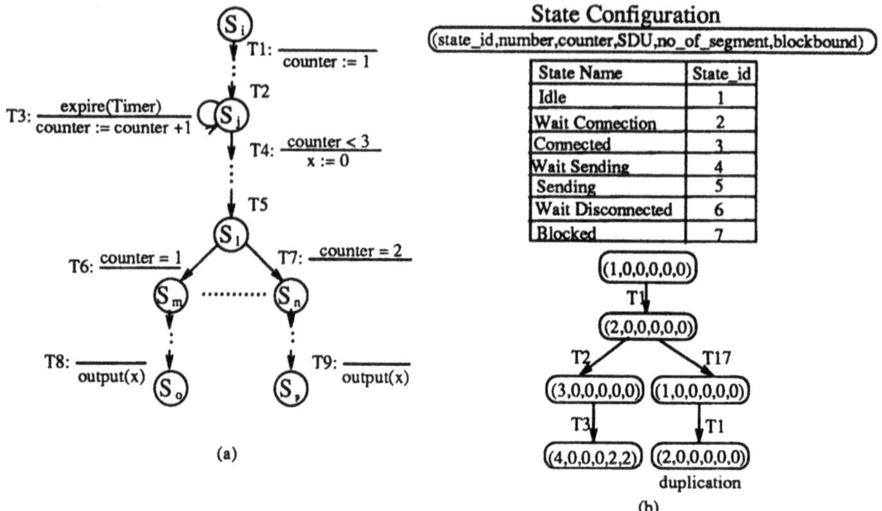

State Name	State_id
Idle	1
Wait Connection	2
Connected	3
Wait Sending	4
Sending	5
Wait Disconnected	6
Blocked	7

Figure 2: (a) Association between ESSs and DO-paths; (b) the TEA tree used for detecting the ESSs with respect to the A-use of variable *number* in transition T_3.

either:
(1) t_n has an O-use of x and the path is *def-clear* path with respect to x, or
(2) t_k contain a *P-use* of x, where $2 \leq k \leq n-1$, the path $t_1,..., t_k,...,t_{n-1}$ is a *def-clear* path with respect to x, such that (i) t_n has an *O-use* of any variable, and (ii) the execution of t_n depends on the execution of t_k, i.e., t_n indirectly depends on the predicate of t_k.

Definition 8: (IS-connected)
An EFSM M is *IS-connected*, where IS represents the initial state of M, if each reachable state configuration S has at least one executable path T, such that S is T's head state and IS in T's tail state.

Definition 9: (bounded)
An EFSM M is bounded, if for each variable x in M, x's legal values is bounded in a finite set of bound(x), i.e., the number of x's legal values are finite.

Definition 10: (TEA tree)
The Transition Executability Analysis (TEA) of an EFSM M generates an expanding tree rooted from a given state configuration of M, such that (1) each circle in the tree represents a reachable state configuration, (2) each arc in the tree represents an executable transition originated from the corresponding head state configuration.

3 Test Sequence Generation for the EFSM Model

In this section, we present our executable test sequences generation method for data flow test. For an EFSM-specified protocol M, such that M is (1) IS-connected and (2) bounded, the generated executable test sequences (ETSs) cover all of the executable DO-paths which exist in M using our method.

Our method contains three main steps. In the first step, executable switch sequences (ESSs), which can reach DO-paths, are generated by expanding a TEA tree rooted from the EFSM's initial state configuration. In the second step, all of the executable DO-paths (EDO-paths), which are generated by expanding TEA trees rooted from the state configurations of the tail states of ESSs, are selected. When all of the EDO-paths have been selected, the third step is invoked. In the third step, an executable back path from each EDO-path's tail state to the initial state of the EFSM is explored. An executable back path for an EDO-path P is derived by expanding a TEA tree rooted from the tail state of P.

Before presenting the derivation procedures in detail, some notations are introduced to simplify the presentation. Let A and B be two transition sequences. Then, the concatenation C of A and B is represented as C = A @ B. TSP1 and TSP2 are temporary state pools storing existing state configurations. IS is the initial state. DE, which identifies the definition use, i.e., A-use or I-use, of a variable in a transition, contains two attributes: (1) Var: the variable identifier, and (2) Tid: the identifier of the transition which contains the definition use of the associated variable *Var*. A-use(T), I-use(T), P-use(T) and O-use(T) are sets, which identify variables that belong to A-use, I-use, P-use and O-use in transition T, respectively. Each element in A-use(T), I-use(T), P-use(T) and O-use(T) has the similar attributes as that of DE. SE, which denotes a state element in TSPi, i=1 or 2, contains three attributes: (1) State: the state identifier and the associated context variables' values, i.e., the state configuration, (2) Level: the level (number) of the associated state configuration (of the SE) in the TEA tree, in which the level of the root state configuration is 1, (3) Path: the path from the root state configuration to the associated state configuration (of the SE) in the TEA tree. Function tailstate(TT) returns the tail state configuration of the transition sequence TT. Function exist(P, PP) returns a path TP from path pool PP such that path P is a postfix sequence of TP; otherwise, an infinitely long sequence, which is represented by the symbol ∞, is returned. Function shorter(P, T) returns the shorter path of path P and path T.

3.1 The Derivation of Executable Switch Sequences (ESSs)

In order to test DO-paths of an EFSM M with respect to a variable x which has either A-use or I-use in a transition T_i, executable switch sequences which are originated from the initial state of M to the tail state of T_i need to be detected. For an A-use or I-use of a variable x in a transition, there may be more than one DO-path and there may be more than one ESS. The key issue is that these DO-paths may need different ESSs to be executable. An example is given in Figure 2-(a). In Figure 2-(a), S_i is the initial state, an A-use of x is in T_4. In order to test the data flow with respect to x in T_4, an ESS must be applied so that T_4 can be executable. In Figure 2, there are two ESSs which can reach T_4, one is (T_1,..., T_2,T_4) and the other one is (T_1,...,T_2,T_3,T_4) if the bound of variable *counter* is restricted from 1 to 2. There are two DO-paths with respect to x in T_4, one is (T_4,...,T_5,T_6,...,T_8) and the other one is (T_4,...,T_5,T_7,...,T_9). The DO-path (T_4,...,T_5,T_6,...,T_8) can be executable only when the ESS (T_1,...,T_2,T_4) is selected, and the DO-path (T_4,...,T_5,T_7,...,T_9) can be executable only when the ESS (T_1,...,T_2,T_3,T_4) is selected. If only one of the above two ESSs is used to search for these two DO-paths, one of the two DO-paths will be lost. Thus, for an A-use or I-use of a variable x in a transition T_i, all of the possible ESSs, which result in different T_i's head state configurations, need to be detected. The following procedure generates all of the ESSs for a DE.

```
Procedure ESS(IS, DE)
root.State := IS
root.Level := 1
root.Path := ∅
TSP1, TSP2 := ∅
ESS-pool := ∅
add root to TSP1 and TSP2 /* TSP2 is used for duplication check */
```

```
L := 1
while TSP1 is not empty do
  while TSP1 contains elements whose Levels are L do
    remove an SE from TSP1 such that SE.Level = L
    for each executable outgoing transition Ti of SE, such that (tailstate(Ti) is not in TSP2
      and all variables' values in tailstate(Ti) are in their bounds) or (tailstate(Ti) is a
      duplication of a state S that is stored in TSP2 and none of S's incoming transitions
      is Ti) do
      if DE ∈ A-use(Ti) or DE ∈ I-use(Ti)
      then
        ESS := SE.Path @ Ti
        add the ESS to ESS-pool
      endif
      tailstate(Ti).Level := L+1
      tailstate(Ti).Path := SE.Path @ Ti
      add tailstate(Ti) to TSP1 and TSP2
    endfor
  endwhile
  L := L+1
endwhile /* end of procedure */
```

The procedure of ESS(IS, DE) searches for the ESSs of the states whose outgoing transitions contain some definition uses of some variables. Initially, the ESS(IS, DE) procedure sets the initial state IS as the root of the TEA tree. Then, expand the TEA tree one state by one state and one level by one level, i.e., expand the TEA tree using the Bread-First-Search (BFS) way, until the TSP1 pool becomes empty, i.e., all of the possible ESSs have been explored for the definition use of DE.Var in DE.Tid. The *for* loop explores all of executable outgoing transitions of a state element. An outgoing transition T_i is checked whether (1) T_i's tail state configuration doesn't exist in the TEA tree, where TSP2 stores all of currently existing state configurations in the TEA tree, and all of the variables' values are in their bounded values after T_i having been executed, or (2) T_i's tail state configuration S has been reached previously in the TEA tree, but the previous S is not generated by executing T_i. If the answer is negative, then stop expanding this path; otherwise check whether A-use(T_i) or I-use(T_i) contains the DE element. If DE is not contained in A-use(T_i) or I-use(T_i), then continue to expand the TEA tree; otherwise, the path from root to T_i's head state is an executable switching sequence to be obtained. For a transition T_i that has an A-use or I-use of variable x, and a transition T_o that has an O-use of variable x, there may be two executable switching sequences, ESS_l and ESS_s, such that (1) T_i is the tail transition of both ESS_l and ESS_s, (2) the length of ESS_l is greater than that of ESS_s, (3) the tail state configuration of ESS_l is different from that of ESS_s, and (4) the length from ESS_l to T_o is shorter than that from ESS_s to T_o. That is, a longer executable switching sequence may result in a shorter executable DO-path. Thus, the tail state of T_i, which contains an A-use or I-use of a variable x, still needs to be added in TSP1 to explore the other possible executable switching sequences. Repeat the above procedure, until all of the ESSs have been detected. Figure 2-(b) shows the TEA tree used for detecting the ESS with respect to the A-use of variable *number* in transition T_3. Table 1-(b) shows all of the ESSs for each A-use or I-use of the protocol depicted in Figure 1.

Theorem 1:
The TEA tree's state space for the ESS(IS, DE) procedure is finite. That is, the ESS(IS, DE) procedure can be terminated in finite steps.

Proof:
Since all variables' values are bounded, the number of possibly generated state configurations is GSC=$|S| * |V_1| * ... * |V_n|$, where $|S|$ is the number of states, and $|V_i|$ is the number

of legal values of variable V_i, i=1,...,n. Let k=maximum(out(S_1), out(S_2),...,out($S_{|S|}$)), where out(S_i) is the number of state S_i's outgoing transitions and i=1,...,|S|. Hence, the state space of the TEA tree is $GSC * k$. In other words, the state space of the TEA tree is finite, i.e., the expanding of a TEA tree can be terminated in finite steps.

3.2 The Derivation of Executable DO-paths (EDO-paths)

For an ESS, which contains an A-use or I-use of a variable x in ESS's tail transition T_i, all of the DO-paths with respect to x need to be checked based on ESS's tail state configuration. The DO-paths to be obtained have two types: (1) Let T_{i+n} have an O-use of x, and T_i,...,T_{i+n} be a def-clear path with respect to x and be executable. The path (T_i,...,T_{i+n}) is a DO-path of type 1. In this condition, x can be directly observed using the output event in T_{i+n}. (2) Let T_{i+n} have a P-use of x and T_i,...,T_{i+n} be a def-clear path with respect to x and be executable, and a nearest O-use be in T_{i+n+k} such that the path (T_i,...,T_{i+n},...,T_{i+n+k}) be executable. The path (T_i,...,T_{i+n},...,T_{i+n+k}) is a DO-path of type 2. In this condition, the effect of the P-use of x can be observed indirectly by observing the output event of T_{i+n+k}.

```
Procedure EDO-path-search (ESS, x)
root.State := tailstate(ESS)
root.Level := 1
root.Path := ∅
TSP1, TSP2 := ∅
add root to TSP1 and TSP2 /* TSP2 is used for duplication check */
L := 1
while TSP1 is not empty do
  while TSP1 contains elements whose Levels are L do
    remove an SE from TSP1, such that SE.Level = L
    for each executable outgoing transition Ti of SE, such that all variables' values in tailstate(Ti)
      are in their bounds do
      Case 1: x ∈ O-use(Ti) and x ∉ I-use(Ti) and x ∉ A-use(Ti)
        DO-path := SE.Path @ Ti
        EDO-path := shorter(ESS @ DO-path, exist(DO-path, EDO-path-pool))
        add EDO-path to EDO-path-pool
        expand-or-discard(tailstate(Ti))
      Case 2: x ∈ O-use(Ti) and x ∉ I-use(Ti) and x ∈ A-use(Ti) and (the execution
        of the statement containing O-use of x is before that of the statement containing A-use of x)
        DO-path := SE.Path @ Ti
        EDO-path := shorter(ESS @ DO-path, exist(DO-path, EDO-path-pool))
        add EDO-path to EDO-path-pool
      Case 3: x ∈ P-use(Ti) and x ∉ I-use(Ti)
        tailstate(Ti).Path := SE.Path @ Ti
        if O-use(Ti) is not empty
        then
          DPO-path = tailstate(Ti).Path
          EDO-path := shorter(ESS @ DPO-path, exist(DPO-path, EDO-path-pool))
          add EDO-path to EDO-path-pool
        else
          DPO-path-search(tailstate(Ti))
        endif
        if x ∉ A-use(Ti)
        then
          expand-or-discard(tailstate(Ti))
        endif
```

```
      Case 4: x ∉ I-use(Ti) and x ∉ P-use(Ti) and x ∉ A-use(Ti) and x ∉ O-use(Ti)
        expand-or-discard(tailstate(Ti))
      Case 5: otherwise, stop expanding this path
    endfor
  endwhile
  L := L+1
endwhile /* end of procedure */

Procedure expand-or-discard(S)
if S is not a duplication of any state configuration in TSP2
then
  S.Path := SE.Path @ Ti
  S.Level := L+1
  add S to TSP1 and TSP2
endif /* end of procedure */

Procedure DPO-path-search(S)
root.State := S
root.Level := 1
TSP := ∅
add root to TSP
L := 1
while TSP is not empty do
  while TSP contains elements whose Levels are L do
    remove an SE from TSP, such that SE.Level = L
    for each executable outgoing transition Tj of SE do
      if O-use(Tj) is not empty
      then
        DPO-path := SE.Path @ Tj
        EDO-path := shorter(ESS @ DPO-path, exist(DPO-path, EDO-path-pool))
        add EDO-path to EDO-path-pool
        exit
      else
        tailstate(Tj).Path := SE.Path @ Tj
        add tailstate(Tj) to TSP
      endif
    endfor
  endwhile
  L := L+1
endwhile /* end of procedure */
```

For a given variable x, and an ESS in which ESS's tail transition has a definition use, i.e., A-use or I-use, of x, the procedure of EDO-path-search(ESS, x) searches for the EDO-paths using the Bread-First-Search (BFS) way. Initially, the tail state of the ESS is set as the root of the TEA tree. All of the state elements in the TEA tree is explored one state by one state and one level by one level. Each executable outgoing transition T_i, in which all of the variables' values are in their bounds after executing T_i, of a selected state configuration in TSP1 is checked. The use of variable x in a transition T_i can be classified into five cases:

- Case 1: T_i has no re-definition use but has an O-use of x. In this case, the path which is from ESS's tail state to the tail state of T_i is a DO-path with respect to x. The DO-path is appended with the ESS to form an EDO-path. However, there may exist some EDO-paths which have the same DO-path but with different ESSs. A shorter EDO-path with respect to the same DO-path should be selected.

Functions *exist* and *shorter* are called to have the shorter EDO-path with respect to the DO-path. When an EDO-path (T_1,...,T_i,...,T_{k_1}), which contains a D-use in T_i and an O-use (or P-use) in T_{k_1} of a variable V, is detected, it is still possible that there is an EDO-path (T_1,...,T_i,...,T_{k_1},...,T_{k_2}), such that T_{k_2} has an O-use (or P-use) of V. Therefore, the tail state configuration of a newly detected EDO-path still needs to have further exploration. The expand-or-discard(tailstate(T_i)) procedure is called to decide whether continue to expand the path or not. The expand-or-discard(tailstate(T_i)) procedure checks whether the tail state S of a path P is a duplication or not. If the answer is positive, then stop expanding path P; otherwise, add S to TSP1 and TSP2 to further expand path P.

- Case 2: T_i has no I-use of x, but has an A-use and an O-use of x, and the execution of the statement containing O-use of x is before that of the statement containing A-use of x in T_i. In this case, the DO-path is valid. Thus, the corresponding actions are the same as that of Case 1, except the expanding of the path is stop at the tail state of T_i, because there is a re-definition use of x in T_i.
- Case 3: T_i has no I-use but has a P-use of x. In this case, if T_i has an output event, then the path which is from ESS's tail state to the tail state of T_i is a DO-path of type 2 with respect to x; otherwise, the DPO-path-search(tailstate(T_i)) procedure is called to search for the DO-path of type 2 with respect to x. In the procedure of the DPO-path-search(tailstate(T_i)), a new TEA tree is expanded until a transition T_j containing an output event is reached[3]. When the DO-path is obtained, the DO-path is appended with the ESS to form the EDO-path. Functions *exist* and *shorter* are called to have the shorter EDO-path with respect to the DO-path. Then, if T_i does not have an A-use of x, the expand-or-discard(tailstate(T_i)) procedure is called to decide whether continue to expand the path or not; otherwise, if T_i has an A-use of x, stop expanding this path.
- Case 4: variable x is not used in T_i. In this case, the expend-or-discard(tailstate(T_i)) procedure is called to decide whether continue to expand this path or not.
- Case 5: The use of variable x is not in above cases, i.e., there is a re-definition use in T_i, stop expanding this path.

Repeat the above steps, until all EDO-paths with respect to x have been detected. Figure 3 shows the TEA tree used for detecting DO-paths with respect to the A-use of variable *number* in transition T_3. Table 2 shows all of the EDO-paths of the protocol depicted in Figure 1.

Theorem 2:
The TEA tree's state space for EDO-path-search(ESS, x) is finite. That is, the EDO-path-search(ESS, x) procedure can be terminated in finite steps.

Proof:
The proof of this theorem is the same as that for **Theorem 1**.

3.3 The Derivation of Executable Back Paths

When an EDO-path is derived, the back path from the tail state of the EDO-path to the initial state IS needs to be generated, such that IUT can be brought to the initial state to have the next test sequence. The following EBP-path-search procedure can search for the back path for an EDO-path.

---

[3]Since the EFSM is IS-connected, an output event can be detected eventually.

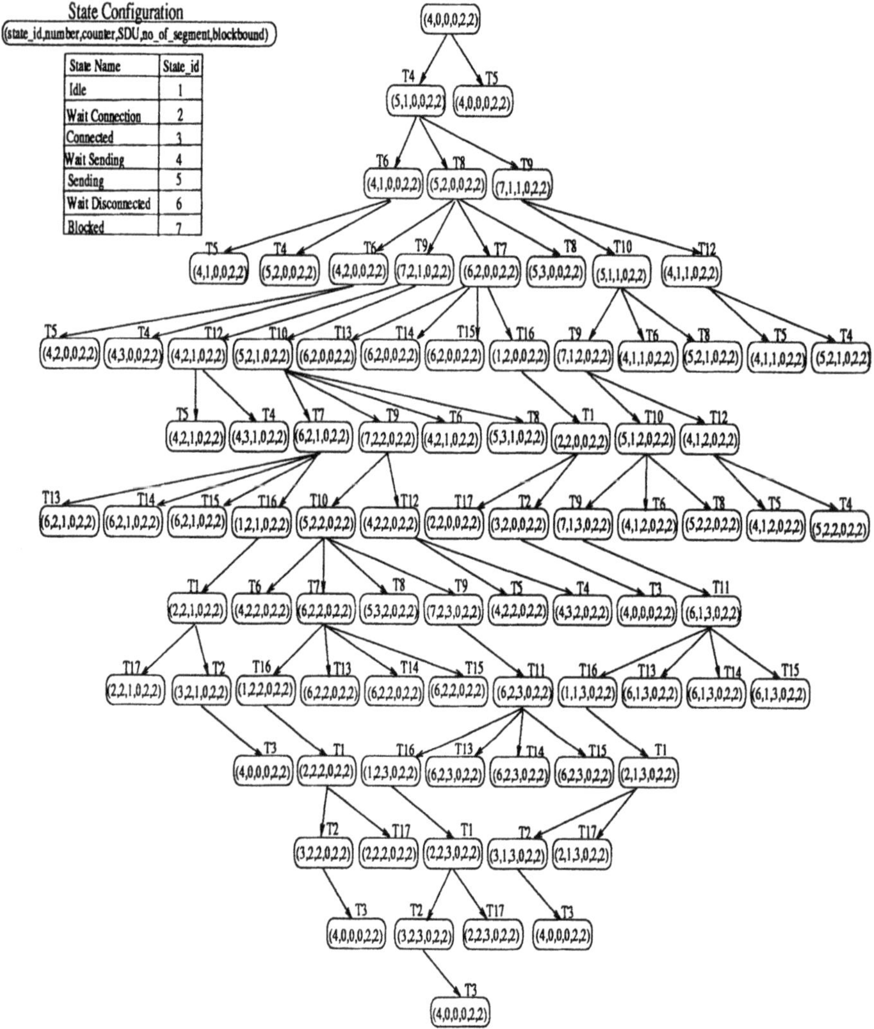

Figure 3: The TEA tree used for detecting DO-paths with respect to the A-use of variable *number* in transiton T_3.

```
Procedure EBP-path-search(EDO)
root.State := tailstate(EDO)
root.Level := 1
root.Path := EDO
TSP := ∅
add root to TSP
L := 1
while TSP is not empty do
  while TSP contains elements whose Levels are L do
    remove an SE from TSP, such that SE.Level = L
    for each executable outgoing transition T_i of SE do
      if tailstate(T_i) = IS
      then
        EBP-path := SE.Path @ T_i
        add EBP-path to EBP-path-pool
        exit
      else
        if tailstate(T_i) is not in TSP
        then
          tailstate(T_i).Path := SE.Path @ T_i
          add tailstate(T_i) to TSP
        endif
      endif
    endfor
  endwhile
  L := L+1
endwhile /* end of procedure */
```

Starting from the tail state of an EDO-path, the procedure of EBP-path-search(EDO) expands one state by one state and one level by one level. If the initial state is not reached at level L, level L+1 is explored. The procedure is continued until the initial state of the EFSM is reached[4]. In this way, a complete executable test sequence with respect to a DO-path is derived.

Theorem 3:
The TEA tree's state space for the EBP-path-search(EDO) procedure is finite. That is, the EBP-path-search(EDO) procedure can be terminated in finite steps.

Proof:
Since the EFSM is *IS-connected*, there is an EBP-path for each EDO-path according to **Definition 7**. Using the same calculation as that in the *proof* for **Theorem 1** (and 2), the EBP-path can be detected in finite steps.

3.4 Generation of Executable Test Sequences (GETS)

Using the ESS(IS, DE), EDO-path-search(ESS, x), and EBP-path-search(EDO) procedures, all of the executable test sequences covering all of the executable DO-paths can be generated. The executable test sequence generation algorithm is as follows:

Algorithm Generating-Executable-Test-Sequence (GETS)

[4] For simplicity, it is assumed that all variable values can be reset to their initial values respectively. It can be achieved as follows. For each $S \xrightarrow{T} S_{init}$, in which S_{init} is the initial state and T is an incoming transition of S_{init}, a split state S' and a reset transition T_r are added, such that $S \xrightarrow{T} S' \xrightarrow{T_r} S_{init}$. Each variable is reset to its initial value in T_r. Otherwise, the EBP-path-search procedure keeps expanding until an ESS's head state configuration is reached.

Define-Element (variable, transition_id)	ESS	TYPE (DO/DPO)	(EDO_PATH; #PATH_ID)
(number, 3)	T1,T2,T3	DO	(ESS,T4 ; #1), (ESS,T4,T6,T4 ; #2), (ESS,T4,T8 ; #3), (ESS,T4,T8,T9,T10,T9,T10,T9,T11 ; #4), (ESS,T4,T9,T12,T4 ; #5), (ESS,T4,T9,T10,T8; #6), (ESS,T4,T9,T10,T9,T12,T4 ; #7), (ESS,T4,T9,T10,T9,T10,T9,T11 ; #8).
		DPO	(ESS,T4,T8,T9,T10,T7 ; #9), (ESS,T4,T8,T9,T10,T9,T10,T7 ; #10), (ESS,T4,T8,T7 ; #11).
(counter, 3)	T1,T2,T3	DO	(ESS,T4,T8,T7 ; #12).
		DPO	(ESS,T4,T8,T9,T10,T6 ; #13), (ESS,T4,T8,T9,T10,T7 ; #14), (ESS,T4,T8,T9,T10,T9,T10,T6 ; #15), (ESS,T4,T8,T9,T10,T9,T10,T7 ; #16), (ESS,T4,T8,T9,T10,T9,T10,T9,T11 ; #17), (ESS,T4,T8,T9,T10,T9,T12 ; #18), (ESS,T4,T9,T10,T6 ; #19), (ESS,T4,T9,T10,T8 ; #20), (ESS,T4,T9,T10,T9,T10,T9,T11 ; #21). (ESS,T4,T9,T10,T9,T12 ; #22), (ESS,T4,T9,T12 ; #23), (ESS,T4,T9,T10,T9,T10,T6 ; #24), (ESS,T4,T9,T10,T9,T10,T8 ; #25).
(SDU, 3)	T1,T2,T3	DO	(ESS,T4 ; #26), (ESS,T4,T6,T4 ; #27), (ESS,T4,T8 ; #28), (ESS,T4,T9,T12,T4 ; #29), (ESS,T4,T9,T10,T8 ; #30), (ESS,T4,T9,T10,T9,T10,T8 ; #31), (ESS,T4,T9,T10,T9,T12,T4 ; #32).
(no_of_segment, 3)	T1,T2,T3	DPO	(ESS,T4,T8 ; #33), (ESS,T4,T8,T9,T10,T7 ; #34), (ESS,T4,T8,T9,T10,T9,T10,T7 ; #35), (ESS,T4,T8,T7 ; #36).
(blockbound, 3)	T1,T2,T3	DPO	(ESS,T4,T8,T9,T10,T6 ; #37), (ESS,T4,T8,T9,T10,T7 ; #38), (ESS,T4,T8,T9,T10,T9,T10,T6 ; #39), (ESS,T4,T8,T9,T10,T9,T10,T7 ; #40), (ESS,T4,T8,T9,T10,T9,T10,T9,T11 ; #41), (ESS,T4,T8,T9,T10,T9,T12 ; #42), (ESS,T4,T9,T10,T9,T12 ; #43), (ESS,T4,T9,T10,T8 ; #44), (ESS,T4,T9,T10,T6 ; #45), (ESS,T4,T9,T10,T9,T10,T6 ; #46), (ESS,T4,T9,T10,T9,T10,T8 ; #47), (ESS,T4,T9,T10,T9,T10,T9,T11 ; #48), (ESS,T4,T9,T12 ; #49).
(number, 4)	T1,T2,T3, T4	DO	(ESS,T6,T4 ; #50), (ESS,T8 ; #51), (ESS,T8,T9,T10,T9,T10,T9,T11 ; #52), (ESS,T9,T12,T4 ; #53), (ESS,T9,T10,T8 ; #54), (ESS,T9,T10,T9,T12,T4 ; #55), (ESS,T9,T10,T9,T10,T9,T11; #56).
		DPO	(ESS,T8,T9,T10,T7 ; #57), (ESS,T8,T9,T10,T9,T10,T7 ; #58), (ESS,T8,T7 ; #59).
	T1,T2,T3, T4,T6,T4	DO	(ESS,T9,T10,T9,T10,T9,T11 ; #60).
		DPO	(ESS,T7 ; #61), (ESS,T9,T10,T7 ; #62), (ESS,T9,T10,T9,T10,T7 ; #63).
	T1,T2,T3, T4,T9,T12, T4	DO	(ESS,T9,T10,T9,T11; #64).
		DPO	(ESS,T7 ; #65), (ESS,T9,T10,T7 ; #66).
	T1,T2,T3, T4,T9,T10, T9,T12,T4	DO	(ESS,T9,T11 ; #67).
		DPO	(ESS,T7 ; #68).
(number, 8)	T1,T2,T3, T4,T8	DO	(ESS,T9,T10,T9,T10,T9,T11; #69).
		DPO	(ESS,T7 ; #70), (ESS,T9,T10,T7 ; #71), (ESS,T9,T10,T9,T10,T7 ; #72).
	T1,T2,T3,T4, T9,T10,T8	DO	(ESS,T9,T10,T9,T11 ; #73).
		DPO	(ESS,T7 ; #74), (ESS,T9,T10,T7 ; #75)
	T1,T2,T3,T4, T9,T10,T9, T10,T8	DO	(ESS,T9,T11 ; #76).
		DPO	(ESS,T7 ; #77).
(counter, 9)	T1,T2,T3,T4, T8,T9	DPO	(ESS,T10,T6 ; #78), (ESS,T10,T7 ; #79), (ESS,T10,T9,T10,T7 ; #80), (ESS,T10,T9,T10,T6 ; #81), (ESS,T10,T9,T10,T9,T11 ; #82).
	T1,T2,T3, T4,T9	DO	(ESS,T12,T4,T7 ; #83), (ESS,T10,T9,T12,T4,T7 ; #84).
		DPO	(ESS,T12 ; #85), (ESS,T12,T4,T9,T10,T6 ; #86), (ESS,T12,T4,T9,T10,T7 ; #87), (ESS,T12,T4,T9,T10,T9,T11 ; #88), (ESS,T10,T6 ; #89), (ESS,T10,T8 ; #90), (ESS,T10,T9,T12 ; #91), (ESS,T10,T9,T12,T4,T9,T11 ; #92), (ESS,T10,T9,T10,T6 ; #93), (ESS,T10,T9,T10,T8 ; #94), (ESS,T10,T9,T10,T9,T11 ; #95).
	T1,T2,T3,T4, T9,T10,T9,	DO	(ESS,T10,T8,T7; #96), (ESS,T12,T4,T7; #97).
		DPO	(ESS,T10,T6 ; #98), (ESS,T10,T8 ; #99), (ESS,T10,T8,T9,T11 ; #100), (ESS,T10,T9,T11 ; #101), (ESS,T12,T4,T9,T11 ; #102).
	T1,T2,T3,T4, T8,T9,T10,T9	DPO	(ESS,T10,T6 ; #103), (ESS,T10,T8 ; #104), (ESS,T10,T9,T11 ; #105).
	T1,T2,T3,T4, T9,T10,T9, T10,T9	DPO	(ESS,T11 ; #106).
	T1,T2,T3,T4, T8,T9,T10, T9,T10,T9	DPO	(ESS,T11 ; #107).

Table 2: The EDO-paths for the protocol depicted in Figure 1.

NOTE: In the field of "Test Target", (*variable*, [*PATH_ID*]) represents the *PATH_ID*s and the corresponding defined *variable* that are contained in the associated ETS.

ETS No.	Executable Test Sequnece (ETS)	Test Target (variable,[PATH_ID])
1	(T1.T2.T3.T4.T6.T4.T7.T16)	(number, [1,2,50,61]), (SDU, [26,27])
2	(T1.T2.T3.T4.T6.T4.T9,T10,T7.T16)	(number, [1,2,50,62]), (SDU, [26,27])
3	(T1.T2.T3.T4.T6.T4.T9,T10,T9,T10,T7.T16)	(number, [1,2,50,63]), (SDU, [26,27])
4	(T1.T2.T3.T4.T6.T4.T9,T10,T9,T10,T9,T11,T16)	(number, [1,2,50,60]), (SDU, [26,27])
5	(T1.T2.T3.T4.T8.T9.T10.T7.T16)	(number, [1,3,9,71]), (counter, [14,79]), (SDU, [26,28]), (no_of_segment, [33,34]), (blockbound, [38])
6	(T1.T2.T3.T4.T8.T9.T10.T9.T10.T7.T16)	(number, [1,3,10,51,58,72]), (counter, [16,80,104]), (SDU, [26,28]), (no_of_segment, [33,35]), (blockbound, [40])
7	(T1.T2.T3.T4.T8.T9.T10.T9.T10.T9.T11.T16)	(number, [1,3,4,51,52,69]), (counter, [17,82,105,107]), (SDU, [26,28]), (no_of_segment, [33]), (blockbound, [41])
8	(T1.T2.T3.T4.T8.T7.T16)	(number, [1,3,11,51,59,70]), (counter, [12]), (SDU, [26,28]), (no_of_segment, [33,36])
9	(T1.T2.T3.T4.T9.T10.T9.T10.T9.T11.T16)	(number, [1,8,56]), (counter, [21,95,101,106]), (SDU, [26]), (blockbound, [48])
10	(T1.T2.T3.T4.T8.T9.T10.T6.T4.T9.T10.T9.T11.T16)	(number, [1,3]), (counter, [13,78]), (SDU, [26,28]), (no_of_segment, [33]), (blockbound, [37])
11	(T1.T2.T3.T4.T8.T9.T10.T9.T10.T6.T4.T9.T11.T16)	(number, [1,3,51]), (counter, [15,81,103]), (SDU, [26,28]), (no_of_segment, [33]), (blockbound, [39])
12	(T1.T2.T3.T4.T8.T9.T10.T9.T12.T4.T9.T11.T16)	(number, [1,3,51]), (counter, [18]), (SDU, [26,28]), (no_of_segment, [33]), (blockbound, [42])
13	(T1.T2.T3.T4.T9.T10.T6.T4.T7.T16)	(number, [1]), (counter, [19,89]), (SDU, [26]), (blockbound, [45])
14	(T1.T2.T3.T4.T9.T10.T9.T10.T6.T4.T7.T16)	(number, [1]), (counter, [24,93,98]), (SDU, [26]), (blockbound, [46])
15	(T1.T2.T3.T4.T9.T10.T8.T9,T10,T7.T16)	(number, [1,6,54,75]), (counter, [20,90]), (SDU, [26,30]), (blockbound, [44])
16	(T1.T2.T3.T4.T9.T10.T8.T9,T10,T9.T11,T16)	(number, [1,6,54,73]), (counter, [20,90]), (SDU, [26,30]), (blockbound, [44])
17	(T1.T2.T3.T4.T9.T10.T9.T10.T8.T7.T16)	(number, [1,77]), (counter, [25,94,96,99]), (SDU, [26,31]), (blockbound, [47])
18	(T1.T2.T3.T4.T9.T10.T9.T10.T8.T9,T11.T16)	(number, [1,76]), (counter, [25,94,100]), (SDU, [26,31]), (blockbound, [47])
19	(T1.T2.T3.T4.T9.T12.T4.T7.T16)	(number, [1,5,53,65]), (counter, [23,85]), (SDU, [26,29]), (blockbound, [49])
20	(T1.T2.T3.T4.T9.T12.T4,T9.T10.T6,T4,T9,T11,T16)	(number, [1,5,53]), (counter, [23,85,86]), (SDU, [26,29]), (blockbound, [49])
21	(T1.T2.T3.T4.T9.T12.T4,T9.T10.T7,T16)	(number, [1,5,53,66]), (counter, [23,85,87]), (SDU, [26,29]), (blockbound, [49])
22	(T1.T2.T3.T4.T9.T12.T4,T9.T10.T9,T11,T16)	(number, [1,5,53,64]), (counter, [23,85,88]), (SDU, [26,29]), (blockbound, [49])
23	(T1.T2.T3.T4.T9,T10,T9.T12.T4,T7.T16)	(number, [1,7,55,68]), (counter, [22,102]), (SDU, [26,32]), (blockbound, [43])
24	(T1.T2.T3.T4.T9,T10,T9.T12.T4,T9,T11.T16)	(number, [1,7,55,67]), (counter, [22,92,102]), (SDU, [26,32]), (blockbound, [43])

Table 3: Final ETSs for the protocol depicted in Figure 1.

```
Step 1:
  D-pool := ∅, EDO-path-pool := ∅
  for each transition T do
    generate the sets of A-use(T), I-use(T), P-use(T), and O-use(T)
    add A-use(T) and I-use(T) to D-pool
  endfor
Step 2:
  remove an element DE from D-pool
  ESS(IS, DE) /* in Section 3.1 */
Step 3:
  for each sequence ESS in ESS-pool do
    remove an ESS from ESS-pool
    EDO-path-search(ESS, DE.variable) /* in Section 3.2 */
  endfor
Step 4:
  If D-pool is not empty, go to Step 2; otherwise, go to Step 5.
Step 5:
  for each sequence EDO in EDO-path-pool do
    EBP-path-search(EDO) /* in Section 3.3 */
  endfor
```

The Generating-Executable-Test-Sequence (GETS) algorithm is explained as follows: In Step 1, the four sets of A-use(T), I-use(T), P-use(T) and O-use(T) for each transition T are generated. Additionally, all of the A-use and I-use elements are added to D-pool in order to find the DO-paths. In Step 2, each DE element in the D-pool is explored. ESSs for each DE element are derived using the ESS(IS, DE) procedure. In Step 3, all of the shortest EDO-paths with respect to a DE element are generated. In Step 4, repeat Step 2 and Step 3 until all of the EDO-paths for all DE elements have been detected. Step 5 searches for the back path from the tail state of each EDO-path to the initial state IS. Appending the back path to an EDO-path forms an executable test sequence. Table 3 shows all of the executable test sequences for the protocol depicted in figure 1.

4 Conclusion

In this paper, we have presented a new data flow protocol test sequence generation method for EFSM-specified protocols. An executable test sequence (ETS) contains three paths: (1) the executable switch sequence (ESS); (2) the executable DO-path (EDO-path); and (3) the executable back path (EBP-path). For solving the executability problem, we have proposed the Transition Executability Analysis (TEA) technique to analyze paths' executability. Using the TEA technique, which is expanded in the Bread-First-Search (BFS) way, (1) each shortest executable switch sequence, which connects the initial state to the head state of the transition containing an A-use or I-use of a variable x, is derived based on the initial state's configuration; (2) the executable DO-paths (EDO-paths) with respect to a variable x are derived based on the associated ESSs' tail state configurations, then the EDO-paths are derived by concatenating the associated ESSs and DO-paths; (3) the executable back paths that connect EDO-paths' tail states to the initial state are generated, and (4) all of the executable test sequences (ETSs) are derived by concatenating the associated EDO-paths and their back paths, respectively.

References

[Chanson 93] S. T. Chanson and J. Zhu, "A Unified Approaches to Protocol Test Sequence Generation", *Proc. of IEEE INFOCOM*, pp. 106-114, 1993.

[Koh 94] L. S. Koh and M. T. Liu, "Test Path Selection Based on Effective Domains," *Proc. of International Conference on Network Protocols*, pp. 64-71, 1994.

[Miller 92] R. E. Miller and S. Paul, "Generating Conformance Test Sequences for Combined Control and Data Flow of Communication Protocols," *Proc. of International Symposium on Protocol Specification, Testing and Verification, XII*, pp. 13-27, 1992.

[Sarikaya 87] B. Sarikaya, G. V. Bochmann, and E. Cerny, "A Test Design Methodology for Protocol Testing," *IEEE Transactions on Software Engineering*, VOL. 13, NO. 5, pp. 518-531, 1987.

[Ural 91] H. Ural and B.Yang, "A Test Sequence Selection Method for Protocol Testing," *IEEE Transactions on Communication*, VOL. 39, NO. 4, pp.514-523, 1991.

Biography

Chung-Ming Huang received the B.S. degree in electrical engineering from National Taiwan University in 1984/6, and the M.S. and Ph.D. degrees in computer and information science from The Ohio state University, in 1987/12 and 1991/6, respectively. He is currently an associate professor in Institution of Information Engineering (IIE) National Cheng Kung University (NCKU), Taiwan, R.O.C. He is also the Chairman of ISO Special Interested Group (SIG) of Open Document Architecture (ODA) in Taiwan, R.O.C. He was the General Secretary of the Chinese Image Processing and Pattern Recognition Society from 1993/1 to 1994/12. His research interests include protocol testing and verification, multimedia networking protocols, and multimedia document systems.
Yuan-Chuen Lin received the B.S. degree in applied mathematics from Chung Yuan University in 1993/6, and had his master degree in IIE NCKU on 1995/6. His research interests include protocol testing and multimedia networking protocols.
Ming-Yuhe Jang received the B.S. degree in information computer engineering from Chung Yuan University in 1990/6, had qualifications to study directly Ph.D. degree in IIE NCKU in 1994/9, and is currently a Ph.D. student in IIE NCKU. His research interests include protocol testing, multimedia networking protocols, and ATM networking.

3

Transformation of Estelle modules aiming at test case generation

O. Henniger[a], *A. Ulrich*[b], *and H. König*[c]

[a]*GMD – German National Research Center for Information Technology*
Rheinstr. 75, 64295 Darmstadt, Germany
e-mail: henniger@darmstadt.gmd.de

[b]*Dept. of Computer Science, University of Science and Technology*
Clear Water Bay, Kowloon, Hong Kong
e-mail: ulrich@cs.ust.hk

[c]*Dept. of Computer Science, Technical University of Cottbus*
P.O. Box 101344, 03013 Cottbus, Germany
e-mail: koenig@informatik.tu-cottbus.de

Abstract

This paper presents a method for transforming an extended finite state machine (EFSM) given as an Estelle normal form module into an equivalent expanded EFSM without control variables, i.e. an Estelle normal form module free of provided-clauses. The transformed EFSM allows to apply methods based on the finite state machine (FSM) model for test case generation. Using this approach, it is possible to cope with test sequence generation for control and data flow and with test data selection. The transformation is feasible if the variables that occur in provided-clauses have finite, countable domains. For realistic protocol specifications, this condition is fulfilled most of the time.

Keywords

Conformance testing, formal specifications, Estelle

1 INTRODUCTION

Early work on automatic test sequence generation for communication protocols has been based on the model of finite state machines (FSMs) (e.g. [NT81, SD88, ADLU88, SLD89, CVI89]). A main problem of these test sequence generation methods is that FSMs usually model only the control aspect of a system. To model the data aspect as well as the control aspect, extended finite state machine (EFSM) models, which are based on an FSM extended by variables, are applied in many cases. EFSM models form the basis of

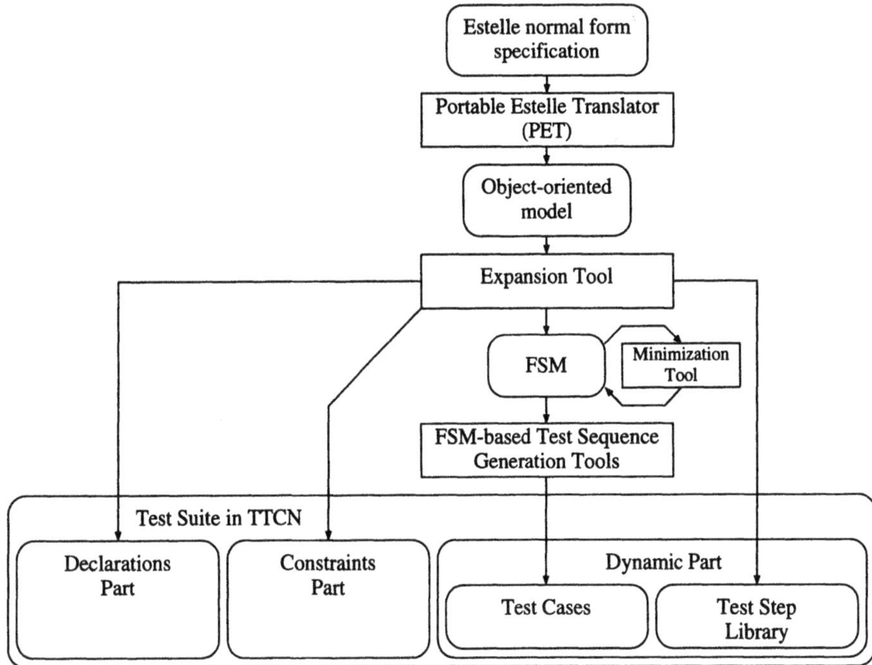

Figure 1 Test generation approach.

the standardized formal description techniques (FDTs) Estelle [ISO89] and SDL [ITU92] that increasingly come into use.

Theoretically, provided that all variables of an EFSM have a finite, countable domain, an EFSM can be transformed into an equivalent FSM. The transformation leads to the removal of variables from the state machine and an increase in the number of states. Practically, since the number of states of the resulting FSM may get very large, the transformation of an EFSM into an equivalent FSM is not feasible.

This paper discusses an approach to bridge the gap between an EFSM and an FSM avoiding inordinate increase in the number of states. Our aim is to make FSM based methods for test generation and for validation applicable to EFSM. The approach is discussed on the basis of an one-module Estelle normal form specification [Sar93] representing an EFSM.

Our approach is based on the observation that variables influencing the control flow usually have a small, finite domain (Boolean type, enumerated type, or subrange type), and that only these variables cause serious problems in applying FSM based methods. We transform an EFSM into an equivalent expanded EFSM where enabling of transitions depends only on the current state and the input, i.e. variables do not influence the control flow. That means we obtain an Estelle normal form specification without provided-clauses. The resulting specification still contains variables which, however, are not used in provided-clauses. Therefore, it still represents an EFSM and not a pure FSM. The expanded EFSM can be interpreted in terms of FSM.

From the expanded EFSM, test sequences can be generated using classic methods, such as the transition tour method [NT81] or a UIO method [SD88, ADLU88, SLD89, CVI89]. The transformation algorithm does not cause loss of information. This allows to generate test sequences that cover control and data aspects of the original EFSM.

Beside test generation, the expanded EFSM is also useful for analyzing the original EFSM and for detecting specification errors.

The transformation algorithm has been implemented as a prototype expansion tool based on the PET&DINGO tool set [SS90]. Input to the expansion tool is the object-oriented model of an one-module specification produced by the Portable Estelle Translator (PET). Output of the expansion tool is an FSM representation of the source specification, a test step library and parts of the declarations and constraints parts in the test notation TTCN.MP [ISO92]. Figure 1 shows a scheme of our test generation tool set.

The rest of this paper is organized as follows: Section 2 defines the prerequisites necessary for the transformation algorithm. It gives formal definitions of an FSM and an EFSM, a definition of an Estelle normal form specification, its link to EFSMs, and a classification of EFSM variables. Section 3 introduces the algorithm for transforming a given EFSM into an expanded EFSM, discusses the interpretation of the expanded EFSM in terms of FSM and FSM based test sequence generation methods. Section 4 discusses issues arising from test generation in the context of multi-module specifications. In Section 5 our approach is demonstrated for the Inres protocol specification. Section 6 shortly reviews related work dealing with test generation from EFSM, and Section 7 gives some concluding remarks.

2 PRELIMINARIES

2.1 Finite state machines and extended finite state machines

Definition 1 A *finite state machine (FSM)* is a tuple $\langle S, I, O, T, s_0 \rangle$, where S is a non-empty finite set of states, I is a non-empty finite set of inputs, O is a non-empty finite set of outputs, $T \subseteq S \times I \times O \times S$ is the transition relation, and $s_0 \in S$ is the initial state of the FSM. $\diamond$

A transition $t \in T$ of an FSM is a tuple $\langle s, i, o, s' \rangle$ where $s \in S$ is a current state, $i \in I$ is an input, $o \in O$ is an output related to s and i, and $s' \in S$ is a next state related to s and i.

To enhance the descriptive power, additional variables are introduced into the mathematical model. These variables are used in programming language constructs specifying conditions for the execution of transitions and calculations carried out during transitions. For extending a conventional finite state machine by variables, such model is called extended finite state machine.

Definition 2 An *extended finite state machine (EFSM)* is a tuple $\langle S, C, I, O, T, s_0, c_0 \rangle$ where S is a non-empty finite set of main states, $C = \text{dom}(v_1) \times \ldots \times \text{dom}(v_n)$ is a non-empty countable set of contexts with $v_i \in V$, V is a non-empty finite set of variables and dom(v_i) is a non-empty countable set which is referred to as the domain of v_i, I is a non-empty finite set of inputs, O is a non-empty finite set of outputs, $T \subseteq S \times C \times I \times O \times S \times C$ is the transition relation, $s_0 \in S$ is the initial main state, and $c_0 \in C$ is the initial context of the EFSM. $\diamond$

A context is a concrete assignment of values to the variables. A transition $t \in T$ of an EFSM is a tuple $\langle s, c, i, o, s', c' \rangle$ where $s \in S$ is a current main state, $c \in C$ is a current context, $i \in I$ is an input, $o \in O$ is an output, $s' \in S$ is a next main state, and $c' \in C$ is a next context.

The mathematical structure EFSM can be expressed using different syntactic constructions, e.g. using modules or processes of the FDTs Estelle [ISO89] or SDL [ITU92].

2.2 Estelle normal form specification

Estelle uses a subset of ISO Pascal which is complemented by special constructs for expressing the elements of EFSM transitions, for structuring, and for expressing communication concepts. In Estelle, a system is specified as a hierarchy of module instances that communicate with each other via FIFO channels. The behavior of a single module instance is characterized in terms of an EFSM.

Starting point of our test generation approach is a specification in a specification style that makes the interpretation of module instances in terms of EFSM easy. A specification of this style is referred to as a normal form specification. In certain cases, specifications based on other specification styles may be transformed into normal form specifications by means of syntax-directed transformation rules [SB85, Sar93]. Main characteristics of a normal form specification are:

- The influence of variables on the control flow is specified only in provided-clauses, i.e. the specification contains no conditional-statements ("if", "case", or "forone" statements) and no repetitive-statements ("repeat", "while", "for", or "all" statements).
- The specification is complete, i.e. it contains no "any" constant-definitions, no "..." type-definitions, and no "external" and "primitive" directives.
- The specification has a static structure, i.e. it contains "init", "connect", "attach", "release", "terminate", "disconnect", and "detach" statements only in the initialization-part of module definitions.
- Shorthand notations, like nested transitions, "provided otherwise", state sets etc., are expanded.

Furthermore, we assume that the specification is deterministic and does not contain priority-clauses.

2.3 Classification of variables in an Estelle normal form specification

The variables occurring in a module definition contained in an Estelle normal form specification can be classified as context variables and interaction variables according to the place of their declaration.

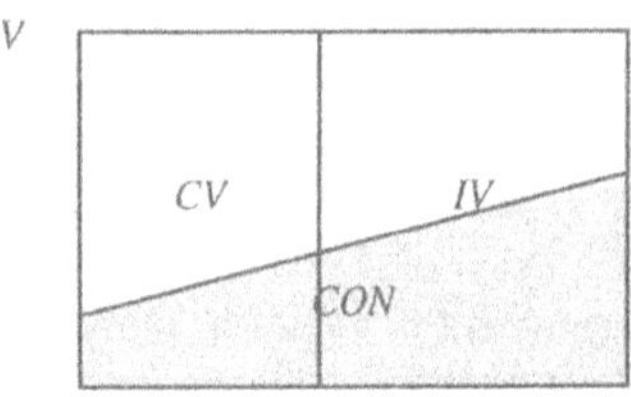

Figure 2 Visualization of the variable classes.

Definition 3 A variable that is declared within the declaration-part of a body-definition is called *context variable.* ◇

Definition 4 A variable that is declared within a value-parameter-specification of an interaction-definition contained in a channel-definition is called *interaction variable.* ◇

The set of all context variables and the set of all interaction variables of an Estelle module are called CV and IV respectively. Each variable belongs either to CV or to IV; i.e., CV and IV together form the set of all variables V of an Estelle module: $V = CV \cup IV$.

Furthermore, context variables as well as interaction variables can belong to the class of control variables. Control variables influence the selection of transitions.

Definition 5 A variable that occurs in a provided-clause is a *control variable.* A variable that is used to assign a value to a *control variable* is a *control variable* itself. ◇

The set of all control variables of an Estelle module is called CON. The set of all control variables can be found using a recursive algorithm based on the given recursive definition of a control variable. CON is a subset of the set of all variables: $CON \subseteq CV \cup IV$. Figure 2 depicts the relationship between the classes of variables.

2.4 Interpretation of an Estelle module in terms of an EFSM

The EFSM described by a module definition in an Estelle normal form specification is the tuple $\langle S, C, I, O, T, s_0, c_0 \rangle$ with $S = SID$ where SID is the set of state-identifiers introduced within the declaration-part of the body-definition, $C = \text{dom}(v_1) \times \ldots \times \text{dom}(v_n)$ with $v_i \in CV \cup IV$ where CV and IV are the sets of context variables and interaction variables as explained in Section 2.3, $I = \{iref \in IREF \mid iref \text{ is used in a when-clause}\}$ and $O = \{iref \in IREF \mid iref \text{ is used in an output-statement}\}$ where $IREF$ is the set of interaction-references, T represented by the set of transition-declarations TR, $s_0 = initial$ where *initial* is the state-identifier specified in the from-clause contained in the initialization-part, and c_0 represented by the values assigned in the transition-block contained in the initialization-part.

An interaction-reference $iref_i \in IREF$ may be associated with interaction variables as parameters: $iref_i(iv_{i,1}, \ldots, iv_{i,m_i})$. The interaction variables associated with $iref_i$ make up the set $IV_i \subseteq IV$.

A transition-declaration $tr \in TR$ has the following form:

trans from *current* **to** *next* **when** *input* **provided** *predicate*
begin *assignments*; *outputs* **end;**

where

- *current* $\in$ *SID* and *next* $\in$ *SID* are state-identifiers specifying the start state and the end state of a transition;
- *input* $\in$ *IREF* is an interaction-reference (possibly associated with interaction variables as parameters) specifying the input of a transition; alternatively to the when-clause, a delay-clause could be given to designate a time-out event as input;
- $predicate(v_1, \ldots, v_k)$ is a Boolean expression of the variables $v_1, \ldots, v_k$ specifying the enabling condition for a transition;
- *assignments* is a (possibly empty) sequence of assignment-statements specifying that variables are set to new values during execution of a transition;
- *outputs* is a (possibly empty) sequence of output-statements specifying the output caused by execution of a transition.

Since the Boolean expression in the provided-clause may be true for several contexts, a single Estelle transition-declaration comprises in general several EFSM transitions.

3 EXPANSION OF EFSM

3.1 Outline of the transformation approach

The FSM based test sequence generation methods are not easily applicable to EFSMs. In contrast to the transitions of an FSM, the transitions of an EFSM depend not only on the input and the current state, but also on the actual values of variables, which may depend on the whole record of previous inputs. Application of FSM based test sequence generation methods taking into consideration only the main states, but neglecting provided-clauses and the variables contained in them would lead to infeasible subtours.

In order to apply FSM based methods, the EFSM should be transformed into an equivalent FSM. An EFSM $\langle S, C, I, O, T, s_0, c_0 \rangle$ could theoretically be transformed into an equivalent FSM $\langle S', I, O, T', s'_0 \rangle$ by means of Cartesian multiplication, i.e. $S' = S \times C$. Carrying out this transformation, one is faced with the state-explosion problem, i.e. with a very large or even infinite set of states. If all variables have a finite domain, S' is a finite set, and an equivalent FSM exists. If any of the variables does not have a finite domain, S' is an infinite set, and the EFSM can not be transformed into an equivalent FSM.

Since not all variables must have a finite domain, the transformation of an EFSM into an equivalent FSM is in general not applicable. However, taking into consideration the different classes of variables in Estelle, an approach can be established which limits the growth of the number of states.

We studied example specifications (e.g. [Hog92]) and realistic protocol specifications in Estelle (e.g. [HHP93, GHLP93, Häh94]) and observed that in many cases

- only a subset of all variables are control variables, and

- the control variables have small finite domains, i.e. they have the Boolean type, an enumerated type, or a subrange type with few values.

Our transformation approach is based on this observation, and the transformation algorithm is applicable only for specifications that satisfy the prerequisite that each control variable $c \in CON$ has a finite domain. Though not all variables must have a finite domain, at least the variables from the set of control variables usually have a finite domain.

Our approach is to eliminate only the control variables since the existence of control variables and provided-clauses is the main problem that hinders us from applying FSM based methods. Variables that do not influence the control flow can still remain in the transformed state machine; they do not bother us. The control variables are eliminated by shifting them to a set of new states S' or to a set of new inputs I'.

For determining the new set of states S', only the context variables that are control variables need to be taken into consideration. A new state is a combination of a main state of the original EFSM and of a concrete assignment of values to all context variables that are control variables.

For determining the new set of inputs I', interaction variables that are control variables are taken into consideration. A new input is a combination of the original input and of a concrete assignment of values to all parameters associated with this input that are control variables.

Once S' and I' are determined, we work out the final states and outputs of the new transitions by symbolic evaluation of the original transitions. The output parameters are partly replaced by concrete values during symbolic evaluation.

After carrying out the transformation, the transformed EFSM can be interpreted in terms of an FSM.

3.2 Transformation algorithm

This section describes the transformation algorithm in detail. Input to the algorithm is a module definition contained in an Estelle normal form specification and representing an EFSM as defined in Section 2.4. Output is an Estelle module definition without provided-clauses representing an expanded EFSM without control variables. The expanded EFSM is semantically equivalent to the original EFSM; no information is lost during the transformation.

Algorithm *Expansion*

Input: Estelle module definition representing an EFSM M with the set of state-identifiers SID, the set of interaction-references $IREF$, the subsets of variables CV, IV, and CON, and the set of transition-declarations TR.

Output: Estelle module definition without provided-clauses representing the expanded EFSM M'.

begin

{Computation of new states}

$ST = SID \times \text{dom}(cv_1) \times \ldots \times \text{dom}(cv_k)$ such that $cv_i \in CV \cap CON$ for $1 \leq i \leq k$;

{Computation of new input interactions}
$INT = INT_1 \cup \ldots \cup INT_l$ such that
$INT_i = \{iref_i\} \times \text{dom}(iv_{i,1}) \times \ldots \times \text{dom}(iv_{i,m_i})$, $iref_i \in IREF$ for $1 \leq i \leq l$, and
$iv_{i,j} \in IV_i \cap CON$ for $1 \leq j \leq m_i$;

{Computation of new transitions}
for each $tr \in TR$ **do**
  for each $st = \langle sid, \text{value}(cv_1), \ldots, \text{value}(cv_k)\rangle \in ST$ **do**
    for each $int = \langle iref_i, \text{value}(iv_{i,1}), \ldots, \text{value}(iv_{i,m_i})\rangle \in INT$ **do**
      if $((sid = current) \wedge (iref_i = input) \wedge$
        $(predicate(\text{value}(cv_1), \ldots, \text{value}(cv_k), \text{value}(iv_{i,1}), \ldots, \text{value}(iv_{i,m_i})) = \text{true}))$
      then
        begin
          Create a new transition-declaration $tr' \in TR'$ such that
          - $current' = \text{stateid}(st)$;
            {stateid(st) is a unique state-identifier assigned to st}
          - $input' = \text{intref}(int)$;
            {intref(int) is a unique name assigned to int}
          - $predicate' = \text{true}$;
            {i.e., the provided-clause can be omitted}
          - $next' = \text{stateid}(\langle next, \text{newvalue}(cv_1), \ldots, \text{newvalue}(cv_k)\rangle)$;
            {newvalue(cv_i) is the value of cv_i after evaluation of *assignments*}
          - *outputs'* = *outputs* with actual parameters replaced by their symbolic values after evaluation of *assignments*;
          - *assignments'* = *assignments* with assignment-statements having a control variable on their left-hand side omitted
        end;
end.

ST is the generalized Cartesian product of the set of state-identifiers SID and of the domains of all context variables that are control variables. Each ordered n-tuple $st \in ST$ consists of a state-identifier $sid \in SID$ and of a concrete value from the domain of each variable $cv_i \in CV \cap CON$. Because of the prerequisite that each control variable has a finite domain, the cardinality of ST (i.e. the number of states of the expanded EFSM) is a finite number: $|ST| = |SID| \cdot |\text{dom}(cv_1)| \cdot \ldots \cdot |\text{dom}(cv_k)|$.

INT is a set of ordered n-tuples that consist of an interaction-reference $iref_i \in IREF$ and of concrete values from the domains of all control variables iv_j that are declared in the interaction-definition for $iref_i$.

Since concrete values are given for all control variables, it is no problem to compute a concrete value for $next'$. If variables that are not control variables are used in a transition-block, the computation of actual output parameters and of new values for these variables will lead to symbolic values, i.e. expressions with variable names, arithmetic operators, and constants instead of concrete values.

All control variables are taken into account while computing the sets of new states and of new interactions; therefore, they are removed from the expanded EFSM. The transition-declaration-part TR' of the expanded EFSM consists only of transition-declarations that do not contain a provided-clause, and that contain assignments, if any, only to variables that do not influence the control flow.

3.3 Interpretation of the expanded EFSM in terms of FSM

A transition-declaration of the transformed module definition $tr' \in TR'$ has the following form:

trans from *current'* **to** *next'* **when** *input'*
begin *assignments'*; *outputs'* **end;**

Only remaining variables and the statement-part between "begin" and "end" offend against the definition of an FSM. Because there are no provided-clauses, the output of the transformation algorithm can easily be interpreted in terms of an FSM $\langle S, I, O, T, s_0 \rangle$.

The state-identifiers correspond to the states of the FSM. Input interaction-references, possibly still associated with interaction parameters that were not used in the provided-clauses of the original specification, are mapped to FSM inputs by disregarding the parameters. The statement-parts of transition-blocks are mapped to FSM outputs as a whole. Statement-parts with equivalent sequences of output-statements correspond to the same FSM output.

Since no information should be lost by interpreting the expanded EFSM as an FSM, our expansion tool generates a declarations part, a constraints part, and a test step library at the same time as the FSM is generated.

3.4 Minimization of FSM

Given an FSM $\langle S, I, O, T, s_0 \rangle$, two states $s_1 \in S$ and $s_2 \in S$ are referred to as equivalent states if each sequence of inputs produces identical sequences of outputs in s_1 and s_2. An FSM without equivalent states is a minimal FSM. Each FSM can be reduced to a minimal FSM by applying minimization algorithms merging equivalent states to one state [HU79].

The FSM obtained by interpreting the expanded EFSM as an FSM may contain equivalent states. Minimization reduces the number of states and makes the state-explosion problem less severe.

An FSM $\langle S, I, O, T, s_0 \rangle$ is called a complete FSM if the transition relation is defined for each pair $\langle s, i \rangle \in S \times I$, otherwise it is a partial FSM. In case of partial FSMs, the classic minimization algorithms fix a next state and an output for pairs $\langle s, i \rangle$ for which no next state and no output have been defined originally. In the context of conformance testing, this means that additional conformance requirements are introduced that have not been given in the original specification. Since this must be avoided, minimization is carried out only for complete FSMs.

3.5 Test generation based on FSM

The different test sequence generation methods based on FSMs have a common basic idea [BU91]: A test sequence is a preferably short sequence of consecutive transitions that contains every transition of the FSM at least once and allows to check whether every transition is implemented as defined. To test a transition, one has to apply the input for the transition in the starting state of the transition, to check whether the correct output occurs, and to check whether the correct next state has been reached after the transition. Checking the next state might be omitted (transition tour method [NT81]) or be carried out by means of distinguishing sequences (checking experiments method), characterizing sequences (W-method), or unique input/output sequences (UIO methods [SD88, ADLU88, SLD89, CVI89]).

Different FSM based test sequence generation methods may be applied to the FSM interpretation of the expanded EFSM. In case of large expanded EFSM, the transition tour method is most appropriate.

4 MULTI-MODULE SPECIFICATIONS

So far, our presented approach is able to deal with single-module specifications only. On the other hand, a system may be specified as a set of communicating modules. To apply our approach in this case, it is necessary to combine the communicating modules together. The composition is normally carried out by assuming synchronous communication between modules [SLU89]. If we use multi-module Estelle specifications as input to our approach, we slightly change the semantics of Estelle, which uses asynchronous communication over unbounded buffers. The changes in the semantics may be accepted if only modules are combined which will be implemented locally on a single processor system.

While combining modules to a composite machine, we are faced with another kind of state space explosion caused by interleaving events. To alleviate this problem, two different approaches are suggested in the literature: The first approach takes advantage of the hierarchical structure of the specification [SLU89], and the second one tries to prune the modules of a set of communicating modules before combining them according to a certain test purpose [LSKP93]. Whereas the first approach can not guarantee a feasible size of the composite module, in the second approach the module composed from a set of pruned machines may not represent the correct behavior under some circumstances.

In case of truly concurrent modules, execution of test cases generated from a composite machine becomes difficult because of weak controllability of the implementation under test. The test suite may require to test a certain order of interleaving events, but this order depends on a concrete test run and is in general non-deterministic. Thus, the entire concept of testing concurrent systems must be reconsidered. Currently, we investigate a general approach to test generation and execution of concurrent systems. The test generation will be based on partial order semantics [UC95].

5 RESULTS OF THE APPLICATION TO THE INRES PROTOCOL

We demonstrate the transformation algorithm on the Estelle specification of the Inres protocol [Hog92]. The Inres protocol is used as demonstration example for many FDT based methods for test case generation [FMC95]. The Inres protocol provides a simple data transfer service over an unreliable medium. Only the initiator side of the Inres protocol is considered. The initiator side consists of two modules, "Initiator" and "Coder". The two module definitions were merged into one, and transformation rules for obtaining a normal form specification were applied.

In the declaration-part of the module definition three context variables are introduced:

```
var olddata : ISDUType;
    counter : 0..4;
    number : Sequencenumber;
```

The context variables `counter` and `number` are used in provided-clauses. Searching for variables that are used to assign values to these two variables, only `counter` and `number` themselves are found. Thus, these two context variables are control variables and will be eliminated by the algorithm. `olddata` is used for buffering data and will not be eliminated. It does not matter whether or not such a variable has a finite domain.

In the channel-definitions, some interaction variables are introduced:

```
channel ISAPchn(User, Station);
    by User :
        ICONreq; ICONresp; IDATreq(ISDU : ISDUType); IDISreq;
    by Station :
        ICONconf; ICONind; IDATind(ISDU : ISDUType); IDISind;
channel MSAPchn(Station, Medium_Service);
    by Station :
        MDATreq(id : PduType; num : Sequencenumber; data : ISDUType);
    by Medium_Service :
        MDATind(id : PduType; num : Sequencenumber; data : ISDUType);
```

Only the interaction variables `id` and `num` belonging to the interaction `MDATind` are used in provided-clauses. Thus, these two interaction variables are control variables and will be eliminated as well.

With the following type-definitions, all control variables have a finite domain:

```
type PduType = (CR, CC, DT, AK, DR);
type Sequencenumber = 0..1;
```

Thus, the prerequisite for the applicability of our transformation algorithm is satisfied for the Inres example.

The first step of the algorithm delivers a set of new states by combining the original states with concrete values for all context variables that are control variables. The following set of states is given:

```
state DISCONNECTED, WAIT, CONNECTED, SENDING;
```

Combining the original states with concrete value assignments for `counter` and `number`, we obtain the following new set of states*:

```
state
    SENDING_0_0, SENDING_1_0, SENDING_2_0, SENDING_3_0, SENDING_4_0,
```

*The notation `SENDING_0_0` refers to `SENDING` combined with `counter` = 0 and `number` = 0, etc.

```
SENDING_0_1, SENDING_1_1, SENDING_2_1, SENDING_3_1, SENDING_4_1,
CONNECTED_0_0, ..., CONNECTED_4_1,
WAIT_0_0, ..., WAIT_4_1,
DISCONNECTED_0_0, ..., DISCONNECTED_4_1;
```

The next step of the algorithm delivers new interaction-definitions for interactions associated with control variables as parameters by combining these interactions with concrete value assignments for the control variables. So, `MDATind` is modified as follows†:

```
MDATind_CR_0(data : ISDUType);
MDATind_CR_1(data : ISDUType);
..
MDATind_DR_1(data : ISDUType);
```

In the last step of the transformation algorithm, the new states and the new interactions are utilized to replace each transition-declaration by a number of new transition-declarations without provided-clauses. Let us consider the following transition-declaration as an example:

```
trans from SENDING to SENDING
    when MSAP.MDATind
        provided (id = AK) and (num <> number) and (counter < 4)
            begin
                counter := counter + 1;
                output MSAP.MDATreq(DT, number, olddata)
            end;
```

This transition-declaration is replaced by 8 new transition-declarations without provided-clauses:

```
trans from SENDING_0_0 to SENDING_1_0
    when MSAP.MDATind_AK_1(data)
        begin output MSAP.MDATreq(DT, 0, olddata) end;
trans from SENDING_0_1 to SENDING_1_1
    when MSAP.MDATind_AK_0(data)
        begin output MSAP.MDATreq(DT, 1, olddata) end;
...
trans from SENDING_3_1 to SENDING_4_1
    when MSAP.MDATind_AK_0(data)
        begin output MSAP.MDATreq(DT, 1, olddata) end;
```

The variables `data` and `olddata`, which do not influence the enabling of transitions, remain in the newly generated transitions. `data` is an interaction variable, whereas `olddata` is a context variable for buffering the last data package sent. Since there are no provided-clauses, the result of the transformation is suitable for the application of FSM based test

†The notation `MDATind_CR_0` refers to `MDATind` combined with `id = CR` and `num = 0`, etc.

Table 1 Size of the transformed specifications

	Number of main states	Number of transitions
Normalized module	4	34
Expanded EFSM	40	410
Minimized FSM	18	185

sequence generation methods. Table 1 illustrates the influence of the transformation on the size of a specification. We applied test generation tools to the minimized FSM for the Inres initiator part. A transition tour [NT81] derived from the specification of the initiator part consists of 400 transitions.

Analysis of the expanded EFSM led to the revelation of some errors in the original specification, like unintended non-determinism. So, the transformation is also useful for the validation of a specification.

Since the variable `data`, which is left over after the transformation, occurs in the send events of the test suite, the test suite designer has to choose a suitable value for it. What a "suitable" value means, is up to the test suite designer. The variable `olddata` remains in the test suite and is defined as a test suite variable.

6 RELATED WORK

Test sequence generation methods based on FSM have been widely studied. The extension by a data portion complicates the test generation from EFSM: both, the control aspect and the data aspect have to be tested, and besides the generation of a sequence of test events, one has to cope with the selection of test data.

The existing methods for test generation from EFSM can be roughly classified into methods with explicit test purposes and methods with implicit test purposes: methods with explicit test purposes require information about the test purpose or the fault model for the generated test cases as input in addition to the EFSM (e.g. [GHN93, WL93]); methods with implicit test purpose assume test purposes for the generated test cases implicitly and usually do not require supplementary inputs in addition to the EFSM (e.g. [Sar93, CA90, PG90, UY91, MP92, CZ93]).

The methods with explicit test purposes require the test designer to choose what to test. Then the methods ensure that test cases consistent with the specification and the test purposes are generated. These methods offer much flexibility, but on the other hand, they require considerable manual efforts and do not guarantee systematic fault coverage. For methods with implicit test purposes, the picture is reversed: While requiring less manual efforts, they offer less flexibility in choosing what faults to generate test cases for.

The methods with implicit test purposes take a normalized one-module specification as their starting point. Main problem of these approaches is to guarantee that the test sequences are executable. Some subtours may be not executable if the enabling predicates (also called constraints) of transitions along the test sequence can not be satisfied with any input parameter values. In general, the executability problem is undecidable. However, on condition that variables have finite domains, constraint satisfaction techniques may be applied to ensure the executability of test sequences [CA90, CZ93].

Most of the methods base the test suite structure on the EFSM transitions. This leads to a dependence of the resulting test suite on the specification style: For the same system, very different test suites may be generated depending on whether the EFSM is specified in a more state-oriented or a more data-oriented style.

The approach presented in this paper is founded on experiences from earlier work on test generation from EFSM and extends the existing methods in certain aspects: The approach tackles the problem of test generation for both control and data flow by transforming an EFSM into an expanded EFSM representing both control and data aspect of the original EFSM and allowing to apply FSM based test generation methods. FSM based test generation methods assume an implicit fault model and generate shortest possible test sequences with a guaranteed fault coverage with respect to this fault model. The test sequence generated from the expanded EFSM is always executable since the enabling conditions of transitions have been considered in the course of the transformation. Furthermore, the approach copes with selection of test data by exhaustive enumeration of the control variables, which are required to have finite domains. High fault detection power is traded for considerable length of the test sequence. The generated test sequence is to a great extent independent from the specification style.

7 CONCLUSIONS

Although a formal specification is theoretically the best starting point for the automatic generation of conformance test suites, the way from a given formal specification to a conformance test suite is not straightforward in practice. Based on an Estelle normal form specification, we have proposed an algorithm to transform each module body into an equivalent EFSM without provided-clauses. The approach is limited to specifications where variables influencing the control flow range over a small domain. For many realistic protocol specifications, however, this limitation is fulfilled.

The described approach has been implemented as a prototype tool. The tool was applied to a number of example specifications and to realistic protocol specifications developed in the area of fieldbus systems and wireless telecommunication systems, and the feasibility of the method has been shown.

ACKNOWLEDGEMENTS

The authors are grateful to Karsten Nickoll for his assistance in programming the transformation algorithm.

REFERENCES

[ADLU88] A.V. Aho, A.T. Dahbura, D. Lee, and M.U. Uyar. An optimization technique for protocol conformance test generation based on UIO sequences and Rural Chinese Postman Tours. In S. Aggarwal and K. Sabnani, editors, *Protocol Specification,*

Testing, and Verification, VIII, pages 75–86, Atlantic-City, New Jersey, USA, 1988. Elsevier Science B.V. (North-Holland).

[BSV89] E. Brinksma, G. Scollo, and C.A. Vissers, editors. *Protocol Specification, Testing, and Verification, IX*, Enschede, The Netherlands, 1989. Elsevier Science B.V. (North-Holland).

[BU91] B.S. Bosik and M.U. Uyar. Finite state machine based formal methods in protocol conformance testing: from theory to implementation. *Computer Networks and ISDN Systems*, 22(1):7–33, 1991.

[CA90] W. Chun and P.D. Amer. Test case generation for protocols specified in Estelle. In Quemada et al. [QMV90], pages 191–206.

[CVI89] W.Y.L. Chan, S.T. Voung, and M.R. Ito. On test sequence generation for protocols. In Brinksma et al. [BSV89], pages 119–130.

[CZ93] S.T. Chanson and J.-S. Zhu. A unified approach to protocol test sequence generation. In IEEE INFOCOM'93 [IEE93], pages 106–114.

[FMC95] FMCT guidelines on "Test generation methods from formal descriptions", 1995. ITU-T Q.8/10 and ISO/JTC1/SC21/Project 54.2.

[GHLP93] F. Graner, O. Henniger, B. Lehnert, and A. Pöschmann. Estelle specification of the Lower Layer Interface (LLI) based on the ISP fieldbus specification. Technical report, Institute of Automation and Communication, Magdeburg, Germany, 1993.

[GHN93] J. Grabowski, D. Hogrefe, and R. Nahm. Test case generation with test purpose specification by MSCs. In O. Færgemand and A. Sarma, editors, *6th SDL Forum*, pages 253–266, Darmstadt, Germany, 1993. Elsevier Science B.V. (North-Holland).

[Häh94] J. Hähniche. Estelle specification of the PIA protocol. Technical report, Institute of Automation and Communication, Magdeburg, Germany, 1994.

[HHP93] J. Hähniche, E. Hintze, and A. Pöschmann. Portable implementation for PROFIBUS layer 7. Technical Report DFAM AIF No. 357D, University of Magdeburg, Magdeburg, Germany, 1993. In German.

[Hog92] D. Hogrefe. OSI formal specification case study: the Inres protocol and service, revised. Technical Report IAM-91-012, University of Bern, Bern, Switzerland, 1992. Update 1992.

[HU79] J.E. Hopcroft and J.D. Ullmann. *Introduction to automata theory, languages, and computation.* Addison-Wesley, Reading, Mass., USA, 1979.

[IEE93] *IEEE INFOCOM'93 Conference on Computer Communications*, volume 1, 1993.

[ISO89] Information processing systems – Open Systems Interconnection – Estelle: A formal description technique based on an extended state transition model. International Standard ISO 9074, 1989.

[ISO92] Information technology – Open Systems Interconnection – Conformance testing methodology and framework – Part 3: The Tree and Tabular Combined Notation (TTCN). International Standard ISO/IEC 9646-3, 1992.

[ITU92] Specification and description language SDL '92. ITU-T Recommendation Z.100, 1992.

[LSKP93] D. Lee, K.K. Sabnani, D.M. Kristol, and S. Paul. Conformance testing of protocols specified as communicating FSMs. In IEEE INFOCOM'93 [IEE93], pages 115–127.

[MP92] R.E. Miller and S. Paul. Generating conformance test sequences for combined control and data flow of communication protocols. In R.J. Linn and M.U. Uyar, editors, *Protocol Specification, Testing, and Verification, XII*, Lake Buena Vista, Florida, USA, 1992. Elsevier Science B.V. (North-Holland).

[NT81] S. Naito and M. Tsunoyama. Fault detection for sequential machines by transition-tours. In *11th IEEE Fault Tolerant Computing Conference*, pages 238–243, 1981.

[PG90] M. Phalippou and R. Groz. From Estelle specifications to industrial test suites, using an empirical approach. In Quemada et al. [QMV90], pages 175–190.

[QMV90] J. Quemada, J. Mañas, and E. Vázquez, editors. *Formal Description Techniques, III*, Madrid, Spain, 1990. Elsevier Science B.V. (North-Holland).

[Sar93] B. Sarikaya. *Principles of protocol engineering and conformance testing.* Ellis Horwood Ltd., Hemel Hempstead, Herts., G.B., 1993.

[SB85] B. Sarikaya and G. v. Bochmann. Obtaining normal form specifications for protocols. In L. Csaba, K. Tarnay, and T. Szentiványi, editors, *Conference on Computer Networking, COMNET'85*, pages 601–612, Budapest, Hungary, 1985. Elsevier Science B.V. (North-Holland).

[SD88] K.K. Sabnani and A.T. Dahbura. A protocol testing procedure. *Computer Networks and ISDN Systems*, 15(4):285–297, 1988.

[SLD89] Y.-N. Shen, F. Lombardi, and A.T. Dahbura. Protocol conformance testing using multiple UIO sequences. In Brinksma et al. [BSV89], pages 131–143.

[SLU89] K.K. Sabnani, A.M. Lapone, and M.U. Uyar. An algorithmic procedure for checking safety properties of protocols. *IEEE Transactions on Communications*, 37(9), 1989.

[SS90] R. Sijelmassi and B. Strausser. NIST integrated tool set for Estelle. In Quemada et al. [QMV90].

[UC95] A. Ulrich and S.T. Chanson. An approach for testing distributed software systems. In *Protocol Specification, Testing, and Verification, XV*, Warsaw, Poland, 1995.

[UY91] H. Ural and B. Yang. A test sequence selection method for protocol testing. *IEEE Transactions on Communications*, 39(4):514–523, 1991.

[WL93] C.-J. Wang and M.T. Liu. Automatic test case generation for Estelle. In *International Conference on Network Protocols*, pages 225–232, 1993.

PART TWO

Test Environments

4

Application of a LOTOS based Test Environment* on AAL5

J. Burmeister and A. Rennoch

GMD FOKUS, Research Institute for Open Communication Systems
Hardenbergplatz 2, D – 10623 Berlin, Germany
phone: +49 30 25499 - 241, fax: +49 30 25499 - 202
email: burmeister@fokus.gmd.de

Abstract

This paper provides a first experience report on a formal based testing methodology, which we applied on the ATM Adaptation Layer Type 5 (AAL5). A feasibility study is given, which shows the applicability of a formal description technique, namely LOTOS [2], to be suitable as a testing notation. In this approach it is one of our intentions to be able to check a test suite against a formal specification without leaving the language (and its formal semantics) domain.

The methods and tools presented in this paper are addressing the following aspects:

- validation of (possibly manual written) abstract test cases with formal data specification,
- the generation of the corresponding executable test cases,
- the test operation, and
- the visualized animation of the test execution trace.

As a prerequisite, a formalized, but simplified, protocol or service specification is expected, which will be also used for the analysis of the test execution. The generation (creation) of the LOTOS test cases is not addressed here.

Keywords: AAL5 Testing, Protocol Conformance Testing, Quality of Service Verification, Formal Description Techniques, LOTOS, TTCN, ASN.1.

1 Introduction

It was our intention to develop a platform which supports the standardized conformance testing methodology (CTMF) on the basis of a formal description technique (LOTOS).

*This testing environment has been partially funded by the European RACE Project R2088 TOPIC (Tool Set for Protocol and Advanced Service Verification in IBC Environments).

Although the ISO 9646 [4] standard provides its own testing notation, a formal description technique (FDT) has been selected due to some reasons presented in the following. The platform should be open for experiments which go beyond PCT (Protocol Conformance Testing) and TTCN [4, 1]. The use of an FDT like LOTOS makes a link to FMCT (Formal Methods in Conformance Testing, ISO/IEC JTC1/SC21 P54), a joint ISO/ITU-T project, which has already identified the needs of formal methods in protocol and service (conformance) testing.

Future requirements for ODP testing and/or QoS verification (addressed in QoS Framework Project ISO/IEC JTC1/SC21 P57) needs more powerful expressiveness than TTCN is currently able to provide. Unfortunately, TTCN does not have a sound formal semantics, which makes its extension nearly impossible. On the other side, any new test notation has also to integrate the features of the CTMF standard.

Our approach represents an alternative to the development of TTCN compiler which is currently lacking of a sound formal basis for the relationship between formal (FDT) protocol and (TTCN) test specifications. We've avoided any additional model for the test specification and can validate LOTOS test and protocol specifications without leaving the underlying semantics model.

In the following, if LOTOS is referred to as a test notation, only a certain subset of this language will be addressed w.r.t our intention to derive the executable test cases (ETC) automatically as far as possible. Specially, the abstract data types (ADT) are reduced to those, which can be derived from an ASN.1 [3] specification.

The rest of this paper is divided as follows: first, the used methodology is defined (chap. 2), then, the supporting tool environment is introduced (chap. 3). In chapter 4 the applicability of the methods and tools are demonstrated, followed by some further application ideas. And finally, the conclusions are given.

2 Methodological Approach

The objective of the LOTOS based test environment is to automate the generation of the executable test case (ETC) or suite (ETS) from a given abstract test case (ATC) or suite (ATS). Therefore, it is also intended to provide some support in the adaption of the different IUT (Implementation Under Test) interfaces, which are in general distributed, and which need coordination. To be flexible as possible, the test environment is designed to the distributed test method, which is also of important relevance for ODP service testing. For test analysis purposes, post-animation of the behavior of the test system is proposed to get a detailed view of the observations, and to identify those parts of the formal specification which are fulfilled or not by the IUT.

Due to a consequent formal approach in the code generation from an abstract test case, LOTOS and ASN.1 has been chosen as the front end to the user. Data specified in ASN.1 can be mapped into both some implementation languages like C, and also into LOTOS abstract data types, which then allows an implementation semantics for ADTs. Furthermore, an extension of ASN.1 is selected (RO-Notation from ISODE [10]) to define the abstract interfaces, i.e. the data exchange parameters at the abstract communication points.

LOTOS has been chosen due to its programming language design (although it is not a programming language), its checking and simulation features, and its parameterization capability. It has been already recognized, that the application of full LOTOS is not required to use it as a test notation, but some restrictions on that language enables the application of algorithm and tools to make LOTOS fit for being a test notation like TTCN.

Not at least, the LOTOS based test environment shows the integration of different available tools towards a powerful test tool set, which also integrates platform features like remote procedure calls (RPC under SunOS) and possibly ROSY/PEPSY (ISODE).

Figure 1 gives an overview of our methodological approach in technical terms, i.e. by illustrating the logical relationship between relevant documents (specifications and programs). The rectangle in the middle of the figure comprises the executable programs, which realizes the test execution phase:

- the test system, i.e. a master tester with its upper/lower tester agents and individual timer servers, and
- the system under test (SUT), which includes the IUT and its environment (e.g. communication services).

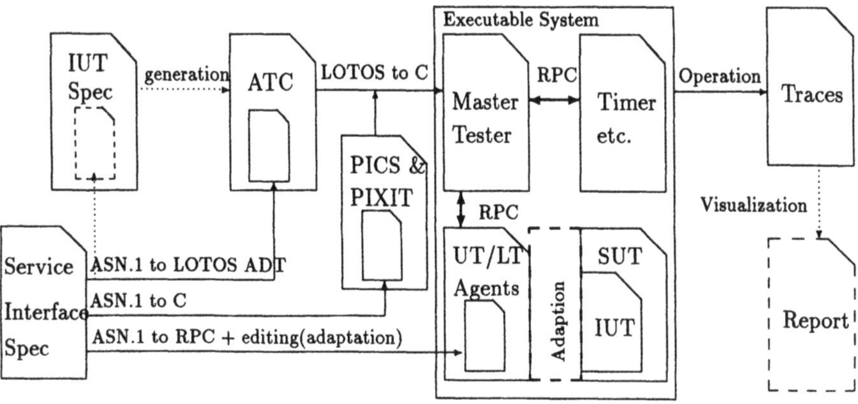

Figure 1: LOTOS based Test Approach

On the left side, which presents the preparation phase, the IUT specification and the service interface specification constitutes the starting point for the production and completion of the ATC and the PICS/PIXIT documents, which are input for the test system compilation.

The right side of the figure illustrates the test traces from the test operation, which will be used for visualization and test report production.

2.1 Preparation Phase

In this paper the term preparation refers to the tester preparation w.r.t. the IUT interfaces and the integration of additional context information (like PICS and PIXIT, Protocol Information Conformance Statement and eXtra information for Testing). It is not referred to the generation of abstract test cases, which is out of the scope of this paper.

In the following three aspects of preparation are discussed.

Abstract Service Interface Definition

For each test case (suite) an abstract service interface definition has to be specified formally, which in fact will be defined once. This task may be part of standardization bodies in the future.

The definition, usually specified in ASN.1, comprises the used data types and the communication points. And using the RO-Notation, the location of the data exchange can be described in an abstract style. An example is given later in chapter 4.

Adaptation of the IUT Interface

It is obvious, that for each different implementation of an IUT some implementation code must be generated manually for its adaptation. Since the LOTOS based test environment is designed to be an open tool, which should be able to adapt to a wide range of communication interfaces (in C), it uses and works only on (a subset of) abstract data types. The approach here is to provide a basic implementation representation (in C) of the abstract interface specification to reduce the manual involvement only to the implementation of the mapping between this (C) representation and the individual IUT interface.

The creation of the (C) representation has to be done only once for each protocol/service specification, but the final mapping to the concrete service primitive calls has to be done for each implementation.

Parameterization of a Test Case

Abstract test cases are abstract in the sense, that they do not refer to concrete objects like machines, implementation languages, and communication interfaces. If an abstract test case will be made executable, some specific values from the testing environment and the IUT must be taken into account. Normally, all these information are collected in the PIXIT document. Host names of the machines, time limitations, buffer and data constraints, and other information and requirements are to be provided.

Using LOTOS as a test notation, the user can easily integrate all requested parameters in the specification, as it can be seen later in chapter 4.

2.2 Execution Phase

The interpretation semantics of TTCN is weakly formalized in a procedural program notation like PASCAL. Based on a snapshot of the queues at the PCOs, an interpretation along the events is defined. It is not clear what will happen in case of events to be ignored.

If they are not removed from the input queues, any stored event might match to a future expectation, which was not intended. On the other hand side a queue might contain too many events due to the fact, that the frequency of taking a snapshot can be too low. In fact, the snapshot semantics is firstly to take a snapshot, secondly, to try to match the events, and thirdly to take a next snapshot. It is not based on what to do with the most recent observed event.

Other features of TTCN are the OTHERWISE construct, and that timers are dealt separately. Again, it is not clear how these features can be reflected properly in the executable code, e.g. where to locate the timers in a distributed environment. This comes along with the problem of the evaluation of the distributed input queues of the different PCOs. Although timeouts are defined, in time critical applications the time for the coordination and transmission of the observed events via e.g. a ferry clip must be taken in to account.

In this paper a slightly different execution method for abstract test cases is proposed. FDTs, which are already available since 1988, can be used to derive a so-called master tester, which is primarily independent from any test method. With some restrictions on the FDTs it is possible to compile the specification without loosing their usability as a test notation.

In our approach, LOTOS has been selected for the test specification notation (as well as for protocol and service specification). The following concepts and requirements for test notations are supported by LOTOS:

parameterization: The definition of external gates corresponds to the PCOs. Other specification parameters, which have unknown values during the test case generation, must be instantiated in the executable test, which are e.g. concrete host names (locations), timer values or variances of them, data to fill buffers etc. .
If a test specification consists of a test suite, the LOTOS parameterization technique can be also used to select and/or exclude certain test cases with respect to a given PICS document.

verdict assignment: At certain places within a specification, verdicts are required. This can be expressed by the `exit` statement (together with an appropriate `verdict` data type) and the functionality of processes and the specification itself. In case of a conformance test, also the final verdict can be directly specified.

structuring: The process definitions are suitable to build hierarchies for test groups, test cases, and test steps. Required loops are presented by recursive instantiations of processes.

constraints: Like in procedural programming languages, temporary variables can be declared, and values can be assigned. The **guards** in combination with the events express constraints (conditions) on parameter values.

test event: In general, an event is expressed by an `action` in LOTOS. Normally, a test event is composed of the PCO, where the event must occurs, and a data unit to be send or to be await. In LOTOS, the PCO is defined by the **gate**, and similar to TTCN, the sending and the expectation (receiving) of an event is expressed by '!' and '?' resp. .
The semantics of LOTOS is based on synchronous communication of events. Using LOTOS here as a test notation, we can retain this semantics in the sense, that the communication via PCOs (and thus with the environment) must be implemented synchronously. That means, that a sending event is always successful (synchronization with the environment), and a receiving event is then successful, if it has gotten the required (and possibly by a **guard** constrained) data.
By this concept, we define the basic different interpretation of an abstract test case in opposite to TTCN. A master tester has to react on each exchange of data at any PCO immediately.

test sequence: The sequential ordering of events is provided by the `action denotation` of LOTOS.

alternative: Alternate behavior of an abstract test step is supported by two operators of LOTOS. The disable operator `[>` is suitable to express the interruption of a test step due to some more general events like timeouts or disconnecting events. And the choice operator `[]` will be used in the sense of an **if-then-else** construct, either constrained by the environment, which offers an appropriate event, or conditioned by **guard** expressions.

timer: Timers are not directly supported by LOTOS. Nevertheless, on the abstraction level of abstract test cases, a timer can be seen as a service, which provide appropriate functionalities at certain interfaces. The event at these interfaces can be expressed exactly in the same style as it will be done for the PCOs.

otherwise: Unless reliable timer can be used, timeouts in combination with the disable operator will catch every deadlock situation, which might come from a blocking of receiving events which do not get anything. Thus, it's just a question of specification style.

data presentation: As to the behavior description feature of LOTOS, its expressiveness of abstract data must be restricted. Since ASN.1 is one of the most popular data description techniques for communication protocols and services, the LOTOS ADTs are reduced to those which are related to ASN.1 data types.
The advantage of using ASN.1 is the abstraction from programming languages, and it is suitable for the expression of concrete structures necessary for protocol and service data unit descriptions. Several relationship definitions between ADTs and ASN.1 are already defined.

The above listed requirements for a test notation are completely supported by a well-defined subset of LOTOS, which supports therefore the automated generation of an executable test case from an abstract one.

2.3 Analysis Phase

Each ATC should be related to a specific path in the given specification. Verdicts are assigned with respect to the preamble, the test steps, and the test events of the test case. A concrete verdict will be given depending on the actual behavior of the IUT. The relationship between the test sequence and the corresponding path in the specification has to be provided manually and may expressed via comments in the ATC. After the test run, an appropriate test report has to be provided. In case of IUT failures the implementation supplier is interested in those parts of the specification, which have not been satisfied by the IUT.

If the protocol or service has been specified formally, the observed trace (also expressed in an FDT) can be combined with the formal protocol or service specification. Using appropriate visualization tools, a post-animation of the test operation is possible, which shows the final result, and which also indicates the test steps, which has been undertaken, and which part of the specification has been either fulfilled (finishing with `pass`) or couldn't be verified (finishing with `inconclusive` or `fail`).

Similar to the validation of the abstract test suite against the specification, the obtained traces identify the specification parts, which have been verified in the test operation. The post-animation of the test run helps to reconsider the test without its repetition. The methodological approach here is to combine the specification together with the obtained traces plus some testing environment services like e.g. timers and verdict processors.

3 Tool Set Implementation

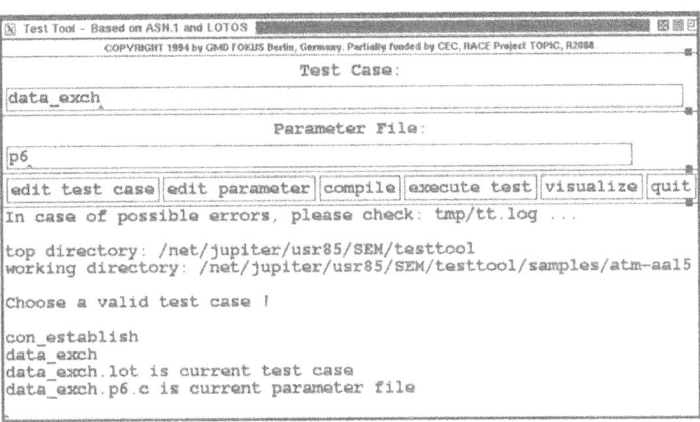

Figure 2: Graphical User Interface of the Test Tool `tt`

The current status of the test tool set (`tt`) supports the user in various ways in the three testing stages of preparation, execution, and analysis. Namely, three tools have been developed, and some tools of the TOPO tool set have been selected (previous versions are part of LITE [8]). A user interface is provided, which provides all these tool functionalities without presenting them explicitly. This user interface is a rapid prototyping of an X11 window application, and can be replaced easily by other (modified) user interface implementations (see fig. 2). In fact, a command line interface would fulfill the same tasks, so this tool environment could also run independent of any window surface.

In the following, the tools and their functionalities are briefly explained w.r.t. the three testing stages. They are following the methods given in the previous chapter. The automation process requires a certain file subsystem strategy. Since the executable test system will be derived from a selected abstract test case, all required information must be located in certain directories under particular names. Here, we abstract from file name conventions, but indicate the logical file categories.

3.1 Preparation Tools

The preparation phase is supported by two tools dealing with the data specifications, the interface specification, and the data transmission between distributed hosts. The presented tool environment supports the presentation of abstract data using ASN.1. More concrete, the RO-Notation from the ISODE platform has been identified as a suitable abstract description of data together with the interface description. Such an abstract description of data and interfaces allows the use of tools which provides encoding and decoding facilities to exchange data between machines (hosts), hence, using a distributed platform like ISODE or ANSAWare [12].

In our approach, we use an ASN.1 to C Compiler (a tool called `asn`), whose output is suitable for `rpcgen` [9], a tool under SunOS. The same ASN.1 compiler generates the LOTOS ADTs as required for the abstract test case. The used mapping mechanism have been reported in the TOPIC deliverable *Verification Tools (V2)* [6].

The current status of the tools does not support the automated generation of appropriate operations on data yet (e.g. comparison), neither in LOTOS ADT nor in C. They must be provided manually, but can be automated in the future.

Another task, which must be provided manually, but which cannot be automated, is the mapping between abstract represented data in C and the concrete data types used by the IUT at the PCOs.

3.2 Execution Tools

The transformation of the ATC into in ETC is supported by a set of tools, managed by a user interface, which combines the different functionalities (see fig 2). The testing tool set (`tt`) needs a directory for the configuration of the executable test of the IUT. This directory must be divided into the following subdirectories:

- a temporary (`tmp`) directory, which contains all the generated and copied files to produce the master tester;
- a PIXIT (`pixit`) directory, which contains the generated (RPC) coordination interfaces and the abstract data types to be used in the abstract test cases;

- a test suite (**testcases**) directory, which consists of the different test cases for a certain protocol or service specification, and which also contains a LOTOS specification of the testing environment, which will be completed and used later for the visualization of the test run (see next section);

- a binary (bin) directory containing the executable interfaces (RPC demons) to the IUT and/or underlying service (PCOs), which are generated from the interface specifications given in the PIXIT directory; other binaries may also located here;

- optional directories (include and lib) which refer to additional implementation details of the environment (libraries and header files) and the IUT to access the PCOs. Further directories are omitted, which store the source code (src) of the IUT (if available at all), documentations (doc), and/or manuals (man).

The main feature of the testing tool is to generate and run the Master Tester. For that purpose, the PIXIT documentation must be available from the preparation stage, i.e. the LOTOS abstract data types are generated, and the interface routines of the PCOs are prepared. Then, tt combines the selected test case with the data types, checks the obtained specification with TOPO tools, and takes the generated so-called *common representation* of LOTOS to produce the Master Tester using a LOTOS to C compiler. In fact, basically the behavior part of LOTOS is compiled as follows, and the data elements are mapped to those C data types and operations already generated by the ASN.1 tool.

The generated code implements the test events with RPCs, which are in fact implemented as synchronous calls. A racing strategy has been implemented, which discards all other events, if one of the set of alternative events has been occurred. Note, that the parallel operator is not implemented; firstly, it is not a requirement for conformance test notations as listed above (but may be one in future to test e.g. platforms and distributed services), and secondly, only the interleaving operator ||| and the partial synchronized parallel operator |[..]| make sense, whereby the latter one uses internal (hidden) gates to synchronize.

The generated trace consists of a sequence of actions together with the verdict of the test case, which can be easily transformed into a LOTOS behavior expression (see log file in chapter 4).

3.3 Analysis Tools

The analysis of the test run is supported by a visualization tool. The input specification for the DEMON [11] visualization tool is composed of the IUT specification, the observed trace, and some additional specification parts like e.g. used timers. Furthermore, some visualization information must be integrated as annotations. In this approach, the test operator and the IUT supplier can analyze the observed communication events. In case of a failure, a backwards reference to the specification is feasible to indicate the part of the specification, which has not been implemented correctly in the IUT. Note, that the visualization information does not change the semantics of the specification itself like e.g. other annotations will do for code generation of e.g. a state machine.

An excerpt of the visualization specification as used for our AAL5 experimentation looks as follows:

```
specification aal_env_spec[U (*# region 2 #*), ... , T2 (*# region 5 #*),
                           info (*# region 5 #*) ]:noexit
  (*# user_defined_icons
   region_definitions
   GRID R3 50 665 8 8 1.0; # timer & verdict
...
   GRID R7 140 30 8 8 2.0; # aal
  #*)

   type number is ...   endtype

   type primitives is
     sorts primitive
     opns
        a_connect_req, ... , a_close_req,
        pass, ..., tt_ti_start, tt_ti_expired :-> primitive
   endtype

   behavior
       test_system[U,L,T1,T2,info]
     ||
       (aal[U,L] (1) ||| timer1[T1] ||| timer2[T2] ||| verdict[info])
  where
     process aal[U,L](PN:number):noexit:=
       (*# region 7 #*)
     (  U ! 1 ! a_connect_req ; (* The appropriate AAL5 Specification *) ... )
    endproc

   process test_system[U,L,T1,T2,info]:noexit:=
       (*# region 1 #*)
#include "auto.lot"
       (* Integration of the observed trace expressed in LOTOS *)
     stop
   endproc

   (* Specification of the environment: *)

   process verdict [info]:noexit:=
         (*# region 3 #*)
        info ! 1 ! pass ; stop
     [] info ! 1 ! fail ; stop
     [] info ! 1 ! inconclusive ; stop
   endproc

   process timer1[T1]:noexit:=
         (*# region 3 #*)
        T1 ! 1 ! tt_ti_start ; timer1[T1]
     [] T1 ! 1 ! tt_ti_expired ; timer1[T1]
   endproc
   process timer2[T2]:noexit:= ... endproc
endspec
```

Another LOTOS compiler will be used, which generates the appropriate input code for DEMON, using the annotations '`(*#`' and '`#*)`' for visualization support. The current state of the art of this tool supports only very simple data types, but the tool is powerful enough to express at least the used abstract primitives.

In case of the AAL testing environment, we consider four independent processes (AAL5, 2 timers, verdict processor), which are synchronized by the test system, whose behavior is defined by the actual observed trace (`auto.lot`, derived from the log file). The visualization is presented in the following chapter.

4 Experiments with AAL5

First experiences have been made within the RACE Project TOPIC. Here, the service over XTPX [7] has been tested. After extending the testing tool set, tests have been applied on the AAL5 service [13].

We have used Sun work stations under SunOS 4.1.x. Two machines were equipped with FORE ATM adapter cards, together with the AAL software. The machines were connected via an ATM FORE ASX 200 switch (a third machine). One of the applied tests just only checks the sending and receiving of data:

```
specification test_data_exchange [T1,T2,L,R]
  (remote: CHARACTERSTRING, any_qos, sel_qos: DL_qos,
   data_flow: DL_dataflow, data: CHARACTERSTRING, data_len: INTEGER,
   to1, to2: INTEGER) : exit(verdict)

library Verdict endlib
library Timer endlib
library tt__primitive endlib
#include "../pixit/datatypess.adt"

behaviour
 test_data_exch [T1,T2,L,R]
 (remote,any_qos,sel_qos,data_flow,data,data_len,to1,to2)
 where
  process test_data_exch [T1,T2,L,R]
   (remote: CHARACTERSTRING, any_qos: DL_qos, sel_qos: DL_qos,
    data_flow: DL_dataflow, data: CHARACTERSTRING, data_len: INTEGER,
    to1, to2: INTEGER) : exit(verdict) :=
  T1 ! tt_ti_start ! to1 ;
  ( R ! a_connect_req ! SEQ_A_Con_Send(dest      (remote),
                                       qos       (any_qos),
                                       dataflow  (data_flow));
      L ! a_connect_ind ! data_flow ? re: A_Con_Send;
       L ! a_connect_resp ! SEQ_A_Con_Receive(qos      (sel_qos),
                                              dataflow (data_flow));
        R ! a_connect_conf ? re: A_Con_Receive;
         T2 ! tt_ti_start ! to2 ;
         ( R ! a_data_req ! SEQ_A_Data_Send(buf (data), len(data_len)) ;
             L ! a_data_ind ? re: A_Data_Send;
             exit(pass)
         [>
            T2 ! tt_ti_expired  ? x:BOOLEAN ;
```

```
          exit(fail)
        )
  [>
    T1 ! tt_ti_expired ? x:BOOLEAN ;
    exit(inconclusive)
  )
 endproc
endspec
```

The identifiers (e.g. `a_data_request`) of this specification have been taken in particular from the abstract service interface definition (after its mapping into LOTOS ADTs (`datatypes.adt`)). A small excerpt of this interface definition looks as follows:

```
AAL5-CT DEFINITIONS ::=
BEGIN ...
DL-dataflow ::= ENUMERATED
 { simplex(0),
   duplex(1),
   multicast(2) }
...
A-Data-Send ::= SEQUENCE
 { buf        CHARACTERSTRING,
   len        INTEGER }
...
a-data-req                        -- Abstract Data Request Primitive
    OPERATION
    ARGUMENT  A-Data-Send
  -- RESULT None
    ERRORS    { ... }
    ::= 5
...
END
```

The abstract test specification has been parameterized with different values for time out and data formats. Arbitrary user data have been taken, because no specification was available during the experimentation phase. An example parameterization for selected PIXIT values is given in the sequel, provided in ASN.1. The identifier names like `DL-dataflow` have been taken from the above abstract service interface definition:

```
PIXIT-NO-6 DEFINITIONS ::=
BEGIN
  IMPORTS DL-qos, DL-dataflow, simplex FROM AAL5-CT ;
remote CHARACTERSTRING ::= "\000\000\000\001\361\044\015\013"
any-qos DL-qos ::= { { 256 , 128 } , { 128 , 64 } , { 2 , 1 } }
sel-qos DL-qos ::= { { 256 , 128 } , { 128 , 64 } , { 2 , 1 } }
data-flow DL-dataflow ::= simplex
data-len INTEGER ::= 4096
data CHARACTERSTRING ::= "\
0123456789012345678901234567890123456789012345678901234567890123\
. . .
0123456789012345678901234567890123456789012345678901234567890123"
to1 INTEGER ::= 400  -- ms
to2 INTEGER ::= 200  -- ms
END
```

Note, that at the current state of the art the ASN.1 tool does semantics checks on data type definitions only, and therefore data values must be compiled manually into C, e.g. the `remote` value:

```
CHARACTERSTRING remote = { 8 , "\000\000\000\001\361\044\015\013" } ;
```

In the PIXIT example above the general timeout (`to1`) has been set to 400 ms, whereby the data transmission phase (`to2`) has been limited to 200 ms. As you can simply identify, the ATM host address is given as an extra information for testing, since the abstract test case does not know concrete addresses of the testing environment.

Furthermore, the parameterization of the PCO (`L, R`) and timer gate (`T1, T2`) locations shows the independence of these interfaces from the generated (master) tester location. Due to the generation of RPCs, the (master) tester could be located on a fourth machine.

Assuming, that the first timeout value `to1` has been chosen to small, the verdict will lead to an inconclusive verdict. The appropriate log file looks as follows:

(* Started: Mon Mar 27 18:41:21 (723 ms) 1995 *)
T1!tt_ti_start!to1
R!a_connect_req!SEQ_A_Con_Send(dest(remote), qos(any_qos), dataflow(data_flow))
T1!tt_ti_expired?x:BOOLEAN
exit(*inconclusive*)
(* Finished: Mon Mar 27 18:41:22 (190 ms) 1995 *)
(* Finished with verdict inconclusive *)

This log file can be combined with the formal service specification or, if not available, with a simplified version, and run by the visualization tool DEMON. The post-animation demonstration is presented in figure 3. The example illustrates a timeout during the connection establishment, i.e. here it results with an inconclusive verdict. Therefore, the test operator and the IUT supplier may conclude, that either the preamble does not work properly, or the connection establishment is not correct implemented.

The presented test approach is not fully applicable to do performance measurements. E.g. specially in the realm of AAL testing, the exchange of data within the IUT is much more faster as it is provided by the tester coordination procedures (like RPCs). Furthermore, as long as timers have to be integrated like additional services, a certain delay must be taken into account. Hence, like in TTCN, timers should be used only to limit the duration of a set of events.

On the other hand side, long time statistical measurements are feasible. E.g. the transmission of data by an appropriate loop within a certain time frame can be expressed easily.

5 Conclusions and future Plans

This paper has shown the feasibility of using an FDT (LOTOS) as a suitable test notation. It has focussed on the distributed testing platform which implements certain features of a restricted subset of LOTOS. The problem of test case generation has been left out; and to validate a test specification against the protocol or service specification, certain constraints like e.g. a common data type part specification must be fulfilled. In this paper, ASN.1 is proposed to define the minimal set of (abstract) data types.

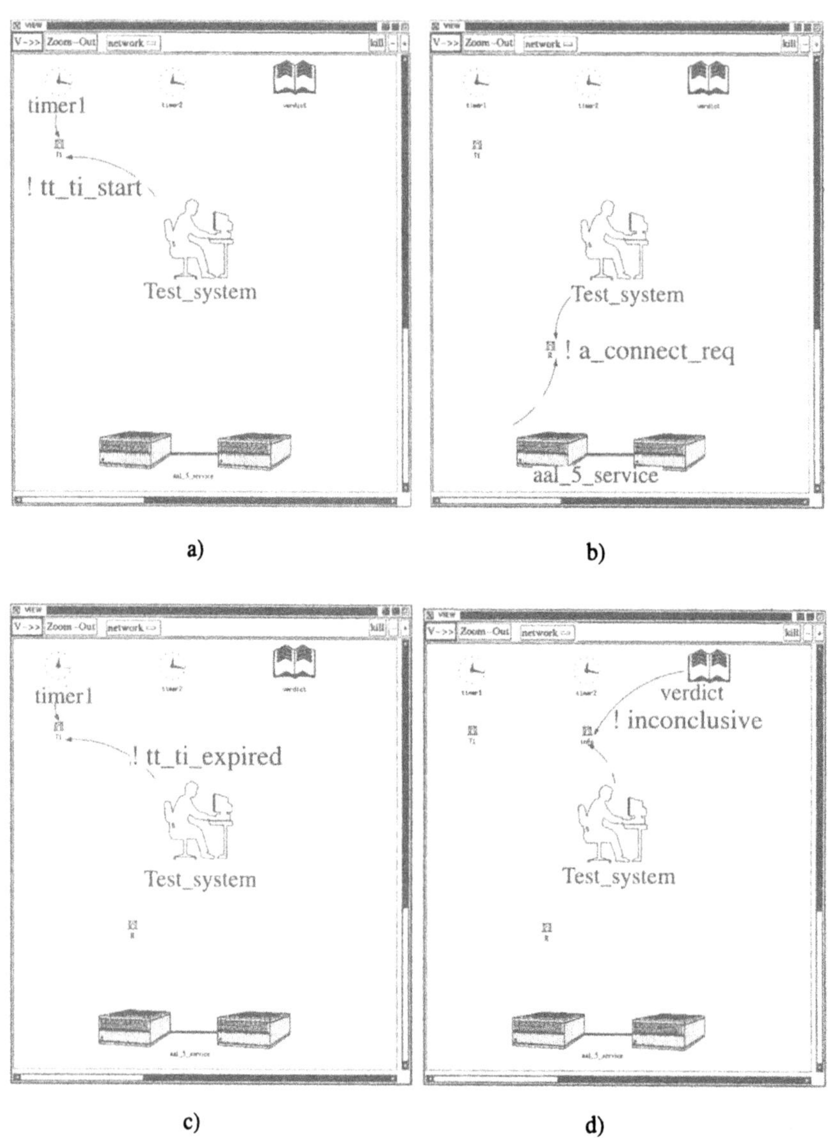

Figure 3: AAL5 Test Trace Visualization

A set of advantages makes LOTOS to be an alternative to TTCN:

1. it is based on a sound formal method and its semantics is well defined;
2. the relationship between an abstract test case and the corresponding specification parts (if it is also formalized by an FDT) can be verified (specially in case of manual generation);
3. the derivation of the executable test suite is supported due to the sound semantics definition of LOTOS;
4. the parameterization of an abstract test case is an integrated part of the notation;
5. PCOs are not restricted to service access points of the IUT, and may also include other test environment components like timer, and, in future, e.g. traffic generator and CPU loader.

Since this feasibility study has been realized by a prototype, certain limitations have to be accepted. E.g. the availability of formal specification and its corresponding (formal) test suites are very rare, especially on new protocols and services like ATM and the adaptation layers. In fact, very rudimentary ATM-AAL specification and test specifications have been developed manually to demonstrate the distributed testing platform.

The current testing tool set covers the conformance testing methodology from generated abstract test cases, their execution and analysis, up to a preliminary test report generation.

Furthermore, extensions are feasible, which make this tool set and its testing approach also suitable for QoS testing. PCOs can be introduce in an abstract test suite to control any traffic generator or CPU load, and recursive process instantiations can be used to specify loops for statistical measurements. On the other hand side, language extensions of the FDTs are desirable w.r.t. their testing purpose and for their applicability to quality testing aspects including time and resource constraints on events.

Acknowledgment

This testing environment has been partially funded by the European RACE Project R2088 TOPIC (Tool Set for Protocol and Advanced Service Verification in IBC Environments, see also WWW: `http://www.fokus.gmd.de/step/topic`). We would like to thank J. de Meer, technical leader of the TOPIC project, who invented and discussed with us the basic ideas of the distributed test system, and also B. Stepien as a project consultant form the University of Ottawa, Canada, who worked on the implementation of the visualization concepts.

References

[1] B. Baumgarten, A. Gießler; OSI Conformance Testing Methodology and TTCN, North-Holland, Amsterdam, 1994.

[2] ISO, IS 8807, Information Processing Systems, Open Systems Interconnection, *LOTOS–A Formal Technique Based on the Temporal Ordering of Observational Behaviour*, July, 1988.

[3] ISO/IEC DIS 8824-1, *ASN.1. Abstract Syntax Notation One – Part 1, Specification of Basic Notation*, 1992.

[4] ISO/IEC IS 9646, Information Retrieval, Transfer and Management for OSI: *Conformance Testing Methodology and Framework*, (1991).

[5] RACE Project R2088 TOPIC, *Integration of the Toolset Prototypes (V2)*, Deliverable 16, Ref. R2088/GMD/SEM/DS/L/016/b4, 1994.

[6] RACE Project R2088 TOPIC, *Verification Tools (V2)*, Deliverable 18, Ref. R2088/GMD/SEM/DS/L/018/b1, 1994.

[7] RACE project R2088 TOPIC, *Experience Report of the Verification Demonstrator*, Deliverable 20, Ref. R2088/CLE/TEE/DS/P/020/b1, December 15th. 1994.

[8] ESPRIT Project 2304 LOTOSPHERE, *LITE User Manual*, Ref. Lo/WP2/N0034/V08, 1992.

[9] Sun Microsystems, *Network Programming Guide*, Part Number 800-3850-10, 1990.

[10] ISODE, The ISO Development Environment: User's Manual Vol. 1, *Application Services*, & Vol. 4, *The Applications Cookbook*, 1991

[11] DEMON, *Reference Manual V3.0*, Mari Computer Systems Ltd, 1993.

[12] Architecture Projects Management Limited, *ANSAware Version 4.0 Manual Set*, Cambridge, UK, March 1992.

[13] FORE Systems Inc. , *ATM Devices and Network Interfaces, Manual Rel. 3.0*, June 1994.

5

Stable Testers for Environment Sensitive Systems

Mohammed Ghriga
Department of Computer Science, Long Island University,
1 University Plaza, Brooklyn, NY 11201 USA
E-mail: mghriga@comsci.liunet.edu

Abstract

We present a new testing technique for a restricted class of nondeterministic systems, we qualify as *environment sensitive.* Such systems are specified and implemented such that all nondeterministic transitions, whether internal or not, are conditionally driven function of a system's *environment conditions.* The rationale and the practical aspects of this class will be given. The intention is to provide a *pragmatic* framework to improve the testability aspects of nondeterministic systems. We show that our restricted class of systems coupled with a new testing postulate *Wait and Lock Stability* lead to testing conditions similar to those of traditional sequential/deterministic systems; the testing postulate is analyzed and an algorithmic characterization for the construction of testers, that are both *sequential* and *stable*, from system specifications provided. Finally, we establish the correlation of our results and testing approach to the related work in failure and failure trace semantics when applicable.

Keywords

Conformance Testing, Labelled Transition Systems, Testing with Deadlock Detection.

1 Introduction

Conformance testing is known to be a problem of considerable complexity. Nevertheless, it is crucial in that it helps ensures *interoperability* among multi-vendor products. Significant efforts have been expended to establish testing theories and frameworks, formulate methodologies for the generation of conformance tests, and develop techniques to optimize such tests. Yet, many problems and issues remain open as indicated by Cavalli, Favreau and Phalippou (1994). This paper attempts to: (1) investigate notions of implementation in conjunction with a *restricted* class of nondeterministic systems, we will qualify as *environment sensitive*; and (2) link *test execution procedures* to current practices in testing sequential/deterministic systems. We use ***notion of implementation*** to simply refer to the conditions under which an implementation is said to satisfy its specification.

Communicating systems can be described as processes whose dynamic behaviors are formalized as labelled transition systems (LTS). A system is said to be ***environment sensitive*** if all nondeterministic transitions in the system, whether internal or not, are conditionally driven by current environment conditions rather than being implicitly driven. Such environment sensitive systems can be of particular interest to industry in various domains, e.g.: client-server applications, communicating systems, critical-safety systems, etc. These issues will be made clearer in Section 2.

Testing LTS-based implementations has concentrated on notions of implementation based on trace semantics (exemplified by Cavalli, Kim and Maigron (1993)), failure semantics (exemplified by Brinksma (1988); Brinksma, Scollo and Steenbergen (1987); Tretmans (1994); Drira (1994); Fujiwara and Bochman (1992); Pitt and Freestone (1990)), and failure trace semantics (Langerak, 1990). The most detailed work on failure semantics is due to Brinksma (1988). Unfortunately, the implied notions of implementation may not be satisfactory as indicated by Langerak (1990). This same criticism of inadequate discrimination applies with respect to the class of environment sensitive systems. On the other hand, we argue that Langerak's notions of implementation are too expensive to be of practical interest. This paper proposes a more practical and pragmatic notion of testers that preserve the intuitive behavior induced by nondeterminism.

The paper is organized as follows. In Section 2, we formally define the class of environment sensitive systems and project on their practical aspects. In section 3, we present the elements of our testing approach and the concept of ***stability locks*** as well as the feasibility of testing. In Section 4, we give a characterization of finite LTSs. In Section 5, we propose an algorithmic treatment of internal events based on a simple ***fragmentation process***. In section 6, we provide an algorithmic characterization of our testers and define their induced notion of implementation. In Section 7, we examine/establish the relationships between our work and the related work in both failure and failure trace semantics. Section 8 concludes the paper.

2 Environment Sensitive Systems

In this section, we formally define the notion of environment sensitive systems, illustrate their basic concepts, and discuss their practical aspects.

We first give the definition of a ***labelled transition system*** (LTS). An LTS S is a 4–tuple $S = (Q, \sum, \Delta, s_0)$, where Q is a countable non-empty set of states, $\sum$ a countable set of observable actions, Δ a set of transitions which is a subset of $Q \times (\sum \cup \{\tau\}) \times Q$ with $\tau \notin \sum$ being the unobservable or internal action, and $s_0 \in Q$ the initial state of S.

An element $(s, \mu, s') \in \Delta$ is interchangeably written as $s \xrightarrow{\mu} s'$. The main notational conventions are given in Figure 1, where the μ_is and μ are in $\sum \cup \{\tau\}$ and S is an LTS with s_0 as an initial state. A state s is said to be ***stable*** if $\neg(s \xrightarrow{\tau})$ (that is, there is no outgoing transition with label τ at s), and ***unstable*** otherwise (Petrenko, Bochmann and Dssouli, 1994). An LTS is ***stable*** if all its states are stable, and ***unstable*** otherwise.

Notation	Meaning
$s \stackrel{\mu_1\mu_2\ldots\mu_n}{\longrightarrow} s'$	$\exists s_i (1 \leq i \leq n)\ s \stackrel{\mu_1}{\longrightarrow} s_1 \stackrel{\mu_2}{\longrightarrow} \ldots \stackrel{\mu_n}{\longrightarrow} s_n = s'$
$s \stackrel{\epsilon}{\Longrightarrow} s'$	$s \equiv s'$ or $\exists n \geq 1\ s \stackrel{\tau^n}{\longrightarrow} s'$
$s \stackrel{\mu}{\Longrightarrow} s'$	$\exists s_1, s_2\ s \stackrel{\epsilon}{\Longrightarrow} s_1 \stackrel{\mu}{\longrightarrow} s_2 \stackrel{\epsilon}{\Longrightarrow} s'$
$s \stackrel{\mu_1\mu_2\ldots\mu_n}{\Longrightarrow} s'$	$\exists s_i (1 \leq i \leq n)\ s = s_0 \stackrel{\mu_1}{\Longrightarrow} s_1 \stackrel{\mu_2}{\Longrightarrow} \ldots \stackrel{\mu_n}{\Longrightarrow} s_n = s'$
$s \stackrel{\mu_1\mu_2\ldots\mu_n}{\Longrightarrow}$	$\exists s'\ s \stackrel{\mu_1\mu_2\ldots\mu_n}{\Longrightarrow} s'$
$s \stackrel{\mu_1\mu_2\ldots\mu_n}{\not\Longrightarrow}$	$\neg\exists s'\ s \stackrel{\mu_1\mu_2\ldots\mu_n}{\Longrightarrow} s'$
$out(s)$	$= \{\mu \in \sum \mid s \stackrel{\mu}{\Longrightarrow}\}$
$Tr(s)$	$= \{\sigma \in \sum^* \mid s \stackrel{\sigma}{\Longrightarrow}\}$
$Tr(S)$	$= Tr(s_0)$
$s\ ref\ A$	$\forall \mu \in A:\ s \stackrel{\mu}{\not\Longrightarrow}$ (where $A \subseteq \sum$)
$S\ after\ \sigma$	$= \{s \mid s_0 \stackrel{\sigma}{\Longrightarrow} s\}$ (where $\sigma \in \sum^*$)
$(S\ after\ \sigma)\ ref\ A$	$\exists s \in (S\ after\ \sigma):\ s\ ref\ A$
$N_\tau(S)$	number of τ labelled transitions in S

Figure 1: Notations for Labelled Transition Systems.

Definition 1 *An environment sensitive system E is a 3-tuple $E =< S^E, env_p^E, env_{ip}^E >$, where S^E is an LTS specification of E, env_p^E a countable set of (measurable) environment parameters of E, and env_{ip}^E a finite (ordered) list of predicates that are boolean functions of the environment parameters of E.* □

Let E be an *environment sensitive system* (ESS). The set env_p^E contains basically the parameters of E's environment that can influence the dynamic behavior of E. These parameters are used as *variables* in the predicates of env_{ip}^E that are used to describe the interaction policies of E. Thus, we use the term *interaction policy condition instance* (IPCI) to refer to a sequence of 0s and 1s, where the ith *value* (either 0 or 1) in the sequence is the value of the ith *predicate* in the list env_{ip}^E. It is assumed that all nondeterministic transitions in S^E, whether internal or not, are conditionally driven by the predicates in env_{ip}^E. E is said to be stable iff S^E is stable. Throughout this paper, we assume a tree representation of S^E. These trees, which are generally infinite, will be called *tree labelled transition systems* (TLTS). We let $TLTS(S^E)$ denote the tree representation of S^E. We assume that these processes do not contain *divergences* (i.e. no infinite chains of internal actions/transitions). Moreover, we assume that $Event(s \longrightarrow s')$ returns the label of the edge $s \longrightarrow s'$ and $parent(s)$ returns the parent of node s in the tree representation. Note $parent(s_0)$ is **nil**.

Inspired by Langerak (1990), we give two simple examples from the realm of coffee/tea vending machines. We view tea (t) and coffee (c) as measurable resources in these machines. Let A_t (for tea) and A_c (for coffee) be parameters that indicate the availability of such resources.

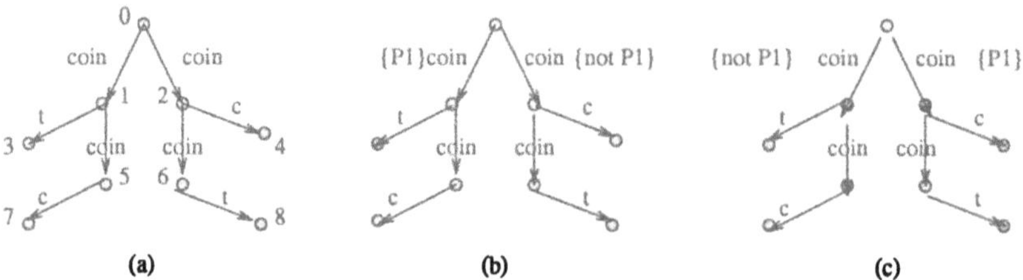

Figure 2: LTS Specification of E_1 and its Interpretations with $P_1 = A_t \geq A_c$. (a) Specification S^{E_1}. (b) An Interpretation of E_1. (c) Another Possible Interpretation of E_1.

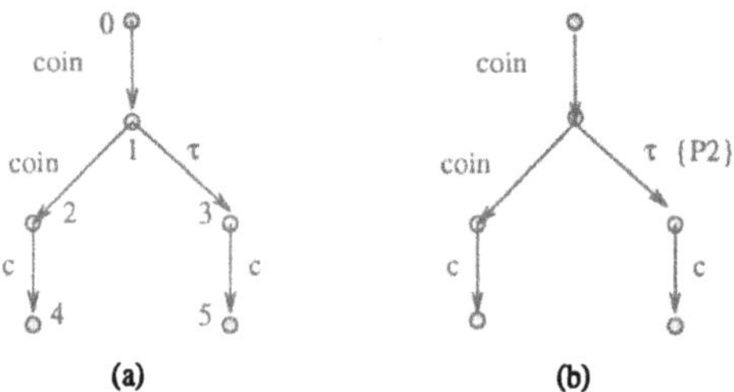

Figure 3: LTS Specification of E_2 and its Interpretation with $P_2 = A_c \geq \kappa$. (a) Specification S^{E_2} of E_2. (b) Interpretation of E_2.

Example 1 *Let* $E_1 =< S^{E_1}, env_p^{E_1}, env_{ip}^{E_1} >$*, where* S^{E_1} *is depicted in Figure 2,* $env_p^{E_1} = \{A_t, A_c\}$*, and* $env_{ip}^{E_1} = [P(A_t, A_c), \neg P(A_t, A_c)]$ *with* $P(A_t, A_c) = A_t \geq A_c$. □

As in (Langerak, 1990), E_1 always gives us what we want (Figure 2(a)). However, the cost for a desired choice (i.e. drink) depends on the availability of resources (i.e. tea and coffee). The possible IPCIs of E_1 are the sequences [01] and [10]; the other possible sequences are infeasible given $env_{ip}^{E_1}$. Figure 2(b),(c) give the possible interpretations of E_1, which differ in the attachment of the predicates to the nondeterministic transitions. In Figure 2(b) and under IPCI [10] (for example), if one desires coffee, he/she has to insert an additional coin to get coffee after the *refusal* of the first trial for getting coffee. Similarly, other scenarios (that may or may not contain refusals) can be derived.

Example 2 *Let* $E_2 =< S^{E_2}, env_p^{E_2}, env_{ip}^{E_2} >$*, where* S^{E_2} *is depicted in Figure 3,* $env_p^{E_2} = \{A_c\}$*, and* $env_{ip}^{E_1} = [P(A_c)]$ *with* $P(A_c) = A_c \geq \kappa$ *(κ is a constant).* □

Again, E_2 always give us coffee, but the cost is always enforced to be *two coins* if the availability of coffee is below certain threshold κ (a constant). Otherwise, the cost is either one or two coins. Notice that the possible IPCIs of E_2 are [0] and [1]. Suppose that under IPCI [0], one inserts a coin and then tries the coffee button. Regardless of the waiting time before pressing the button after the first coin, the request will be refused. In order to get coffee, he/she has to insert an additional coin. On the other hand, transition τ may occur under IPCI [1] after an insertion of a coin. After this first coin, there are two states s_1 and s_3 that are reachable, where the vending machine might *lock its stability* (i.e. commit) after waiting. If the lock occurs at s_1 (s_3, respectively), the cost for coffee will be two coins (one coin, respectively). The underlying motivations of this *relaxed* interpretation are: 1) to avoid the concept of "quickly offer" (Langerak, 1990) an event which may deemed to be impractical in real applications, and 2) to bring a robust notion of stability (through stability locks) to unstable ESSs, which allows to bring closer our systems to the traditional sequential/deterministic systems.

Coffee and tea having been treated as resources, it is easier now to project on the practical aspects of ESSs. In practical situations, one may include several system parameters (as environment parameters) to *anticipate* and *formulate* interactions policies under various environment conditions, to *establish* discriminating views of the system with respect to the classes of potential users, and to *enforce* such views to maintain/converge towards a normal operational mode of the system or to meet certain reliability requirements. The underlying concepts of ESSs are of particular interest to client-server applications, communication systems, critical-safety systems, etc- which can be applied towards the specification and implementation of systems with improved reliability.

3 The Elements of Our Testing Approach

Given an ESS E, the problem of testing E in our framework translates into the following: 1) *environment set-up* which consists of adjusting E's environment to *satisfy* a feasible IPCI based on env_p^E and env_{ip}^E; and 2) *derivation and application of tests* using S^E as a *reference specification*, which is the result of our weak characterization of ESSs. Throughout the paper, it is assumed that a test is applied under a feasible IPCI. Since the attachment of the predicates in env_{ip}^E to S^E is not known, the second aspect of the problem has been reduced to resolving the problems of test derivation and the procedural application of tests based on $< S^E, \emptyset, \Lambda >$ (Λ denotes an empty list). For readability, such system notation will be referred to -interchangeably- as S^E or S.

Our approach is conceptually based on the (intuitive) basic idea of n-testers (Pitt and Freestone, 1990) and a weaker- but modified- notion of traces than *failure traces* (Langerak, 1990) called *Quasi-Refusal (QR) traces*. It is qualified as such as we make only partial use of the unspecified interactions at some states. We shall extend the notion of n-testers to allow for the observation of deadlocks (or refusals) and continue testing afterwards, which leads to what we call n-QRtesters and their corresponding formal notion of implementation $QRimpl_n$ (short for n-$QRimplementation$)- for better conforming implementations that capture our intuitiveness of nondeterminism. For practical testing, we limit the deadlocks that appear in our traces to those that can discriminate between states that are nondeterministically reachable.

For testing ESSs, we advance a more relaxed and concrete version of *Wait for Stability* (Langerak, 1990), called *Wait and Lock Stability*: it could be possible to postulate a maximum response time for a given system; after the maximum time, the system is assumed to have *locked its stability*. In the presence of internal actions, the system may elect to take an arbitrary sequence- including the empty sequence- of internal actions based on its current IPCI, but the stability is *locked* afterwards. We assume *fairness* between such arbitrary choices so that one can get all possible observations under a given feasible IPCI. Our detection of deadlocks is as feasible as Langerak's because the only element of difference is that we do not assume that waiting leads implicitly to a stable state. For this end, suppose that the "locked stability" were waved off and that the system has reached stability in a given state after the postulated maximum time; if an internal event occurs afterwards because stability is not locked this would be a violation of the maximum response time for stability, and thus, it is also a violation of Langerak's postulate. Most importantly, *locked stability* can be easily realizable in practice since the action *lock stability* can be implemented as *commit to a state*. The consequences of this postulate will be apparent in Section 5. Moreover, our testing postulate is of practical significance in that testing unstable systems will translate into testing a stable system with *unknown stability locks*, which gives rise to the notion of *stable testers*. Throughout, the use of the expression "wait(ing) for stability" should be understood in the context of our postulate, i.e. "wait(ing) for stability to be locked". Also, any subsequent reference to an LTS or one of its variant forms must be taken in the context of being a component to an ESS. Finally, it is assumed that there are no internal actions that result in timeouts for clarity of exposition. We defer the treatment of such events to a later stage.

4 Finite Tree Labelled Transition Systems

The testing process is finite, and so must be the test sequence interactions. Accordingly, we reduce the number of executed events in the LTS component of an ESS to a certain maximum bound n.

Definition 2 *Let S be an LTS. A state s in $TLTS(S)$ is said to be further-expandable if $\exists \mu \in \sum \cup \{\tau\}$ such that $s \xrightarrow{\mu}$ in $TLTS(S)$.* □

Definition 3 *Given an LTS S and an arbitrary number n, we define $TLTS_n(S)$ as the finite tree obtained from $TLTS(S)$ by truncating every path $s_0, s_1, \ldots, s_k, s_{k+1} \ldots$ of $TLTS(S)$ to path $s_0, s_1, \ldots, s_k$ such that the following conditions hold:*

1. $s_0 \stackrel{\sigma}{\Longrightarrow} s_k \wedge \sigma \in \sum^* \Longrightarrow |\sigma| \leq n$,
2. $|\sigma| < n \Longrightarrow s_k$ *is not further expandable in $TLTS(S)$, and*
3. s_k *is further expandable in $TLTS(S)$* $\Longrightarrow Event(parent(s_k) \longrightarrow s_k) \in \sum$. □

The first condition indicates that all traces of $TLTS_n(S)$ are of length less than or equal to n (action τ is considered of length 0). The second condition indicates that such traces are of length strictly less than n only if the ending states of the corresponding paths are not further expandable. The last condition of the definition ensures the unicity of the paths in $TLTS_n(S)$. Clearly, one can always add chains of internal events to the leaves of a $TLTS_n()$ without affecting the satisfiability of the first two conditions. Such ambiguity is avoided in our definition. Note these chains of internal events will be considered in $TLTS_{n+1}()$.

We do not use the *distance* from the root for the derivation of finite trees on the basis that there are varied degrees of nondeterminism induced by internal events in specifications, and that the assessment of conformance (testing confidence) implied by n–QRtesters should not be dependent on such factor. The construction process of $TLTS_n()$ is finite under the assumption of no divergences in specifications. For a given LTS S that describes a finite process, we note that there exists k such that $TLTS_k(S) = TLTS(S)$.

5 Treatment of Internal Events by Fragmentation

In this section, we propose a new technique for the treatment of internal events based on the fragmentation of initial LTS specifications. We first provide its primitive form and then its closure process. The primitive form is viable in that the validity of transformations with respect to our postulate becomes intuitive. The goal is to generate ***stable weakly-initial*** labelled transition systems that simulate (step by step) the initial LTS specification and vise-versa, under our testing postulate.

Recall that we have used $S = (Q, \sum, \Delta, s_0)$ to denote an (*initial*) LTS S with initial state s_0. Similarly, we write $S = (Q, \sum, \Delta, IQ)$, where $IQ \subseteq Q$, to denote a *weakly-initial* LTS S with IQ prescribed as the set of all possible initial states of S. This qualification of LTSs by *initial* and *weakly-initial* has been borrowed from Starke (1972), used in the context of nondeterministic automata. The notions of *stability/unstability* of a state or the LTS as a whole are the same for both models. A weakly–initial LTS is thus stable if all its states are stable. To accommodate for such weakly–initial LTS, we define $Tr(S) = \{\sigma \in \sum^* \mid s_i \stackrel{\sigma}{\Longrightarrow} \wedge s_i \in IQ\}$ and $S\ after\ \sigma\ =\ \{s \mid s_i \stackrel{\sigma}{\Longrightarrow} s\ \wedge\ s_i \in IQ\}$ with $\sigma \in \sum^*$.

5.1 Primitive Fragmentation Process

Contrary to other mechanisms for the removal of internal transitions (Cavalli, Kim and Maigron, 1993) (Drira,1994), we propose a fragmentation based technique of *initial* LTS specifications. In this section, we focus on the primitive form of this process and some of its properties.

For weakly-initial LTSs, we define the *forest* counterpart of tree LTS representation: a forest LTS (FLTS) is a collection of components, where each is a TLTS. A *partial ordering by level* of a set of states Q of an FLTS is a sequence $L_0 L_1 \dots L_k$, where L_i $(0 \leq i \leq k)$ is a set of the states at level i in the FLTS (notice the roots are considered to be of level 0).

Definition 4 *Let S be an unstable FLTS with a set of states Q. Let $d^+(s)$ denote the number of outgoing edges for $s \in Q$. The primitive form of the fragmentation process of S is described by the following sequence of actions:*

1. *Partially order by level the set Q of FLTS S;*
2. *Select the first unstable state s in the above partial order;*
3. *Perform one of the following actions if its applicability conditions are satisfied:*
   - *If $parent(s) \neq nil$ and $d^+(s) > 1$, apply the operation described in Figure 4.*
   - *If $parent(s) \neq nil$ and $d^+(s) = 1$, apply the operation described in Figure 5.*
   - *if $parent(s) = nil$ and $d^+(s) > 1$, apply the operation described in Figure 6.*
   - *if $parent(s) = nil$ and $d^+(s) = 1$, apply the operation described in Figure 7.* □

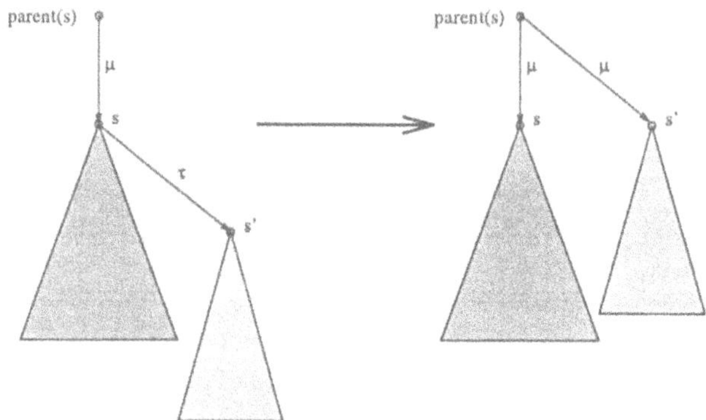

Figure 4: Primitive Form of Fragmentation: $parent(s) \neq nil$ and $d^+(s) > 1$; $\mu \in \sum \cup \{\tau\}$.

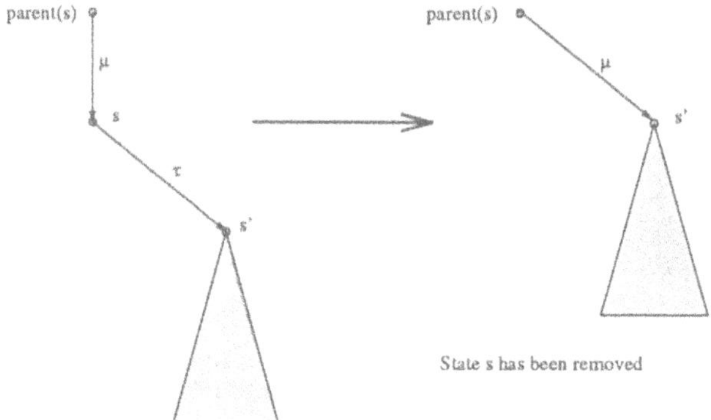

Figure 5: Primitive Form of Fragmentation: $parent(s) \neq nil$ and $d^+(s) = 1$; $\mu \in \sum \cup \{\tau\}$.

Dark tinted triangles in Figures 4 and 6 denote *non–empty* subtrees. Figure 4 depicts the fragmentation process that occurs at an unstable state s that has a parent. The form of the FLTS at s before fragmentation is straightforward since s is unstable and $d^+(s) \geq 2$; if there are two or more internal transitions at s then arbitrarily choose one. Similar reasoning applies to the other cases. In Figure 4, the removal of transition $s \xrightarrow{\tau} s'$ is coupled with the addition of transition $parent(s) \xrightarrow{\mu} s'$ with μ being the label of the transition from $parent(s)$ to s. The subtree rooted at s' remains unchanged. The intuition captured by this transformation is that if the flow of execution has reached s by some trace then it is also possible that this flow of execution may have reached state s' (which is different from s). Moreover, if stability (after waiting) might have been locked at state s then the internal event would not happen after this lock (otherwise, it is going to be a violation of the maximum response time). This information would be lost if states s and s' were collapsed.

Figure 5 describes the degenerate case of Figure 4 when the subtree T (the dark tinted triangle) is empty. In such case, we have similar basic steps coupled with the removal of

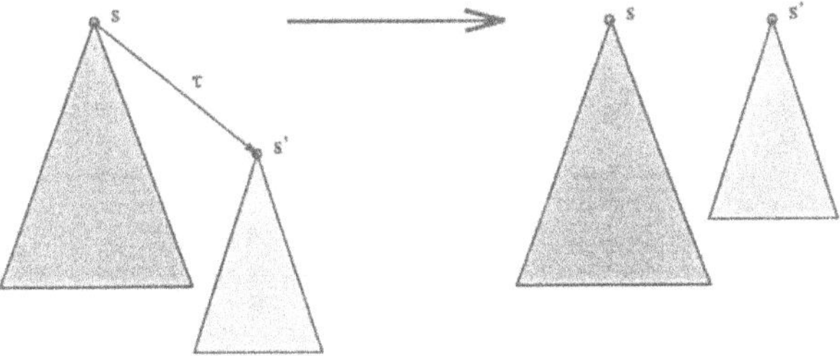

Figure 6: Primitive Form of Fragmentation: $parent(s) = nil$ and $d^+(s) > 1$; $\mu \in \sum \cup \{\tau\}$.

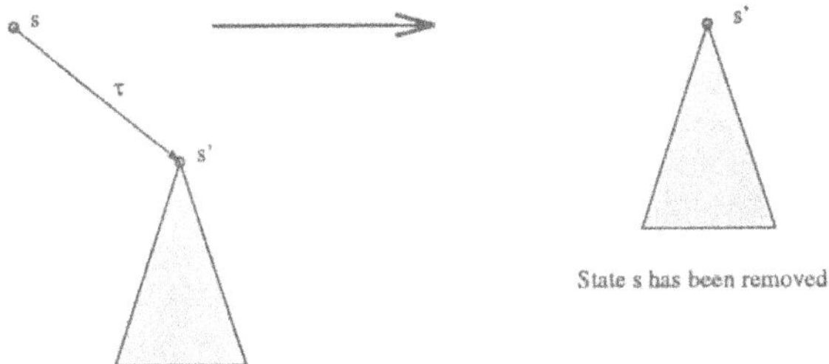

Figure 7: Primitive Form of Fragmentation: $parent(s) = nil$ and $d^+(s) = 1$; $\mu \in \sum \cup \{\tau\}$.

state s (notice that transition $parent(s) \xrightarrow{\mu} s$ is deleted as the result of removal of s). The intuition is best described by the notion of "waiting for stability" which is part of "wait and lock stability". Under the conditions of Figure 5, state s is a *transient* state and if we do wait for stability the flow of execution must reach state s', otherwise the lock of stability at s would result in a "deadlocked" system (i.e. no subsequent progress).

Figure 6 describes an intuition similar to that of Figure 4 when state s is a root. Upon the start of the system, it is possible that the system may reside in state s'. However, Figure 7 is a degenerate case of Figure 6 in which the root is a transient state and if we do wait for stability the system may reside in state s'.

Given an FLTS S, let fp be an algorithm such that $fp(S) = S$ if S is stable, otherwise $fp(S)$ is the result of the application of the primitive form of the fragmentation process described by Definition 4. Hence, we have

Lemma 1 *Let S be an FLTS. $fp(S)$ is an FLTS.* □

Lemma 2 *Let S be an unstable FLTS. $N_\tau(fp(S)) = N_\tau(S) - 1$.*

Proof: For any of the operations in Figures 5 to 7, the lemma is trivial. For the operation described by Figure 4, it is sufficient to observe that the number of internal transitions decreases by one in $fp(S)$ iff $\mu \neq \tau$. By partial ordering of the set of states, it follows that $\mu \neq \tau$. Thus, $N_\tau(fp(S)) = N_\tau(S) - 1$ in all cases. □

5.2 Fragmentation Closure Process

For some given LTS S, let $TLTS_n(S) = (Q, \sum, \Delta, s_0)$. $TLTS_n(S)$ is an initial finite LTS: it is a particular case of FLTSs. By virtue of Lemma 1, we can recursively define $fp^{[k]}(TLTS_n(S))$ for a nonnegative k as follows: $fp^{[0]}(TLTS_n(S)) = TLTS_n(S)$ and $fp^{[k]}(TLTS_n(S)) = fp^{[k-1]}(fp(TLTS_n(S)))$ when $k > 0$.

The fragmentation closure process is simply $fp^*(TLTS_n(S)) = fp^{[\infty]}(TLTS_n(S))$. In the remainder of this paper, we write $fp^*(TLTS_n(S)) = (Q', \sum, \Delta', IQ')$ where Q' is the set of states and IQ' the set of initial states of $fp^*(TLTS_n(S))$.

Lemma 3 *$fp^*(TLTS_n(S)) = fp^{[k]}(TLTS_n(S))$, where k is $N_\tau(TLTS_n(S))$.*

Proof: It is trivial by mathematical induction on k: use Lemma 2. □

Lemma 4 *$fp^*(TLTS_n(S))$ is a stable FLTS.*

Proof: $fp^*(TLTS_n(t))$ is an FLTS by Lemma 1 and stable by definition of fp^*. □

We now refine the notion of unstability of states so that we can account for the states where the system might lock its stability.

Definition 5 *Let s be an unstable state. State s is said to be definitely-unstable (d-unstable) if all its immediate transitions are internal transitions. A state s that is unstable but not d-unstable is said to be potentially–unstable (p-unstable).* □

We note that p-unstable states (not d-unstable states) are legitimate states where a system might lock its stability after waiting for stability; this is not allowed under Langerak's "wait for stability". Some properties of fp^* are given below; see Ghriga (1995) for details.

Lemma 5 *The following properties hold:*

1. *$Q' = Q \setminus U_d(TLTS_n(S))$, where $U_d(TLTS_n(S))$ is the set of d-unstable states in $TLTS_n(S)$.*
2. *$IQ' = (TLTS_n(S)\ after\ \epsilon) \cap Q'$.*
3. *$Tr(fp^*(TLTS_n(S))) = Tr(TLTS_n(S))$.* □

Theorem 1 *$fp^*(TLTS_n(S))$ is unique.*

Proof: Let d_{max} be the number of branches in the longest path in $TLTS_n(t)$. fp^* works in stages due to the partial ordering of states. Hence, we can write $fp^* = fp^{[k]} = fp^{[k_l]} \circ fp^{[k_{l-1}]} \circ \ldots \circ \circ fp^{[k_0]}$ where $k_0 + \ldots + k_{l-1} + k_l = k$ $(= N_\tau(TLTS_n(S)))$ and $fp^{[k_i]}$ represents k_i applications that are needed to make level i stable (that is, all states of level i stable) given that levels $i-1$ down to 0 are stable. Let fp_i denote $fp^{[k_i]} \circ fp^{[k_{i-1}]} \ldots \circ fp^{[k_0]}$. Now, one can easily show- by induction on i- that $\forall\ i,\ (0 \le i \le l < d_{max}) :\ fp_i(TLTS_n(S))$ is unique. □

Given a TLTS or an FTLTS, a path is said to be complete if the originating state (or node) is a root and the ending state is a leaf.

Theorem 2 *Given an LTS S, let* $TLTS_n(S) = (Q, \Sigma, \Delta, s_0)$. *(1)* $fp^*(TLTS_n(S)) = (Q', \Sigma, \Delta', IQ')$ *is a unique stable weakly-initial LTS with* $Q' = Q \setminus U_d(TLTS_n(S))$ *and* $IQ' = (TLTS_n(S)\ after\ \epsilon) \cap Q'$ *that is trace-equivalent to* $TLTS_n(S)$*; and (2) there is a bijection between the set* Π_1 *of complete paths in* $TLTS_n(S)$ *and the set* Π_2 *of complete paths in* $fp^*(TLTS_n(S))$.

Proof: (1) See Lemmas 4, 5, and Theorem 1. (2) Consider the following *characteristic behavior function* χ on paths such that $\chi(p) = \chi(t_1)\chi(t_2)\ldots\chi(t_k)$, where $p = t_1t_2\ldots t_k$ (a sequence of transitions) and $\chi(t_i) = (s_i, \mu_{i+1})$ iff ($t_i = (s_i, \mu_{i+1}, s_{i+1})$and $\mu_{i+1} \neq \tau$); otherwise $\chi(t_i) = \epsilon$. Now, it is trivial to see that there is a bijection f from Π_1 to Π_2 such that $f(p) = p'$ iff $\chi(p) = \chi(p')$. (Notice that this bijection is maintained after each individual application of a primitive fragmentation operation throughout the steps taken to reach fp^*). □

Accordingly, $TLTS_n(S)$ and its fp closure simulate one another (in a step-by-step fashion) under our testing postulate. It follows that

Corollary 1 $\forall \sigma \in \Sigma^* :\ fp^*(TLTS_n(S))\ after\ \sigma\ =\ (TLTS_n(S)\ after\ \sigma) \cap Q'$. □

6 Derivation of Conformance Testers

In this section, we propose an algorithmic characterization for the construction of stable testers and provide their induced notion of implementation with an inherent *procedural requirement* for testers.

Definition 6 *Let* $S = (Q, \Sigma, \Delta, IQ)$ *be a stable FLTS with a set of initial states* $IQ \subseteq Q$. *We define* $\xi(S)$ *as an FLTS that satisfies the following conditions:*

1. *Each state (or node) s in the FLTS S has an extra label "RS(s)" which is a refusal set at s with* $RS(s) \subseteq \Sigma$.
2. *For each state* $s \in Q$, $\mu \in RS(s)$ *iff there exists* $s' \in Q$ *and a trace* $\sigma \in Tr(S)$ *such that* $s,\ s' \in (S\ after\ \sigma)$ *and* $s \overset{\mu}{\not\longrightarrow}$ *but* $s' \overset{\mu}{\longrightarrow}$. □

The elements of a refusal set are the deadlocks that may be observed during testing. Transformation ξ extends a stable FLTS with only the deadlocks that can discriminate a state from others that are simultaneously reachable. The refusal sets help check the nondeterministic structure of implementations. Clearly, $\xi(S)$ is unique for any stable FLTS.

Definition 7 *Given an LTS S,* $\xi \circ fp^*(TLTS_n(S))$ *is said to be n-QRtester of S.* □

Such *n-QRtesters* are unique for any LTS; they will be run in parallel with implementation processes under test and observe the traces with refusals. Testing an LTS implementation process of an ESS translates into testing in the presence of the *unknown stability locks* taken by the implementation process. This is due to i) the unknown nature of the nondeterministic transitions that are not internal that are taken under a feasible

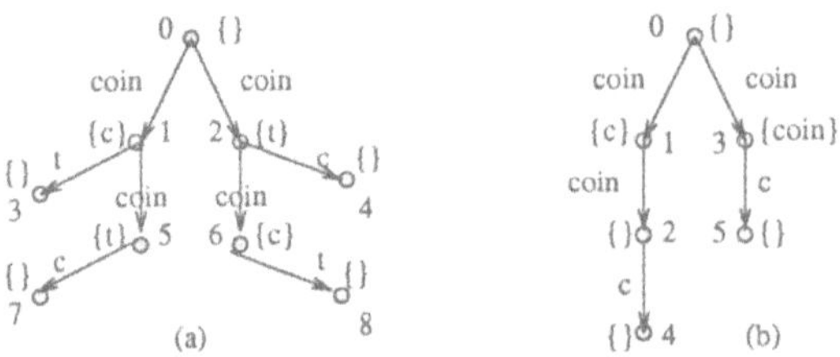

Figure 8: Examples of *n-QRtesters*. (a) 3-*QRtester* of S^{E_1}. (b) 3-*QRtester* of S^{E_2}.

IPCI, and ii) the unknown arbitrary sequences of internal transitions taken by an implementation process. Under our postulate, the problem of dealing with these two latter items becomes equivalent to finding out what stability locks (after waiting for stability) the implementation has chosen. Since the locks are unknown, we define the notion of (finite) quasi-refusal traces of an LTS component, whether stable or unstable. These traces will allow us to discriminate between the states that holds the stability locks.

Definition 8 *Let S be a LTS. Let $\xi \circ fp^*$ $(TLTS_n(S)) = (Q', \Sigma, \Delta', IQ')$. Let $\varrho \in Tr(TLTS_n(S))$. This means that $\exists s_1, s_2, \ldots, s_k \in Q'$: $w_0 \xrightarrow{a_1} s_1 \xrightarrow{a_2} s_2 \ldots \xrightarrow{a_k} s_k$ in $\xi \circ fp^*(TLTS_n(S))$ where $w_0 \in IQ'$ and $\varrho = a_1 a_2 \ldots a_k$ with $a_i \in \Sigma$, $1 \leq i \leq k$.*

A trace $\rho = A_0 a_1 A_1 a_2 A_2 \ldots a_k A_k$ obtained from ρ such that $A_i \subseteq RS(s_i)$ is called a quasi-refusal trace of $TLTS_n(S)$. The set of all quasi-refusal traces of $TLTS_n(S)$ will be denoted by $QRTr_n(S)$. □

These *quasi-refusal traces* allow us also to test whether the implementation exhibits the nondeterministic structure of the specification or not. This effect results from capturing the refusals (observed during a run of an *n-QRtester* with an implementation) that help us gain discrimination between the states that hold the stability locks. For a formal treatment of this issue and detectability analysis of structural abnormalities in implementations, see Ghriga (1995). As an example, Figure 8 depicts the 3-*QRtester* of S^{E_1} and the 3-*QRtester* of S^{E_2} that have been introduced in Section 2. Notice that $TLTS_3(S^{E_1}) = S^{E_1}$ and the same holds for S^{E_2}. To find the quasi-refusal traces of S^{E_1}, one has to augment its traces at any given state by sets of refusals which are subsets of their corresponding $RS()$. For example, the quasi-refusal traces associated with the trace *coin coin c* in S^{E_1} are as follows. Notice this trace corresponds to the traversal of $s_0 \xrightarrow{coin} s_1 \xrightarrow{coin} s_5 \xrightarrow{c} s_7$ in $\xi(S^{E_1})$; notice that S^{E_1} is initially stable. We can construct a quasi-refusal trace by inserting a subset of $RS(s_0)$ at s_0, a subset of $RS(s_1)$ at s_1, and so on. For example, $\{\}coin\{c\}coin\{\}c\{\}$ is a quasi-refusal of S^{E_1}: the $\{\}$s (or their equivalent representation $\emptyset$) are important and denote the fact that the system has to wait for stability. The same semantics is attached to non-empty refusal sets. Semantically (and by Definition 8 also), the trace $\{\}coin\{c\}coin\ c\{\}$ is not a quasi-refusal trace of S^{E_1}. We would like to point out that the 3-*QRtester* of S^{E_2} has a different structure than its initial specification because of fp^*: fp has been applied once to remove the initial internal transition.

Based on this notion of quasi-refusal traces under *wait and lock stability*, the sets (whether empty or not) are indications that our *n-QRtesters* behave in a step by step fashion and before each step (i.e. before offering an event) the tester has to wait for stability, after which the stability of the system is assumed to be locked and then offer an event. Notice that the empty set at the end of our sequences should not be removed as there are particular instances where these sets are not empty. The semantics associated with these empty sets are consistent with Langerak's. We do not allow for the possibility of replacing $RS()$ by an ϵ (generally speaking, "quickly offer" an event according to Langerak's TLOTOS framework) as a result of our testing assumptions.

We also would like to point out that the behavior of an *n-QRtester* can be improved. Suppose that all nodes of $TLTS_n(S)$ have been labelled by "d-unstable", "p-unstable" and "stable" and all transformations are made to preserve such labelling. After the required transformations, the *n-QRtester* has stable states only labelled by "p-unstable" and "stable". If the current set of states at any stage of testing contains only states labelled "stable", then the *n-QRtester* can safely override the waiting for stability of our postulate.

Before we define our notion of implementation (or, notion of conformance), it is important to note that a successful execution of a quasi-refusal trace $\{\}a_1\{\}a_2\{\}\ldots\{\}a_k\{\}$ by an *n-QRtester* establishes that the implementation exhibits the trace $a_1a_2\ldots a_k$. Similarly, a successful execution of a quasi-refusal trace $A_0a_1A_1a_2A_2\ldots a_kA_k$ establishes that the implementation exhibits the trace that is obtained from $A_0a_1A_1a_2A_2\ldots a_kA_k$ by removing the empty refusal sets.

Before we define the notion of *n-QRimplementation*, we need to define the concatenation of traces with sets of refusals. Let $\sigma = \sigma_1a_1A_1$ and $\sigma' = A_2a_2\sigma_2'$ where $a_1, a_2 \in \sum$ and $A_1, A_2 \subseteq \sum$. The concatenation of σ and σ' is $\sigma\sigma' = \sigma_1a_1Aa_2\sigma_2'$ where $A = A_1 \cup A_2$.

Definition 9 *Let S be a LTS and I its implementation process. Implementation I is said to be an n-*QRimplementation *of S, written I $QRimpl_n$ S, if*

1. $QRTr_n(S) \subseteq Tr(I)$, *and*
2. $\forall\ \sigma\mu\{\} \in QRTr_n(S):\ (I\ after\ \sigma)\ ref\ \{\mu\} \implies \sigma\{\mu\} \in QRTr_n(S)$. □

The first condition expresses trace inclusion: all quasi-refusal traces $QRTr_n(S)$ must be exhibited by the implementation. The second condition, however, states that the implementation should not deadlock more often than $\xi \circ fp^*(TLTS_n(S))$ when placed in the *n-QRtester* environment whose traces are limited to $QRTr_n(S)$. In other words, all refusals that occur during testing for the traces of the *n-QRtester* must be contained within its traces. This same condition makes explicit the *sequentiality* aspect of the application of tests. It places a *procedural requirement* on any tester that checks for $QRimpl_n$. By complying to this requirement, the procedural step-by-step behavior of a *n-QRtester* that is run in parallel with an implementation under test is formally described. Let $\sigma\mu\{\} \in QRTr_n(S)$. If there were a successful interaction with σ then the tester would offer μ and wait for observation; if this μ is refused by the implementation (formally, $(I\ after\ \sigma)\ ref\ \{\mu\}$) then this refusal must be within the behavior of the tester (formally, $\sigma\{\mu\} \in QRTr_n(S)$). We now would like to show how one would test for a set of refusals *sequentially*. The trace σ indicated above can be written as $\sigma'A$ where A is a set of refusals which may eventually be empty. By definition of the concatenation operation, $\sigma\{\mu\}$ is basically $\sigma'A'$ with $A' = A \cup \{\mu\}$; we use such process to gather the set of refusals. The validity of such operation follows directly from the concept of "locked stability": once stability is locked at a stable state or a p-unstable state, a process progresses (or moves from its current state) only by one of its specified interactions. In the context of stable states, this coincides with the usual argument in failure trace semantics.

Given an ESS $E =< S^E, env_p^E, env_{ip}^E >$, our notions of n-*QRtesters* and $QRimpl_n$ allow us to *check* for conformance and *define* a notion of implementation using a reference ESS $< S^E, \emptyset, \Lambda >$. Testing for our desired containment of the intuitive behavior of E in an implementation under test would be impossible without consideration of the *environment set-up* for test runs. Several runs under a feasible IPCI will be scheduled to get all possible observations under such *environment conditions.* This same approach for gathering observations (and applying them towards our desired coverage as indicated by $QRimpl_n$) will be followed for all feasible IPCIs.

7 Related Work

We relate our induced notion of conformance (or implementation) to $\underline{conf}$, $\underline{ext}$ and examine the correlation between our traces and failure traces (Langerak, 1990). For comparative analysis, we subject testing for $\underline{conf}$ and $\underline{ext}$ to our testing postulate to mean that the set of tests of these relations remain the same but the waiting and the stability locks are an integral part for testing with respect to these relations.

Theorem 3 *Let S be a finite LTS and I its implementation process. Let n be the smallest nonnegative number such that $TLTS_n(S) = TLTS(S)$.*

$I\ QRimpl_n\ S \implies I\ \underline{conf}\ S$ (under "wait and lock stability").

Proof: Let an implementation process I be such that $I\ QRimpl_n\ S$. Suppose that $I\ \neg\underline{conf}\ S$. It follows that $\exists\sigma \in TLTS_n(S)$, $\exists A \subseteq \sum$ such that $(I\ after\ \sigma)\ ref\ A$ but $(S\ after\ \sigma)\ ref\ A$ does not hold. This means that $\forall s \in (S\ after\ \sigma)$, $\exists \mu \in A : s \stackrel{\mu}{\Longrightarrow}$. It follows that $S\ after\ \sigma$ contains at least a state s' which is not d-unstable. By Corollary 1, we have $\forall s \in (fp^*(TLTS_n(S))\ after\ \sigma), \exists\mu \in A : s \stackrel{\mu}{\longrightarrow}$ and $fp^*(TLTS_n(S))\ after\ \sigma)$ is not empty (it contains at least s'). Therefore, $\sigma'\mu\{\} \in QRTr_n(S)$ and $\sigma'\{\mu\} \notin QRTr_n(S)$ where $\sigma' = \{\}a_1\{\}a_2\{\}\ldots\{\}a_k\{\}$ with $a_1a_2\ldots a_k = \sigma$. Since $\mu \in A$, $(I\ after \sigma')\ ref\ \{\mu\}$ which is a contradiction with I being an n-QRimplementation of S (as $(I\ after\ \sigma')\ ref A$ since testing I for $\underline{conf}$ is done under "wait and lock stability" and $(I\ after\ \sigma')\ ref A \implies (I\ after\ \sigma')\ ref\ B$ for any subset B of A). □

However, the converse does not hold. For example, consider the following implementation process described in LOTOS: I = (*coin*; (c[](*coin*; c)))[](*coin*; (*t*[](*coin*; *t*))). Clearly, $I\ \underline{conf}\ S^{E_1}$ holds; see Example 1 for S^{E_1}. However, $I\ QRimpl_3\ S^{E_1}$ does not hold. Amongst the quasi-refusal traces of S^{E_1} that trigger the *fail* verdict, we list e.g.: $\{\}coin\{t\}coin\{\}t\{\}$, $\{\}coin\{c\}coin\{\}c\{\}$, etc.

Corollary 2 *Let S be an LTS and I its implementation process. Let n be the smallest nonnegative number such that $TLTS_n(S) = TLTS(S)$.*

$I\ QRimpl_n\ S \implies I\ \underline{ext}\ S$ (under "wait and lock stability").

Proof: We have $Tr(S) \subseteq Tr(I)$ (under "wait and lock stability") since $QRTr_n(S) \subseteq Tr(I)$. The result is then immediate by Theorem 3. □

The converse, however, does not hold. Note that I $\underline{ext}$ S^{E_1} holds, but I $QRimpl_3$ S^{E_1} does not hold; use the same example as above. Therefore, both $\underline{conf}$ and $\underline{ext}$ do not lead to adequate notions of conformance (or implementation) for ESSs.

Theorem 4 *Let S be a finite LTS process. Let n be the smallest nonnegative number such that $TLTS_n(S) = TLTS(S)$. Let $FTr(S)$ be the set of failure traces of S. Let U_p be the set of "p-unstable" states in S. $QRTr_n(S) \subseteq FTr(S)$ iff $U_p = \emptyset$.*

Proof: It is trivial by definition of *failure traces* (Langerak, 1990), Definition 8, Theorem 2(2), and the extension implied by our postulate for locking stability at "p-unstable" states. □

Even though that our set of tests is not always a subset of the set of failure traces, it is significantly smaller because we make only a partial use of the unspecified interactions (i.e. deadlocks) of processes. Most importantly, the concept of *stability locks* has allowed us to introduce a robust notion of stability in the specification and testing of ESSs, which draws closer this class of systems to that of sequential/deterministic systems. The "stable" characteristic of our testers is a crucial attribute in gaining confidence in the observations during testing and improving the testability aspects of nondeterministic systems. Furthermore, our notion of implementation provides practical means for the conformance testing of infinite processes, which are consistent with current practices in testing.

8 Conclusion

Environment sensitive systems have been formally introduced. We have shown how the concept of environment sensitiveness can be used for the specification and implementation of systems with improved reliability. This particular class of systems coupled with our testing postulate *Wait and Lock Stability* allowed us to translate the complex problem of testing unstable systems to an equivalent problem of testing stable systems with *unknown stability locks*. Based on this, we have proposed new testers (that are both stable and sequential), characterized their construction, and established the correlation of their notion of implementation (with an inherent procedural requirement) to the related work in failure and failure trace semantics (when applicable). Our testers and their notion of implementation do not have an equivalent in testing theories. Most importantly, we have made an attempt to identify some of the practical aspects of nondeterminism in specifications and implementations and draw testing of LTS based implementations closer to current practices in testing; the concept of *stability locks* allowed us to create testing conditions similar to those of sequential and deterministic systems.

There are many directions for future research, foremost among them being to *extend* our testing approach to deal with data parameters and *timeouts*. Some other topics include the investigation of stronger notions of environment sensitive systems and development of validation tools for such systems.

Acknowledgement

The author would like to thank Phyllis G. Frankl (Polytechnic University, NY) and the anonymous reviewers for their helpful comments and suggestions.

References

Brinksma, E. (1988) A Theory for the Derivation of Tests. *Protocol Specification, Testing, and Verification VIII*, (North-Holland), Elsevier Science Publishers B.V., 63–74.

Brinksma, E., Scollo, G. and Steenbergen, C (1987) LOTOS Specifications, their Implementations and their Tests. *Protocol Specification, Testing, and Verification VI*, (North-Holland), Elsevier Science Publishers B.V., 349–360.

Cavalli, A.R., Favreau, J.P. and Phalippou, M. (1994) Formal Methods for Conformance Testing: Results and Perspectives. *Protocol Test Systems VI*, (North-Holland), Elsevier Science Publishers B.V., 3–19.

Cavalli, A.R., Kim, S.U. and Maigron, P. (1993) Automated Protocol Conformance Test Generation Based on Formal Methods for LOTOS specifications. *Protocol Test Systems V*, (North-Holland), Elsevier Science Publishers B.V., 212–222.

Drira, K. (1994) The Refusal Graph: a Tradeoff between Verification and Test. *Protocol Test Systems VI*, (North-Holland), Elsevier Science Publishers B.V., 301–316.

Fujiwara, S. and Bochmann, G.V. (1992) Testing non-deterministic state machines with fault coverage. *Protocol Test Systems IV*, (North-Holland), Elsevier Science Publishers B.V., 267–280.

Ghriga, M. (1995) Conformance Testing of Nondeterministic Communication Systems. *PhD thesis*, Computer Science Department, Polytechnic University, NY.

Langerak, R. (1990) A Testing Theory for LOTOS using Deadlock Detection. *Protocol Specification, Testing, and Verification IX*, (North-Holland), Elsevier Science Publishers B.V., 87–98.

Petrenko, A., Bochmann, G.V. and Dssouli, R. (1994) Conformance relations and test derivations. *Protocol Test Systems VI*, (North-Holland), Elsevier Science Publishers B.V., 157–178.

Pitt, D.H. and Freestone, D. (1990) The derivation of Conformance Tests from LOTOS Specifications. *IEEE Transactions on Software Engineering*, **16**, 1337–1343.

Starke, P.H. (1972) Abstract Automata. North-Holland: Elsevier Science Publishers.

Tretmans, J. (1994) A Formal Approach to Conformance Testing. *Protocol Test Systems VI*, (North-Holland), Elsevier Science Publishers B.V., 261–280.

Bibliography

Mohammed Ghriga received the Dipl. d'Ingénieur d'état in computer science from the University of Sciences and Technology at Algiers (USTHB), Algeria; and the M.S. and Ph.D. degrees in computer science from Polytechnic University, Brooklyn, NY, USA.

He has been an Assistant Professor of Computer Science at Long Island University, NY, since September 1994. His current research interests include software and protocol testing, conformance testing, formal specification techniques, software verification and validation, and software testability.

PART THREE

Theoretical Framework

6

Timed systems behaviour and conformance testing – a mathematical framework

B. Baumgarten
GMD
Rheinstr. 75, D-64295 Darmstadt, Germany
Tel +49 6151 869 263, Fax +49 6151 869 224
baumgart@darmstadt.gmd.de

Abstract

A formal framework for conformance testing is a prerequisite for the verification as well as for the correct generation of test cases. In this paper, we develop a formal view of systems, behaviour, and testing, that includes aspects of time. It deals on a semantic level with systems cooperating via timed input/output rendezvous and was developed with a view to the OSI architecture, service and conformance testing concepts. We outline a theoretical framework for testing, in which many important informal notions of conformance testing are reconstructed as formal notions with clear relationships among them. In the process, some of these notions are refined and new ones are added. The verdict concept is clarified by the introduction of the notions of evidence function, verdict strategies, and additional parameters.

Keywords

Conformance testing, system behaviour, time, observation, test verdicts, system parameters

1 INTRODUCTION

While OSI protocols reportedly are falling back behind other protocol families in the number of installations, the architectural and conformance testing (CT) concepts of OSI seem to have a lasting impact on theory and practice of protocols, even outside of OSI. The OSI Conformance Testing Methodology and Framework (CTMF) standard [ISO91, ISO94a] provides one of the most comprehensive sources for practical concepts of CT, of which it gives informal definitions. Terms and notions of CT can be roughly divided into

- administrative (document formats, mandatory texts and the like),
- procedural (rules for human beings and institutions), and
- behavioural, dealing with the validity of implementation and tester behaviour, their specification and assessment.

In this paper, we outline a theory of systems and testing, clarifying many behavioural CT notions, in particular with respect to the operational semantics of TTCN [BG94, ISO91, ISO94a]. Its starting point was the time-free framework described in [BW95].

1.1 Overview

In Section 2, we develop a formal view of cooperating systems taylored to architectural conventions in the protocol world. This is a semantic view, not a formal specification language. Appropriate formal specifications in many languages can be interpreted in it. We discuss why this view is fairly general. In Section 3, we give a precise and practically useful meaning to behavioural terms related to protocol testing, even to some that were originally not very clearly defined in standards. Moreover, we develop some new formal concepts, such as the evidence of a test outcome and the correctness of a test case. Our approach solves some of the well-known problems in CT specification [Bau94].

Due to space limitations, some of our definitions are given merely in the guise of an informal summary and no specification examples in any of the current specification languages are given. Natural language terms being defined explicitly or implicitly are italicized.

1.2 Comparison with other approaches

There exist a number of formal and informal frameworks for black box testing of protocols, such as [Bri89, ISO95, ISO91, Pha94, Pha94a, Tre92, Tre94]. Available space does not suffice to compare extensively those texts with the present one. We confine ourselves to the observation that in each of the following points our framework differs from one or several of the cited approaches:

- It is not assumed that the IUT (implementation under test) behaviour can theoretically be specified in the same formal language as the specification ('test assumption'). Instead we assume that the specification in the chosen formal language can be, and is, interpreted in the semantic framework of Section 2, and that the testing related real systems' behaviour can be modelled in this framework.
- The implementation relation is not considered to be arbitrary or depending on circumstances. Rather, we attempt to formalize a single implementation relation which appears to prevail implicitly in TTCN.
- We explicitly do without the 'PCO queues' of CTMF. In [Bau94] it was shown that they are ambiguously described, and that some obvious ways to define them more clearly result in their being either superfluous, non-implementable, or contradicting other standards. Therefore, we also permit Send events to be unsuccessful.
- We formalize the intuitive notion of the test verdict INCONCLUSIVE to the effect that it applies to observations which, even when fully exploiting all the knowledge the tester has, leave it open whether the IUT behaved externally as specified or not.
- Tests presupposing restrictions of the non-determinism originally granted by the specification are interpreted as presuming an agreement on different, more restrictive, specifications – a procedural question. The applicability of INCONCLUSIVE is considered to be independent from non-determinism.
- Our approach involves a notion of timed rendezvous with explicit enabling and disabling events. It gives a precise meaning to the observation of refusals of actions: refusals are observed by a disabling action after a specified waiting period, corresponding to the use of the TTCN timeout mechanism.

1.3 Mathematical preliminaries

For any set A, the set of all *finite words* over A is $A^* := \{a_1 \dots a_n \mid n \geq 0 \wedge \forall 1 \leq i \leq n: a_i \in A\}$, while the sets of all (countably) *infinite* and of all *countable words* over A are $A^\omega := \{a_1 a_2 \dots \mid \forall 1 \leq i: a_i \in A\}$ and $A^\infty := A^* \cup A^\omega$, respectively. We use 'countable' in the sense of 'finite or countably infinite.'

For every $w \in A^\infty$, *length*(w) is the length of w, a natural number or ω. ε denotes the empty word, i.e. *length*$(\varepsilon) = 0$.

Any infinite sequence $w_1w_2 \ldots$ of finite words such that each w_i is a prefix of w_{i+1} and $\lim_{i\to\infty} length(w_i) = \infty$ defines a unique infinite limit word $\lim_{i\to\infty} w_i$. To each (finite-word) language over A, $L \subseteq A^*$, we can associate the limit language L_ω and the countable-word closure $L_\infty = L \cup L_\omega$, obtained by constructing, or adding to L, respectively, the limit words of all suitable sequences of words in L. In the same vein, any prefix-order-preserving mapping from $L \subseteq A^*$ to $K \subseteq B^*$ can be canonically extended to a mapping from L_∞ to K_∞.

2. SYSTEMS AND SYSTEM COOPERATION

We develop a behaviour-oriented view of discrete systems performing in Newtonian physical time. We also show how our view can be used to model other views.

2.1 Actions and system signatures

Systems can perform actions, which are either internal, intermediate, or external. An external action is either an input or an output action. Intermediate actions enable or disable inputs and output actions. Each action is associated with a data type. Each performance of an action is associated with a data object of that type and occurs at some moment in time. A system may perform behaviour sequences, i.e. time-ordered countable sequences of action occurrences.

Throughout this paper, the term 'data type' may be interpreted in an intuitive sense. If more formality is desired, all definitions should be considered to be given relatively to a fixed model M of some many-sorted abstract data type [EM85] that is rich enough to encompass all of the finitely many sorts of interest. A *data type* is then a sort domain of M.

A *system signature*, describing types of actions and data objects, is an octuple

$$\Sigma = (\ IntActs(\Sigma), IEnActs(\Sigma), InpActs(\Sigma), IDisActs(\Sigma), OEnActs(\Sigma), OutActs(\Sigma), ODisActs(\Sigma), Type_\Sigma\ )$$

such that (cf. Figure 1)

- $IntActs(\Sigma)$, $IEnActs(\Sigma)$, $InpActs(\Sigma)$, $IDisActs(\Sigma)$, $OEnActs(\Sigma)$, $OutActs(\Sigma)$, and $ODisActs(\Sigma)$, are mutually disjoint sets with fixed bijections

  En_Σ: $InpActs(\Sigma) \cup OutActs(\Sigma) \to IEnActs(\Sigma) \cup OEnActs(\Sigma)$

  such that $En_\Sigma[InpActs(\Sigma)] = IEnActs(\Sigma)$ and $En_\Sigma[OutActs(\Sigma)] = OEnActs(\Sigma)$, and

  Dis_Σ: $InpActs(\Sigma) \cup OutActs(\Sigma) \to IDisActs(\Sigma) \cup ODisActs(\Sigma)$

  such that $Dis_\Sigma[InpActs(\Sigma)] = IDisActs(\Sigma)$ and $Dis_\Sigma[OutActs(\Sigma)] = ODisActs(\Sigma)$.
- $Type_\Sigma$ assigns to each action a data type.

We call the elements of $IntActs(\Sigma)$, $IEnActs(\Sigma)$, $InpActs(\Sigma)$, $IDisActs(\Sigma)$, $OEnActs(\Sigma)$, $OutActs(\Sigma)$, $ODisActs(\Sigma)$, *internal*, *input enable*, *input*, *input disable*, *output enable*, *output*, and *output disable actions*, respectively. The set

$ExtActs(\Sigma) := InpActs(\Sigma) \cup OutActs(\Sigma)$

comprises all *external actions*. Other systems may participate in them, as described in 2.4. The set

$ItrmActs(\Sigma) := IEnActs(\Sigma) \cup IDisActs(\Sigma) \cup OEnActs(\Sigma) \cup ODisActs(\Sigma)$

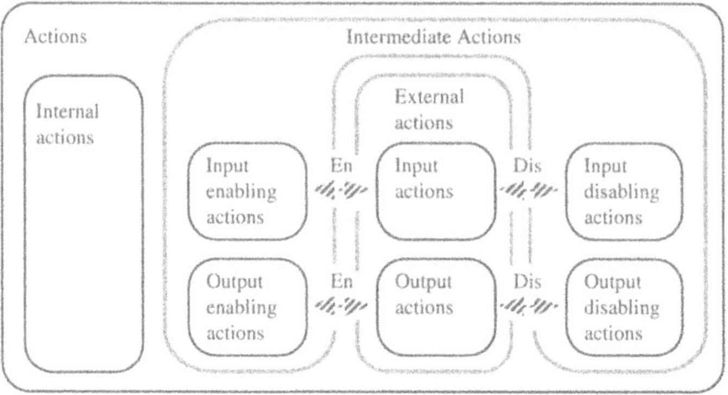

Figure 1 Action sets.

comprises all *intermediate actions*. $VisActs(\Sigma) := ExtActs(\Sigma) \cup ItrmActs(\Sigma)$ is the set of all *visible actions*, and $Actions(\Sigma) := IntActs(\Sigma) \cup VisActs(\Sigma)$ is the set of all *actions*. In natural language terms, we will often omit the attribute 'over Σ', as long as we are not dealing with more than one system signature.

In 2.3, systems will be defined by pairing system signatures with behaviours.

Application to CTMF

In TTCN test cases, after the expansion of constructs, Receive and Send at a PCO are input and output actions, respectively. More precisely, the entering of a list of Receive alternatives is an input enable action for each pair (PCO,service primitive) appearing in this list, while the success of one alternative on this list is the corresponding input action. The immediate success of Send lines claimed by TTCN would mean that the matching input action in the accepting entity must always be enabled before the Send line is enabled. Timeout is internal or disabling, depending on the context, and pseudo-events are internal actions. Send constraints are the data objects of send occurrences, while Receive constraints are subtypes of the input action type. They are used to determine subsequent behaviour, represented by the subtree below the successful alternative, cf. 2.2.2.

2.2 Timed behaviour

In this subsection, we consider a fixed system signature Σ.

Time is counted in seconds having passed after 1.1.1900, 0h00, GMT, for example. The actual choices of the zero point and the time unit do not really matter, of course, but it is necessary to choose in order to be unambiguous. Purely relative times can be expressed easily by using sets of absolutely timed traces, as we will see in 2.3.1.

Occurrences and occurrence sequences

An action may be performed repeatedly and with various data objects; we speak of various possible occurrences. For example, an action *message_reception* may occur at several points in time and with various, sometimes even identical, messages. In each occurrence *occ* of an action, the unique action $Act(occ)$ is associated with a data object $Obj(occ)$ of type $Type_{\Sigma}(Act(occ))$ and with a real number $Time(occ)$. The data object may be trivial, as in the case of a synchronization event, or it may encompass several parameters, such as message type identifier, sender and receiver addresses, and user data in a message. We assume that actions are instantaneous and that $Time(occ)$ denotes the point in time at which *occ* 'happens.' 'Time-consuming activities' can be represented by two instantaneous actions each, representing start and end.

Mathematically, we define the set of all *occurrences* over Σ as

$$Occs(\Sigma) := \{(act,obj,t) \mid act \in Actions(\Sigma),\ obj \in Type_\Sigma(act),\ t \in \mathbb{R}\},$$

entailing $\forall\ occ \in Occs(\Sigma):\ occ = (Act(occ), Obj(occ), Time(occ))$.
For a finite word over the occurrences, $occseq \in Occs(\Sigma)^*$, and an occurrence *occ*, we define the logical expression

$$After(occseq,occ) :\Leftrightarrow occseq = \varepsilon \vee (occseq \neq \varepsilon \wedge Time(last(occseq) \leq Time(occ)).$$

We partition $Occs(\Sigma)$ according to their *Act*-values, thus defining $IntOccs(\Sigma)$, $IEnOccs(\Sigma)$, $InpOccs(\Sigma)$, $IDisOccs(\Sigma)$, $OEnOccs(\Sigma)$, $OutOccs(\Sigma)$, $ODisOccs(\Sigma)$, and $ExtOccs(\Sigma)$ in the obvious way.
The set of all *finite occurrence sequences over* Σ, $FOccSeqs(\Sigma)$, and the *sets of enabled actions* after these sequences are simultaneously inductively defined by the following rules:

$$\varepsilon \in FOccSeqs(\Sigma) \wedge Enabled(\varepsilon)=\emptyset$$

$$\begin{aligned}&occseq \in FOccSeqs(\Sigma) \wedge occ \in IntOccs(\Sigma) \wedge After(occseq,occ)\\ \Rightarrow\ & occseq \circ occ \in FOccSeqs(\Sigma)\\ &\wedge Enabled(occseq \circ occ) = Enabled(occseq)\end{aligned}$$

$$\begin{aligned}&occseq \in FOccSeqs(\Sigma)\\ &\wedge occ=(En_\Sigma(act),obj,t)\\ &\wedge act \in ExtActs(\Sigma) \setminus Enabled(occseq)\\ &\wedge After(occseq,occ)\\ \Rightarrow\ & occseq \circ occ \in FOccSeqs(\Sigma)\\ &\wedge Enabled(occseq \circ occ) = Enabled(occseq) \cup Act(occ)\end{aligned}$$

$$\begin{aligned}&occseq \in FOccSeqs(\Sigma)\\ &\wedge occ \in ExtOccs(\Sigma)\\ &\wedge Act(occ) \in Enabled(occseq)\\ &\wedge After(occseq,occ)\\ \Rightarrow\ & occseq \circ occ \in FOccSeqs(\Sigma)\\ &\wedge Enabled(occseq \circ occ) = Enabled(occseq) \setminus Act(occ)\end{aligned}$$

$$\begin{aligned}&occseq \in FOccSeqs(\Sigma)\\ &\wedge occ=(Dis_\Sigma(act),obj,t)\\ &\wedge act \in Enabled(occseq)\\ &\wedge After(occseq,occ)\\ \Rightarrow\ & occseq \circ occ \in FOccSeqs(\Sigma)\\ &\wedge Enabled(occseq \circ occ) = Enabled(occseq) \setminus Act(occ).\end{aligned}$$

These rules ensure the 'local rendezvous order' of enable, external, and disable action occurrences, cf. either side of Fig. 2. This order concerns the (rendezvous-) *related action set* of an external action $act \in ExtActs(\Sigma)$, defined by

$$Rend_\Sigma(act) := \{En_\Sigma(act), act, Dis_\Sigma act)\}.$$

Now, the set of possible *occurrence sequences* is defined by

$$OccSeqs(\Sigma) := FOccSeqs(\Sigma)_\infty.$$

The *action restriction* operation on occurrence sequences is defined inductively, and via subsequent canonical extension to infinite words, by :

$\forall w \in Occs(\Sigma)^*, a \in Occs(\Sigma), A \subseteq Actions(\Sigma)$:

$$\varepsilon|_A := \varepsilon \text{ and } (wa)|_A := \text{IF } Act(a) \in A \text{ THEN } (w|_A)a \text{ ELSE } w|_A.$$

Behaviour

A *behaviour* over Σ is defined as a subset *beh* of all occurrence sequences over Σ, i.e. $beh \subseteq OccSeqs(\Sigma)$, fulfilling the following requirements, which are presented in an informal manner, due to limited space:

- While an external action is enabled, it can happen at any moment, but only once and unless it is disabled (*rendezvous interval* requirement). It is possible that disabling is not intended, in which case the external action will remain enabled forever if it does not occur. Occurrences of other actions can happen in the interval between the enabling and actual performance or disabling of the action.
  For example, if some occurrence sequence in *beh* consists of an enabling of external action a at time 0 and a disabling of a at time 1, then for all $0 \le t \le 1$, *beh* contains a sequence consisting of $En_\Sigma(a)$ at time 0 and of a itself at time t.
- Whenever the behaviour permits an input action *get*, it is prepared to receive any data object of $Type_\Sigma(get)$ (*free-input* requirement).

The elements of a behaviour are called *behaviour sequences*. Intuitively spoken, *beh* represents the set of 'behaviourally maximal occurrence sequences' and thus gives information on termination: the system may stop working after such a behaviour sequence, even if *beh* contains continuations of it. This corresponds to the distinction of "potentially last actions" in a tree-representation, or the distinction of "potentially maximal words" in a prefix-closed language representation of the behaviour.

2.3 Systems and Conformance

A *system* (over *Signature(S)*) is a pair $S := (Signature(S), Behaviour(S))$ such that *Signature(S)* is a system signature and *Behaviour(S)* is a behaviour over *Signature(S)*. $Systems(\Sigma)$ is defined as the class of all systems over the signature Σ.

System descriptions usually define behaviour sequences only indirectly, e.g. by logical conditions on sequences or by operational models that generate or accept the desired sequences. Examples of operational system description languages that permit to express systems in our sense are time(d) Petri nets [Mer74, Ram74], timer nets [Bau90], timed automata [AD94], and TTCN. These languages have, for example, much more compact ways of specifying that an action may ocur in a time interval, than our semantic model.

Concrete examples

Let us model simple timers that can be used at most once (to keep things simple). Ideally, a timer can be set and started (say, together, in one atomic action) at any time $t\_set \ge 0$ and will then ring after the chosen duration. Apart from noting that *Start* is input and *Ring* is output, signature definitions are omitted. We use trees, written by means of indentation in the style of TTCN, to denote several occurrence sequences with common prefixes.

- *Behaviour(ExactTimer)* consists of the (uncountably many!) sequences

(*EnableStart*, *duration*, 0)
(*Start*, *duration*, *t_set*)
(*EnableRing*, *"Ring"*, *t_set+duration*)
(*Ring*, *"Ring"*, *t_set+duration*)
(*DisableRing*, *dummy*, *t_set+duration*),

where *duration* > 0, and $t\_set \geq 0$.

- *Behaviour*(*InexactTimer*) comprises the sequences

(*EnableStart*, *duration*, 0)
(*Start*, *duration*, *t_set*)
(*EnableRing*, *"Ring"*, *t_set+duration+allowance*)
(*Ring*, *"Ring"*, *t_set+duration+allowance*)
(*DisableRing*, *"Ring"*, *t_set+duration+allowance*),

where *duration*>0, $t\_set \geq 0$, and $-1 \leq allowance \leq 1$.

- *Behaviour*(*UnreliableTimer*) comprises the sequences

(*EnableStart*, *duration*, 0)
(*Start*, *duration*, *t_set*)
(*InternalOK*, *dummy*, *t_set+duration*)
(*EnableRing*, *"Ring"*, *t_set+duration*)
(*Ring*, *"Ring"*, *t_set+duration*)
(*InternalBreakdown*, *dummy*, *t_set+duration*),

where *duration*>0 and $t\_set \geq 0$.

2.4 Interactions, system composition, and observations

Interactions and compatibility

Interactions are common external actions of two system signatures Σ_1 and Σ_2,

$Interacts(\Sigma_1,\Sigma_2) := ExtActs(\Sigma_1) \cap ExtActs(\Sigma_2)$.

Σ_1 and Σ_2 are called *interaction compatible* if they have only interactions, and no other actions, in common, if these interactions have the same type in both signatures, and if these interactions consist of input-output pairs, i.e. if

$$Actions(\Sigma_1) \cap Actions(\Sigma_2) = Interacts(\Sigma_1,\Sigma_2)$$
$$\wedge\ \forall\ int \in Interacts(\Sigma_1,\Sigma_2):\ Type_{\Sigma 1}(int) = Type_{\Sigma 2}(int)$$
$$\wedge\ int \in InpActs(\Sigma_1) \Leftrightarrow int \in OutActs(\Sigma_2),$$

We call a finite set of system signatures $\{\Sigma_1,\ldots,\Sigma_n\}$ *compatible* if its members are pairwise interaction compatible and if, for any three different Σ_i, Σ_j, Σ_k,

$$ExtActs(\Sigma_i,) \cap ExtActs(\Sigma_j) \cap ExtActs(\Sigma_k) = \varnothing.$$

Thus, within a compatible set of signatures, each single interaction is bilateral.

Cooperations and Composite systems

Compatible systems can be composed to form cooperations and larger systems. First we look at the cooperation of systems, for which we define a behaviour, but not a signature; then we define composed systems and a default signature for them.

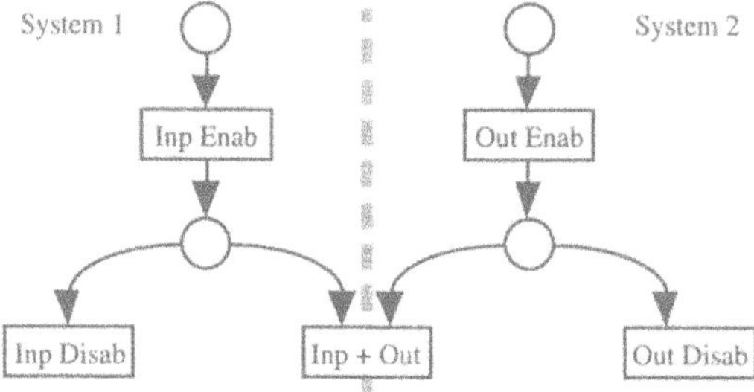

Figure 2 Rendezvous structure.

A *cooperation* is a finite set $\{S_1,\ldots,S_n\}$ of systems with compatible signatures. We prepare the definition of its cooperation behaviour:

$$Occs(S_1,\ldots,S_n) := Occs(S_1) \cup \ldots \cup Occs(S_n)$$

$$PreBehaviour(S_1,\ldots,S_n) := \{\, w \in Occs(S_1,\ldots,S_n)^{\infty} \mid \forall S \in \{S_1,\ldots,S_n\}: w|_{Actions(S)} \in Behaviour(S) \wedge \forall\, 1 \leq i < j \Rightarrow Time(w_i) \leq Time(w_j) \,\}.$$

Again omitting the full formalism, $Behaviour(S_1,\ldots,S_n)$, the *cooperation behaviour* of $\{S_1,\ldots, S_n\}$, is the subset of those *PreBehaviour* sequences in which interactions must occur at the earliest moment, i.e. at the same time as, and in the order behind, the matching enabling by the second partner. This way, interactions shall always occur in both systems in a common, atomic and synchronous step, with the same data object, and such that original local effects (i.e. how behaviour can continue) are achieved in both systems, cf. Figure 2. Internal, intermediate and unmatched external actions happen without any partner, as if the system were running alone.

A *composite system* is a system

$$S_1 \oplus_{\Sigma} \ldots \oplus_{\Sigma} Sn := (\Sigma, Behaviour(S_1,\ldots,S_n)), \text{ where } \Sigma \text{ is a system signature such that}$$

$$Actions(\Sigma) = Actions(\Sigma_1) \cup \ldots \cup Actions(\Sigma_n) \text{ and}$$

$$Type_{\Sigma} = Type_{\Sigma 1} \cup \ldots \cup Type_{\Sigma n}.$$

In principle, input, output, and internal actions can be chosen freely. For our purposes, we choose in the remainder of this paper the *default composite signature* $\Sigma_1 \oplus \ldots \oplus \Sigma_n$ as follows: internal actions and interactions between component systems, as well as their related intermediate actions, are internal. Unmatched external actions form the external actions of the composite signature, and their related intermediate actions play the same role in the composite signature.

Remark:

$Behaviour(S_1,\ldots,S_n)$ is a behaviour over $\Sigma_1 \oplus \ldots \oplus \Sigma_n$.

Application to CTMF

Examples of systems and external actions are the protocol entities and service primitives in OSI terminology [ISO92, ISO94]. For user and provider entities cooperating at the same interface (called service access point) service primitives are executed in common atomic actions. Service primitives usually have an initiator system, which is also the source of an information exchange encompassing the data object associated with the interaction (output action). Input service primitives may occur with any content of the type specified for the service primitive. Intermediate actions are hidden in the protocol mechanisms: both protocol specifications and TTCN test cases require that e.g. an 'incoming primitive' will only be performed if the entity or test case is ready for it (enabled). TTCN semantics take care that enabled primitives are performed as soon as possible, by continuously repeated attempts. Zero delay is our approximation to a very small delay. Disabling is not uncommon: protocol entities and test cases avoid getting stuck (especially in receive primitives) by taking other action after a timer-controlled period of unsuccessful attempts or waiting.

2.5 Discussion

Our restriction of interactions to identical actions could easily be relaxed by the introduction of name associations or a renaming operator; our conventions merely lead to simpler definitions.

Our restriction of synchronous interactions to be two-sided covers many practical applications, such as OSI modelling, and excludes various implementation problems. Principally, many-sided interactions can be simulated by properly coordinated two-sided interactions. Non-default signatures permit many-sided interactions by repeated system composition.

Some system modelling techniques associate 'partial behaviour descriptions' to interfaces, maybe even different ones to 'each side of an interface.' A priori, systems connected by this interface can only execute a particular subset of the possible sequences of interaction occurrences. Examples are service definitions in OSI. Without loss of generality, we will consider such occurrence sequence restrictions as integrated into the behaviour descriptions of the systems.

Some approaches favour asynchronous interactions between systems S_1 and S_2 via some intermediate machinery such as queues etc. Usually, this machinery can be modelled as a separate system C performing synchronous interactions with S_1 (e.g. put into queue) and with S_2 (e.g. take out of queue). Alternatively the queue in question can be treated as an integral part of S_1 and S_2. Hence, synchronicity does not impose essential restrictions.

Typing restrictions could be relaxed without changing system behaviour. The type associated to an interaction on the 'sending' side could be permitted to be a true subset of the type associated at the 'receiving' side.

If 'action refusal' is to be decided by timeouts, this presupposes exact timing specifications, because otherwise there will always be uncertainty whether specifications are met by an implementation or not. If, on the other hand, refusal is communicated explicitly, this will happen in an interaction occurrence.

Rendezvous, with its earliest timing policy, is a natural cooperation mechanism, e.g. for operating on queues: pushing succeeds as soon as both the queue user pushes an object and the queue has the available space, and pulling succeeds as soon as both the user pulls and the queue offers an object.

3. CONFORMANCE TESTING

3.1 Observations

We consider 'direct' observations first. In 3.3 we will introduce indirect observations by means of direct observations in three-sided cooperations. An *observation* of a system S by a system T is a 'finite common trace' of S-T-interaction occurrences, enriched by the observing system's intermediate action occurrences associated to the interactions, allowing to observe that an enabled interaction does not succeed within an appropriate time. Letting τ := *Signature(T)*, the set of all observations is defined as

$$Obs(S,T) \quad := \quad Behaviour(S,T)|_{\cup Rend_\tau[Interacts(S,T)]} \cap Occs(S,T)^* .$$

Note that *Obs(S,T)* is a subset of $Behaviour(S,T)|_{VisActs(T)}$, the 'visible behaviour' of the observer.

We say that a system *S'* *cannot be distinguished from* system *S*, or *S' nodif S*, if, for any system *T*:

$$Obs(S',T) \subseteq Obs(S,T).$$

S' nodif S means that no finite observation of *S'* can ever reveal that it is not *S* that is being observed.

Even though we do not pursue the topic any further, we could call *S* and *S'* *observation-equivalent* if neither of them can be distinguished from the other:

$$S' \ obseq \ S \quad :\Leftrightarrow \quad S' \ nodif \ S \wedge S \ nodif \ S'.$$

For non-deterministic systems, a possible distinction generally cannot be guaranteed (i.e. enforced) by any number of arbitrarily long tests, even if both their number and their length were infinite. Here, we are using the term non-determinism/tic in an informal sense, but leaving aside time, it amounts to the testing non-determinism defined in [Pha94, Pha94a].

An observation example

Let us consider observations of all three example systems in 2.3.1. The system *A_TimerTester* with the behaviour (*sometime* being an arbitrary fixed real number)

(*EnableStart*, 10, *sometime*)
  (*Start*, 10, *sometime*)
    (*EnableRing*, *"Ring"*, *sometime*+9)
      (*Ring*, *"Ring"*, *sometime+x*)
      (*DisableRing*, *dummy*, *sometime*+12)
  (*DisableStart*, 10, *sometime*),

where $9 \leq x \leq 12$. Here, in contrast with the timers, *Start* is output and *Ring* is input. *A_TimerTester* may distinguish *UnreliableTimer* from *InexactTimer*, because *Obs(UnreliableTimer, A_TimerTester)* \ *Obs(InexactTimer, A_TimerTester)* contains the observation

(*EnableStart*, 10, *sometime*)
  (*Start*, 10, *sometime*)
    (*EnableRing*, *"Ring"*, *sometime*+9)
      (*DisableRing*, *dummy*, 12).

It can be shown that both *ExactTimer* and, of course, *InexactTimer* cannot be distinguished from *InexactTimer*, to name but two more relationships.

3.2 Tester systems

Obviously, no real test is ever performed through an infinite time period or with infinitely many observed events. Therefore we define a *tester system T* as a system whose occurrence sequences are all finite and take place within uniformly bounded time (see also 2.2), the latter meaning that either *occ* is empty, or

$$\sup\{\ Time(occ_n) - Time(occ_1) \mid (occ_1, \ldots, occ_n) \in Behaviour(T) \wedge n \in \mathbb{N}\ \} \in \mathbb{R}.$$

3.3 Test configuration

Figure 3(a) shows the common situation in protocol testing [ISO91]. Three systems interact in a compound:

- an examinee system *Ex*, in CTMF the 'implementation under test' (IUT),
- a tester system *T*, and
- a testing context *Co*, in CTMF the 'underlying service provider.'

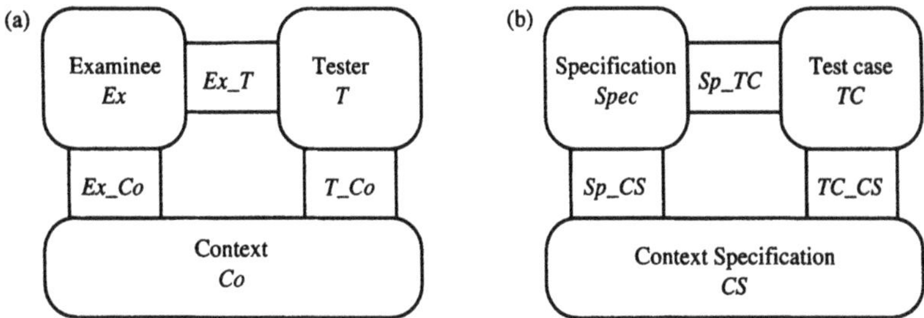

Figure 3 The test configuration (a) and the reference configuration (b).

Ex_T, *Ex_Co*, and *T_Co* represent *Interacts* (*Ex*,*T*), *Interacts*(*Ex*,*Co*), and *Interacts*(*T*, *Co*). By the definition of composite systems, the three are pairwise disjoint. *Ex* may have more external actions than those contained in $Ex\_T \cup Ex\_Co$ – in the cases of 'hidden buttons' or a multi-purpose implementation: a clock radio can pass for a clock (testers will look at the time display, manipulate the clock buttons, listen to the alarm sounding) and for a radio (testers will look at the frequency display, manipulate the radio buttons, listen to the programs). A multi-purpose protocol implementation may have an operator command interface – different from a service access point – to select a protocol or profile or parameter set.

Application to CTMF and non-standardized testing

Two special cases of Figure 3(a) are single-party OSI testing and direct testing.

In single-party testing [BG94, ISO91, ISO94a, Kni93], the tester is considered as a composite system consisting of Upper Tester, Lower Tester, and Test Coordination Procedures. 1992 TTCN, however, treats all three together as a single system. *Ex_Co*, the lower IUT boundary is considered as inaccessible.

In particular, in the Distributed Test Method, *Ex_T* is the Upper Tester PCO, and *T_Co* is the Lower Tester PCO. The standard postulates two opposing queues at each PCO. For the reasons explained in [Bau94], we omitted these queues.

In non-standardized *direct testing* (e.g. first party, development testing), *Co*, *Ex_Co*, and *T_Co* are empty, i.e. *Signature*(*Co*)=(∅,∅,∅,∅,∅,∅,∅,∅), etc.

3.4. Test cases

OSI conformance testing is black box testing comparing *Ex* in the test configuration with its specification *Spec* in the reference configuration shown in Figure 3(b). Specifications are assumed to be semantically interpreted as systems in our sense.

The tester would like to collect information as to whether *Ex* 'looks to *T* and *Co* as *Spec* would.' Unfortunately, it can usually only observe the composite subsystem $Ex \oplus Co$ at *Ex_T* and *T_Co*, and cannot observe what is happening at *Ex_Co*.

The tester specification is called the *test case TC*. A set of test cases is called a *test suite*. In real testing it is usually assumed that only the specified service primitives are used and that the context and the tester operate as specified. In our model this amounts to $Ex\_T \subseteq Sp\_TC$, $Ex\_Co \subseteq Sp\_CS$, $T\_Co \subseteq TC\_CS$, *Co nodif CS*, and *T nodif TC*.

For the sake of simplicity, we assume that every test case is run with a new copy of *Ex*, such that we do not have to deal with restoring *Ex* 'back to its original state.'

3.5 Test outcomes, evidence and verdicts

Test outcomes and conducts

In the following, let us consider the three specifications *Spec,TC,* and *CS* as fixed. A *test outcome* of a tester *T* performed on *Ex* is an observation of *Ex⊕Co* by *T*:

$$Outcomes(Ex,Co,T) := Obs(Ex \oplus Co,T).$$

A test outcome on *Ex* is an indirect observation, recorded at the surface of the tester, providing generally less information about *Ex* than a direct observation of the visible surface of the examinee, a *conduct*, as we call it here. A *valid test outcome* is an observation of *Spec⊕CS* by *TC*:

$$ValidOut := Obs(Spec \oplus CS,TC).$$

A *valid conduct* is 'what should happen at the surface of *Ex*' while it cooperates with *Co* and *T*, i.e. what could happen at the surface of *Spec*:

$$ValCond := Obs(Spec,CS \oplus TC).$$

Due to possible non-determinism in *Ex*, valid test outcomes may be produced (fortuitously) by a non-conforming examinee. Even worse, due to the possible masking of errors in *Co*, valid test outcomes may even be produced by a (non-conforming) examinee performing invalid conduct.

If the examinee is always assumed to operate in conjunction with *Co*, as shown in Figure 3, then we can speak of *relative conformance* w.r.t. *Co*, which amounts to direct conformance of *Ex⊕Co* to *Spec⊕CS*.

Assessing test outcomes: the evidence function

From test outcomes the tester may try to draw conclusions about the validity of 'what happened at the surface of *Ex*.' Of course, from an invalid test outcome it can be concluded that *Spec* was not observed, i.e. that *Ex* does not conform. We say, that the *evidence* of this test outcome is FAIL. If a particular valid test outcome implies that the conduct of *Ex* was valid, then we assign the evidence PASS. In all other cases, we assign the evidence INCONCLUSIVE. Let for example *Co* be a channel that delivers all messages (interactions at *Sp_CS*) truly (in order, in time, without duplication etc.), until it possibly loses one, in which case it immediately apologizes to the tester. As long as the tester obtains a valid sequence of messages it can infer that the examinee's actual conduct in the run of the composite system was valid: the evidence is PASS. If the tester receives an apology after a hitherto valid sequence, it knows that a message was lost, and that this message may have been valid or invalid. In this a case, the evidence is INCONCLUSIVE. Formally,

$$
\begin{array}{llll}
Results := & \{\text{PASS, FAIL, INCONCLUSIVE}\} & & \\
Evidence: & Outcomes(Ex,Co,T) \to Results & & \\
Evidence(out) := & \text{IF} & out \not\in ValidOut & \text{THEN FAIL} \\
 & \text{ELSE IF} & Orig(out) \subseteq ValCond & \text{THEN PASS} \\
 & \text{ELSE} & & \text{INCONCLUSIVE,}
\end{array}
$$

where *Orig* maps, informally spoken, each test outcome to the set of all observations that can be made by *CS⊕TC* and may lead to this outcome. Therefore, we complete the definition of *Evidence* as follows:

$$Orig(out) := \{w|_{VisActs(Signature(CS \oplus TC))} \mid w \in Behaviour(CS \oplus TC), w|_{VisActs(Signature(TC))} = out\}.$$

Of course, one can replace the unknown function domain *Outcomes(Ex,Co,T)* by any known superset, such as $Behaviour(TC)|_{VisActs(Signature(TC))}$, cf. 2.3.

Verdicts and verdict strategies

One part of the story that we have not told yet is that a test case has one additional external action *Verdict*, which it performs exactly at the end of each behaviour sequence. *Verdict* is typically an interaction with a human test operator or a test case driver and logging process. Occurrence events of *Verdict* have a data object of type *Results* which is determined by the outcome up to this moment. We continue, however, to ignore verdict actions and represent verdict occurrences instead by a *verdict function*

$$Verdi(T): Outcomes(Ex,T) \to Results,$$

viewing a *tester* as 'a system plus a verdict function.'

Ideally, a test case would use the evidence of the test outcome as the verdict, but often it is difficult or impossible to calculate the evidence. A *verdict strategy* is a mapping *Strat* from *Results* to *Powerset(Results)*. *Verdi(T) complies* with *Strat* if

$$\forall\, outc \in Outcomes(Ex,T): Verdi(T)(outc) \in Strat(Evidence(outc)).$$

Of course, the *perfect* strategy is $PerfStrat(x)=\{x\}$. Considering *Results* as ordered by FAIL < INCONCLUSIVE < PASS, *Strat* is

- *acceptable* if $\forall x \in Results, y \in Strat(x): \; y \geq x$,
- *wary* if $\forall x \in Results, y \in Strat(x): \; y \leq x$, and
- *strict* if $Strat(\text{FAIL}) = \{\text{FAIL}\}$

Non-acceptable verdicts amount to false accusations. Strict verdicts discover all 'errors,' i.e. invalid outcomes. Non-strict test cases are often much easier to write: at the extreme, the tester could immediately terminate with PASS. We assume that all test cases of a test suite follow the same verdict philosophy, because otherwise test results are very hard to interpret for anyone without detailed knowledge of the test suite.

As the present paper is apparently among the first to fully formalize the evidence and verdict strategy concepts, research on how to approach perfect verdicts is only beginning. In [ISO91], the notion of verdict was based on a vaguely described concept of test purpose [Bau94, CL93].

3.6 System parameters and non-determinism

Specification parameters

Protocol specifications often contain 'static nondeterminism' in the form of protocol *parameters* and options, where the latter can be considered a special cases of the former. Parameters are replaced (i.e. values assigned to them) either rigidly by the implementor or controllably by the operator of the implementation. In CTMF, parameter values are assigned in ICS and IXIT documents. Usually a test case has a subset of the specification parameters as its own parameters.

A *parameter set* (over *M*, cf. 2.1) is a triple

Params = (*ParNames*, *ParTypes*, *ParDecl*),

where *ParNames* is a finite set, *ParTypes* a set of types, and *ParDecl* is a mapping from *ParNames* to *ParTypes*. A *parameter assignment* for *Params* is a mapping

$Assig\colon ParNames \to \bigcup ParTypes$ with $\forall parm \in ParNames\colon Assig(parm) \in ParDecl(parm)$.

Let *Assigs*(*Params*) be the set of all parameter assignments for *Params*. A *parameterized* (by *Params*) *system* over Σ is a mapping

$ParSys\colon Assigs(Params) \to Systems(\Sigma)$.

Parameterized systems are usually defined inductively as terms with variable names in *ParNames*, i.e. syntactically. Then *ParSys* is also inductively defined by the evaluation function on terms canonically associated with the parameter assignments. Once the parameters have been replaced by values, we are back to the ordinary systems. Still, the generation of parameterized test cases is a challenge.

Additional parameters

Dynamic non-determinism (DND) refers to internal behaviour alternatives remaining in *Spec*, after all parameters have been assigned. As shown in 3.1, DND decreases the diagnostic powers of testing. An incorrectly implemented alternative may just 'happen not to occur' during the performance of a test suite, and the error will remain undetected.

A way out of this dilemma is the introduction of additional parameters, which allow to specify on a voluntary basis, but with a binding effect, in which way the examinee has less DND than *Spec*. This additional information increases testability and the value of positive test results, but also the chance of failing test cases.

In CTMF, IXITs are the documents in which additional parameters are assigned values. A typical case is assigning a time limit to reaction times of the examinee, if such limits are unreasonably high for practical testing, or were plainly forgotten in *Spec*.

An *additional parameterization* of *Spec* with *AddParms* is a parameterized system

$$SpecPlus\colon Assigs(AddParms) \to \{S \in Systems(\Sigma) \mid S\ nodif\ Spec\} \text{ such that} \quad (1)$$

$$Spec \in SpecPlus(Assigs(AddParms)). \quad (2)$$

Practically expressed, assignment shall not violate *Spec* (1), and the test client may always choose not to restrict DND at all (2).

Test case behaviour may depend on the additional parameters, and test verdicts must respect the additional parameter assignment *assig*. Test cases whose verdicts do not follow the test suite verdict strategy w.r.t. *Spec'*= *SpecPlus*(*assig*) are not *selected* for execution. This is often the case where the test case 'concentrates on' a special DND-alternative, but the examinee is allowed to avoid this alternative, even according to *assig*.

With the appropriate procedural measures, additional parameter assignments might even be permitted to vary from test case to test case, but we omit the details here.

3.7 Correctness criteria for test cases

We summarize some of the mentioned or implicit criteria that a test case *TC* must meet in order to be considered correct. We assume that a verdict strategy has been defined for the entire test suite. We consider *Spec* and *CS* as given. We add corresponding informal requirements in parentheses.

- The signatures of *Spec*, *CS* and *TC* must be compatible. (In protocol testing: *TC* must use the right set of service primitives.)
- The parameter set of *TC* must be a (parameter) subset of the parameter set of *Spec*, if applicable, united with the standardized additional parameter set. (*TC* must not have any undefined parameters.)
- The verdict strategy must be acceptable and strict. *Verdi*(*TC*) must comply with the verdict strategy. (*TC* must deliver valid verdicts.)

4 CONCLUSION AND OUTLOOK

We have defined a theoretical framework for conformance testing, giving precise and consistent meanings to many practically motivated CT notions. More of them will be formalized in forthcoming papers. Moreover, we have introduced formal correctness criteria for test cases. We have contrasted our approach with comparable existing approaches.

A clean and comprehensive mathematical framework for behaviour testing is the necessary foundation for

- the generation of correct test cases,
- the validation of existing test cases,
- coverage optimization, and
- protocol design rules for testability.

In this paper, we have contributed to the first two objectives. We have neither dealt with the last two, nor with the topics of

- conformance requirements and test purposes,
- alternatives between regular and exceptional behaviour,
- concurrent TTCN,
- probability in conformance testing.

We expect to do so in the future.

5 ACKNOWLEDGEMENTS

The author gratefully acknowledges numerous hints and improvements suggested by Helmut Wiland and the anonymous referees, as well as fruitful discussions with Alfred Giessler, Christa Paule, Gunther Gattung, and Olaf Henniger.

6 REFERENCES

[AD94] R. Alur, D.L. Dill: A theory of timed automata, *Theoretical Computer Science* 126, 1994, 183-235

[Bae94] U. Bär: *OSI-Konformitätstests: Validierung und qualitative Bewertung*, VDI Verlag, 1994

[Bau90] B. Baumgarten: *Petri-Netze, Grundlagen und Anwendungen*, B.I. Wissenschaftsverlag, 1990

[Bau94] B. Baumgarten: Open Issues in Conformance Test Specification, *7th IFIP WG6.1 International Workshop on Protocol Test Systems*, Proceedings, Tokyo, 1994

[BG94] B. Baumgarten, A. Giessler: *OSI Conformance Testing Methodology and TTCN*, North-Holland, 1994

[Bri89] E. Brinksma et al.: A Formal Approach to Conformance Testing, *Protocol Test Systems, North-Holland*, 1989

[BW95] B. Baumgarten, H. Wiland: What is a correct test case? Elements of a Petri net oriented theory of protocol testing, *Petri Nets applied to Protocols*, Proceedings of a Workshop of the 16th ICATPN, Torino, 1995

[CL93] S. T. Chanson, Qin Li: On Inconclusive Verdict in Conformance Testing, *Protocol Test Systems, V*, North-Holland, 1993, 81-92

[EM85] H. Ehrig, B. Mahr: *Fundamentals of Algebraic Specifications, Vol. 1*, Springer, 1985

[ISO91] ISO/IEC IS 9646: Information Technology – Open Systems Interconnection – Conformance Testing Methodology and Framework, 5 parts, 1991/2

[ISO92] ISO/IEC DIS 7498-1: Information Technology – Open Systems Interconnection – Reference Model, Part 1: Basic Reference Model, 1992 (Revision of 1984 IS 7498)

[ISO94] ISO/IEC JTC 1, IS 10731: Information Technology – Open Systems Interconnection – Basic Reference Model – Conventions for the definition of OSI services , 1994

[ISO94a] ISO/IEC IS 9646: Information Technology – Open Systems Interconnection – Conformance Testing Methodology and Framework, 7 parts, 1994–19XX

[ISO95] ISO/IEC JTC1/SC21/P.54.1: Framework: Formal Methods in Conformance Testing, Interim version, 1995

[Kni93] K. Knightson: *OSI Protocol Conformance Testing*, McGraw-Hill, 1993

[Mer74] P.M. Merlin: A Study of the Recoverability of Computing Systems, Irvine; Univ. California, Dept. of Information and Computer Science, TR 58, 1974

[Pha94] M. Phalippou: *Relations d'implantation et hypothèses de test sur des automates à entrées et sorties*, Ph.D. thesis, Université de Bordeaux, no. d'ordre 1112, 1994

[Pha94a] M. Phalippou: Executable testers, *Protocol Test Systems, VI*, North-Holland, 1994

[Ram74] C. Ramchandani: *Analysis of Asynchronous Concurrent Systems by Timed Petri Nets*, MIT, Project MAC, Technical Report 120, Feb. 1974

[Tre92] J.G. Tretmans: *A Formal Approach to Conformance Testing*, CIP - Gegevens Koninklijke Bibliotheek, Den Haag, 1992

[Tre94] J.G. Tretmans: A Formal Approach to Conformance Testing, *Protocol Test Systems, VI*, North-Holland, 1994

7 BIOGRAPHY

B. Baumgarten received the degree in mathematics in 1973 and the doctorate in 1976 at the Technical University of Darmstadt, Germany. In 1977, he joined GMD – German National Research Center for Information Technology, where he has taken part in several research and development projects in the fields of distributed systems, formal description techniques, and testing. Since 1988, he has been involved in OSI conformance testing standardization. He is author and co-author of books on Petri nets and on conformance testing.

An adaptive test sequences generation method for the users needs

Richard Castanet
Christophe Chevrier
Ousmane Koné
Bertrand Le Saëc
La*boratoire* **B***ordelais de* **R***echerche en* **In***formatique*
351 crs de la liberation, 33405 Talence cedex, France
{castanet, chevrier, kone, lesaec}@labri.u-bordeaux.fr

Abstract
We present a test sequences generation method according to users needs. Our method is based on formal languages and monoid computations. The testing users needs are expressed by extended regular expressions. A combination between the specification and the test criteria is computed, providing a specific target sub-specification. Finally, a set of test sequences is generated from this target sub-specification.

Keywords
Conformance testing, formal languages, target sub-specification, faults model, users needs.

1 INTRODUCTION

Test sequences generation is an important area of protocol engineering. The generation methods can be applied on the one hand on finite state automata: Distinguishing Sequences [Gon70], W method [Cho78] (and some improvements with Wp [Fa91]), Transition Tour [NT81], Unique Input Output (UIO) sequences [SD85, SD89] (and some improvements with UIOv [VCI89]), on the other hand on Formal Description Technique (FDT) specifications (Estelle, LOTOS ...) [Fav87, CKM92]. The same methods are used with finite state machines and FDT. Empirical methods can be used to produce automatically test sequences. These methods provide sequences similar to those produced by hand by protocols experts [PG90].

All the methods are based on faults model applied to finite state machines (transfer fault between two states, output fault on a transition, $\cdots$)[BDD+91]. The reader can find the presentation for the formal language point of view in [YL95].

The aim of all these methods is to produce the shortest sequences for a type of fault, but these methods make some assumptions for the finite state machine of the specification: determinism, completeness, reset mechanism, equality between the state number of the specification and the implementation... Some extensions have been proposed, for instance, with partially-specified non deterministic state machines [LPB93].

The industrial users [CCC92] want sets of test sequences which are accorded to the execution of their applications using the tested protocol. So it appears that their needs do not always correspond to the general behaviour of their application. The conformance sequences are not sufficient enough to verify the behaviour of a protocol in these cases. So, additional sequences have to be built and tested.

We propose the following general goals :

- to consider finite state machine with minimal restrictions (for example without reset mechanism, non deterministic machine, $\cdots$)
- to take a larger fault model into account (output error, transfer error, additional state, missing state).
- to propose an intensive test of subsets of the specification (parts of the specification considered critical for the application). This approach, that can be named best effort generation, will give a large coverage of the subset, with, of course, the inconvenience of an enlarge set of test sequences.

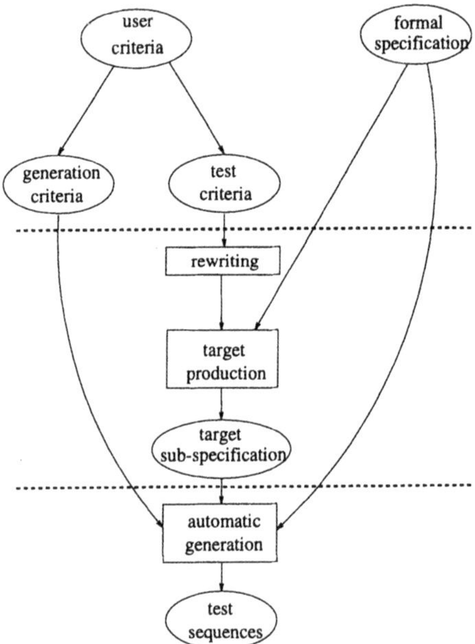

Figure 1 Global architecture of the method

As we use IOSM, we can use formal languages theory and in particular the monoid theory. Each possible calculus in the automaton is associated with an element of a monoid. The considered monoid is finite and interesting elements are the idempotent one. Such an element satisfies the property that the consecutive execution of representatives of this element is equivalent to the execution of ones of them. This is the base of our method and we will only consider a finite number of consecutive computations of given idempotents. This method is described in section 4 following the description of the considered fault model in section 3.

We propose a generation method which is able to cover a large fault model. This method produces a great number of test sequences. So it is not clear to identify the part of the specification which is tested by it. We propose then an overlayer. This tool offers the user the possibility to control the fault model and moreover a criteria specification language. The control of the fault model is based on rules in the monoid to filter the test sequences generation. The criteria specification language is based on generalized regular expressions (predefined operator, references to the states and transitions of the specification). This criteria language correspond to the definition of sub-specifications which are still IOSM. This part is described in section 5.

The next section is devoted to preliminaries and the last one contains examples.

2 PRELIMINARIES

- Regular Expressions

Let Σ be a set of symbols and ε, the empty word . Classical operators with regular expressions are: $\cup$ (union), . (concatenation), $*$ (iteration) and the syntax of an expression E is:

$$E ::= \emptyset|\varepsilon|a|E \cup E|E.E|E^* \qquad \forall a \in \Sigma$$

$\mathcal{L}(E)$ denotes the language, the set of sequences (or words) over Σ associated with E, $\mathcal{L}(\emptyset) = \emptyset$, $\mathcal{L}(\varepsilon) = \{\varepsilon\}$, $\mathcal{L}(a) = \{a\}$, $\mathcal{L}(E_1 \cup E_2) = \mathcal{L}(E_1) \cup \mathcal{L}(E_2)$, $\mathcal{L}(E_1.E_2) = \{u_1.u_2|u_1 \in \mathcal{L}(E_1), u_2 \in \mathcal{L}(E_2)\}$, $\mathcal{L}(E^*) = \{\varepsilon\} \cup \{u_1 \ldots u_n | \forall\, 1 \leq i \leq n, u_i \in E \; and\, n \geq 0\}$.

- Automata

Let Σ be a finite alphabet (Σ is a set of input and output symbols). Σ^+ denotes the words built over the alphabet Σ and $\Sigma^* = \Sigma^+ \cup \{\varepsilon\}$. An automaton $\mathcal{A}$ is a quadruple (Q, q_0, δ, F) where Q is a finite set of states, q_0 the initial state, δ the transfer function: $Q \times \Sigma \rightarrow Q$, F is the set of final states, $F \subset Q$. $\mathcal{L}(\mathcal{A})$ denotes the language recognized by the automaton $\mathcal{A}$. A word $w \in \Sigma^*$ belongs to $\mathcal{L}(\mathcal{A})$ if and only if $q_0 \xrightarrow{w} q_f$ where $q_f \in F$. Figure 6 depicts an automaton which specifies a part of the INRES protocol.

- The transitional monoid [Ber79, Eil76, Lal79, Pin86]

With any automaton $\mathcal{A}(Q, q_0, \sigma, F)$, we can associate a congruence $\approx$ defined on Σ^* (i.e., an equivalence relation $\approx$ satisfying: $u, v \in \Sigma^*, u \approx v \Rightarrow \forall w, w' \in \Sigma^*, wuw' \approx wvw'$) in order to describe the computation in $\mathcal{A}$ in a global way:

$$\forall u, v \in \Sigma^*, u \approx v \; iff \; \forall q, q' \in Q, (q \xrightarrow{u} q') \Leftrightarrow (q \xrightarrow{v} q')$$

This congruence, named the transitional congruence of $\mathcal{A}$, is of finite index. $u_\approx$ denotes the class of u, i.e. the set of words equivalent to u. Two different words in a same class have same computations from any state q to a same state q' of $\mathcal{A}$. In the example of the figure 2, the congruence $\approx$ has 4 classes:

$$\varepsilon_\approx = \varepsilon, a_\approx = a^+, b_\approx = b^+, ab_\approx = a^+b\Sigma^* \cup b^+a\Sigma^*.$$

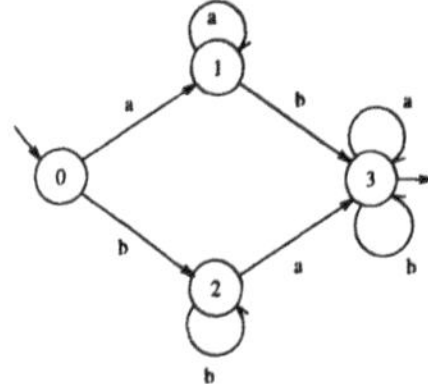

Figure 2 Example of Finite State Machine

$\circ$	m_ϵ	m_a	m_b	m_{ab}
m_ϵ	m_ϵ	m_a	m_b	m_{ab}
m_a	m_a	m_a	m_{ab}	m_{ab}
m_b	m_b	m_{ab}	m_b	m_{ab}
m_{ab}	m_{ab}	m_{ab}	m_{ab}	m_{ab}

Table 1: Composition table

With any congruence $\approx$, we can associate a "transitional monoid"$\mathcal{M}$ (i.e. a set equipped with an associative operation denoted by $\circ$ and a neutral element $\epsilon_\approx$) in bijection with the set of classes of $\approx$. In order to clarify the presentation, we denote by m_u, the element of $\mathcal{M}$ in bijection with $u_\approx$. For our example, the monoid is defined by $\mathcal{M} = \{m_\epsilon, m_a, m_b, m_{ab}\}$

The operation $\circ$ associated with $\mathcal{M}$ is defined by: $m_u \circ m_v = m_{uv}$. We can also define a composition table of the elements of $\mathcal{M}$ (Table 1 for the previous monoid).

If the congruence is of finite index, the monoid $\mathcal{M}$ is obviously finite. So we have a finite representation of a language (the one recognized by $\mathcal{A}$) which has an infinite number of words.

The test sequences generation method, we propose in the following, is closely linked to a notion of "strong periodicity". In monoid theory, this notion can be found in the notion of idempotence:

$$f \in \mathcal{M} \text{ is an idempotent if } f \circ f = f^2 = f$$

The neutral element of a monoid is always an idempotent and if the monoid has a zero, it is also an idempotent.

Any element of a finite monoid admits a power which is an idempotent (see, for instance, [Pin86]):

Theorem 2.1 *For all finite monoid $\mathcal{M}$, $\exists k_\mathcal{M} \in \mathbb{N}$ such that $\forall m \in \mathcal{M}$, we have $m^{k_\mathcal{M}} = m^{2k_\mathcal{M}}$.*

The number $k_\mathcal{M}$ is characteristic of the monoid, it does not depend on the considered element m.

We also use the famous Ramsey theorem [GRS80]:

Theorem 2.2 *Let E be a set, we denote $P_p(E)$ the set of subset of E with p elements. Let θ be a partition of $P_p(E)$ in k classes. If, for any n, there exist $R(n,p,k)$ such that $Card(E) \geq R(n,p,k)$ then there exist a subset $F \subset E$ with n elements such that $P_p(F)$ is included in a class of θ.*

Applying this result to the computation of sequences in the monoid $\mathcal{M}$ having k elements (with $p = 2$ and $n = n' * k_\mathcal{M}$), we can easily obtain the following corollary:

Corollary 2.3 *For any finite monoid $\mathcal{M}$ with k elements, there is $R(n'k_{\mathcal{M}}, 2, k) \in \mathbb{N}$ such that any sequence longer than R can be factorized in $n' + 2$ parts:*

$$s : m_1 \underbrace{m^{k_{\mathcal{M}}} \ldots m^{k_{\mathcal{M}}}}_{n'} m_2$$

where, according to theorem 2.1, $m^{k_{\mathcal{M}}}$ is an idempotent of the monoid $\mathcal{M}$ and $m_1 \circ m^{k_{\mathcal{M}}} = m_1$.

3 FAULT DETECTION POWER

3.1 Protocols as automata

Specifications of protocols are represented by inputs outputs state machines (IOSM). In these machines, the alphabet is the union of input alphabet, output alphabet and internal alphabet (see example in figure 6). IOSM can always be reduced in a deterministic minimal automata. Our tool tool contain an algorithm wich provide minimal automata. So in the sequel we suppose that the used automata are deterministic and reduced. Moreover protocols modelize iterations of periodical prefix processes. So, in the sequel, we also assume that the used automata are strongly connected (with eventually an additional sink state) and they have only one final state which is the initial one.

3.2 Faults model

Several types of implementation faults linked to specification modelized by IOSM can be defined [BDD+91]. We have mainly two basic faults: *"output faults"* and *"transfer faults"* (respectively bad output event and bad final state for a transition of the automaton). Moreover, some other faults can be considered: missing state, missing transition and additional state or transition.

Generally, in order to express the covered fault model, the generation method assumes that the tested implementation is complete and has the same number of states than the implementation. In this case the fault model is only composed by the basic faults.

In our approach, we make also supose that the implementation is complete, but we do not need any assumptions on the size of the implementation. So we have to take into account two other types of faults:

- *"Simple additional state fault"*: An additional state is considered as a fault if the calculus from it in the specification do not correspond to the ones from any states of the specification. For the specification described in the figure 3, the *b*) is a correct implementation of *a*) but *c*) is a faulty one (the transition $(3', e, 5)$ is omitted.
- *"Linearization fault"*: The others additional state type of faults are due to linearizations of loops. A n-linearization of a loop is the substitution of this loop by n consecutive paths with the same label. For example, figure 4a points out a specification that contains a loop on the state 1. The implementation of figure4b is a correct one but the implemention 4c and 4d are faulty ones due to a 1-linearization, respectively a 2-linearization, of the loop b.

The method described in the sequel allows to capture, in a certain limit, this fault model. This limit is completely controled by the user.

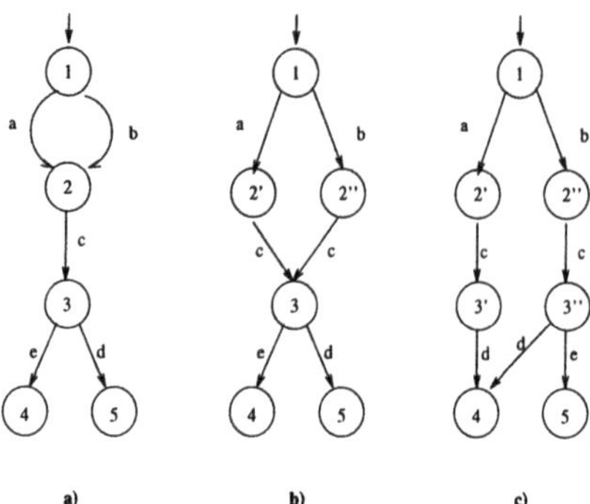

Figure 3 simple additional state fault

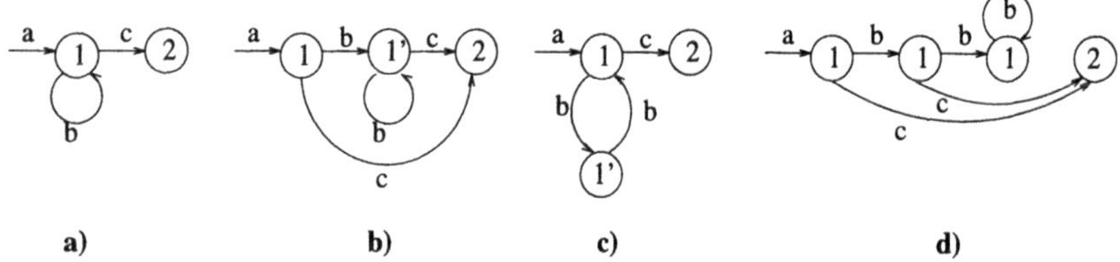

Figure 4 linearization fault

4 Test sequences generation

The basic method, we enlarge here, has been described in [CCSC95]. It can be viewed as a general method that incrementally develops sequences for a given goal (initial state to final one). The sequences are kept or rejected with respect to the generation rule that consider the global behaviour of the transitions in the specification.

4.1 Application to communication protocol test

In a "well written" specification, two elements of the alphabet have different behaviours; i.e. their transfer functions on the automaton are different. So, in the transitional monoid of a specification, two different elements of the alphabet correspond to two different elements. We call basic element, an element of the monoid which corresponds to a letter of the alphabet.
For the transitional monoid, the idempotent property $f_u^2 = f_u$ means that:

$$\forall u', u'' \in u_{\approx}, \forall q \in Q, \delta(q, u') = q' \Rightarrow \delta(q', u'') = q'$$

Our method consist in controlling the number n' of consecutive iterations of a same idempotent: For any $n' \geq 0$, we will consider only factorization of $m \in \mathcal{M}$ of the form (α, f^i, β) where $0 \leq i \leq n' - 1$ with $\alpha, f, \beta \in \mathcal{M}$, $\alpha \neq f$, $\beta \neq f$, $f^2 = f$, $\alpha f = \alpha$. This set of factorization is finite. So for each n', Corollary 2.3 guaranties that the set of sequences, obtained by factorization of an element of the monoid, is finite.

We first introduce the method on the example of the figure 2. The first step of the method leads us to calculate the classes of the congruence $\approx$ and the composition table of the corresponding monoid (cf table 1). In this case, we want to produce some sequences which lead form q_0, the initial state to q_3, the final one. The corresponding element in the monoid is m_{ab} so the sequences will be deduced from the factorization of this element. For $n' = 3$, the obtained set of factorization is:

$$\begin{aligned} m_{ab} \rightarrow \{ \quad & m_{ab}, \\ & m_{ab}m_{ab}, m_\varepsilon m_{ab}, m_{ab}m_\varepsilon, m_a m_b, m_b m_a, \\ & m_\varepsilon m_\varepsilon m_{ab}, m_a m_a m_b, m_b m_b m_a, \ldots\} \end{aligned}$$

In this example, all the elements of the monoid are idempotent and so no generated factorization contains terms like $m_\varepsilon^3, m_a^3, m_b^3, m_{ab}^3$. Then, we replace the elements of the monoid by their representatives in the alphabet. We obtain, extracting from this set all the sequences that are prefixes of another the following set of sequences:

s_1 : $abaaba$
s_2 : $abaabb$
s_3 : $abbabb$
s_4 : $aabaabaa$
s_5 : $aabaabb$
s_6 : $aabbaa$
$s_7, s_8, s_9, s_{10}, s_{11}, s_{12}$: symmetrical from $s_1 \ldots s_6$ by interchanging a and b.

4.2 Sequences generation algorithm

The monoid associated with a specification is computed using the AMORE system (computing Automata, MOnoid, and Regular Expression)[JPTW90], developed at the Aachen university since 1986, is a tool for finite states automata, syntactic monoids [Pin86][Eil76] and regular languages computations. It offers different procedures that convert regular expressions or automata into complete deterministic minimal automata, moreover it compute the associated syntactic monoids. The syntactic monoid is the transitional monoid of the minimal automaton.

We select the set of elements which correspond to calculus from the initial state to itself and we only develop test sequences for this set of elements. So, we obtain sequences which can be chained with no need of a reset.

Each class admits a factorization (cf table 1) whose first term is a basic element. We take into account only such factorizations. We compute in an incremental way (developing at each step the left term of the factorization of a class) the test sequences σ verifying the property:

Generation rule: If a sequence σ is factorizable in the form:

$$\sigma = a_1 \ldots a_{i_1} a_{i_1+1} \ldots a_{i_2} a_{i_2+1} \ldots a_{i_3} a_{i_3+1} \ldots a_{i_{n'}} a_{i_{n'}+1} \ldots a_j$$
with $f = (a_{i_1+1} \ldots a_{i_2}) = \ldots = (a_{i_{n'-1}+1} \ldots a_{i_{n'}})$
then $f^2 \neq f$.

(i.e. not any chosen sequences contain n' consecutive identical idempotent factors.)

We refine the previously built set T_1, avoiding any sequence of T_1 which is strict prefix (operator $\leq$) of another one in T_1. So we produce the set T_2:

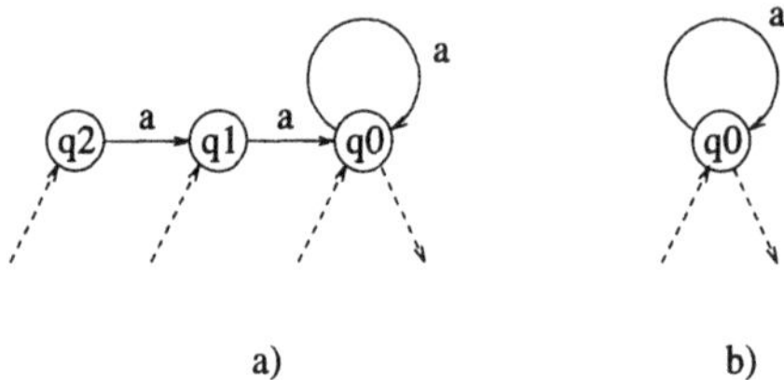

Figure 5 Example of behaviour lost

Prefix rule: $\sigma \in T_2\ iff$:

- $\sigma \in T_1$
- $\forall \sigma, \sigma' \in T_2, \sigma \leq \sigma' \Rightarrow \sigma = \sigma'$
- $\forall \sigma' \in T_1, \exists \sigma \in T_2$ *such that* $\sigma' \leq \sigma$

In fact, the implementation of our algorithm directly works on the words corresponding to basic elements.

Corollary 2.3 guaranties the termination of the algorithm and so, our set of sequences is finite.

4.3 Generating from a sub-specification

In the following section we give the user the possibility to express a sub-specification (i.e. a part of the initial specification). We have seen that a monoid is able to express some properties in a global way. When we have a sub-specification, some informations about the global behaviour of some inputs or outputs are lost. Let us consider the figure 5 that describes two parts (*a*) and *b*)) of a same specification where transition labelled by "*a*" appears. In the complete specification, the sequence "*aa*" corresponds to an idempotent, it is not the case for the sequences "*a*". If we consider a sub-specification where only the configuration b) appears, then the basic element "*a*" becomes an idempotent. If we directly extract sequences from this sub-specification, factors "*aaa*" and "*aaaa*" are not allowed. But in the complete specification, this factors may occurs in the test sequences. So, in this case we have lost an information about the behaviours of the transitions labelled by "*a*" in the sub-specification.

We have to take into account both the monoid which corresponds to the target specification and the monoid which corresponds to the total specification in our test sequence generation. The first monoid is used to develop the sequences and the second one to compute the possible reductions of the sequences.

5 USER CONTROL

In this section, we propose to the user two ways to control the generation of the test sequences. The first control concerns the possibility to define targets to test in the specifications. In order to obtain this, we propose a language to manipulate test criteria. The second allows the user to increase or decrease the fault model. A convenient interface will be provided in the sequel.

5.1 Formal expression of test criteria

The aim of the section is to provide a formal way to express test criteria. A *criterion* is a user requirement for a particular test. A criterion being defined, our method selects a set of sequences that matches the criterion.
As we use automata theory, regular expressions (see below) will be the basis to express test criteria.

5.1.1 User Requirements

Let $(\Sigma, Q, q_0, \delta, F)$ be an automaton specification. The automaton is defined with some "characteristic objects" that are

- the set of states,
- the symbols from the alphabet,
- the transfer function or the transitions.

It is obvious that user requirements for a particular test are defined over these objects.
(1) *Testing a particular transition* is a widely used requirement (see the survey by [SL89]). Such a test consists of executing some preamble which is a sequence that leads the IUT (Implementation Under Test) to the origin of the transition, executing the transition, and eventually executing some postamble (the sequence after the transition). This is also known as the "single transition" test method and it is basically used in many tests requirements. For instance,
(2) *Testing a transition with all possible preambles,*
(3) *Testing a given sequence of transitions*
(4) *Performing the previous test* (3) *n consecutive times* $(n = 2, 3, ...)$.
We consider the automaton specification of figure 6 and the previous requirements are used in the sequel to illustrate the proposed formalism for test criteria.

5.1.2 (Test) Criteria Expressions

The reader may report to [Aut87, Pin86, BS86] for more details on regular expressions.

A test criterion is defined by an expression. We define a test criterion expression or simply criterion expression (CE) by means of (additional) operators that are more suitable to facilitate criteria description and to improve the readability. These operators are $+$, $*$, $\wedge$, $\cap$, $\cup$, $-$, $LENGTH$, $length$, MIN, min, $NOLOOPS$.
E stands for a regular expression, CE stands for a criterion expression, q_k stands for state k and i, j, n stand for integers.
The expressions syntax is defined in a recursive way by the grammar:

$$CE ::= (q_0 S) | CE^+ | CE^* | CE \wedge n | (CE \cap CE) | (CE \cup CE) | (CE - CE)$$
$$S ::= F q_k S | F.q_0 \qquad \forall k \in Q$$
$$F ::= MIN(E) | min(E) | NOLOOPS(E) | E$$
$$E ::= \emptyset | \varepsilon | a | E \cup E | E.E | E^* | E^+ | LENGTH(E, i, j) | length(E, i, j)$$

The grammatical rule CE acts as a generator and a combiner of criterion expressions. The rule S permits to describe the general structure of a criterion expression. The operation on the extended regular expressions are defined by the rule F and extended regular expressions by the rule E.
The criteria described by users are combinations of criterion CE in the following form:

$$q_0.F_1.q_{i_1}.F_{i_2}.q_2 \cdots F_n.q_0$$

by means of the operators $+$, $*$, $\wedge$, $\cap$, $\cup$, $-$ whose semantic is:

- $\mathcal{L}(CE*) = \mathcal{L}(CE)^*$
  $*$ defines iteration.
- $\mathcal{L}(CE+) = \mathcal{L}(CE^*) - \{\varepsilon\}$
  $+$ defines iteration without the empty word, ε.
- $\mathcal{L}(CE_1 - CE_2) = \mathcal{L}(CE_1) - \mathcal{L}(CE_2)$
  $-$ defines the difference of the 2 languages associated with 2 criterion expressions.
- $\mathcal{L}(CE \wedge n) = \mathcal{L}(CE.CE...CE)$ n times
  $\wedge$ (power) defines the number of iterations of a criterion expression.
- $\mathcal{L}(CE_1 \cap CE_2) = \mathcal{L}(CE_1) \cap \mathcal{L}(CE_2)$
  $\cap$ defines the intersection of the 2 languages associated with 2 criteria expressions.
- $\mathcal{L}(CE_1 \cup CE_2) = \mathcal{L}(CE_1) \cup \mathcal{L}(CE_2)$
  $\cup$ defines the union of the 2 languages associated with 2 criteria expressions.

The language associated with a criterion expression CE of the form:

$$CE = (q_0.F_1.q_{i_1}.F_2.q_{i_2} \cdots F_n.q_0)$$

is equal to:

$$\mathcal{L}(q_0.F_1.q_{i_1}).\mathcal{L}(q_{i_1}.F_2.q_{i_2}).\cdots.\mathcal{L}(q_{i_{n-1}}F_n.q_0)$$

where F_i is $MIN(E)$, $min(E)$, $NOLOOPS(E)$ or simply an extended regular expression E (using length). Each language $\mathcal{L}(q_i.F_i.q_j)$ describes the words defined by F_i which lead from q_i to q_j. In particular, $\mathcal{L}(q_i.S.q_j) = \mathcal{L}_{ij}$ denote the set of word u in the specification such that $q_i \xrightarrow{u} q_j$

- $\mathcal{L}(q_i.E.q_j) = \mathcal{L}_{ij} \cap \mathcal{L}(E)$
- $\mathcal{L}(q_i.MIN(E).q_j) = \{u \in \mathcal{L}(q_i.E.q_j) \mid \forall\, v \in \mathcal{L}(q_i.E.q_j), |u| \leq |v|\}$ where $|u|$ is the length of u
  MIN defines the set of words of an extended regular expression with minimal length.
- $\mathcal{L}(q_i.min(E).q_j) = u \in \mathcal{L}(q_i.E.q_j) \mid \forall\, v \in \mathcal{L}(q_i.E.q_j), |u| \leq |v|\}$
  min is a word of an extended regular expression with a minimal length.

- $\mathcal{L}(q_i.NOLOOPS(E).q_j) = \{u \in \mathcal{L}(q_i.E.q_j) \mid \forall vwv' \in \Sigma^+ | u = vwv', q_i \xrightarrow{v} q_j \xrightarrow{w} q_k \xrightarrow{v'} q_f \Rightarrow q_j \neq q_k\}$
  $NOLOOPS$ denotes the set of words of an extended regular expression which do not use loops.
- $\mathcal{L}(LENGTH(E, n, m)) = \{u \in \mathcal{L}(E) |\ n \leq |u| \leq m\}$
  $LENGTH$ denotes the set of words of an extended regular expression with a specified length.
- $\mathcal{L}(length(E, n, m)) = u \in \mathcal{L}(E) |\ n \leq |u| \leq m$
  $length$ denotes a word of an extended regular expression with a specified length.

In next section, we give some examples of criteria expressions that use these operators.

Criteria Expression Examples

(1) $CE_1 = q_0.\Sigma^*.q_5.IDATreq.\Sigma^*.q_0$ send an $IDATreq$ in state q_5 with any preamble and any postamble.

(2) $CE_2 = q_0.min(\Sigma^*).q_5.IDATreq.\Sigma^*.q_0$ same as (1) but with a preamble of minimal length.

(3) $CE_3 = q_0.\Sigma^*.q_5.(IDATreq.\Sigma^*) \wedge 3.q_0$ brings the IUT in state q_5 with any preamble, then send an $IDATreq$ followed by some postamble, 3 consecutive times.

(4) $CE_4 = q_0.\Sigma^*.q_5.LENGTH(IDATreq.\Sigma^*, 3, 5).q_0$ brings the IUT in state q_5 with any preamble, then sends a sequence starting with an $IDATreq$ which length is between 3 and 5.

5.1.3 From test criteria expression to target sub-specification

It is not difficult to translate the test criteria expressions into regular expressions and then to combine the criteria with the specification in order to obtain the part of the specification that the user want to test (the target sub-specification). The first step of the method allows to transform all the additional operators used by the criteria. In the second step, we eliminate the state notation to merge the alphabet of the specification.

When the criterion expression has been transformed in the form of a regular expression the final step, before the generation step, consists of the generation of the target sub-specification. It corresponds to the computation of the intersection between the criteria expression and the specification. If the intersection is empty, then it means that the expressed criteria is not suitable with the specification.

All the operations over automata and regular expressions we made are computed using the software AMORE [JPTW90].

5.2 Formal expression of the fault model

Basicaly, the generation method works with $n' = 3$. In this case, the set of sequences contains a distinguishing sequences if it exists, the UIO sequences if they exist, a set of caracterisation as produce by the W method. Moreover, we can test all the faults due to a 1-linearization of loops and the faults due to simple additional state fault. The user can modify this model either increasing or decreasing it

5.2.1 Test weakness

We can reduce the number of sequences and so decrease the fault detection power by adding some test rules. For example, the user can specify the following rules:

Orderness rule: If two sequences S_1 et S_2 are built in the following form:

$S_1 : a_0 \dots a_h \dots a_i \ a_{i+1} \dots a_j \ a_{j+1} \dots a_g \ a_{g+1} \dots a_p$

$S_2 : a_0 \dots a_h \dots a_i \ a_{j+1} \dots a_g \ a_{i+1} \dots a_j \ a_{g+1} \dots a_p$

with $m = (a_h \dots a_i)_{\approx}, f = (a_{i+1} \dots a_j)_{\approx}, g = (a_{j+1} \dots a_g)_{\approx}$ and $mog = m$ and $mof = m$ then we only keep one of them.
This rule intuitively expresses that the order of loops appearance is not important in this case.

Repetition rule: If a sequence σ is built in the following form:

$\sigma : a_0 \dots a_h \dots a_i \ a_{i+1} \dots a_j \ a_{j+1} \dots a_g \ a_{g+1} \dots a_p$

with $m = (a_h \dots a_i)_{\approx}, f = (a_{i+1} \dots a_j)_{\approx} = (a_{j+1} \dots a_G)_{\approx}$, $m \circ f = m$ and $f \circ f = f$ then we only keep one f factor.

This rule expresses that a sequence never contain two consecutive idempotent factors. In particular, a loop which belongs to an idempotent class is tested only once.

These reduction rules can be expressed in a global way or in a local way (specifying the elements on which the rules can be applied).

Comments: These reduction rules do not garantie the basic fault model.

5.2.2 Test increase

The basic method is able to detect the faulty implementation of figure 4c ("*b*" correspond to an idempotent), but it cannot detect the faulty implementation of figure 4d due to a 2-linearization. Intuitively, this detection is linked to the number of iterations of the faulty loop specified in the generation rule. If we enforce the generation rule, we are able to detect faults hide by n-linearizations. It correspond to the parametrization of the generation rule by n': $mf^{n'+1} \longrightarrow mf^{n'}$ with $m \circ f = m$ and $f \circ f = f$.

5.3 Advantages of the method

5.3.1 Industrial needs in communication protocol testing

We can mention the following requirements:

- Each sequence has to begin in the initial state and return to the initial state (this is the usual representation of a test sequence obtained in an empirical way).
- To lead the specification to a given state and to go to another given state to cover a part of the specification.

- To have a large choice of sequences selections to cover given areas of the specification.
- To maximize the number of possible testable errors.

5.3.2 Advantages of the sequences generated with our method

- Production of sequences whatever required initial and final states, and in particular the possibility to produce sequences for inopportune events: If the specification has an error state where leads all an inopportune event transition and a reset to return to the normal behaviour, the general method produce some sequences to test these transitions. But if the error state is a sink state, we have to produce unchainable sequences which lead from q_0 to the sink state.
- Ability of criteria expression by means of the test criteria languages.
- Taking into account of the required faults model by means of the generation rules (orderness, repetition rules $\cdots$).
- Preservation of the fault detection power of the sequences even if they are not applied to the initial state. For instance, if the IUT is not in its initial state, the used sequences are still valid.

6 EXAMPLES

We illustrate our generation method on the example of the INRES protocol [Hog92]. This protocol provide a simple asymmetric data transfer over an unreliable medium. The connection asked by the initiator is established by means of the primitives: $ICONreq$, $ICONind$, $ICONres$, $ICONconf$ and then the data transfer by the primitives $IDATreq$ and $IDATind$. At any time the connection can be disrupted by the responder by means of the primitives $IDISreq$ and $IDISind$. A specification for the initiator part is given figure 6. We give below some examples of criteria and the corresponding sequences for this specification. We use for these examples a local test architecture.

To this automaton corresponds a monoid with 1198 elements. If we apply our method on this monoid, the produced set of sequences will be too big to be used to test the implementation. Also, with the test criteria expression language, we cut the test problem in different test objectives. The examples below are inspired from the basic test of the 9646 standard [ISO].

- Test purpose: Basic connection test - this test aims at verifying if the implementation under test (IUT) is able to connect to another entity. The criterion is expressed as follows:

  $$CE_{basic\_connection} = q_0.ICONreq.(\Sigma - (IDATreq \cup T1))^*.IDISind.q_0$$

  This criterion means that the upper tester asks for a connection ($ICONreq$) and waits for a connection end ($IDISind$). Not any data transfer ($IDATreq$) or reconnection timer ($T1$) are authorized at this level of test.

  The monoid which corresponds to the intersection with the specification have 38 elements. The test set is composed by only 4 sequences.

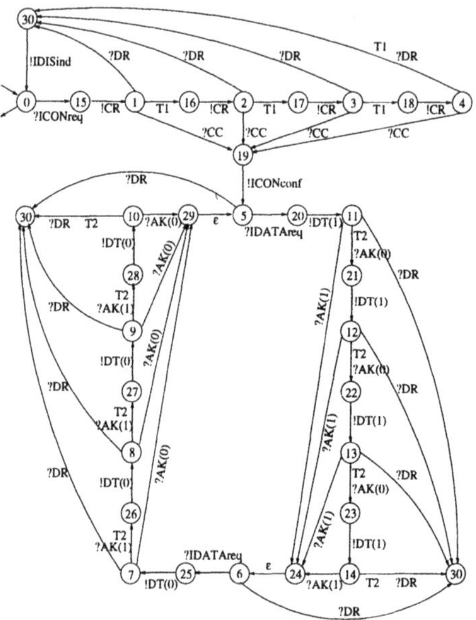

Figure 6 IOSM for the initiator part of the InRes protocol

```
ICONreq.CR.DR.IDISind.ICONreq.CR.DR.IDISind
ICONreq.CR.DR.IDISind.ICONreq.CR.CC.ICONconf.DR.IDISind
ICONreq.CR.CC.ICONconf.DR.IDISind.ICONreq.CR.DR.IDISind
ICONreq.CR.CC.ICONconf.DR.IDISind.ICONreq.CR.CC.ICONconf.DR.IDISind
```

- Test purpose: Elementary data transfer - this test aims at verifying the ability of the IUT to exchange data with another entity. The criterion is expressed as follows:

$$CE_{elementary\_data\_transfer} = q_0.min(\Sigma^*).q_5.IDATreq.q_{20}.min(\Sigma^*.AK(1).\Sigma^*).q_0$$

This criterion means that we want to take a minimal way to lead the implementation in the state (q_5) in which data transfers are allowed. Then we asked for a data transfer ($IDATreq$) and we terminate the connection in the shortest way with a positive acknowledgement.

The corresponding monoid have 83 elements and it allows us to produce one sequence:

```
ICONreq.CR.CC.ICONconf.IDATreq.DT(1).AK(1).DR.IDISind.
          ICONreq.CR.CC.ICONconf.IDATreq.DT(1).AK(1).DR.IDISind
```

- Test purpose: Behaviour with respect to connection timer - this test aims at verifying the connection protocol with respect to the connection timer. The criterion is expressed as follows:

$$CE_{connection\_behaviour} = q_0.ICONreq.(\Sigma - (DR))^*.ICONconf.(\Sigma - (IDATreq))^*.q_0$$

This criterion means that, after a connection request, ($ICONreq$) all events, and in particular the event corresponding to a connection timer end, are authorized except a disconnection request (DR). After a connection confirm ($ICONconf$), not any data transfer ($IDATreq$) is accepted.

The target sub-specification is associated with a monoid which contains 155 elements and we produce a set of 16 sequences:

```
ICONreq.CR.CC.ICONconf.DR.IDISind.ICONreq.CR.CC.ICONconf.DR.IDISind
ICONreq.CR.CC.ICONconf.DR.IDISind.ICONreq.CR.T1.CR.CC.ICONconf.DR.IDISind
ICONreq.CR.CC.ICONconf.DR.IDISind.ICONreq.CR.T1.CR.T1.CR.CC.ICONconf.DR.IDISind
ICONreq.CR.CC.ICONconf.DR.IDISind.ICONreq.CR.T1.CR.T1.CR.T1.CR.CC.ICONconf.DR.IDISind
ICONreq.CR.T1.CR.CC.ICONconf.DR.IDISind.ICONreq.CR.CC.ICONconf.DR.IDISind
ICONreq.CR.T1.CR.CC.ICONconf.DR.IDISind.ICONreq.CR.T1.CR.CC.ICONconf.DR.IDISind
ICONreq.CR.T1.CR.CC.ICONconf.DR.IDISind.ICONreq.CR.T1.CR.T1.CR.CC.ICONconf.DR.IDISind
ICONreq.CR.T1.CR.CC.ICONconf.DR.IDISind.ICONreq.CR.T1.CR.T1.CR.T1.CR.CC.ICONconf.DR.IDISind
ICONreq.CR.T1.CR.T1.CR.CC.ICONconf.DR.IDISind.ICONreq.CR.CC.ICONconf.DR.IDISind
ICONreq.CR.T1.CR.T1.CR.CC.ICONconf.DR.IDISind.ICONreq.CR.T1.CR.CC.ICONconf.DR.IDISind
ICONreq.CR.T1.CR.T1.CR.CC.ICONconf.DR.IDISind.ICONreq.CR.T1.CR.T1.CR.CC.ICONconf.DR.IDISind
ICONreq.CR.T1.CR.T1.CR.CC.ICONconf.DR.IDISind.ICONreq.CR.T1.CR.T1.CR.T1.CR.CC.ICONconf.DR.IDISind
ICONreq.CR.T1.CR.T1.CR.T1.CR.CC.ICONconf.DR.IDISind.ICONreq.CR.CC.ICONconf.DR.IDISind
ICONreq.CR.T1.CR.T1.CR.T1.CR.CC.ICONconf.DR.IDISind.ICONreq.CR.T1.CR.CC.ICONconf.DR.IDISind
ICONreq.CR.T1.CR.T1.CR.T1.CR.CC.ICONconf.DR.IDISind.ICONreq.CR.T1.CR.T1.CR.CC.ICONconf.DR.IDISind
ICONreq.CR.T1.CR.T1.CR.T1.CR.CC.ICONconf.DR.IDISind.ICONreq.CR.T1.CR.T1.CR.T1.CR.CC.ICONconf.DR.IDISind
```

7 REFERENCES

[Aut87] J.M. Autebert. *(Algebraic Langages) Langages Algébriques.* Etudes et Recherches en Informatique. Masson, 1987.

[BDD+91] G.v. Bochmann, A. Das, R. Dssouli, M. Dubuc, A Ghedamsi, and G Luo. Fault models in testing. In 4th *IWPTS*, Leishendam, The Netherlands, october 1991.

[Ber79] J. Berstel. *Transductions and context free Languages.* Terbner Studienbucher, Stuttgart, 1979.

[BS86] G. Berry and R. Sethi. From regular expressions to deterministic automata. *Theoritical Computer Science*, (48):117–126, 1986.

[CCC92] R. Castanet, R. Casadessus, and P. Corvisier. Industrial experience on test suite coverage management. In 5th *IWPTS*, Montréal, Canada, october 1992.

[CCSC95] C. Chevrier, R. Castanet, B. Le Saëc, and R. Casadessus. Génération de séquences de test par calcul de monoïdes. In *CFIP'95*, Rennes, France, may 1995. Hermes.

[Cho78] T.S. Chow. Testing software design modelled by finite state machines. *IEEE Transactions on software engineering*, SE-4(3):178–187, may 1978.

[CKM92] A. Cavalli, S. Kim, and P. Maigron. Automated protocol conformance test generation based on formal methods for lotos specifications. In 5th *IWPTS*, Montréal, Canada, october 1992.

[Eil76] S. Eilenberg. *Automata, langages and machines*, volume A, B. Academic Press, New York, 1974-1976,.

[Fa91] S. Fujiwara and all. Test selection based on finite state models. *IEEE Transaction on software engineering*, 17(6), june 1991.

[Fav87] J. P. Favreau. Automatic generation of test scenario skeletons from protocol specifications written in estelle. In 6th *PSTV*. North Holland, 1987.

[Gon70] G. Gonenc. A method for the design of fault detection experiment. *IEEE Transaction on computer*, C-19(6), june 1970.

[GRS80] Ronald L. Graham, Bruce L. Rothschild, and Joel H. Spencer. *Ramsey theory*. Wiley interscience series in discrete mathematics. John Wiley and Sons, 1980.

[Hog92] D. Hogrefe. Osi formal specification case study: the inres protocol and service, revised. Universität Bern", 1992.

[ISO] ISO. Information processing systems - open systems interconnection - conformance testing methodology and framework- part 1: general concept - part2: abstract test suite specification - part3: the tree and tabular combined notation (ttcn) - part4: test realization - part5: requirement on test laboratories and clients for the conformance assessment process. ISO 9646.

[JPTW90] V. Jensen, A. Potthof, W. Thomas, and U. Wermuth. A sort guide to the amore system (computing automata, monoids and regular expressions). Technical report, RWTH Aachen, 1990.

[Lal79] G. Lallement. *Semigroups and combinatorial Applications*. J. Wiley and Sons, New York, 1979.

[LPB93] G. Luo, A. Petrenko, and G.v. Bochmann. Selecting test sequences for partially-specified non deterministic finite state machines. Technical report, Université de Montréal, février 1993.

[NT81] S. Naito and M. Tsunoyama. Fault detection for sequential machines by transition tours. In *IEEE Fault Tolerant Computing Conference*, pages 238–243, , 1981.

[PG90] M. Phalippou and R. Groz. From estelle specifications to industrial test suites, using an empirical approach. In *FORTE'90*, Madrid, Spain, november 1990.

[Pin86] J.E. Pin. *Varieties of formal languages*. Oxford Univ. Press, 1986.

[SD85] K. Sabnani and A.T. Dahbura. A new technique for generating protocols test. *ACM Computer Communication Review*, 15(4):36–43, 1985.

[SD89] K. Sabnani and A.T. Dahbura. A protocol test generation procedure. *Computer Network and ISDN Systems*, 15, 1989.

[SL89] D. Sidhu and T. Leung. Formal methods for protocol testing : A detailed study. *IEEE Trans. on Software Engineering*, 4(15), April 1989.

[VCI89] S.T. Vuong, W.Y.L. Chan, and M.R. Ito. The uiov method for protocol test sequence generation. In 2th*IWPTS*, Berlin, october 1989.

[YL95] M. Yanakakis and D. Lee. Testing finite state machines: Fault detection. *Journal of Computer and System Sciences*, 50(2):209–227, april 1995.

8

A FRAMEWORK FOR TESTING TELECOMMUNICATION SERVICES

G.S. Vermeer, M.F. Witteman, J. Kroon.
KPN Research
P.O.Box 421
2260 AK Leidschendam
Netherlands

Email:
G.S.Vermeer@research.ptt.nl
M.F.Witteman@research.ptt.nl
J.Kroon@research.ptt.nl

ABSTRACT

Service testing is a new development pushed by the rapid increase of the number of services offered by telecom operators. This paper defines a framework for the test process related to the operation of new services. First an overview is presented of terms and concepts in the service life cycle. Then a procedure is defined that covers all testing activities that should be considered before and during service operation.

Keywords

Service testing

1. INTRODUCTION

In the turbulent telecommunications world, the rapid introduction of new telecommunication services can supply Service Providers with a competitive edge. The introduction of an Intelligent Network architecture will enable operators to shorten the introduction of new services from many years down to a number of months, or even to a couple of weeks. Furthermore operators will have the possibility to create new services in their own Service Creation Environments.

In telecommunications, on the other hand, service providers can hardly afford failures of services. Failures of the network may have a catastrophic effect on the operators' image as a reliable service provider. This threat provides the need for 'Service Testing'. Before a telecommunication service is deployed into the network, it should be ascertained that the service is functionally correct, and that the service does not harm the already installed base.

This paper aims at providing a framework for the test activities considering validation of new telecommunication services. Rather than presenting new and ambitious test techniques we will investigate the role of testing in the service life cycle. This framework is largely based on the Intelligent Network architecture. The framework however is fit for application on other architectures as well.

Section two will explain the various terms and concepts related to the service life cycle. Section three will introduce a procedure for all testing activities that should be considered before, and during service operation. Section four will conclude the paper.

2. SERVICE LIFE CYCLE

This section summarises some terms and concepts that are used in the world of service creation. Understanding these concepts is important in order to see the relation with the testing activities presented in the next section. Readers that are very familiar with service design may wish to continue with the next section immediately.

2.1. What is a service?

A very basic aspect is the notion of what a service actually is. It appears that many different definitions are used over the various documentation. This is due to the fact that services can be seen from commercial, technical, and user perspectives. As testing is merely focusing on the technical perspective a technical definition would probably be most useful.
For this reason we introduce the following definition of a service:

A (telecommunication) service is a stand-alone commercial offering, which can be identified by all parts of hardware and software which are especially designed, installed or configured, to enable the service.

We will use this definition of a service for the service testing activities suggested in this paper.

An example may clarify the meaning of this definition. Consider a service, commercially defined as: "A time dependent call forwarding service that connects all calls to B from 17.00 till 9.00 to another number C". The software that enables this service comprises code that checks the B-number, and decides upon its re-routing. The required hardware could consist of the clock mechanism that provides the appropriate time signals for the service.

2.2. Stages in the service testing life cycle

In order to understand the various aspects that contribute to the realisation of a service it is useful to observe the stages that are traversed during the service development.
The Eurescom P103 (Eurescom, 1992) project has defined a life cycle that we consider quite appropriate (see Figure 1).

In this paper we like to position most of the service testing activities after the service implementations tasks are completed. The service testing tasks take place before installing the service into the network. The result of service testing therefore should be a service which exhibits correct behaviour and is ready for being installed in the operational network.

A minor part of the service testing activities however will take place during service operation, which comprise the phases after the service installation, and before service de-installation. We will come back to this anomaly in section three.

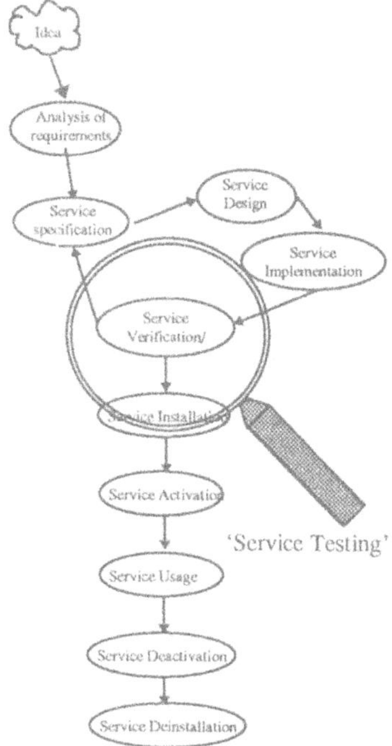

Figure 1 Service Testing in the Service Life Cycle

3. THE INTELLIGENT NETWORK ARCHITECTURE

Services can be provided using the 'Intelligent Network' architecture. In Figure 2 a possible configuration of an intelligent network is pictured. In this architecture an IN-(service) call is detected at the Service Switching Point (SSP). The SSP will consult the Service Control Point (SCP) for instructions. The SCP contains the service logic and service data that are used to provide IN services. The SCP can access data in a Service Data Point (SDP) either directly, or through a signalling network.

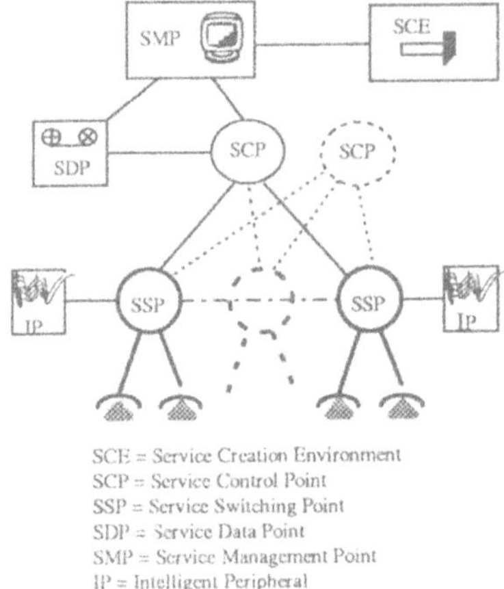

Figure 2 Intelligent network architecture

The intelligent Peripheral (IP) provides specialised resources for customisation of services, and supports flexible interaction between a user and the network.
An IP can be used, for example, to play announcements like 'type your PIN' in a credit card validation service. The Service Management Point performs service management control, service provision control, and service deployment control. The Service Creation Environment (SCE) is used to define, develop and test an IN service and input it into the SMP.

4. SERVICE TESTING

At a first glance, the testing activities that should be performed with the introduction of a new service should be quite limited. One could suggest that a simple check on the availability of the functionality would satisfy. There are however a number of reasons to apply a careful installation procedure:

- Realisation of a new service may require modification to the service platform, or it may affect the platform functionality unintentionally;
- Realisation of a new service may require modification to existing services, or it may change the behaviour of the existing services unintentionally;
- Realisation of a new service may seem to be successful but a malfunction may show up (long) after installation, due to complex relations with the various network elements.

To assure the proper working of service the following 5 tasks should be performed:

1. Test the service platform
2. Test the 'new' service
3. Perform regression tests
4. Perform Service Interaction tests
5. Perform Service Management

Before installing a service into a IN structured network, the network itself has to be tested; This first task is in this paper referred as platform testing. The second step considers the new service. It has to be tested whether the functionality of the new service conforms to its specification. When a new service is installed and activated, it has to be tested (task 3) whether the already installed services are still operational. Then, before making the service operational, task 4 checks for unwanted service interactions.

When the service is operational still tests are to be performed during service management (task 5). These tests consider aspects like 'is the service still working?', 'is the performance still satisfactory', etc. In the following subsections the test steps will be discussed into more detail.

4.1. Platform Testing

Before installing new services into a service platform, the platform itself has to be tested. In the IN-standards (CCITT,1992), a service is composed of so-called Service Independent Building Blocks (SIBs).

It can be said that these building blocks are part of the platform and should be subject of platform testing. It is possible to test a SIB in some artificial environment (like simulation). However, to test a single SIB implemented in a SCP will be difficult.
For this purpose a 'test service' may be useful. Such a service exhibits not necessarily any functionality employable by end-users, but supports testing the relevant part of the service logic.

The communication between the network elements SSP and SCP (and in some cases SDP) will make use of the standardised core-INAP protocol. For testing this protocol stack, conformance test standards are (being) developed within ETSI. Before deploying these network elements into the network, these elements should have passed these conformance tests.

Besides the functional behaviour, non functional aspects like performance should be tested as well. It should be verified that the platform exhibits sufficient performance to run the desired services. This can be done by either performance calculations (including prediction algorithms) or by performance benchmark tests.

4.2. Testing the new service

Once the service is created the new service will be tested in a test environment. Naturally, this environment should be representative for the operational environment. Of course measures should be taken to ensure that errors which occur have no effect on the operational environment.

Several approaches can be taken to test whether the service complies to the specification, i.e. the service subscriber's needs. These approaches differ in costs, ease-of-use, and fault detection capabilities. A straight-forward technique applies simple end-to-end testing of the service behaviour. A more advanced approach provides in monitoring network or protocol traffic on a number of strategic locations. The most complex and exhaustive testing can be performed by stand-alone testing of a single network element. Below these approaches are discussed in detail.

4.3. The end-to-end service testing approach

Figure 3 pictures the most basic approach towards the testing of a new service.

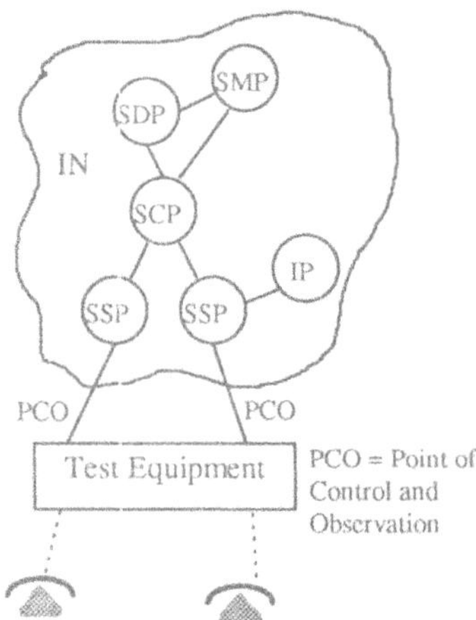

Figure 3 End-to-end service testing approach

In this approach the service is downloaded into the (test)platform. The tests are controlled at the user-end of the platform, i.e. by simulating the service user behaviour using DTMF (terminal key pad buttons) access. The tests are observed at this interface as well. This type of testing can be referred to as 'black-box' testing.

For example, testing a Call-Forwarding service can be done using three phones. Suppose phone B is forwarded to phone C. Subsequently phone A dials B's number. Phone C rings and when picked up a speech channel is established between A and C. The advantage of this approach will be the low costs and short time needed to set up the test configuration. The description of the tests could even be done using natural language. In that case the specifications would be easy to write.

The major drawback of such an approach is the limited coverage of the tests. Applying the terminal access as a PCO implies that part of the service logic that deals with exception situations cannot easily be tested. This is caused because the terminal access is not, or hardly, able to provide in network failures or resource limitations. Another drawback of this approach is the lack of fault localisation. If errors occur it will be very difficult to determine which part of the network caused the error.

4.4. The monitored end-to-end service testing approach

One of the drawbacks of the approach described above are the possibilities of fault localisation. A more refined approach is the so called 'Monitored end-to-end testing' (Figure 4).

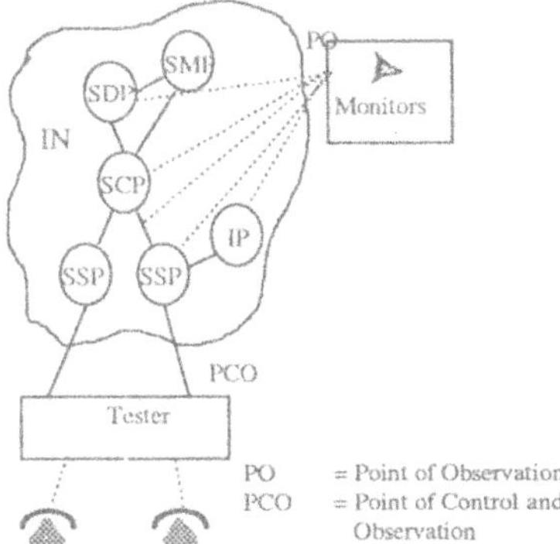

Figure 4: Monitored end-to-end testing approach

In this approach, again the tests are controlled from a user point of view. To observe the network's 'internal behaviour' monitors can be installed at several places. Obvious observation points for monitor equipment are the communication links. These links are often based on international standards (core-INAP). These monitors are (or will be) therefore commercially available. In some cases network elements will provide possibilities to monitor the inside behaviour of the network elements as well.

For the test description two approaches can be used. One approach is to use the same kind of test specification as used by the 'end-to-end' test approach. In this case the monitors are only used when errors occur in order to locate the error sources .

Another approach can be found in (Gabrielli, 1992). In this approach all expected observations visible at the monitors are fully described. In the test specification not only the external behaviour (like in the end-to-end testing approach) will be described, a description of the internal events in the SSP and SCP, the INAP messages exchanged over the SS7 interface, and the queries in the SDP are to be specified as well. On the one hand this provides detailed information on what is tested. On the other hand the test specification will tend to become very large and difficult to read.

Looking at the advantages/disadvantages, using the monitored end-to-end testing approach, information is gained about the IN internal events. This provides the opportunity to locate the source of the error. The testing port (the terminal access) is still the same as described for the previous approach. Therefore the testing power will still be limited to the possibilities of the user interface. We will refer to this type of testing as 'grey-box' testing.

4.5. The Service Conformance Testing approach

The most powerful test method namely the Service Conformance Testing approach can be found in Figure 5.

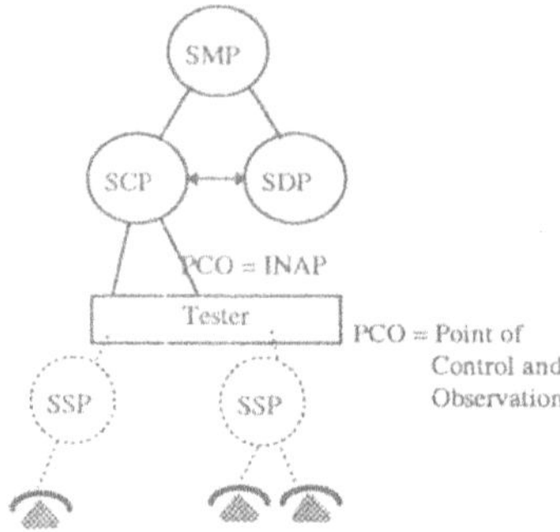

Figure 5 Service Conformance Testing approach

Using this method the tests are mainly focused on one network element, in this configuration the SCP. The service specific software largely resides in the SCP. The part of the service in the SSP will in most cases only affect some parameter settings (like the update of trigger detection points)

In Figure 5, the PCO is located at the communication link, which exhibits the standardised core-INAP protocol. Tests are controlled by this access. Using the core-INAP interface provides extensive possibilities for testing, that go beyond the level of valid behaviour testing. The core-INAP interface provides possibilities to perform tests for invalid behaviour and data encodings as well.

This testing approach has a number of similarities with 'traditional protocol conformance testing' which is defined in ISO9646 (ISO,1994). It is therefore likely that the TTCN language will be suitable for this approach to Service Testing.
The test specification for the call forwarding service will look like:

[1]! IntialDetectionPoint (CF,B,C.........)
"forward phone B to phone C"
    !IntialDetectionPoint(A,B,....)
    "A dials B number"
        ?Connect (A,C,...)
        "Phone A is connected to phone C"

In this example the action of forwarding a phone B to phone C is translated in the INAP message 'InitialDetectionPoint(Cf,B,C...)'. Subsequently phone A dials B's number (InitialDetectionPoint(A,B,...) and the SCP will provide the instructions to forward the call to phone C (Connect(A,C,..) As the PCO is now located inside the Intelligent Network, we will refer to this type of as 'white-box' testing.

The advantage of such an approach lies in the extensive testing power. This testing power provides possibilities to perform a number of robustness tests which are not possible with the approaches described before. However, using such an approach will require still a number of end-to-end tests to ascertain interoperability. The disadvantage of this an approach clearly lies in the effort needed to set up a test campaign and to produce the test

In protocol conformance testing the ! stands for 'Send' while the ? stands for 'receive' suites. In a 'time-critical' provision schedule the large effort may be a bottle-neck in the testing process.

4.6. Which approach?

The choice of which approach should be used, depends on a number of factors. On one hand we have to deal with factors like 'required quality of service', c.q. required level of confidence'.

On the other hand we are limited by the available resources and time. At this moment there is little core-INAP conformance test equipment, and no complete test specifications available. Furthermore the lack of experience in service conformance testing may cause unacceptable delays in the tight testing schedule.

It seems therefore reasonable to start gaining experience in the field of service testing by applying manual end-to-end tests. Along the way of practice the process can be improved by automation and the use of more powerful methods. The generation and execution of tests could be automated, while the test methods could shift from end-to-end testing towards service conformance testing.

4.7. Regression tests

Even when a newly installed service is not activated yet, it may interfere with the functionality of other services; the addition of the new service to the installed base may have caused changes to the existing services. Therefore it is necessary to perform tests on existing services as well. These type of tests are called regression tests.

In Figure 6 a scenario for service regression tests is pictured. Before starting the regression tests, the new service should be completely tested using one of the approaches discussed in the previous paragraphs. To assure the other services are still working correct, regression tests have to be executed, by performing a subset of the functional tests for each services.

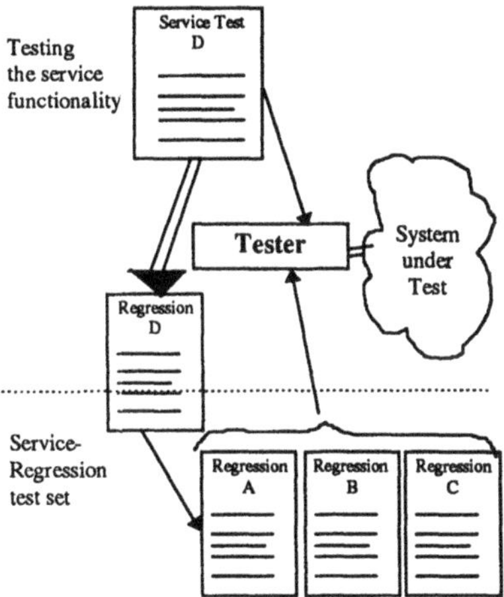

Figure 6 Service Regression tests

After the execution of all regression tests the standard set of regression tests can be extended by a subset of the functional tests for the new service. (In Figure 6, the service D.)

From the Figure 6, it can be seen that regression testing may involve a serious management problem. Like pictured, the set of standard regression tests will grow with the introduction of each new service.

For reducing the regression test set two alternative approaches can be used. Instead of using a subset of the functional tests, one can figure that it might be possible to use some 'test service' for the regression test. This service should contain features of all already installed services. Whether such an approach is feasible is hard to say.

Another possibility might be the use of categories and priorities for the selection of a regression set. This principle requires all functional tests to be categorised(e.g. according to the involved SIBS), and prioritised.(e.g. according to the impact of the failure). Although the complete set of available tests grows with the addition of new services, it would be possible to select a constant number of regression tests based on categories and priorities.

4.8. Interaction tests

In the previous paragraphs it was stated that after testing the service on functionality, regression tests are to be performed to verify that the already installed services are still working correct. Therefore a subset of the functional tests of each installed service is executed. This is however not enough for deploying the service into the network. When each individual service is working correct, this gives no guarantee yet that the services operating concurrently will work correctly together. Services may conflict with each other and even with (other instances of) them-selves.

Two types of service interaction can be identified. The first type is inherent to the service logic and can (should) be detected in the specification. Smart algorithms can detect these service interactions, and alteration to the service can prevent the interactions to occur. The second type however, is depending on dynamic service data values, and is difficult to recognise at specification time.

In practice however it is yet not possible to detect undesirable service interactions in the specifications. Therefore some kind of 'service interaction' test must be executed to prevent the network from service interactions. These tests can be performed either in a simulation environment or in a real test environment.

An interesting approach can be found in (Razol, 1994). A simultaneous execution of functional tests can detect the second kind of interactions. When all functional tests executed simultaneously pass, it can be argued the network is free of service interactions.

4.9. Service management tests

Once the service is operational, test activities should still be performed. These test activities fall into the category of fault management. Within the Telecommunication Management Network (TMN) (CCITT,1992), fault management is defined. Fault (or maintenance) management consists of a set of functions which enables the detection, isolation and correction of abnormal operation of the telecommunication network and its environment.

4.10. TMN Fault Management

Within the TMN standard the fault management is divided into the following four types of actions:

- Alarm surveillance (some Network Element (NE) or link is not working)
- Fault Localisation (where did the fault occur?)
- Fault Correction (resolve the fault)
- Testing (is a service still working)

Because this paper aims at service testing, the first three types will not be discussed. These types of actions consider errors, which will cause notifications. This error is subsequently localised and corrected.

In this report we concentrate on the testing type of fault management. Within M.3400 (CCITT, 1994) two types of testing activities are identified. Either the TMN requests the Network Element to test a service, and send back the report, or the TMN requests access to a Network Element and performs some testing itself. These types of testing can be used either preventive (testing if a service still is operational) or for fault location (after complaints about the service are received). Although Service Management is very important in keeping the service alive, unfortunately this area is still quite unexplored.

5. ACKNOWLEDGEMENTS

We like to thank the people within Eurescom, KPN Research and PTT Telecom who have contributed to the ideas that were presented in their papers. We owe our colleague Wilbert Schelvis for carefully reviewing this article.

6. CONCLUSIONS

This paper presented a framework for Service Testing. Before testing the new service itself, it should be assured that the platform is operating correctly. The new service should be tested in three steps. Within this frame several possibilities are available to ascertain that both the new service and the installed base of services, are operating correctly. Finally, when the service is in operation tests can be performed to assure that the services are still working and to locate error sources in case of failures.

Although the area of service testing is relatively new, a number of already available testing techniques can be reused for service testing. Already for basic telephony services testing call generators and ordinary telephone equipment is used. These equipment can be reused for service testing as well.

Also concepts and methodology from the conformance testing world can be applied for service testing. The really new aspect on service testing will be the time available for testing. To meet market goals the pressure for a rapid service introduction will grow. Therefore the time left for service testing will be under pressure more and more.

This time factor will become more and more important in the choice for a test methodology. It is therefore a challenge for the test research community to find time efficient solutions

7. REFERENCES

Eurescom-P103 (1992) 'Evolution of the Intelligent Network, deliverable No.2'.

H. Velthuijsen et al. (1993), ' A feature Interaction Benchmark in IN and beyond', Bellcore Memorandum TM-SV-021982

CCITT, Q1201 (1992), 'Principles of Intelligent Network Architecture'.

ISO/IEC 9646: April 1994, Information Technology, OSI Conformance Testing Methodology and Framework.

S.P. van de Burgt, J.Kroon, E.Kwast, H.J.Wilts (1990), The RNL Conformance Kit, Protocol Test Systems, IFIP.

E. Anders, J. Ellsberger, A. Wiles (1994), Experiences with Computer Aided Test Generation, proceedings IWPTS conference 1994.

L. Gabrielli, P. Marchese, (1992) 'Intelligent Network Service Testing', ICCC, Conference on IN, Tampa, USA.

R.J. Heijink (1994), 'FAITH, a general purpose protocol test system for ISDN', Computer Networks and ISDN Systems (1994), 1581-1593.

P.Razol et al (1994), 'Service Interaction and Test Generation', proceedings ICIN conference.

CCITT, 'M.3010 (1992), principles for a telecommunications management network'.

CCITT, 'M.3400 (1992), TMN Management Functions'.

8. BIOGRAPHY

Gert Vermeer graduated in 1990 at the department Informatics at the University of Twente. After his study he joined KPN Research, where he started working in the field of testing. He participated in several test projects for FTAM and other OSI protocols. Apart from test development he contributed to the methodology for interoperability testing. Currently he is responsible for the development of various IN test suites.

Marc Witteman graduated in 1989 at the department Electrotechnical Engineering at the University of Delft. In the same year he joined KPN Research, where he started working in the field of conformance testing. He participated in several test projects for GSM, IN and ATM protocols. One of his favorite subjects is the investigation of testability and the design of test architectures. Currently he is mainly involved in the validation of IN Services.

Jan Kroon studied Mathematics and Physics at the University of Utrecht in the Netherlands. He joined KPN Research in 1986. He has been a project leader of research projects in the field of conformance testing, with focus on automated test generation from formal specifications. In 1992 Jan Kroon worked for the European Telecommunications Standard Institute. His group produced a standard on the use of the specification techniques SDL. In june 1995 he completed a thesis with the subject "Specification and Testing of Telecommunication Systems" at the University of Bern.

PART FOUR

Algorithms and Languages

9

Conformance testing of protocols specified as labeled transition systems*

P. V. Koppol and K. C. Tai
Department of Computer Science
North Carolina State University
Raleigh, North Carolina 27607-8206, USA
Tel: (919) 515 7146, Fax: (919) 515 7896
e-mail: gr-kvp@aristotle.csc.ncsu.edu, kct@csc.ncsu.edu

Abstract

In this paper, conformance testing of protocols specified as sets of labeled transition systems (LTSs) is considered. LTSs serve as the semantic model for a number of languages such as CCS, CSP and LOTOS. A straightforward approach to conformance testing of such a protocol is to generate the LTS corresponding to the global behaviour of the set of component LTSs and to apply known test generation procedures to the global LTS. In general, construction of the global behaviour is difficult due to the state explosion problem. Also, testing is complicated due to internal (nonobservable) events and nondeterminism. In this paper, we propose a novel approach to alleviate state explosion in the construction of the global LTS. We also propose how the reduced LTS is used in conformance testing.

Keywords

Protocol conformance testing, adaptive testing, labeled transition systems, reachability analysis

*This work was supported in part by US National Science Foundation under grant CCR-9320992

1 INTRODUCTION

Conformance testing of communication protocols has been an active area of research for almost two decades. Given a protocol specification S and an implementation I (referred to as the **implementation under test or IUT**), conformance testing is to show by testing, while treating I as a black box, that I conforms to S. Test generation is therefore based on the specification S. Traditional methods for test generation assume that the specification S is in the form of a strongly connected, minimal and deterministic finite state machine (FSM). Some of these techniques also require the FSM to be completely specified. The most widely studied formal methods based on these assumptions are the T-method, D-method, W-method and UIO method (Sidhu and Leung, 1989). More recently, test generation for protocols specified as labeled transition systems (LTSs) has received much attention (Arkko, 1994) (Cavalli, Favreau and Phalippou, 1994) (Cavalli and Kim, 1992) (Cavalli et al., 1993) (Cavalli, Kim and Maigron, 1994). LTSs serve as a semantic model for the standardized FDT LOTOS and other languages such as CCS (Milner, 1989) and CSP (Hoare, 1985). The semantics of the other standardized FDTs, Estelle and SDL, can also be partly expressed in LTSs (Tretmans, 1994). Details on standardized FDTs may be found in (Turner, 1993) and references therein.

Test generation methods for LTSs may be broadly classified into two categories (Cavalli, Favreau and Phalippou, 1994). The first category corresponds to the testing theory presented in (Brinksma et al., 1990). These methods are based on the coverage of all traces (Arkko, 1994). The second category consists of those methods that are built upon the traditional test generation methods such as UIO method. Methods in the first category have certain practical limitations as has been pointed out in (Cavalli, Favreau and Phalippou, 1994) (Arkko, 1994) and hence are not considered any further in this paper. Methods in the second category assume that the specification is in the form of a single (monolithic) LTS or that such an LTS can be obtained from the protocol specification (Arkko, 1994) (Cavalli and Kim, 1992) (Cavalli et al., 1993) (Cavalli, Kim and Maigron, 1994) (Fujiwara and Bochmann, 1992). In this paper, we adopt the terminology of (Cavalli, Favreau and Phalippou, 1994) and refer to the methods in the second category as **checking experiments based** methods.

In this paper, protocols specified as sets of (communicating) LTSs are considered. Examples of such specifications include CCS and CSP specifications involving parallel composition and constraint oriented LOTOS specifications. In general, the following three-step process may be used for test generation based on such specifications. First, derive the global LTS corresponding to the set of component LTSs (e.g. using expansion law in CCS). Second, transform the global LTS into a form such that a test generation technique may be applied, and third, apply a known test generation technique. Note that once the first and second steps of this general procedure are completed, this procedure coincides with the checking experiments based approaches. However, the following problems need to be addressed: 1) state space explosion during global LTS construction, 2) preserving the expected behaviour of the implementation while performing LTS transformation, and 3) assuring the executability of the generated test cases. Executability becomes a problem due to the lack of control over the nondeterminism in the IUT. Another important concern is that the length of the generated test cases could be exponential in the number of component LTSs (Lee et al, 1993).

To cope with the above problems, an adaptive testing approach was presented in (Lee et al, 1993) wherein the need for global LTS generation is obviated and testing is based on an adaptive guided random walk. However, due to the nonexistence of the global LTS, the fault detection ability of the approach is not clear. In this paper, we propose an algorithm for the construction of a reduced global LTS, while alleviating state space explosion, such that the reduction preserves the expected behaviour of the implementation and all necessary information that will enhance the fault detection capability of the adaptive testing approach. We show how to construct testers and also identify testability issues that need to be considered in the context of adaptive testing.

The organization of this paper is as follows. Section 2 gives basic definitions. Section 3 provides the background related to checking experiments based conformance testing and the adaptive testing strategy presented in (Lee et al, 1993). It also discusses the motivation for the results of this paper. Section 4 presents an algorithm that, for a given set L of LTSs, incrementally composes and reduces subsets of L and finally produces a reduced LTS whose behavior is equivalent to the global behaviour of L. Section 5 addresses the problem of testing based on the reduced global LTS and addresses issues related to tester construction and testability. Section 6 discusses related work and section 7 concludes this paper with a discussion on future work.

2 PRELIMINARIES

In this paper, we consider communication protocols specified as sets of communicating processes with synchronous communication and direct naming. Each process is modeled as a rooted labeled transition system (LTS) which is defined as a quadruple $< Q, E, \rightarrow, q_0 >$, where Q is a finite set of states, E is a set of events (including a special event τ), $\rightarrow \subseteq Q \times E \times Q$ is the transition relation and q_0 is a designated start state. For readability, a transition $(s, a, s') \in \rightarrow$ is denoted as $s \xrightarrow{a} s'$.

Each event in the set of events for each component process denotes a send (receive) statement where the receiver (sender) is explicitly identified. For example, let L_1 and L_2 be two LTSs corresponding to processes P_1 and P_2. An event denoting sending of a message m from P_1 to P_2 is specified as $P_2!m$ and the corresponding receive event is specified as $P_1?m$. This kind of specification is similar to that of CSP and has also been used in (Lee et al, 1993). A matching operation between a send and receive is referred to as a synchronization event and is denoted by τ. We assign a unique identifier 'u' to matching send and receive events. As a convention, we denote the send event as u and the receive event as $\bar{u}$. Thus, only u or $\bar{u}$, but not both, may belong to the event set of a component LTS. A receive (send) event in L is said to be an **input (output)** of L if the corresponding sender (receiver) is not in L. Events that are neither inputs nor outputs are referred to as **internal** events. Also, transitions labeled by internal (input, output) events are referred to as internal (input, output) transitions. Classification of events in an LTS into input and output events has also been used in (Phalippou, 1994).

Given a set L of LTSs, the LTS representing the global behaviour of the set is referred to as the **composite** LTS of L. Each LTS $L_i \in L$ is referred to as a **component** LTS of L. The composite LTS of L may be obtained as the parallel composition of the component LTSs of L while hiding the internal events in L. For $L_1 =< Q_1, E_1, \rightarrow_1, q_1 >$ and $L_2 =< Q_2, E_2, \rightarrow_2, q_2 >$, the composite LTS L of L_1 and L_2 is given by the quadruple $< Q, E, \rightarrow, q_0 >$, where,

- $Q \subseteq Q_1 \times Q_2$;
- $E = \{\tau\} \cup (E_1 \cup E_2) \setminus (E_1 \cap E_2)$;
- $\rightarrow \subseteq Q \times E \times Q$. A transition $(x, y) \xrightarrow{a} (x', y')$ iff
    - $(a \in E) \wedge (y = y' \wedge x \xrightarrow{a}_1 x')$ or
    - $(a \in E) \wedge (x = x' \wedge y \xrightarrow{a}_2 y')$ or
    - $a = \tau$ and $\exists e. ((x \xrightarrow{e}_1 x' \wedge y \xrightarrow{\bar{e}}_2 y') \vee (x \xrightarrow{\bar{e}}_1 x' \wedge y \xrightarrow{e}_2 y'))$.
- $q_0 = (q_1, q_2)$.

Figure 1 shows two LTSs L_1 and L_2 (events i1 through i7 are internal events). The composite LTS L of L_1 and L_2 is shown in Figure 2(a). The above definition of a composite LTS for a set of two LTSs may be extended to a set of n LTSs in a straightforward manner. The composition operation is both associative and commutative. Each state of the composite LTS is a collection of states of the component LTSs. For example, the start state of the composite LTS of L_1 and L_2, whose start states are q_1 and q_2 respectively, is (q_1, q_2). The states of the composite LTS are referred to as **global** states. For a global state s, *outdegree*(s) is the number of transitions leaving s and $S_\tau(s) = \{s' \mid s \xrightarrow{\tau} s'\}$. The global states are classified into three categories. A state s for which *outdegree*$(s) = \mid S_\tau(s) \mid$ and $\mid S_\tau(s) \setminus \{s\} \mid = 1$ is referred to as a **hidden** state. The states which have only input transitions leaving them are referred to as **stable** states. All other states are referred to as **transient** states. It is assumed that states that have outdegree equal to zero do not exist (if they do, then it will be considered a design error and will not be relevant in the context of conformance testing).

Given an LTS $L = < Q, E, \rightarrow, q_0 >$ and $s_1, s_2 \in Q$, s_1 and s_2 are said to be **strongly equivalent**, denoted by $s_1 \sim s_2$, if the following condition and its symmetric condition hold: $\forall a \in E, s_1 \xrightarrow{a} s_1' \Rightarrow \exists s_2'.s_2 \xrightarrow{a} s_2' \wedge s_1' \sim s_2'$. Informally, $s_1 \sim s_2$ if whenever s_1 has an a transition to s_1', then s_2 also has an a transition to some state s_2' such that $s_1' \sim s_2'$ and the symmetric condition holds. States s_1 and s_2 are said to be **observationally equivalent**, denoted by $s_1 \approx s_2$, if the following conditions and their symmetric conditions hold:

(1) $\forall a \in E \setminus \{\tau\},\ s_1 \xrightarrow{\tau^* a \tau^*} s_1' \Rightarrow \exists s_2'.s_2 \xrightarrow{\tau^* a \tau^*} s_2' \wedge s_1' \approx s_2'$, and
(2) $s_1 \xrightarrow{\tau^*} s_1' \Rightarrow \exists s_2'.s_2 \xrightarrow{\tau^*} s_2' \wedge s_1' \approx s_2'$.

Informally, $s_1 \approx s_2$ if whenever s_1 has a **weak** a transition (i.e., an a transition which subsumes zero or more of its preceding and succeeding τ transitions) to s_1', then s_2 also has a weak a transition to some state s_2' such that $s_1' \approx s_2'$ and the symmetric condition holds. Details concerning these equivalences may be found in (Milner, 1989).

3 BACKGROUND AND MOTIVATION

3.1 Checking experiments based approaches to conformance testing

Conformance testing based upon checking experiments involves testing the behaviour of each transition in a global LTS. A transition, (s, a, s'), is tested in the following manner:

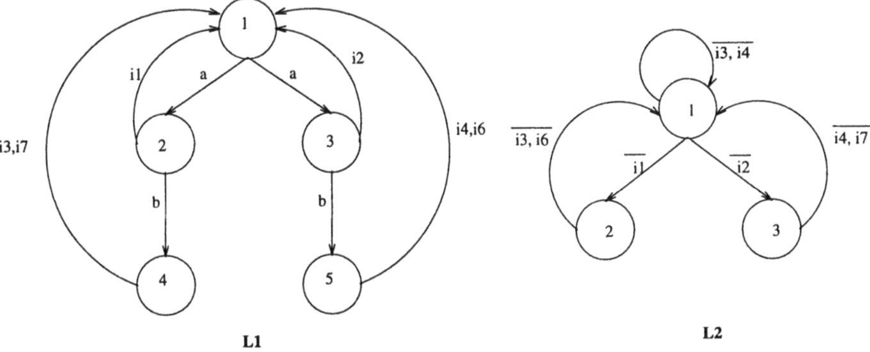

Figure 1 Set of LTSs, $L = \{L_1, L_2\}$

- Step 1: From the current state of the IUT, bring the IUT to state s. This is usually done in two stages: 1) the implementation is brought into a known state, e.g. the initial state, using a sequence of events referred to as the **synchronizing sequence** and 2) a **transferring sequence** is then applied to bring the IUT to state s.
- Step 2: Check if the IUT can synchronize with a matching event for a; if the IUT fails, a fault is detected, else proceed to step 3.
- Step 3: Check if the implementation has reached state s'. This is usually done using a sequence of events referred to as the **state identification sequence.**

The above three step procedure, however, is difficult to achieve due to limited controllability over the IUT. A number of test generation techniques assume the existence of a synchronizing sequence for some or all states of the LTS. For example, the **reset** capability was assumed in (Cavalli, Kim and Maigron, 1994). In the presence of nondeterminism in an LTS, it is difficult or even impossible to find transferring sequences for certain states. Generation of state identification sequences has been studied in several papers. In (Fujiwara and Bochmann, 1992), it is assumed that the LTS does not contain any *tau* transitions. In (Cavalli et al., 1993) (Cavalli, Kim and Maigron, 1994), the LTS is assumed to be deterministic. If it is not, then it is transformed into a deterministic LTS with no τ transitions. As pointed out in (Arkko, 1994), this transformation does not take the inherent nondeterminism in the IUT into consideration. (Arkko, 1994) considers LTSs containing nondeterminism and also τ transitions. Algorithms are provided to generate **state identification machines** that can be used to perform adaptive testing to check if the IUT was in an expected state. However, the complexity of these algorithms is not known. Also, to our knowledge, no such algorithms exist for generation of transferring sequences.

When a set of LTSs is considered, a test generation technique cannot make assumptions of determinism and nonexistence of τ transitions. The reason for this is that, whenever a protocol is specified as a set of (communicating) LTSs, there are bound to be internal (nonobservable) synchronizations that result in τ transitions in the composite LTS.

3.2 The adaptive approach to conformance testing

In this section, we give an overview of the adaptive guided random walk approach (Lee et al, 1993) to conformance testing of protocols specified as a set of communicating processes where each process may be modeled as an LTS.

The basis of this approach is the claim that a test set that checks the conformance of all component processes is also a conformance test for the combined behaviour (the proofs may be found in (Lee et al, 1993) and/or references therein). An implication of this is that it suffices to test every transition in each component machine at least once rather than to test all transitions in the global behaviour. In this context, **testing** a transition means observing the starting and ending states of that transition. A transition is said to be **tested** if both its starting and ending states have been observed, **weakly tested** if its starting state is observed but the ending state is not observed, and **untested**, otherwise.

The following assumptions are made: 1) inputs can be applied to the IUT only when the IUT is in a stable state and 2) at each stable state, the state of each component process can be observed (this assumption may be realized, for example, using a status message for each component process). Assumption about being able to observe every stable state has also been made in (Arakawa and Soneoka, 1992) where it is assumed that for each stable state, the system appearance can be generated using some hardware specifications. The complexity results presented in (Lee and Yannakakis, 1994) make this assumption more interesting. With the goal of testing every transition in each component process, conformance testing is performed as follows. At each stable state reached during testing, the states of all component processes are observed and the input transitions are grouped into three categories: tested, weakly tested and untested. If untested transitions exist, one of them is selected at random. If there are no untested transitions, then one of the weakly tested transitions is selected at random. If no such transitions exist either, then one of the tested transitions is selected. The input associated with the selected transition is then applied and the next stable state reached is observed. Assume that there are n component processes. When a stable state $s' = (s'_0, s'_1, ..., s'_{n-1})$ is reached from stable state $s = (s_0, s_1, ..., s_{n-1})$, coverage of transitions of component processes and faulty behaviour are ascertained as follows:

- If no outputs are observed during the transition from s to s', then a check is performed if, $\forall i$, s'_i is reachable from s_i only through internal transitions. If the check fails, then a fault has been detected. Otherwise, a positive probabilistic measure is added to the coverage probability of each internal transition in each component process that could potentially have been on the path from s to s'.
- Let the set of outputs observed be OO. Then from each component process a set OE of potentially observable outputs during the transition from s to s' is generated and the two sets are compared. If $OO \subseteq OE$ then it is assumed that there was no fault. Otherwise, a fault has been detected.

The testing procedure stops when a reasonable terminating condition is reached. An example of such a condition would be that the probabilities of coverage of all internal transitions have reached a certain minimum threshold and all input/output transitions have been tested (Lee et al, 1993). Efficient techniques for keeping track of coverage of external transitions and the probability of coverage of internal transitions can be found in (Lee et al, 1993).

3.3 Motivation

Checking experiments based testing approaches, when applied in the realm of protocols specified as sets of LTSs, have the following limitations:

(i) State explosion during the construction of the composite LTS.
(ii) It is not necessary, based on results obtained in (Lee et al, 1993), to test all transitions of the composite LTS.
(iii) Generation of state identification sequences (machines) is a difficult problem. Also, such sequences (machines) are difficult to apply to the IUT because of the lack of controllability over the IUT.
(iv) Driving the IUT from the initial state to a given reachable state s is difficult due to the limited controllability over the IUT.

On the other hand, the adaptive testing approach has the following limitations:

(i) When a stable state s is reached during adaptive testing, it is not known if s is reachable according to the specification or was reached due to a fault in the implementation.
(ii) The expected output is in the form of a set of outputs rather than sequences of outputs; this may limit the fault detecting capability of the approach considerably.

These limitations can be alleviated by generating the global LTS. This, however, is a difficult task due to the state space explosion problem. These observations constitute the motivation for the results of this paper, which are:

(i) Generation of a reduced composite LTS such that state explosion is alleviated and all stable states in the composite LTS are preserved.
(ii) Applying the adaptive testing strategy to conformance testing with a greater fault detection capability due to the existence of the reduced composite LTS.
(iii) Identifying testability related issues that need to be addressed during the specification if the adaptive approach to conformance testing is to be applied.

4 GENERATION OF REDUCED GLOBAL LTS

A straightforward approach to generating a reduced global LTS is to first generate the composite LTS L_c for a given set of LTSs L and then apply some reduction technique to L_c. This, however, does not alleviate the state explosion problem. To alleviate state space explosion, we use the incremental technique given in (Tai and Koppol, 1993). This technique works as follows:

- Step 1: Select a subset of L.

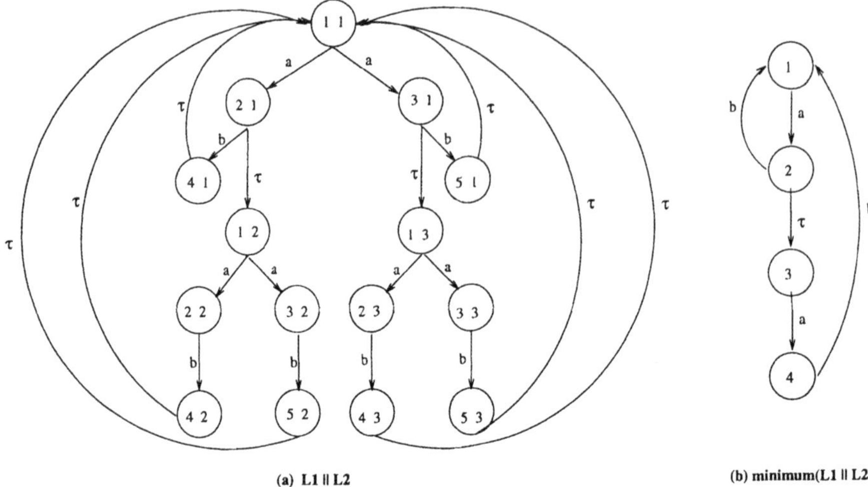

(a) L1 || L2

(b) minimum(L1 || L2)

Figure 2 Global LTS for $(L_1||L_2)$ and its minimal LTS

- Step 2: Generate the composite LTS for the selected subset.
- Step 3: Minimize the composite LTS with respect to observational equivalence.
- Step 4: Update set L as L = L \ (subset of step 1) ∪ (minimum LTS from step 3).
- Step 5: If cardinality of L is greater than 1, goto step 1, else terminate.

One of the main concerns of incremental/compositional techniques for reduced LTS generation is the selection of subsets in step 1 above. The reason is that, if a 'wrong' subset is selected, the LTS generated in step 2 may have more states than the global LTS itself! The hierarchy-based approach presented in (Tai and Koppol, 1993) provides heuristic techniques to alleviate this problem. Step 2 is based on the operational semantics of CCS. For step 3, the set of states is partitioned into equivalence classes such that each class contains all states that are observationally equivalent. The minimum observationally equivalent LTS is generated by collapsing the states in each equivalence class into a single state and updating the transitions accordingly. Due to this reason, the correspondence among global states and the states of component processes is lost. The partitioning algorithm used is the one implemented in the Concurrency Workbench (Cleaveland et al., 1993). Consider the following example. Figure 1 shows two LTSs L_1 and L_2. The composite LTS L that represents the combined behaviour of L_1 and L_2 is shown in Figure 2(a). The LTS L_m shown in Figure 2(b) represents the minimal LTS that is observationally equivalent to L. L_m is constructed by collapsing equivalent states in L into a single state and rearranging the transitions accordingly. Note that there is no correspondence between the states of L and L_m. Also, the stable states of L are not preserved in L_m.

To preserve state information of component processes, we propose a new reduction technique for step 3 which is based on the following observations:

- **Observation 1**: For adaptive testing purposes, the correspondence between component process states and global states is needed only in the case of stable states.
- **Observation 2**: Observational equivalence may equate a stable state and a transient state. A hidden state s is observationally equivalent to the state s' $(s \neq s')$ reached by one of its τ transitions.
- **Observation 3**: Strong equivalence never equates two states that belong to different categories, i.e., if partitioning of states is done with respect to strong equivalence, then all states in a certain equivalence class are either all stable, transient or hidden.

The basic idea of the reduction algorithm is to keep stable states intact while gaining as much reduction as possible in terms of the nonstable states. We exploit observations 2 and 3 to reduce the number of hidden and transient states while leaving stable states intact.

Reducing hidden states

Let $L = < Q, E, \rightarrow, q_0 >$ be the given LTS. Let H be the set of all hidden states in L. For $s \in Q$, let $Pred(s) = \{s' \mid s' \xrightarrow{a} s\}$ and let $Succ(s) = \{s' \mid s \xrightarrow{a} s'\}$. The goal is to eliminate the hidden states in L. To achieve this goal, for each state $s \in H$, L is updated according to the following procedure:

- Step 1: $\rightarrow = \rightarrow \setminus \{s \xrightarrow{\tau} s\}$. Informally, if there is a self looping τ-transition at state s, delete it.
- Step 2: Perform the following updates to L in the given order:

  (i) $\rightarrow = \rightarrow \cup \{x \xrightarrow{a} y \mid x \xrightarrow{a} s \wedge s \xrightarrow{\tau} y\}$. Informally, for each $p \in Pred(s)$, if p has an a transition to s then, $\forall q \in Succ(s)$, add a new transition $p \xrightarrow{a} q$.
  (ii) $\rightarrow = \rightarrow \setminus \{x \xrightarrow{a} y \mid x = s \vee y = s\}$. i.e., delete all transitions leading into s and all those leaving s.
  (iii) $Q = Q \setminus \{s\}$.

The above procedure will henceforth be referred to as procedure **h_reduce(L)**. Since $H \subseteq Q$ and due to the fact that Q is finite, it follows that **h_reduce(L)** terminates and has polynomial complexity. Let L^{r1} be the LTS obtained by applying the above procedure to L. L^{r1} does not contain any hidden states. For example, consider the set of LTSs shown in Figure 1. The reduced LTS, L_{p1}, is shown in Figure 3. L_{p1} does not contain any hidden states, but contains all the stable states contained in L.

Property 1 $L^{r1} \approx L$.

Proof. Follows from observation 2. □

Reducing transient states

Procedure **h_reduce(L)** eliminates all hidden states from the LTS L. Further reduction can be obtained by reducing the number of transient states. For example, L_{p1} shown in

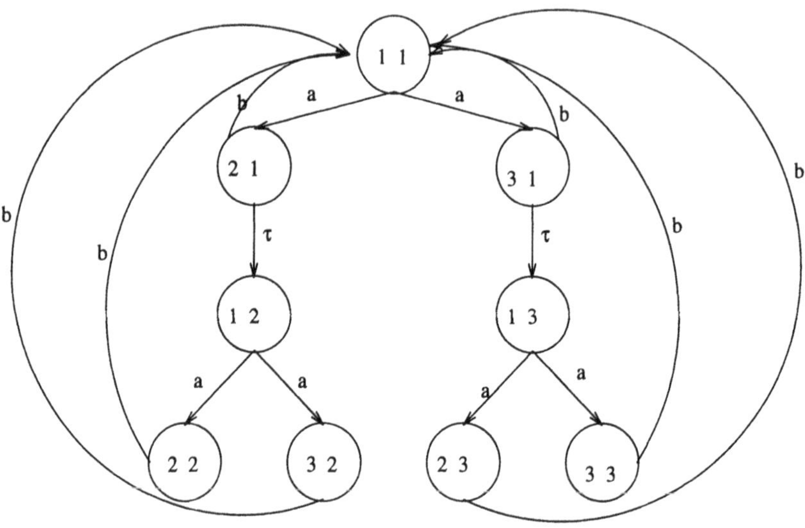

Figure 3 L_{p1}, **the reduced LTS without hidden states**

Figure 3, can be reduced further by collapsing the equivalent transient states $(2,1)$ and $(3,1)$ into a single state. The goal is to achieve this reduction. We exploit observation 3 to gain this reduction. Let $L =< Q, E, \rightarrow, q_0 >$ be the LTS that is to be reduced. It is assumed that L has already been processed by **h_reduce** and hence does not contain any hidden states. The reduction procedure for LTS L, henceforth referred to as **t_reduce(L)**, is as follows:

- **Step 1**: Obtain the partition P of the set of states Q with respect to strong equivalence. This may be done by using well known algorithms given in (Kanellakis and Smolka, 1983) (Paige and Tarjan, 1987). According to observation 3, each block $B \in P$ either contains all stable states or all transient states. A block is said to be stable (transient) if it contains only stable (transient) states. Let $P_s = \{B \in P \mid B\, is\, stable\}$.
- **Step 2**: Refine the partition P as follows. For each $B \in P_s$, do the following:
  - Let $k = \mid B \mid$. Create k sets $B_i, 0 \leq i \leq k-1$, where $B_i = \{s \mid s\, is\, the\, ith\, element\, of\, B\}$.
  - Update P as follows: $P = P \setminus \{B\} \cup \cup_{i=0}^{k-1}\{B_i\}$.
- **Step 3**: Construct the reduced LTS by collapsing states in the same block in P into a single state and arranging the transitions appropriately.

The complexity of the above algorithm is the same as the complexity of the partitioning algorithm. The partitioning algorithms have polynomial complexity. Note that due to step

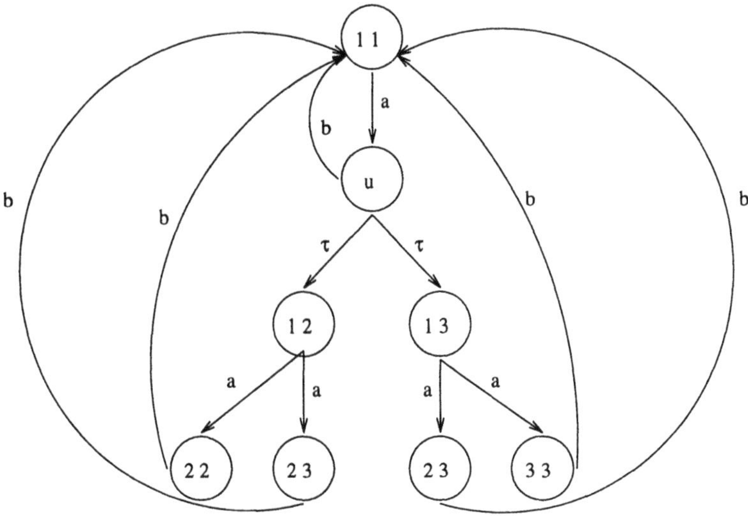

Figure 4 L_{p2}, the reduced LTS without equivalent nonstable states

2 of the algorithm above, each stable block contains exactly one stable state. Hence, when states in a block are collapsed into a single state in step 3, the component process state information may be lost in the case of transient blocks. In the case of stable blocks, due to the fact that no collapsing of states is needed, the component process state information is preserved. Figure4 shows the LTS L_{p2} which is the result of applying the above reduction procedure to the LTS L_{p1}. L_{p2} contains all the stable states of L and does not contain any equivalent transient states.

Property 2 *Let L^{r2} be the LTS obtained by applying the above procedure to L. Then, $L^{r2} \approx L$.*

Proof. Since we use partitioning with respect to strong equivalence, $L^{r2} \sim L$. Due to the fact that strong equivalence implies observational equivalence, $L^{r2} \approx L$. □

Procedure **t_reduce(L)** may, however, introduce hidden states. These hidden states can be eliminated using **h_reduce(L)**. Thus, the overall reduction procedure, referred to as **reduce(L)**, is as shown in Figure 5. In this algorithm, **h_flag** and **t_flag** are boolean variables. It is assumed that the procedures **h_reduce(L)** and **t_reduce(L)** return **true** if reduction is gained and **false** otherwise. Procedure **reduce(L)** has polynomial complexity. Also, it can be made more efficient by making procedures **h_reduce(L)** and **t_reduce(L)**

```
procedure reduce(L)
{
  h_flag = h_reduce(L);
  t_flag = t_reduce(L);
  while (t_flag)
   {
     h_flag = h_reduce(L);
     if (h_flag)
        t_flag = t_reduce(L);
     else
        t_flag = false;
   }
}
```

Figure 5 Algorithm for reduction

to only consider a relevant subset of the states in L. For the LTSs shown in Figure 1, the reduced LTS generated by the above procedure is given in Figure 4.

5 CONFORMANCE TESTING BASED ON REDUCED GLOBAL LTS

In this section, we describe how the reduced global LTS generated in the previous section may be used to perform conformance testing based on an adaptive guided random walk. The following issues need to be addressed: 1) generation of test cases 2) construction of the tester, and 3) issues related to testability.

Since the approach to testing being considered is an adaptive guided random walk, no test cases have to be generated *a priori*. At each stable state reached during testing, the next input to be provided to the IUT is done as described in section 3.2. The construction of testers is more involved. Note that when testing LTSs, the tester needs to synchronize with the IUT.

5.1 Tester generation

At each stable state, a tester can be constructed during adaptive testing as follows:

- Identify the stable state and select the next input to provide to the IUT as discussed in section 3.2.
- In the reduced LTS generated in the previous section, start a depth first traversal at s along the selected input transition(s).
- Construct the subgraph T corresponding to this traversal with the condition that traversal is made only along internal and output transitions.

- If T is nondeterministic, transform it into a deterministic LTS.
- Invert inputs and outputs in T.

5.2 Testability issues

In this paper, we restrict our discussion on testability to the executability of test cases. Issues which are related to coverage need further study and will not be discussed here. When performing conformance testing based on an adaptive guided random walk, the main testability criterion is stable state identification. Stable state identification may become problematic due to **divergence** (a loop consisting of only internal transitions). We assume fairness on the part of the scheduler and deem divergence as not being catastrophic as long as it can be exited. If there is a situation when the IUT stays in an internal loop once it enters it, it is considered a design error and is not relevant during conformance testing - it is assumed that such situations do not exist. Fair divergence can still pose problems in regard to testability. We identify three types of divergence: 1) **hidden divergence** which corresponds to a τ loop that can be exited through only τ transitions, 2) **transient divergence** which corresponds to a τ loop that can be exited using only τ and/or output transitions, and 3) **stable divergence** which corresponds to a τ loop that can be exited only through input transitions.

Hidden and transient divergence do not cause difficulty during state identification, however, when a stable divergence is reached, the IUT cannot proceed any further without an input. In an adaptive guided random walk, an input is provided after observing the current stable state of the IUT. With stable divergence, the IUT would be looping internally among a set of transient states and hence identifying the states of component processes is not possible. Therefore, the adaptive guided random walk fails when stable divergence exists. The necessary testability criterion, hence is not to have stable divergence. The reduced LTS generated in section 4 can be used to detect stable divergence, if exists. The procedure would be as follows:

- Step 1: Identify strongly connected components of τ transitions (i.e., a set of global states where for any pair of states belonging to this set, there is a path consisting of only τ transitions).
- Step 2: Check if each component is such that there exists at least one state that has an output transition or a τ transition leading out of the strong component.
- Step 3: If the check in step 2 fails, stable divergence has been detected.

6 RELATED WORK

Reduction of the LTS while preserving certain information has been studied in different contexts by a number of researchers. In the context of conformance testing, however, the only two approaches that we are aware of are presented in (Cavalli, Kim and Maigron, 1994) and (Arakawa and Soneoka, 1992). In (Cavalli, Kim and Maigron, 1994), specifications given in the form of sets of LTSs are considered. The reduction technique is applied to the composite LTS. This does not alleviate the state explosion problem because the composite LTS has to be constructed in the first place. Also, the reduction is with respect to trace equivalence. This reduction does not take into consideration the inherent nondeterminism in the IUT.

The approach presented in (Arakawa and Soneoka, 1992) considers specifications which are in the form of sets of communicating processes. Each process is represented by its control flow graph. Interprocess communication is assumed to be through FIFO channels. For a specification S, the global behaviour of the set of component processes may be obtained in terms of a reachability graph. A global state is defined as a collection of states of component control flow graphs and the contents of the communication channels. To alleviate state space explosion inherent in the construction of a reachability graph, (Arakawa and Soneoka, 1992) presents an on the fly reduction technique wherein a reduced reachability graph is generated. Each transition in the reduced graph may be associated with more than one event. However, transitions leaving stable states are associated with exactly one input event. This reduced graph contains all stable states that are reachable from the initial global state while adhering to the above requirements on the transitions. The reduction algorithm, however, is based on the following assumptions: 1) the control flow graph for each component process is deterministic, and 2) in the control flow graph for each process, if a node has outdegree greater than 1, then the set of all outgoing transitions from that node must only contain a mixture of either input transitions or internal receive transitions. Due to these assumptions and due to the different semantic model used for the specifications, this reduction technique is not directly applicable in the realm of LTSs. Also, the reduced graph may contain equivalent nonstable states. Hence, procedure **reduce(L)** presented in this paper is more general and is directly applicable in the realm of LTSs.

A reduction technique, referred to as the ICR method, was proposed in (Lapone et al., 1989) in the context of protocol verification. The reduction is with respect to observational equivalence. However, the notion of observational equivalence used in this approach is quite different from the one given in (Milner, 1989). For instance, it equates $a + \tau.b$ and $a + b$. Nevertheless, the reduction does preserve all stable states and hence is applicable in the context of conformance testing using an adaptive guided random walk. It has been demonstrated through a large real life protocol example that this technique produces at least an order of magnitude reduction in the number of states. The rules for reduction in this approach do not collapse equivalent nonstable states into a single state. For this reason, we feel that our approach produces a more significant reduction.

Testing of communicating finite state machines was also considered in (Petrenko et al., 1994). Each component process is represented as a nondeterministic finite state machine (NFSM). Only serial composition of component processes is considered. Test generation is based on a fault model wherein it is assumed that exactly one component process may be faulty. It is our belief that such a fault model may only be suitable when the IUT is embedded within a system under test but may not be suitable when the IUT is directly accessible. Due to these reasons, we refrain from any comparison of the work presented in this paper with that of (Petrenko et al., 1994).

7 CONCLUSIONS AND FUTURE WORK

We have presented a reduction technique which, for a set L of LTSs, generates a reduced LTS observationally equivalent to the composite LTS for L while preserving all the stable states of the composite behaviour of L. The reduction is performed such that the problem of state explosion is alleviated. For a special class of LTSs, we have shown how our reduction technique can be used effectively in conjunction with the adaptive guided random walk approach to conformance testing. This class of LTSs includes LTSs which correspond to CSP style specifications and also to the IOFSM formalism (for the case where only direct naming is considered) of (Phalippou, 1994), for which it has been proved that testing each transition of each component process is sufficient for conformance testing of a set of communicating processes. Also, we have shown how to construct testers and identified criteria for testability in the context of adaptive guided random walk.

While empirical evidence for our reduction technique is not provided in this paper, we have shown with valid reasoning that our reduction approach may perform better than other approaches that have been shown to produce significant reduction. We plan to perform empirical studies with large protocols and study the impact on fault coverage in relation to the adaptive guided random walk. Also, in light of the results presented in (Okazaki et al., 1994), we plan to investigate how our reduction technique can aid test sequence generation for interoperability testing.

Acknowledgement We would like to thank Dr. Rance Cleaveland and V. Natarajan for helpful discussion on equivalences in process algebra.

REFERENCES

Arkko, J. (1994) On the Existence and Production of State Identification Machines for Labeled Transition Systems, *Sixth Int. Conference on Formal Description Techniques - FORTE'93*, pp 351-366, North Holland.

Arakawa, N. and Soneoka, T. (1992) A Test Case Generation Method for Concurrent Programs, *Fourth Int. Workshop on Protocol Test Systems*, pp 95-106, North Holland. Also appears as A Test Case Generation Method for Black Box Testing of Concurrent Programs, *IECE Trans. Commun.*, Vol. E75-B, No. 10.

Brinksma, E., Alderden, R. and Langerak, R. (1990) A Formal Approach to Conformance Testing, *2nd Int. Workshop on Protocol Test Systems*, pp 349-363, North Holland.

Cavalli, R. R., Favreau, J-P. and Phalippou, M. (1994) Formal Methods in Conformance Testing, *Sixth Int. Workshop on Protocol Test Systems*, pp 3-17, North Holland.

Cavalli, A. and Kim, S. U. (1992) Protocol Conformance Test Generation Using a Graph Rewriting System, *Fourth Int. Workshop on Protocol Test Systems*, pp 285-288, North Holland.

Cavalli, A., Kim, S. U. and Maigron, P. (1993) Automated Protocol Conformance Test Generation Based on Formal Methods for LOTOS Specifications, *Fifth Int. Workshop on Protocol Test Systems*, pp 237-248, North Holland.

Cavalli, A., Kim, S. U. and Maigron, P. (1994) Improving Conformance Testing for LOTOS, *Sixth Int. Conference on Formal Description Techniques - FORTE'93*, pp 367-381, North Holland.

Cleaveland, R., Parrow, J. and Steffen, B. (1993) The Concurrency Workbench: A Semantics Tool for the Verification of Concurrent Systems, *ACM Tran. Programming Languages and Systems*, Vol 15, No. 1, pp 36-72.

Fujiwara, S. and Bochmann, G.v. (1992) Testing nondeterministic state machines with fault coverage, *Fourth Int. Workshop on Protocol Test Systems*, pp 267-280, North Holland.

Hoare, C.A.R. (1992) *Communicating Sequential Processes*, Prentice-Hall.

Kanellakis, P.C. and Smolka, S.A. (1983) CCS Expressions, Finite State Processes, and Three Problems of Equivalence, *Proc. of ACM Symposium on Distributed Computing*, pp 228-240.

Lapone, A. M., Sabnani, K. K. and Uyar, M. U. (1989) An Algorithmic Procedure for Checking Safety Properties of Communication protocols, *IEEE Transactions on Communications*, pp 940-948, September.

Lee, D., Sabnani, K., Kristol, D. M., Paul, S. and Uyar, M. U. (1993) Conformance Testing of Protocols Specified as Communicating FSMs, *IEEE INFOCOM'93*, pp 115-127.

Lee, D. and Yannakakis, M. (1994) Testing Finite-State Machines: State Identification and Verification, *IEEE trans. on Computers*, Vol. 43, No. 3, pp 306-320.

Milner, R. (1989) *Communication and Concurrency*, Prentice-Hall.

Okazaki, N., Park, M., Ohta, M. and Takahashi, K. (1994) A New Test Sequence Generation Method for Interoperability Testing, *Proc. 7th Intl. Workshop on Protocol Test Systems*, pp 229-245.

Paige, R. and Tarjan, R. E. (1987) Three Partition Refinement Algorithms, *SIAM Journal of Computing*, Vol. 16, No. 6, pp 973-989.

Phalippou, M. (1994) Executable testers, *Sixth Int. Workshop on Protocol Test Systems*, pp 35-50, North Holland.

Petrenko, A., Yevtushenko, N. and Dssouli, R. (1994) Testing Strategies for Communicating FSMs, *7th Int. Workshop on Protocol Test Systems*, pp 181-196.

Sidhu, D. and Leung, T. (1989) Formal Methods for Conformance Testing: A Detailed Study, *IEEE transaction on software engineering*, vol. 15, no. 4, pp 413-426.

Tai, K. C. and Koppol, P. V. (1993) Hierarchy-Based Incremental Analysis of Communication Protocols, *proc. of Int. Conf. on Network Protocols*, pp 318-325.

Tretmans, J. (1994) A Formal Approach to Conformance Testing, *Sixth Int. Workshop on Protocol Test Systems*, pp 257-276, North Holland.

Turner, K. J. (1993) *Using Formal Description Techniques: An Introduction to Estelle, LOTOS and SDL*, Wiley.

Biography

Pramod V. Koppol is a Ph.D. student in the Computer Science department at North Carolina State University. He received an M.S. degree in Computer Science from Southern Illinois University at Carbondale, IL. His research interests are in Software Engineering, Distributed Systems and Communication Protocols. He is currently working on analysis, testing and debugging of concurrent software and communication protocols.

Kuo-Chung Tai is a professor in the Computer Science department at North Carolina State University. He has published papers in the areas of software engineering, distributed systems, programming languages, and compiler construction. His current research interests include analysis, testing, and debugging of sequential and concurrent software. He received his Ph.D. degree in Computer Science from Cornell University in 1977.

10

Two approaches linking a test generation tool with verification techniques

Marylène Clatin[1], Roland Groz[1], Marc Phalippou[1], Richard Thummel[2]

FRANCE TÉLÉCOM - CNET

Abstract : *This paper presents two methods implemented in a test generation tool to compute significant feasible test paths including parameter values for input-output events. The first method is a kind of symbolic execution. The second method consists in linking the test generation tool with a tool permitting sophisticated types of reachability analysis. Preliminary results on non-trivial protocols are commented.*

Keywords : *conformance testing, protocol verification, Formal Description Techniques, test generation.*

1 Introduction

Conformance testing has long been deemed a major issue of strategic importance for the acceptance and use of OSI and other products based on standards for open systems. ISO has established a methodology in its IS-9646 standard. The central point in a conformance testing process lies in the availability of a test suite which must be closely related to the protocol (or system) specification. Designing such test suites is hard work. Fortunately, the automatic generation of test suites is now becoming a reality, although existing tools are still limited.

Automatic test suite generation for protocols is based on a formal description, in languages such as Estelle, Lotos or SDL. The first goal of Formal Description Techniques (FDT) is to provide a precise and unambiguous description of a protocol. Based on such a description, verification techniques can be applied to check the consistency of a protocol specification with the corresponding service, or simply any protocol property. As a matter of fact, much effort has been devoted to protocol verification techniques, and many tools have been developed which take as input a specification in one of the three standard FDT. Interest in test generation from FDT is more recent, and has followed a different path so far.

Although the aims are different, there is a real interest in integrating verification techniques into test generation. This paper explores two approaches for solving a key problem in test generation. The problem, detailed below, is to find a feasible test path from one protocol state to another. The techniques considered are a kind of symbolic execution, and reachability analysis. Both approaches have been implemented in our test generation tool, called TVEDA, and we have been applying them to significant protocols.

Section 2 introduces the problem and the rationale for considering verification techniques; it also discusses some issues about our test generation environment. Section 3 describes the symbolic execution method used in our tool. Section 4 describes the other method, linking our tool with a powerful tool for reachability analysis. Section 5 discusses the pros and cons of each

---

1. CNET LAA/EIA/EVP, BP-40, F-22301 LANNION Cedex, FRANCE
Tel: 33 96 05 11 11, Fax: 33 96 05 39 45, E-mail: (clatin,groz,phalippo)@lannion.cnet.fr

2. Current address: Direction de l'aviation civile Sud, Aéroport de Toulouse, BP 100, F-31703 Blagnac Cedex
R. Thummel contributed to the work reported here during his stay at Lannion for a collaboration between CNET and DGAC.

method, based on our experiments with a few protocols, and gives some hints on key issues in the use of verification techniques for test generation.

2 Basic issues

2.1 Time is ripe for linking test and verification

Research in test generation techniques focused for a long time on specific issues which did not encompass the whole problem of generating a workable test suite from a formal specification. For instance, many papers in the 80s addressed the test architecture problems, others considered generation of check sequences for FSM etc. However, the development of tools generating test suites (esp. from FDT to TTCN) has led to a broader view of the problems of test generation. We have presented in [Groz 95] a list of problems, all of which are now addressed one way or another in the literature. There is therefore a better understanding of what is needed for generating tests. And some problems can be identified as sub-problems already addressed by verification.

At the same time, protocol verification tools are coming of age. Verification techniques are now implemented in commercial tools, and research prototypes are often strong enough to tackle real-size applications with sophisticated techniques.

Some work has already been done to connect test generation to existing verification tools. For instance, [Chun 90] uses a constraint solver to compute feasible paths traversing an EFSM (Extended Finite State Machine) specification; [Cavalli 92] uses a tool computing the reachability graph for a Lotos specification, in order to apply a modified version of the UIO method [Sabnani 88] on this graph. Both approaches suffer from complexity problems which limit their current applicability to small size specifications. Approaches proposed in commercial tools such as Topic [Montiel 94] or STED [Ek 93] offer also some connection between test generation (from already specified test purposes in this case) with simulation tools (not really verification in this case, but this is a step in the right direction). Although this paper takes a different approach, based on our experience with large scale applications, there is room for many fruitful interactions between test generation and verification tools.

2.2 A key building block for test generation

A test generation tool starting from an FDT-based specification and producing an abstract test suite (ATS) in, say, TTCN, must perform several functions which correspond logically to components of the tool. For instance, we presented in [Groz 95] the following global break up: getting an abstract test-oriented view from the specification, selecting test purposes, computing test case kernels, producing and formatting a complete test suite.

Test selection is specific to test generation, and pertains to research in testing. We do not expect verification techniques to be immediately applicable in that area. The same applies to the production of the final test suite (which takes into account such issues as test suite structure, TTCN format etc.).

Arguably, support could be found in verification techniques for the first step, that is to get an abstraction from the specification. In particular, it is necessary to abstract from events which cannot be observed or controlled from the tester, because of limits in the tester and in the test architecture. We did not (yet) consider verification techniques for this step, because it is not a time-intensive one in the case of our method, which abstracts a specification into an EFSM.

Using efficient techniques might be more crucial in the case of methods translating specifications into less powerful abstract models, such as FSM or LTS [Cavalli 92][Chin 95], because in this case, the resulting abstraction is much larger, and efficiency is required to deal with it.

We have considered verification techniques for that step where efficiency in dealing with the semantics of the application is important, that is to say: computing test case kernels. Given some sort of test purpose, we would like to find which sequences of events are permitted by the specification and fulfil the requirements expressed by that test purpose. This is very similar to a verification problem because we are trying to verify the existence of some property in a specification.

As can be seen, this problem can be well defined (we will describe it in more detail below in section 2.5), and once defined, it is insulated from the rest of test generation procedures. The solver for this problem is a key building block for test generation. Since its interface with the rest of the test generation process can be well defined, solving it can be subcontracted to a verification tool.

Above all, it is important to note that this building block is generic enough to be used in several contexts. It can be used by different test selection strategies; for instance, we use it for computing transition subtours, or test preambles, or test postambles. It could be put to other uses.

2.3 Combinatorial blow-up

The main argument for resorting to verification tools is that the problem considered (identifying sequences of events satisfying a given property) raises combinatorial search with potentially an exponential blow-up. Research in verification techniques has investigated methods to limit that blow-up, and perform efficient searches.

In fact, an EFSM is a heavily factorized view of a protocol. The tests themselves must take into account the precise values of the variables which make up the Extended (E) part of the E-FSM. This means that the execution paths that form the kernel of tests are to be found in the FSM which would result from the full expansion of the EFSM. This FSM is in fact the reachability graph of the EFSM. The problem is that even for very simple protocols, this graph will be gigantic as soon as there are, for instance, several integer variables (e.g. frame numbering at link layer, even with frame numbers limited to a 256 modulo).

Our experience (and common experience) with various protocols and telecommunication services testifies that this problem will appear in most real applications, AND that it has a direct impact on various aspects of test generation. In some cases, finding which test paths are actually feasible can eliminate most proposed paths [Rouger 89]. Section 5 provides data on the complexity of the different protocols which we considered.

2.4 Tool background

Our first tool, called TVEDA-V2 (a previous version 1 - V1, was functionally similar, only the programming environment changed drastically), was based on a single strategy, which we called "test skeleton strategy" (based on the single step method of IS9646). This strategy produced TTCN tables corresponding to test cases, but did not produce the tables for preambles or postambles; also, some constraints (in the sense of TTCN) were not produced adequately in many cases because we considered each transition in isolation from its semantic context, i.e. disregarding the transitions that might have preceded it in a feasible execution path; as a consequence, we could not take into account the values of variables that were dependent on the path leading to this transition. Although this approach may seem limited, it can still generate the bulk of a test suite, as was presented in [Phalippou 90].

Our new tool, called TVEDA-V3 was partially presented in [Phalippou 94]. It incorporates several new features, apart from the fact that the design of the software is completely new from the previous version.

1. A modular architecture, that makes it possible to choose between: specification language (Estelle or SDL), test description language (Menuet [Langlois 89] or TTCN), test selection strategy (single transition, extended transition tour...)
2. A semantic module, which can be called from the strategy modules to compute feasible paths.
3. Sundry functional extensions, such as hypertext links between tests and specification (currently only for Estelle), test coverage analysis...

This paper deals with point 2 above. The idea is that this key point in computing test kernels can be implemented in different ways. We have investigated two approaches to implement this module: symbolic execution, and reachability analysis. Both approaches are actually implemented in TVEDA-V3, and the user of the tool can choose whichever of the two methods seems more appropriate for the protocol considered.

The modular approach to test generation makes it possible to plug into our tool this "semantic module", because the problem it addresses is orthogonal to the tasks addressed by other modules (such as output format, test strategy etc.).

2.5 Precise definition of the problem addressed

Before presenting the two approaches, let us state precisely the problem which they are both aimed at solving. In fact, the two solutions are very different in nature, applicability and efficiency, but they both compute the same result.

Inputs:

1. an EFSM, consisting of a base FSM with extended transitions; those extended transitions identify major from-state, to-state, input-event, provided condition on input parameters + internal variables, output-events, and assignments of new values to internal variables
2. a source domain, consisting of a major (FSM) state, and a condition (boolean expression) on the values of variables
3. a target domain, consisting of a major state, possibly an input event, and a condition on the values of variables (possibly linking them with input parameters)

Goal:

1. Compute at least one path from the source domain to the target domain, chaining transitions in accordance with the conditions specified by the EFSM.
2. Compute the non-determinism inherent in the specification, or due to limited controllability of the IUT (e.g. when the same input can trigger one of several transitions based on the availability of internal resources) or due to limited observability from the tester (when events on non observable channels can interfere). We called these last two types of non-determinism *context non-determinism* and *event non-determinism* respectively, in a previous paper [Phalippou 90]. This is very important to derive correctly the branches leading to *inconclusive verdicts*. In fact, in our modular tool, context and event non-determinism are already included in the EFSM as inherent non-determinism by the module which translates the specification (+ test architecture) into an EFSM.

Outputs:

One solution tree (one = first-found, and shortest-path in fact for both solutions) consisting of

1. a sequence of chainable transitions of the EFSM; let us call this sequence a "path"
2. for each transition containing an input-event, a value for each parameter (such that it is consistent with the chaining)
3. for each transition, observable output events and their associated parameter values
4. for each transition, the list of interfering transitions (those that could fire in the same context as provided by the source domain of the transition AND the parameter values provided in 2 above)

Note: the problem stated above can be defined as the "single target problem". In fact, this basic problem is handled by our tool, but we also address the "multiple target problem": this is the case when the inputs include several targets, and we are trying to compute simultaneously one solution tree for each target. For instance, this happens for one of the most typical occurrence of our problem: compute a preamble for each transition of a specification. In that case, the source domain is the initial state, and we have as many targets as transitions. By computing simultaneously one path for each target, we can factorize the computations done on common prefixes.

3 Symbolic computation technique

In this section we present a first technique used for computing execution paths on extended automata: symbolic computation. We illustrate our explanations with an example based on an Estelle description of INRES protocol [Hogrefe 92]. However, the technique uses the «extended automaton» aspects of the language, and is also valid for SDL: in our tool TVEDA V3, it is implemented by using a preliminary translation of Estelle and SDL subsets into a common model based on EFSM. Figure 1 below describes the part of INRES Estelle specification which will be used (syntax is not strictly respected).

```
state disconnected, wait, connected, sending
var olddata : ...
    counter : 0..4
    number : 0..1

initialize to disconnected begin end;

from disconnected to wait
❶ when user.iconreq
    begin counter := 0 output pdu_access.cr end;

from wait to connected
  when pdu_access.cc
❷   begin number := 1; counter := 0;
    output user.iconconf end;

from wait to same delay(5)
  provided counter<4
❸   begin output pdu_access.cr;
    counter := counter+1 end;

❹ provided otherwise to disconnected
    begin output user.idisind end;

from connected to sending
❺ when user.idatreq(isdu)
    begin output pdu_access.dt(number,isdu)
    olddata := isdu end;

from sending to connected
❻ when pdu_access.ak(num)
    provided num=number
      begin number := succ(number) end;

from sending to same delay(5)
❼ provided counter<4
    begin output pdu_access.dt(number,olddata)
    counter := counter+1 end;

from not_ignore_dr to disconnected
❽ when pdu_access.dr
    begin output user.idisind end;
```

Figure 1 : INRES specification (partial)

An Estelle or SDL specification is a compact description of a big automaton, called the *reachability graph* of the FDT specification. We have explained in section 2.3 why we need to compute execution paths on this big automaton. Symbolic computation is a technique which allows to compute such execution paths without computing the big automaton. This allows to avoid the key problem of managing the size of such an automaton.

3.1 Domain calculus on EFSM transitions

The basic idea of symbolic computation is to avoid the enumeration of data values (as is done in reachability analysis). On the contrary, computation is performed directly on *value domains*: a domain is a compact representation of a set of data values (potentially big).

Estelle data structures (variables and message parameters) are transformed into *elementary data vectors* which can be of two types: integer interval (including the integer type) or enumerated type (including booleans). For doing this, composed types, such as arrays or records, are first expanded into elementary fields. Each elementary data vector has a *name* (built from the name of the corresponding Estelle data structure field) and can store an *elementary value domain*. An elementary value domain models a set of possible values for the elementary vector. The value domains are represented either by the constant value *any* (modelling all possible values of the given type), either by *empty* (no value) either by a list (of values for enumerated types, of intervals for integer interval types). The product of elementary vectors is called a *vector*, and we call *value domain* the product of elementary value domains. We use a vector to represent a set of states of the reachability graph, or a set of events (inputs or outputs). We show on figure 2 the vectors which represent respectively the set of states of the reachability graph of INRES after the *initialize* transition, and the set of *when* inputs of transition ❺ of figure 1. Note that the more *any* values appear in value domains, the more compact the representation is.

states after initialize		*el_vector1*	*el_vector2*	*el_vector3*	*el_vector4*
	name	*state*	*olddata*	*counter*	*number*
	type	*enum.*	*enum.*	*interv.*	*interv.*
	value	*(disconnected)*	*any*	*any*	*any*

when inputs of transition ❺		*el_vector1*	*el_vector2*	*el_vector3*
	name	*pco*	*pdu*	*isdu*
	type	*enum.*	*enum.*	*enum.*
	value	*(user)*	*(idatreq)*	*any*

Figure 2 : Vectors and values

As mentioned above, symbolic computation means that we compute directly on domains. In order to compute transition paths on the reachability graph of the EFSM specification, we must be able to compute the effect of Estelle or SDL transitions on domains. This means:

1. compute the image domain which is reached after executing a transition if started from a given initial state domain and input domain.
2. compute the reverse image of a given final state domain (i.e. which initial state domain and input domain the transition must start from in order to ensure that the final state domain is reached).

In order to do this, for each extended transition of the EFSM structure, we compute the following elements:

1. the guard of the EFSM transition (from state, when event and provided clause) is transformed into a couple of vectors (*state vector* and *when vector*) which represent the complete enabling domains of this extended transition.
2. the body of the EFSM transition (to state, assignment statements and output events) is transformed into a couple of procedures: the first one computes the image domain (final state domain, output domain) of the intersection of any initial domain (start state domain, input domain) with the (*state vector* domain, *when vector* domain) of the Estelle transition. The second one computes the reverse image domain (i.e. a start state domain and an input domain, possibly empty) of any final state domain.

These concepts are illustrated on figure 3. We give two examples of reverse image domains. The first one is a normal case. The second one illustrates the case in which the final domain cannot be reached by the transition: in this case, the initial state and event domains are empty. Of course, this transformation can be done only if the EFSM (Estelle or SDL) satisfies some strong *restrictions* on the constructs which appear:

1. in the provided clauses no coupled constraints on several variables or parameters should appear: a vector domain must be the cartesian product of the domains of its elementary vectors. For instance *provided x<y* where *x* and *y* are variables, is forbidden.
2. in the transition body only constructs for which the image and the reverse image of domains can be computed are allowed. This includes a very limited number of expressions in assignment statements. For instance *x:=y* is allowed (direct image: the domain value of *x* after the assignment is the domain value of *y* before. Reverse image: the domain value of *y* before is the domain value of *x* after, and the domain value of *x* before is *any*). On the contrary, *x:=y+z* is not allowed (impossible to compute the reverse domain as a product of elementary domains for *y* and *z*).

In practice, many real protocol specifications satisfy such constraints, as explained in section 5.

3.2 Path computation

We explain in this section how to use the domain calculus in order to compute the transition paths. As required in section 2.5, our goal is to find an executable transition path from a given start domain D_S to a final domain D_F. We must add the requirement that they are domains in the sense of the previous section, i.e. products of elementary domains. The path computation algorithm has three steps:

1. computation of the shortest sequence of chainable extended transitions such that D_S is included in the start domain of the first extended transition and D_F intersects the final domain of the last extended transition. For this, we consider successively extended transition sequences of growing length, computing a direct propagation of value domains, using the image function mentioned in the previous section, starting from D_S. If we reach a domain which intersects D_F, we compute the intersection D'_F.
2. reverse propagation of D'_F using the reverse image function mentioned in the previous section (this indicates from which subset D'_S of D_S we must start and which subsets of the successive when domains of the extended transitions sequence we must use in order to be sure to reach the final domain D'_F).

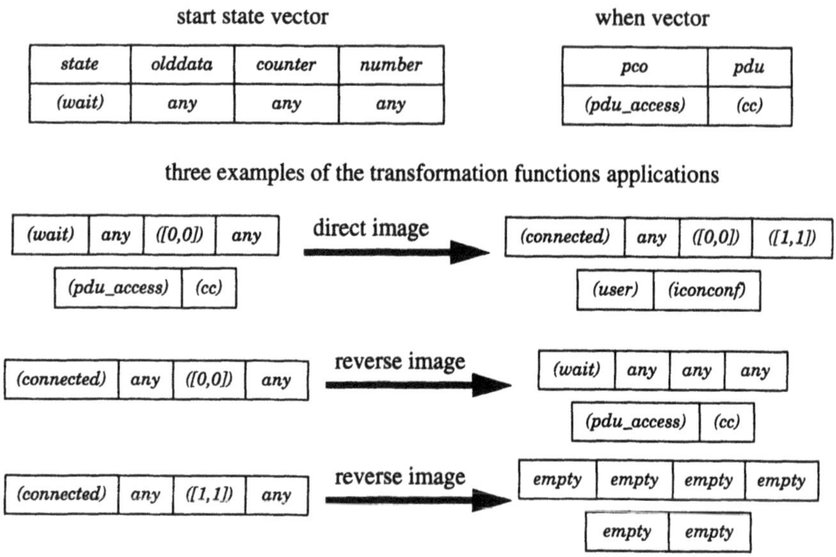

start state vector

state	*olddata*	*counter*	*number*
(wait)	*any*	*any*	*any*

when vector

pco	*pdu*
(pdu_access)	*(cc)*

three examples of the transformation functions applications

(wait)	*any*	*([0,0])*	*any*

(pdu_access)	*(cc)*

direct image →

(connected)	*any*	*([0,0])*	*([1,1])*

(user)	*(iconconf)*

(connected)	*any*	*([0,0])*	*any*

reverse image →

(wait)	*any*	*any*	*any*

(pdu_access)	*(cc)*

(connected)	*any*	*([1,1])*	*any*

reverse image →

empty	*empty*	*empty*	*empty*

empty	*empty*

Figure 3 : Symbolic computation for INRES transition ❷

3. arbitrary choice of values in the domains which result from step 2, in order to obtain an instantiated transition path. Note that such an instantiated path is indeed a path on the accessibility graph which solves the first goal of the problem formulated in section 2.5.

Examples of path which are needed for test generation are preambles. A *preamble* for a given transition is a path which starts from the initial domain of the EFSM (i.e. the domain reached after the *initialize* transition, see figure 2) and reaches the *when* domain of the transition. Some preambles for INRES protocol, obtained by symbolic computation, are shown on figure 4.

Path computation brings no additional restriction (other than the ones which are mentioned in the previous section) on the FDT constructs which are allowed. At this stage, the key point is performance, in order to deal with real protocols. Note that the vectors and the direct and reverse functions are computed from the compact EFSM model of the protocol: the size of such objects corresponds to the size of FDT descriptions, and avoids the size explosion which can be encountered when dealing with accessibility graphs.

The main performance problem comes from the fact that the number of chainable transition sequences grows exponentially with the length of the sequence, and the exponential factor depends on the number of transitions of the EFSM: the computation of long paths for big EFSM is a real challenge. Some optimization techniques have been implemented to improve efficiency: exact computations (e.g. detecting when a domain reached is included in an already reached domain), or heuristics (e.g. first reach a domain with the same major state as the target domain, then reach the exact target domain).

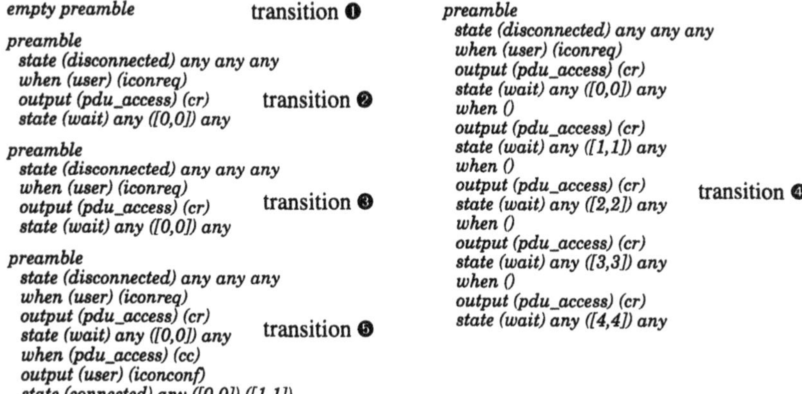

Figure 4 : Some preambles of INRES transitions (partial)

3.3 Non-determinism computation

Finally, the last step is the computation of the effect of non-determinism. According to the requirements of section 2.5, we must compute the effect (output behaviour and state reached) of every transition which can be non-deterministically fired with any transition of the path we have computed. For doing this, we compute the successive domains which are reached starting from the initial domain of the EFSM and using the input values which have been chosen during step 3 of the path computation procedure. For each of these successive domains, we compute the firable transitions (either spontaneous, or with the same input as in the path), and their output and final state domains. Adding non-determinism, the preamble sequences of figure 4 are transformed into the test trees of figure 5. Note that only the preamble of transition ❺ contains a non-deterministic event, leading to an inconclusive verdict.

```
! (user) (iconreq)
  ? (pdu_access) (cr)                    transition ❷

! (user) (iconreq)
  ? (pdu_access) (cr)                    transition ❸

! (user) (iconreq)
  ? (pdu_access) (cr)
    ! (pdu_access) (cc)                  transition ❺
      ? (user) (iconconf)
      ? (pdu_access) (cr) <inconc>

! (user) (iconreq)
  ? (pdu_access) (cr)
    ! ()
      ? (pdu_access) (cr)                transition ❹
        ! ()
          ? (pdu_access) (cr)
            ! ()
              ? (pdu_access) (cr)
                ! ()
                  ? (pdu_access) (cr)
```

Figure 5 : Test trees obtained from the preambles

4 Reachability technique

Another way of computing transition chains consists in using a simulator. In our case, we have coupled TVEDA with Véda. Véda is a tool which has been designed for the specification and verification of systems described with a subset of the ISO Estelle language. Its strength relies mainly on its powerful simulation and verification abilities. Indeed it supports interactive and random simulation techniques as well as formal verification based on reachability state graph analysis. Véda was developed at CNET and industrialized by Vérilog [Algayres 93].

TVEDA is coupled with Véda in the following way. First, TVEDA builds an extended finite state machine (EFSM) from the Estelle or SDL specification given in input. Then Véda is used in three steps:

1. First, this EFSM is transformed so as to reduce its size, while preserving the existence of paths. The transformed EFSM is decompiled into Estelle (the input language of Véda).
2. Then we use three heuristics to address the multiple target problem as stated in section 2.5. Each heuristic allows to compute some of the paths:
   a. an exhaustive simulation is made. Duration of the exhaustive simulation is limited, so as not to spend too much time waiting for the results. During this first step, a major part of the paths is found.
   b. a second step consists, for each transition not reached during the first step, in making an exhaustive simulation using a distance computation between states. When the distance is too large, the simulation is given up. Some other paths may be found during this second step, but possibly not all.
   c. Finally the last step consists in loading a path which brings the specification in a state close to the initial state of the transition, and then in making an exhaustive simulation. The simulation runs until the transition is reached.
3. Finally, these paths have to be completed to integrate non-determinism and constraint values.

We detail now these three steps.

4.1 EFSM transformations

There are two reasons which motivate the EFSM transformations. First, we want to have a specification which can be simulated using Véda, without adding any environment to generate input events. Secondly, we want to eliminate all things which are not really needed for path computation in order to improve the efficiency of the reachability analysis.

The EFSM transformations are the following ones:

1. hiding of the environment (all transitions become spontaneous, output events are suppressed),
2. computation and treatment of the specification «useful part» and suppression of the useless part. The useful part is defined as the specification elements which influence transition firing (variables and parameters which appear in *provided* clauses, and therefore have an influence on transition firing, and transitive closure on parameters and variables for assignments),
3. grouping transitions into equivalence classes, after reductions 1 and 2. A class contains all transitions which become equal when the useless part is suppressed.

These transformations are illustrated on figure 6. From the specification described on figure 1, we obtain this latter after transformations.

The EFSM obtained after transformation is different from the EFSM we had initially, but they both have the same behaviour in terms of possible state sequences (a state is defined here by the global state of the specification and the value of variables computed as «useful»). Input

```
state disconnected, wait, connected, sending
var counter : 0..4
    number : 0..1
    ak_num: 0..1

initialize to disconnected
  begin ak_num := 0 end;

① from anystate to same
  begin ak_num := (ak_num + 1) mod 2 end;

❶ from disconnected to wait
  begin counter := 0 end;

❷ from wait to connected
  begin number := 1; counter := 0 end;

❸ from wait to same
  provided counter<4
  begin counter := counter+1 end;

❹ provided otherwise to disconnected
  begin end;

❺ from connected to sending
  begin end;

❻ from sending to connected
  provided ak_num = number
  begin number := succ(number) end

❼ from sending to same
  provided counter<4
  begin counter := counter+1 end;

❽ from not_ignore_dr to disconnected
  begin end;
```

Figure 6 : INRES specification (partial) after transformations

events have been suppressed, but their useful parameters are kept and modelled as variables which evolves thanks to special transitions added to the EFSM (on the above example, *num* parameter of *ak* interaction is a useful parameter modelled as a *ak_num* variable. The added transition is ①). Therefore, there is no longer interaction with the environment, since the new specification is completely autonomous.

Then, in order to compute paths, we simulate the specification obtained by decompiling the transformed EFSM into Estelle, and the computed paths on this specification are still valid paths for the initial one.

4.2 Path computation heuristics

As indicated in the introduction to section 4, we use three consecutive heuristics to compute a path for each target. This is required because none of the three heuristics is powerful enough to find all paths. We try, successively:

1. *Bounded searching*: this first step consists in making an exhaustive simulation of the specification, restricted by the number of states of the reachability graph. Since the exhaustive simulation is based on building a reachability state graph, when the state number is reached the simulation ends. For simple specifications (such as Inres protocol) all paths are found during this phase, and for more complex ones (such as LAPD protocol) only 80% are found, because the reachability graph is too big to be entirely computed. Due to breadth exploration of the reachability graph, the path found for each transition constitutes the shortest path from the starting state to the target.
2. *Searching based on a distance computation*: for each transition not reached during the first step, a second heuristic is used which consists in performing another exhaustive simulation taking into account a distance between states. Such empiric distance models the «difference of values» between two states. The distance must decrease during the path exploration (more or less strictly). If the distance increases, the current path exploration is given up, and searching goes on another branch of the graph. At the end of this second step, some other paths may have been computed in addition to those found during the first step.
3. *Searching using a path prefix*: finally, for each remaining transition not reached during the last step, a third heuristic consists in loading a prefix which brings the specification in a state close to the transition initial state (with good variable values satisfying the transition guard

predicates). After this prefix loading, an exhaustive simulation happens with no restriction on its duration.

At the end of this phase, we have computed paths in the shape of transition names lists. These paths correspond to transitions of the transformed EFSM. Since the transformed EFSM and the initial one have the same behaviour in terms of possible states sequences, a simple translation on the transition names lists is performed to get paths for the initial specification. Basically, it consists in choosing one original transition for each equivalence class. So far, we have reached the first goal («output» 1 in section 2.5). We must now compute the other outputs (2, 3 and 4).

4.3 Integrating non-determinism and constraint values

Here again, Véda is helpful. Indeed, each path obtained during the last phase is played on the initial specification which is slightly transformed so as to give us the lacking informations. Some of these informations are directly given by Véda, whilst some others have to be computed. This is the reason why we add some extensions to the initial specification.

When playing a path (called scenario in Véda's terminology) on a specification, Véda lists at each scenario step the set of firable transitions. Non-determinism is then obtained at each step from the list of firable transitions and the fired transition name. When playing a scenario, Véda also lists message emissions and the value of each message parameter. Then, informations for constraint values concerning message emissions are directly given par Véda. Concerning message receptions we cannot get directly the informations, because environment has not been modelled. An extension to the specification allows to solve this problem, as described hereafter.

The specification used in this part is obtained from the initial specification, modelled as an EFSM, which is then copied before being transformed. Some reductions are performed so as to simulate the environment; the main reduction makes all transitions spontaneous. Message receptions are found by searching in the initial EFSM the transition thanks to its name (transition names are kept during EFSM copy). So, the last problem concerns reception message parameters. This is solved in the following way. First each parameter is declared as a variable. Then for each parameter belonging to the useful part the same treatment as performed in section 4.1 is done. And for each parameter belonging to the useless part, we give it a value which satisfies the corresponding transition guard by means of another transition (fired just before the one we talk about). All these transformations are illustrated on figure 7.

The reduced and extended EFSM is decompiled in Estelle and simulated with Véda by playing each scenario found during the last phase. A trace file is then produced by Véda, containing all informations relevant to test generation including non-determinism and constraint values.

An example of test generation using this technique is described on figure 8. This figure represents the preamble produced for transition ❻ of the initial specification (cf. figure 1) in a TTCN-like tree notation.

5 Assessment and perspectives

After the presentation of the techniques we used, we conclude now this paper with an assessment of the results that have been reached so far, and we conclude with some words about the future development perspectives.

As we mentioned before, both techniques (symbolic computation and reachability analysis) have been implemented in our tool TVEDA V3, as a means for computing extended transition

```
state disconnected, wait, connected, sending
var olddata : ...
    counter : 0..4
    number : 0..1
    ak_num: 0..1
    idatreq_isdu : 0..1000

initialize to disconnected
   begin ak_num := 0 end;

from anystate to same
   begin ak_num := (ak_num + 1) mod 2 end;

❶ from disconnected to wait
   begin counter := 0; output pdu_access.cr end;

❷ from wait to connected
   begin number := 1; counter := 0;
   output user.iconconf end;

❸ from wait to same
   provided counter<4
   begin output pdu_access.cr;
   counter := counter+1 end;

❹ provided otherwise to disconnected
   begin output user.idisind end;

from connected to connected
   begin forone idatreq_isdu_var:0..1000
      suchthat true do begin
      idatreq_isdu := idatreq_isdu_var end
   end

❺ from connected to sending
   begin
   output pdu_access.dt (number,idatreq_isdu)
   olddata := idatreq_isdu end;

❻ from sending to connected
   provided ak_num = number
   begin number := succ(number) end

❼ from sending to same
   provided counter<4
   begin output pdu_access.dt (number,olddata)
   counter := counter+1 end;

❽ from not_ignore_dr to disconnected
   begin output user.idisind end;
```

Figure 7 : INRES specification (partial) reduced and extended

```
! (user) (iconreq)
  ? (pdu_access) (cr)
    ! (pdu_access) (cc)
      ? (user) (iconconf)
        ! (user) (idatreq) (0)
          ? (pdu_access) (dt) (1,0)
      ? (pdu_access) (cr) <inconc>
```

Figure 8 : Test case obtained for transition ❻

tours [Phalippou 94] on Estelle or SDL specifications. However, the optimization techniques which are listed previously are not completely operational: for this reason the figures we give here will probably be enhanced later on.

5.1 Symbolic computation

We mentioned in section 3 that this technique has two kinds of limitations: it puts some strong restrictions on which constructs are accepted (see section 3.1) and the path computation requires an exponential computation with respect to the length of the path to be computed.

Surprisingly (when we think how strong are the limits on constructs) the first point is not really a problem for many protocols: although many SDL or Estelle specifications considered had some forbidden constructs, in most cases, it has been possible to rewrite the specification to avoid them. This means that the use of such constructs is not linked to the mechanisms of the protocol themselves, but rather to specification styles of the FDT programmer. Among successfully tried protocols, let us mention INRES (Estelle and SDL versions), ISDN D signalling protocol, P1 (a MHS X400 series protocol), FRP and SSCOP (ATM protocols), some supplementary serviced provided on ISDN networks, service descriptions used in Intelligent Network.

As a counter-example, ISDN D link-layer protocol (LAPD) cannot be completely processed, since it incorporates a sliding window mechanism, which cannot be modelled without some provided clauses incorporating coupled variable values: in that case the state vector cannot be represented as a cartesian product of elementary vectors.

About performance during path computation, we obtained the following results:

1. on simple protocols, such as INRES, there is no problem: all paths are computed within some minutes (on a SPARC LX workstation). This result is valid for all small examples we tried, e.g. EFSM up to 50-60 transitions. Let us note however that the exponential time computation is verified: in a very particular case (an EFSM with only 31 transitions, but paths of length up to 15, which is quite big) it took 25 hours on the same machine (in the case of INRES, the maximum length of the paths was 7).
2. on big protocols with (relatively) simple data structure, such as ISDN D protocol, it is possible to compute only short paths (depth 3 in the case of D signalling protocol, which has about 2000 transitions). Of course this is only a very limited subset of the paths to be computed. However, after a (human) analysis of the protocol structure, it is sometimes possible to split the specification into several parts which are independent as far as path calculus is concerned: with such a technique, we succeeded in the computation of 2/3 of the 765 test cases of the LAPD protocol.
3. on big protocols with complex data structures, such as P1, we could not compute the internal representation of vectors, due to memory limitations. As an indication, the state vector has more than 1500 elementary vectors, the when vectors are up to 250 elementary vectors, and there are 115 transitions to be modelled!

In conclusion, the symbolic computation technique, as it is now in TVEDA, is adapted for the treatment of small specifications, but fails when the EFSM become large. The technique by itself seemed promising, since it can be potentially applied to EFSM with big (even infinite) data domains. However, on the studied examples, this did not appear to be a key point (no real protocol has infinite domains!). On the contrary, the weakness of our approach may come from the fact that we developed ourselves an ad-hoc implementation of the method (not at all optimized from a performance point of view), mainly because we could not find such an industrial tool available.

5.2 Reachability analysis

On the contrary, reachability analysis presents two advantages:

1. the technique puts almost no restriction on the Estelle or SDL constructs which are accepted (which means that all our example specifications can be potentially handled by this technique),
2. it is implemented by interfacing our tool TVEDA with a powerful commercial tool for reachability analysis, Véda.

We have not yet finished to implement all optimization features which have been described in section 4.2. However, the results we have so far are promising:

1. on simple protocols such as INRES, all paths are computed within some minutes (in such simple cases, most of the computation time is spent in the compilation step with Véda. Path calculus through Véda execution is almost instantaneous).
2. on LAPD case study, about 80% of the paths needed for the test cases are computed, using only the bounded searching heuristic for path computation (the other heuristics have not been implemented so far). We have good hopes to be able to compute almost all needed

paths once the other heuristics are implemented, as was proved on LAPD in a semi-automatic implementation of heuristics 2 and 3.

From these results we conclude that reachability analysis will probably succeed in solving our path computation problems, at least for some kinds of protocols. However, we must still investigate with different specifications (e.g. P1, which incorporates very big data structures). Moreover, in the case of LAPD, and in spite of all the optimization features that have been incorporated in our technique, we are very close to the limitations of Véda tool (size of the Estelle source that can be accepted at compilation step, for instance). Note that the computation of the preambles took more than 8 hours on our workstation (mainly spent during the compilation step of the simulator).

5.3 About the use of verification techniques

Verification techniques seem to be promising candidates for solving key points in the test generation process. However, their successful use for this purpose will depend on some crucial aspects that have been highlighted by our experiments:

1. the verification technique must have reached a good level of maturity, and should be implemented in a tool of industrial-level efficiency. This was the case for our experiment with reachability analysis, but not with symbolic computation, and this may explain the (relative) failure of the latter.
2. industrial tools for verification have been developed for verifying protocols, not for providing help to test generation tools. Therefore their interfaces must be adapted. For the moment, this leads to tricky interface programming (see the link between TVEDA and Véda described in this paper). In the future, we can imagine that verification tools will evolve towards more open and generic interfaces.
3. even if condition 1 is fulfilled, careful (and sophisticated) optimization must be done in order to be sure that the computation problem submitted to the verification tool is simplified as much as possible (e.g. the computation of the «useful part» of the specification, or the modelling of environment modules in our implementation). Without such optimizations, handling big specification may still be beyond the strength of the tools.
4. a combined use of different techniques (each one with its strengths and weaknesses) may be useful to cope with very different specification styles, and application domains. Several techniques could be potential candidates for further studies in this direction. Let us mention for instance constraint solving techniques, abstract interpretation [Cousot 77], or a mixture of abstract interpretation with reachability analysis.

References

[Algayres 93]
B. Algayres, L. Doldi, H. Garavel, Y. Lejeune, C. Rodriguez, *VÉSAR: a pragmatic approach to formal specification and verification,* Computer Networks and ISDN Systems, special issue on tools for FDTs, vol. 25, n. 7, February 1993.

[Cavalli 92]
A. Cavalli, Sung Un Kim, *Automated protocol conformance test generation based on formal methods for LOTOS specifications,* proceedings of the 5th International Workshop on Protocol Test Systems, Montréal, September 1992.

[Chin 95]
B-M. Chin, A. Cavalli, *Test generation methods for SDL,* tutorial proceedings of SDL forum 95, Oslo, September 1995.

[Chun 90]
W. Chun, P. Amer, *Test Case Generation for Protocols Specified in Estelle*, actes de FORTE'90, Madrid, novembre 1990, Elsevier, pp 191-206.

[Cousot 77]
P. Cousot, R. Cousot, *Abstract interpretation: a unified lattice model for static analysis of programs by construction or approximation of fixpoints*, 4th POPL, January 1977.

[Ek 93]
A. Ek, J. Ellsberger, A. Wiles, *Experiences with computer aided test suite generation*, proceedings of IWPTS 93, Pau, September 1993.

[Groz 95]
R. Groz, M. Phalippou, *La génération automatique de tests est-elle possible ?*, proceedings of CFIP'95, Rennes, May 1995.

[Langlois 89]
C. Langlois, E. Paul, *De la spécification formelle à la validation de systèmes*, actes du deuxième colloque annuel du club FIABEX, INT-Évry, 22-23 novembre 1989.

[Montiel 94]
J. Montiel, R. Roth, A.J.M. Donaldson (Eds), *Methods for QoS Verification and Protocol Conformance Testing in IBC - Results and Further Recommendations*, Race Projet R2088 TOPIC, Deliverable 15, Doc R2088/DAT/TMS/DS/P/015/b1, November 1994.

[Phalippou 90]
M. Phalippou, R. Groz, *Evaluation of an empirical approach for computer-aided test cases generation*, proceedings of the 3rd IWPTS, Washington, October 1990.

[Phalippou 94]
M. Phalippou, *Test sequence generation using Estelle or SDL structure information*, proceedings of FORTE'94, Bern, October 1994.

[Rouger 89]
A. Rouger, P. Combes, *Exhaustive validation and test generation in ELVIS*, proceedings of the 4th SDL forum, North Holland, 1989.

[Sabnani 88]
K. Sabnani, A. Dahbura, *A Protocol Testing Procedure*, Computer Networks and ISDN Systems, vol. 15, n. 4, 1988.

11

PROSPECT - A Proposal for a New Test Specification Language and Its Implementation

Thomas Walter and Bernhard Plattner
Computer Engineering and Networks Laboratory
Swiss Federal Institute of Technology Zürich
ETH Zürich - ETZ Building, 8092 Zürich, Switzerland
e-mail: [walter | plattner]@tik.ee.ethz.ch

Abstract

Although internationally standardized (or are in the process of becoming international standard), the definition of TTCN and concurrent TTCN can still be improved. To solve some of the known problems we propose a test specification language, called *PROSPECT - PROtocol test SPEcification language for Conformance Testing.* The proposed language has a formally defined syntax and has a formally defined semantics. We have introduced *test modules* in our language so that the distribution of test components over real systems can be specified. As we assume that in every test module a main test component exists responsible for creating and terminating other test components, we have been in a position to give our language a precise semantics. In order to prove PROSPECT's suitability we have implemented a compiler and a runtime library. The implementation has been done in a UNIX™ environment. The details of our approach are described in this paper.

Keywords

Conformance testing, test specification, syntax, semantics, compilation

1 Introduction and Motivation

Research and development in test notations have been focussed on *TTCN (Tree and Tabular Combined Notation)* [11] and *concurrent TTCN* [12]. As recommended in [10], TTCN and concurrent TTCN are to be used for the specification of test suites. Whereas in TTCN only one *lower tester (LT)* and optionally one *upper tester (UT)* are active, concurrent TTCN supports the definition of complex test configurations, called *multi party testing context* (Figure 1), which, in general, consists of several concurrently running *test components (TC).* In every multi party testing context, the *lower tester control function (LTCF)* takes responsibility for setting up the test configuration, i.e. creation of lower and upper testers, and the calculation of a final verdict. TCs are connected by channels, called *coordination points (CP)*, through which TCs exchange so-called *coordination messages (CM)*. The rules that govern the exchange of CMs and the coordination of pairs of LT and UT are defined in a *test coordination procedure (TCP)*.

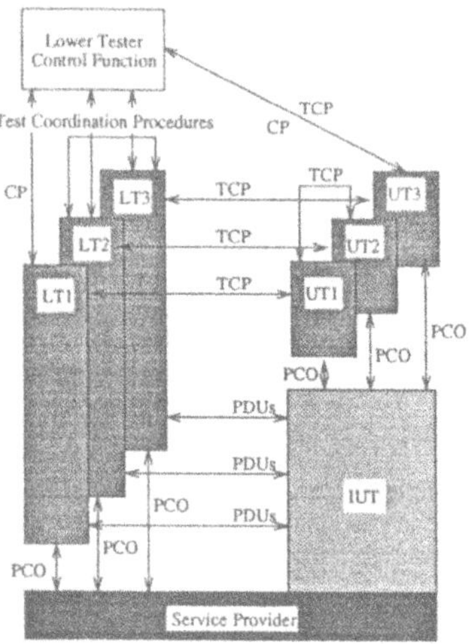

Figure 1: Multi party Testing Context [10].

Communication between lower and upper tester and an *implementation under test (IUT)* is through *points of control and observation (PCO)* directly or via an underlying service provider (Figure 1). PCOs are "modeled as two queues: a) one output queue for control of test events to be sent towards the IUT; and b) one input queue for the observation of test events received from the IUT" [9]. Similarly, CPs "are modeled as two queues, one for each direction of communication" and may be realized "either by local communication or by communication that spans physical boundaries" [12].

Conceptually, a CP abstracts from the real communication infrastructure that connects test components (on either the same system or on different systems). Although CPs "are used to facilitate the exchange of coordination messages between test components", "CPs shall not be used between an LT and a UT ..." [12]. As a consequence, CPs and CMs should be used only for communication between LTs and for communication between UTs. Communication between pairs of LT and UT is supported by different mechanisms. We believe that this is unnecessarily restrictive. Unfortunately, this is not the only flaw in the definition of TTCN and concurrent TTCN (see also [2]). Although LTs and UTs are to be run on different systems, the distribution of test components over systems cannot be made explicit. Having only one lower tester control function (LTCF) responsible for creation and termination of parallel test components (PTC) it is not obvious how the LTCF should perform these operations. One obvious solution is that the LTCF communicates with an "imaginary" entity on the remote system which is instructed by the LTCF to create or to terminate a specific PTC. However, this is not defined in concurrent TTCN and thus is only an ad hoc solution.

To solve the identified problems we propose a new test specification language *PROSPECT - PROtocol test SPEcification language for Conformance Testing*. The main design considerations are summarized as follows: firstly, in order to make the distribution of test components over systems explicit, we introduce *test modules.* A test module comprises all test components running on the same system. Secondly, on every system a lower tester control function exists managing all test components on that system. Thirdly, PROSPECT does not restrict the use of CPs; any two test components may be connected by a CP. The following sections discuss the design of the PROSPECT testing architecture (Section 2), the definition of syntax and semantics of PROSPECT (Section 3) and its implementation (Section 4). We conclude with a summary of related work and identify possible extensions of PROSPECT.

2 PROSPECT Testing Architecture

PROSPECT is a *test specification language.* Its definition is based on a specific testing architecture which is general in the sense that the defined conformance [10] and interoperability [8] abstract test methods (Figures 3 a) and 4 a)) are covered. Our testing architecture (Figure 2) is a slight modification of an existing testing architecture described in [6]. The PROSPECT testing architecture consists of a *test system*, an *implementation under test (IUT)* and a number of *implementation access points (IAP).* A test system consists of *test components* and *links.* Test components are connected by links. *Links* are unidirectional communication channels. Test system and IUT communicate synchronously at IAPs. The difference of our approach with respect to [6] is that we do not distinguish between test components (i.e. LTs and UTs) and *test context*, i.e. an entity relating test actions performed by test components to actions at IAPs. Note that in [6] communication between test context and IUT at IAPs is not assumed to be synchronous. Furthermore, in [6] the test context is not part of a test system, whereas in our approach the test context is realized as a test component too, like UTs and LTs. All components except the IUT but including the test context are considered to be parts of a test system (Figure 2).

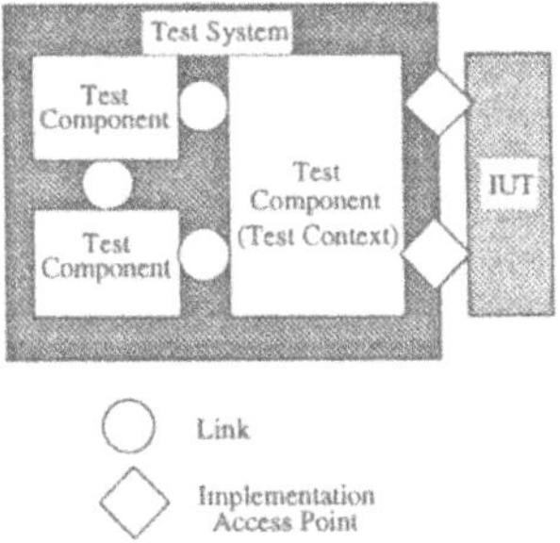

Figure 2: PROSPECT Testing Architecture.

The described testing architecture is the result of the following design considerations:

- It cannot be assumed that a test component can communicate directly with an IUT as, generally, an IUT is embedded in a real system. The test context is an abstraction of this embedding. It comprises the communication infrastructure needed for the interconnection of test components and IUT.

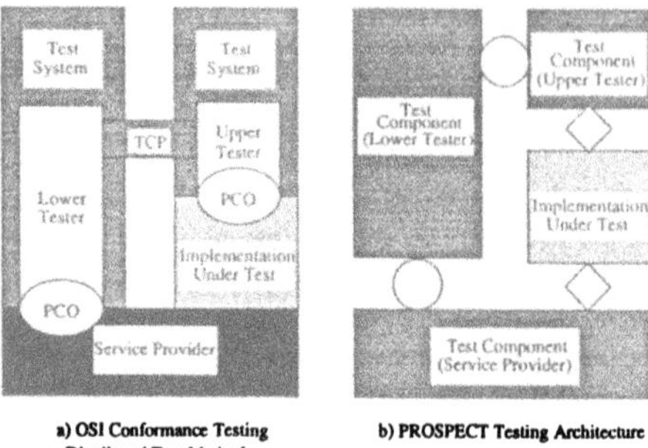

a) OSI Conformance Testing Distributed Test Method

b) PROSPECT Testing Architecture

Figure 3: Distributed Test Method.

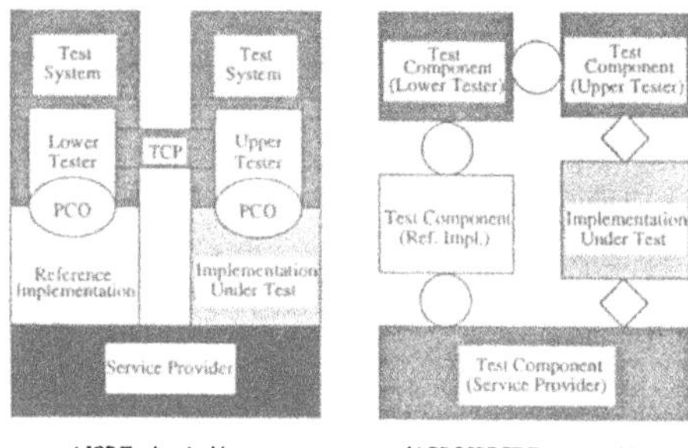

a) IOP Testing Architecture

b) PROSPECT Testing Architecture

Figure 4: Interoperability Testing Architecture.

- In [9] we find the following statement: "It is assumed that the underlying service offered is sufficiently reliable for control and observation to take place remotely." In order to increase the reliability of an underlying service we assume that even the underlying service is part of the test system and is (more or less) under the control of a test system operator.

- In most conformance testing scenarios it is not sufficient to have only one test component. In the *distributed test method* (Figure 3) and in the interoperability testing architecture (Figure 4) at least two components are participating in the execution of a test case. Therefore, in our testing architecture, a test system can consist of several test components.

- Since test execution involves several test components, it becomes essential to provide the instrumentation with which test components can coordinate their behaviour. An important part of this instrumentation are communication channels or links.

- Implementation access points are an abstraction of any mechanism for controlling and observing the behaviour of an IUT. It is for further study how IAPs are to be defined (or even standardized).

Figures 3 and 4 show how the distributed test method [10] and a proposed interoperability test architecture [8] map to our testing architecture: underlying service provider, upper and lower testers simply become test components.

3 Syntax and Semantics of PROSPECT

PROSPECT supports the specification of the behaviour of a test system. The hierarchical structure of the entities of PROSPECT is shown in Figure 5.

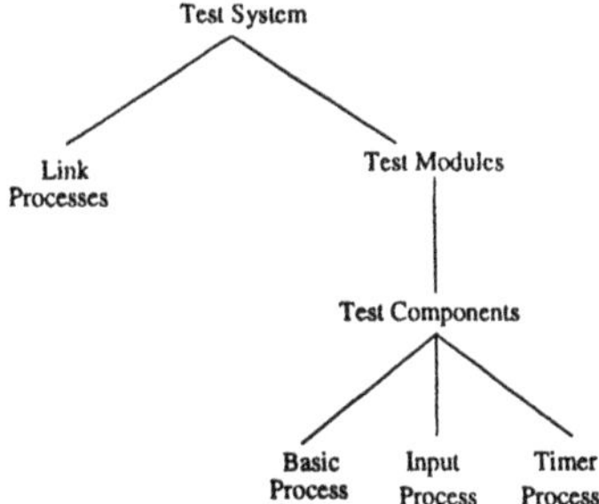

Figure 5: PROSPECT Entities and Structure.

Basic Process (BP): A basic process is a behaviour description of a test component. A behaviour description is an expression over the syntax given in Table 1.

Input Process (IP): An input process stores all messages for a test component which have been received from other test components in a queue-like structure, called *input stream.* Newly received messages are inserted at the end of the input stream. Stored messages can be read and retrieved in any order not just first-in-first-out.

Timer Process (TP): A timer process maintains two finite sequences of timers; one for running or active timers and one for expired timers. The sequence of running timers is ordered according to the timers' expiration times.

Test Component (TC): A test component is composed from a basic process, an input process, a timer process and a storage environment for variables and values. Test components can be created and terminated on demand by a specific test component, called *main test component.*

Test Module (TM): A test module consists of all test components that are supposed to be executed on the same real system. Test components in the same test module communicate asynchronously through communication channels called *routes.* Every test module has its own main test component.

Link Process (LP): A link process connects test components in different test modules. Link processes are unidirectional communication channels. The time required to transmit a message between TCs is a priori unknown but assumed to be finite.

Test System (Sys): A test system consists of link processes and test modules. It communicates synchronously with an IUT at interfaces called *implementation access points (IAP).*

Routes are a concept which has already been used in SDL [22]. As in [22] we assume that routes do not introduce a delay on messages they convey. Every message which is sent on a route to another test component is directly transferred into the receiving test component's input process. Thus, routes and input processes resemble much the semantics of *signal routes* and *input ports* [22]. Note that we distinguish between the reception of a message (in an input process of a test component) and its processing (by a test component). While reception of a message is always possible and is performed *synchronously* with the sending of the message, the processing is, however, performed *asynchronously* with the sending. Therefore we have stated above that communication between test components is asynchronous. With the obvious changes these arguments are also valid for the communication between link processes and test components. Routes and links have been introduced so that the realization of CPs "either by local communication or by communication that spans physical boundaries" [12] can be made explicit.

We assume a discrete time model where time is counted in time units of arbitrary but constant length. A global clock holds the current system time. In addition to the global clock, every test component maintains a local clock which is periodically updated with the current system time.

Example 1 We map the interoperability testing architecture (Figure 4) to PROSPECT (Figure 6) as follows: PROSPECT test components implement test components (Figure 4 b)). Test components that run on the same real system belong to the same test module. We have adopted the approach of mapping the service provider (Figure 4 b)) to several test components that belong to different test modules thus representing the physical distribution of the service provider. □

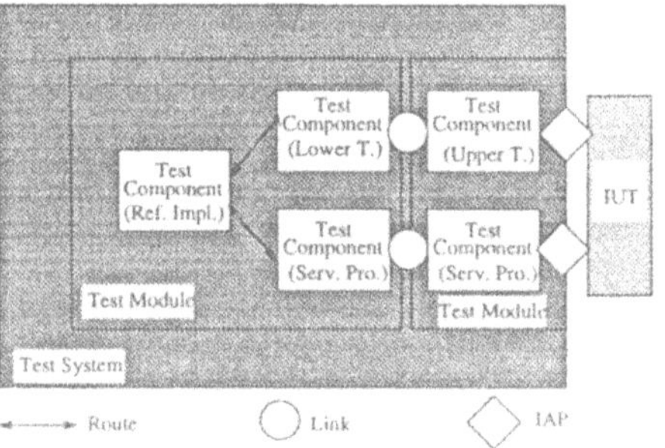

Figure 6: PROSPECT Testing Architecture of Figure 4.

3.1 Syntax of PROSPECT

PROSPECT defines a syntax for basic processes only (Table 1) since all other entities are either composed from lower layer entities (e.g. a test system consists of test modules and link processes) or are completely described by their characteristic property: A link, and this is also valid (with the obvious changes) for input processes and timer processes, is implemented as a FIFO-queue with the following operations pre-defined: the *append* operation to append a new message to the end of the queue, and the *first* operation to remove the first message from the queue.

For the description of data types and data terms we do not give an explicit syntax but assume that an algebraic specification is given from which data terms can be derived. An interpretation of data terms is given in an algebra. Similarly, we assume existence of sets of identifiers for test components, timers etc.

<table>
<tr><td>CommunicationAction</td><td>::=</td><td>“input” “(” lrid “,” dt “)”</td></tr>
<tr><td></td><td>|</td><td>“otherwise” “(” lrid “)”</td></tr>
<tr><td></td><td>|</td><td>“output” “(” lrid “,” dt“)”</td></tr>
<tr><td></td><td>|</td><td>“timeout” “(” tid “)”</td></tr>
<tr><td>BasicAction</td><td>::=</td><td>x “:=” dt</td></tr>
<tr><td></td><td>|</td><td>“start” “(” tid “,” dt “)”</td></tr>
<tr><td></td><td>|</td><td>“cancel” “(” tid “)”</td></tr>
<tr><td></td><td>|</td><td>“stop”</td></tr>
<tr><td></td><td>|</td><td>“return”</td></tr>
<tr><td></td><td>|</td><td>“create” tcid “(” bpid “[” lrid_1 “,” ... “,” lrid_k “]”
“(” dt_1 “,” ... “,” dt_l “)” “)”</td></tr>
<tr><td></td><td>|</td><td>“call” pcid “(” bpid “[” lrid_1 “,” ... “,” lrid_k “]”
“(” dt_1 “,” ... “,” dt_l “)” “(” y_1 “,” ... “,” y_m “)” “)”</td></tr>
<tr><td></td><td>|</td><td>“terminate” “(” tcid “)”</td></tr>
<tr><td>Guard</td><td>::=</td><td>[“[” bt “]”] [CommunicationAction]</td></tr>
<tr><td>ActionSequence</td><td>::=</td><td>BasicAction [“;” ActionSequence]</td></tr>
<tr><td></td><td>|</td><td>PriorityChoiceExpression</td></tr>
<tr><td>GuardedActionSequence</td><td>::=</td><td>Guard “–>” ActionSequence</td></tr>
<tr><td>PriorityChoiceExpression</td><td>::=</td><td>GuardedActionSequence { “+” GuardedActionSequence }</td></tr>
<tr><td>BasicProcess</td><td>::=</td><td>PriorityChoiceExpression</td></tr>
<tr><td>dt, dt_1, ..., dt_l</td><td></td><td>are data terms</td></tr>
<tr><td>bt</td><td></td><td>is a Boolean term</td></tr>
<tr><td>lrid, lrid_1, ..., lrid_k</td><td></td><td>are link, route or IAP identifiers</td></tr>
<tr><td>tid</td><td></td><td>is a timer identifier</td></tr>
<tr><td>bpid</td><td></td><td>is a basic process identifier</td></tr>
<tr><td>tcid</td><td></td><td>is a test component identifier</td></tr>
<tr><td>pcid</td><td></td><td>is a procedure identifier</td></tr>
<tr><td>x, y_1, ..., y_m</td><td></td><td>are variable identifiers</td></tr>
</table>

Table 1: Basic Process Syntax.

PROSPECT distinguishes three types of actions: communication actions, basic actions and guards. Communication actions allow a basic process to communicate with its environment and to determine the status of timers. The communication actions (Table 1) are:

- `input` *and* `otherwise` *actions:* a basic process reads a specific message from the input process or from the IUT by performing an `input` action. A basic process performs an `otherwise` action to read any message that has been received from a specific route or link or the IUT.
- `output` *actions:* a basic process sends a message to another test component or an IUT by performing an `output` action.
- `timeout` *actions:* performing a `timeout` action enables a basic process to check whether a timer has expired.

Basic actions (Table 1) are for the assignment of data terms to variables, for the management of timers, for controlling test components and for dealing with procedures (`call` and `return`). Actions controlling the execution of test components are:

- `stop`: a basic process performs a `stop` action to indicate that it (and implicitly the test component to which it belongs) has successfully terminated.
- `create`: a `create` action instantiates a test component and assigns it a basic process. The instantiated test component runs in parallel with all other active test components.
- `terminate`: a `terminate` action explicitly terminates a test component irrespectively of its current activity. The `terminate` action should be used only by the basic process of the main test component.

A *guard* is a Boolean term followed by a communication action or only a Boolean term or only a communication action. Its meaning (see also Section 3.2) is that the guard is executed if the Boolean term holds and the communication action is possible. If the Boolean term holds and the communication action is possible we say the guard is *fulfilled.* An *action sequence* is either a basic action followed by an action sequence, which means that first the basic action is performed and then the basic process behaves as described by the action sequence; or is a priority choice expression. A *priority choice expression* is similar to a set of alternatives in TTCN and concurrent TTCN: It consists of a finite number of guarded action sequences (i.e. alternatives in TTCN). A guarded action sequence is composed from a guard and an action sequence. If the guard is fulfilled then the guard is executed and the basic process next executes the action sequence. The guards of the guarded action sequences of a priority choice expression are evaluated in sequence which means that the guard of each guarded action sequence is evaluated. The first guard which is fulfilled is executed and execution of the basic process continues with the action sequence that follows the guard. While evaluation of guards takes place no updates to the input stream or expired timers sequence should happen. A *basic process* is a priority choice expression.

3.2 An Operational Semantics of PROSPECT

An operational semantics is defined for every PROSPECT entity. This is done starting with the basic components and proceeding towards the top-level components. This way we emphasize the compositionality of the model (see also [5, 7, 18, 19]).

For every entity an operational semantics is defined by a *labelled transition system (LTS)* [16]:

$$LTS = (S, \epsilon, \rightarrow)$$

where S is a set of states, ϵ is a set of actions and $\rightarrow \subseteq S \times \epsilon \times S$ is a transition relation. An element in $\rightarrow$ is called *transition* and is denoted $s \xrightarrow{e} s'$. A transition is to be interpreted as follows: in state s an entity can perform action e and evolves into state

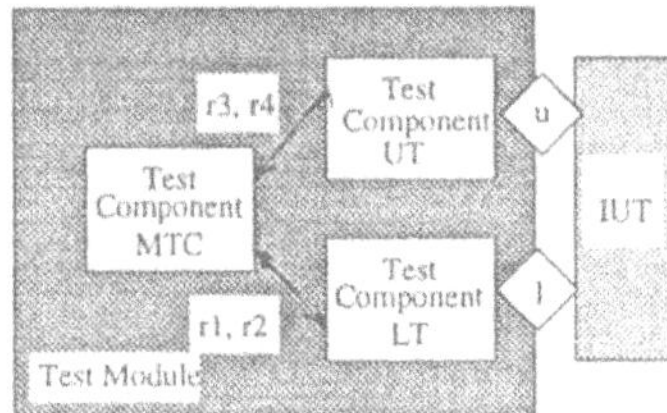

Figure 7: An Example of a PROSPECT Test System.

s'. The behaviour of each entity is given by all transitions which can be derived using inference rules of the form

$$\frac{t_1 \ldots t_n}{t} \quad C$$

where $t_1, \ldots, t_n$ and t are transitions and C is a predicate, called *side-condition.* Transitions $t_1, \ldots, t_n$ can be transitions of the subcomponents of the entity under consideration or are transitions of the entity itself (this is true for all basic entities). Transition t is the inferred behaviour. An inference rule is applicable if the premises and predicate C hold. The concurrent behaviour of components, e.g. test components or links and test components, is represented by interleaving of actions (for more details see also [20]).

Example 2 Assuming we would like to assess conformance of an IUT which is expected to behave as follows: If the IUT receives a character string from its lower layer service provider, then the received character string should be forwarded to the IUT's upper layer service user.

The test system (Figure 7) initiates a service primitive at the IUT's lower layer interface with "`Hello world`" as parameter and then waits for an interaction with the IUT at the IUT's upper layer interface. The character string expected is the string "`Hello world`". The described behaviour is specified using two parallel test components interacting with the IUT and a main test component (MTC) that performs the test coordination. This design is made explicit in the following definition of basic processes:

```
MTC: (* Main Test Component *)
  [true] -> create UT(UT [u, r3, r4]);                    /* line 1 */
    create LT(LT [l, r1, r2]);                            /* line 2 */
    input(r2, cm_r) -> start(timer, now + 5);             /* line 3 */
       input(r4, cm_v(pass)) -> v := pass; stop           /* line 4 */
       + input(r4, cm_v(fail)) -> v := fail; stop         /* line 5 */
       + timeout(timer) -> v := inconc; stop              /* line 6 */

UT: (* Upper Tester *)
  input(u, ind(''Hello world'') ->                        /* line 7 */
    output(r4, cm_v(pass)) -> stop                        /* line 8 */
  + otherwise(u) -> output(r4, cm_v(fail)) -> stop        /* line 9 */

LT: (* Lower Tester *)
  output(l, req(''Hello world'') ->                       /* line 10 */
    output(r2, cm_r) -> stop                              /* line 11 */
```

Test component `MTC` creates parallel test components `UT` and `LT` (lines 1 and 2). Then `MTC` waits for an input from `LT` (line 3). `LT` indicates that it has sent the character string "`Hello world`" to the IUT (line 10) by sending a coordination message `cm_r` to `MTC` (line 11). If `MTC` receives this coordination message (line 3), it starts timer `timer` (line 3) and then it waits for a coordination message from test component `UT` (lines 4 - 6). Upon reception of the character string "`Hello world`" from the IUT (line 7), `UT` sends a coordination message with verdict `pass` to `MTC` (line 8); otherwise (line 9) it sends a coordination message with verdict `fail` (line 9). Test component `MTC` is prepared to accept either the `pass` (line 4) or the `fail` CM (line 5). If due to some unforeseen conditions, `MTC` does not receive any coordination message at all, then, as a third alternative (line 6), `MTC` checks whether timer `timer` has expired in which case the test result is `inconc(lusive)`.

We assume that all actions are atomic, i.e. if two or more actions are performed simultaneously, then this is an indication that entities synchronize and communicate. Although actions are atomic we do not assume that actions are instantaneous in the sense that actions occur "*without consuming time*" as in [3]. Furthermore, the time interval between the end of one action and the start of the next action is unknown but assumed to be finite. Figure 8 represents the execution of test component `LT` under the stated assumptions. Note that after an update of system time, the test component's local clock has to be updated (dotted lines) before a next action is executed.

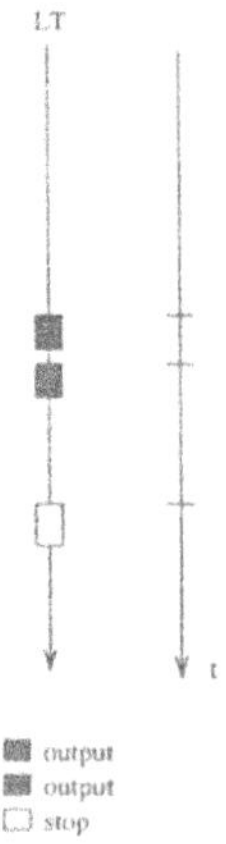

Figure 8: Execution of Test Component `LT`.

Given the following inference rules for basic processes (BP) and test components (TC)

$$\frac{}{GAS \xrightarrow{\text{output}}_{\mathsf{BP}} AS} \qquad \frac{GAS \xrightarrow{\text{output}}_{\mathsf{BP}} AS}{TC \xrightarrow{\text{output}}_{\mathsf{TC}} TC'}$$

we can derive by applying the inference rules that PTC `LT` can perform an `output` action in its initial state as follows:

$$\frac{\overline{GAS \xrightarrow{\text{output}}_{\mathsf{BP}} AS}}{TC \xrightarrow{\text{output}}_{\mathsf{TC}} TC'}$$

Similarly, we infer that PTC LT can perform another output and stop which results in the action sequence shown in Figure 8. Note that before an action is executed the test component performs an internal action due to an update of LT's local clock.

Since we model the parallel execution of actions by interleaving, a specific schedule of the interleaved execution of test components MTC, UT and LT is shown in Figure 9 a). The execution of the test system is shown in Figure 9 b). Communication between test components LT and UT and the IUT is synchronous (dotted lines).

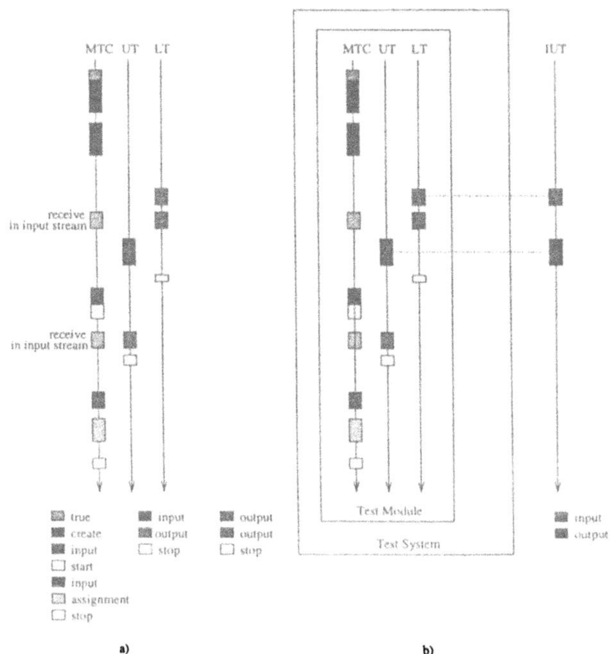

Figure 9: Execution of a Test Module and Test System.

The initial state of a test system is defined as follows: every link process is initialized with an empty message queue; every test module is initialized with the main test component only; the MTC has an empty input stream and empty sequences of running and expired timers, and its storage environment is initialized, i.e. variables are assigned to locations and get their initial values. □

PROSPECT is flexible so that all conformance abstract test methods and interoperability testing architectures can be mapped to PROSPECT test systems. By introducing the notion of a test module we are providing a means that enables test case specifiers to make the distribution of test components over real systems explicit. As we assume that in every test module an MTC exists which is responsible for creating and terminating parallel test components, we are able to define a semantics for the create and terminate basic actions.

In our approach test components communicate through routes (if the test components are in the same test module) or through link processes: No difference exists whether a test component communicates with a local or remote test component. This is of advantage for the definition of test coordination procedures. (In concurrent TTCN test coordination between upper and lower testers and between lower testers and between upper testers are separate issues. Coordination messages and coordination points are used for the communication between lower testers and between upper testers. For the communication between lower tester and upper tester a test management protocol must be defined).

4 PROSPECT Implementation

In order to prove our approach feasible we have implemented PROSPECT in a UNIX™ environment. The implementation consists of a compiler for the behaviour part of test cases and a runtime library.

4.1 Compilation

We have implemented a PROSPECT compiler using *lex* and *yacc* [14]. Compilation of a PROSPECT test case is done in two steps: firstly, programme `P2C (PROSPECT to C)` (Figure 10) is executed that translates PROSPECT source files to C [13] source files. Secondly, the generated C source files are compiled and linked. Besides the generated C sources, the C compiler makes use of the following files (Figure 10):

`_asp.*`: These files contain declarations and definitions of message types, i.e. abstract service primitive types and coordination message types.

`_constraints.*`: These files contain declarations and definitions of test data (or constraints), i.e. instances of abstract service primitives and coordination messages.

`_support.*`: These files contain type definitions, function definitions etc. that are referenced from within a PROSPECT test case. As the previously mentioned files, these files are application specific.

Note that these files are not generated by `P2C` but are assumed to be supplied by a PROSPECT user. As their names suggest the files have to be (manually) derived from the algebraic specifications of *abstract service primitives (ASP)*, coordination messages (CM), *protocol data units (PDU)* and the data terms for ASPs, CMs and PDUs. Furthermore, if necessary, a PROSPECT user has to provide supporting code for encoding and decoding of PDUs and CMs that are exchanged between test components and test system and implementation under test (IUT).

4.1.1 Compilation of a `create`

A `create` is compiled as follows: The MTC first executes a `fork` system call. This system call creates a copy (called the *child* process) of the process that is being executed (called the *parent* process). The child process then executes an `execlp` system call which substitutes the old process image with a new one. As part of the `execlp` system call, actual parameters are passed to the newly instantiated programme. After the MTC has

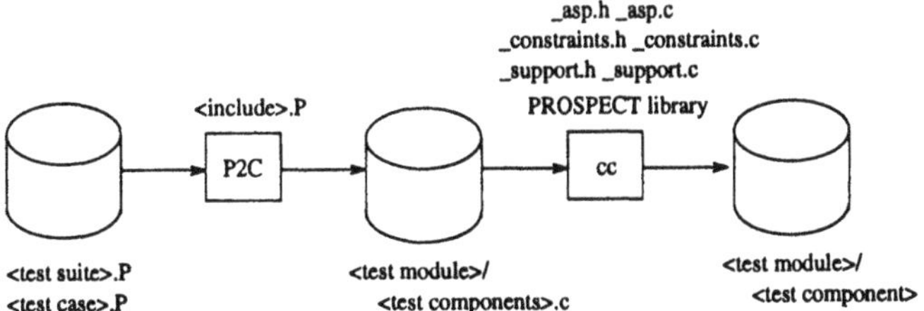

Figure 10: PROSPECT Compilation.

`forked` a PTC, the MTC continues execution of its basic process. The code fragment for the translation of `create LT(LT [1, r1, r2]())` looks as follows:

```
if ((pid = fork()) == -1) {
  /* error handling */
  /* ... */
}
else if (pid == 0) {          /* child process = PTC */
  execlp("LT", "LT", itoa(LT), (char *) 0);
  /* error handling */
  /* ... */
else {                        /* parent process = MTC */
  /* ... */
}
```

where the first `LT` identifies the object code file to be executed and the second `LT` is the parallel test component's name. The third argument passed is the parallel test components unique (with respect to a specific test system configuration) identifier. Some other information needed by the created PTC, e.g. link and route identifiers, is read from a configuration file by the PTC itself during its initialization phase.

4.1.2 Compilation of a `terminate`

In the parent process, upon return from the `fork` system call, the process identifier (an integer) of the child process is known. The parent process maintains an array of process identifiers of child processes. In order to `terminate` a parallel test component, an MTC retrieves the PTC's process identifier and then performs a `kill` system call which terminates the parallel test component unconditionally. For instance, the PROSPECT statement `terminate(LT)` is translated to

```
int tcs[MAXPTC];
/* any other code */
kill(tcs[LT], 9);
/* any other code */
```

where parameter `LT` is used as an index into array `tcs` of process identifiers. The array `tcs` is updated whenever an MTC creates a parallel test component (see also the code fragment above):

```
if                          /* code as shown above */
  /* ... */
else {                      /* parent process = MTC */
  tcs[LT] = pid;            /* <--- change performed here */
}
```

4.1.3 Compilation of Priority Choice Expressions

The evaluation of a priority choice expression is assumed to be atomic (Section 3.1). If a test component evaluates a priority choice expression, no other process (link process, test component or timer process) shall change resources used by the test component, i.e. input stream or expired timers sequence. In order to exclude other processes from accessing critical resources we use semaphores (Section 4.2) to synchronize parallel processes.

Evaluation of a priority choice expression may find all guards unfulfilled. In such a case evaluation of guards is repeated until one guard is fulfilled. Thus, translation of a priority choice expression becomes:

```
for (;;)       /* loop forever until a guard succeeds */
  sem_p();     /* semaphore operation ''P'' */
  if ( /* translation of first guard comes here */ ) {
    /* success */
    sem_v();  /* semaphore operation ''V'' */
    /* ... */
  }
  /* ... */
  if ( /* translation of last guard comes here */ ) {
    /* success */
    sem_v();
    /* ... */
  }
  sem_v();
```

The `sem_v()` function calls are necessary to enable other processes to access critical resources, since otherwise, if all guards are once unfulfilled no progress can ever be made.

4.1.4 Miscellaneous

The translation of other PROSPECT statements is also simple. We consider a few examples: `stop` is compiled to `exit(0)`. `return` (in PROSPECT) becomes `return` (in C). An assignment like `v := pass` is mapped to `v = pass`. An `input` action is translated to a function call in an `if` control statement:

```
input(r2, cm_r);
```

is compiled to

```
if (input(r2, &cm_r) == 0) /* ... */
```

where the constraint, whose declaration and definition are in `_constraints.*`, is passed as a call-by-reference parameter and `0` is returned upon success of the `input` function call.

Timers are implemented as follows: If a timer is started the process performs a `fork` system call and (as the parent process) continues execution. The child process `(u)sleeps` for the specified amount of time. If the `usleep` system call returns, the child process puts a message into its parent's message queue (Section 4.2.1). The message's type identifies the timer that has expired.

4.2 PROSPECT Runtime Support - `PROSPECT library`

The implementation of a runtime library for PROSPECT has been influenced by the structure of a test system (Figure 5) and by our choice of the UNIX™ environment.

4.2.1 Message Queues as an Implementation of Input Processes

Message queues [17] are interprocess communication mechanisms that enable processes to communicate by message passing. A message queue can be shared between several processes. Processes can read and write messages from and to the queue. In our application, link processes or test components write messages to a message queue and the test component for which the message queue is implementing the input process reads messages from the queue. Messages have the following attributes: a message type, an indication of the length of the data part of the message, and the data. The message type is used to identify the link process or route or the timer process which has written the message. The message's data part is, in our context, either an abstract service primitive or a coordination message or empty.

A test component can read a message from its associated message queue by calling the message queue's `receive` function. To identify a specific message the test component has to set the message type appropriately. The mapping from link processes, routes and timer processes to message types is done statically during test case compilation.

There are two advantages of message queues which makes them particularly suitable for our application:

- Sending of a message can be done asynchronously to the receiving and processing of a message.
- Reception of a message of a particular message type (i.e. from a specific link or route) is strictly sequential. This complies with the requirement that only the first message received from a link or route and stored in the message queue should be retrieved.

4.2.2 Sockets as an Implementation of Link Processes

The UNIX™ socket library provides functions to interconnect processes on different systems. Our implementation of link processes is based on *stream sockets* and uses the TCP/IP protocol stack.

A link process is implemented as a client-server pair. The link process client establishes a connection to a link process server and subsequently transmits coordination messages to the link process server. The link process server reads the messages and puts the messages into a test component's message queue.

4.2.3 Semaphores for the Synchronization of Processes

The updating of message queues is critical for the correct evaluation of priority choice expressions. While a test component evaluates a priority choice expression, the content of its message queue should not be changed. Access to a test component's message queues is guarded by semaphores.

In System V semaphores are integer valued variables which are implemented in the UNIX^{TM} kernel. This guarantees that the sequence of operations to update a semaphore is executed atomically. The functions implemented for a semaphore are generalized versions of the P and V operations [4].

To access a resource, a process calls the P operation. If the resource is currently used by another process the calling process is put on a *wait* queue. If the resource becomes available, then the first process in the *wait* queue is activated. A process releases a resource by calling the V operation.

A link process or test component or timer process which intends to update a message queue has to obey the following protocol:

```
P(); /* execute P operation */
/* perform update of message queue */
V(); /* execute V operation */
```

A test component that is about to evaluate a priority choice expression has to perform the protocol steps:

```
P(); /* execute P operation */
/* evaluate the guards of a priority choice expression in sequence */
V(); /* execute V operation */
```

In combination, both protocols prevent the updating of a message queue while evaluation of a priority choice expression is in progress. Furthermore, it is guaranteed that if a process calls a P operation then it will eventually gain access to the message queue.

5 Conclusions

In this paper we have proposed the test specification language PROSPECT and have discussed its implementation. The definition of PROSPECT has been influenced by TTCN and concurrent TTCN [11, 12] and SDL [22]. For instance, main and parallel test components are known from concurrent TTCN and routes are a concept introduced in SDL. PROSPECT extends current approaches for the definition of test notations (see [1, 15] for a definition of *TELL - Test Language LOTOS*): We have introduced test modules in order to enable test case specifiers to define the distribution of test components over real systems. With respect to TELL, PROSPECT supports the use of timers which is not possible in TELL. However, we admit that our time model is rather high level so that it has to be refined to be applicable in e.g. QoS assessment or testing real-time applications.

We have also discussed the implementation of PROSPECT which consists of a compiler for PROSPECT test cases and a PROSPECT runtime library. The implementation has been done in a UNIX™ environment. We have made the experience that the implementation of some parts of PROSPECT has been simple in this environment. Particularly useful have been the different interprocess communication facilities available in UNIX™.

The applicability of PROSPECT is limited since, as pointed out above, timing constraints cannot precisely be specified. With the upcoming deployment of multi-media applications, like video-conferencing, multi-media archiving and retrieval, tele-teaching, etc. not only *functional* properties of an implementation are to be assessed during conformance testing but also *non-functional* properties, e.g. timing constraints (as in real-time applications), quality-of-service (QoS) aspects, synchronization of different data streams (audio, video, textual information). None of the known test notations can deal with these requirements. To close this gap, we have started a research project with the goal to define and implement a test specification language that supports test case specification for multi-media applications [21]. PROSPECT is supposed to be used in this project.

Acknowledgment The authors would like to gratefully acknowledge the anonymous referees for fruitful comments to the previous version of this paper.

References

[1] Ahooja, R., Burmeister, J., de Meer, J., Rennoch, A., *Method for Open Test Sequences and their Implementation generalized in a Conformance Test Tool*, Proceedings PTS II, North-Holland, 1989.

[2] Baumgarten, B., *Open Issues in Conformance Test Specification*, Proceedings PTS VII, 1994.

[3] Bolognesi, T., Brinksma, E., *Introduction to the ISO Specification Language LOTOS*, Computer Networks and ISDN Systems, Vol. 14, North-Holland, 1987.

[4] Dijkstra, E., *Cooperating Sequential Processes*, Genuys, F. (Ed.), *Programming Languages*, Academic Press, 1968.

[5] Walter, T., Ellsberger, J., Kristoffersen, F., Merkhof, P.v.d., *Methods for Testing and Specification (MTS) Semantical relationship between SDL and TTCN A Common Semantics Representation*, European Telecommunications Standards Institute, ETR 071, 1993.

[6] ISO/ITU-T, *Formal Methods in Conformance Testing*, ITU-T TS SG10 Q8 and ISO SC21 WG1 P54 Southampton output, 1994.

[7] Godskesen, J., *An Operational Semantic Model for Basic SDL*, Tele Danmark Research Report TFL RR 1991-2, 1991.

[8] Hogrefe, D., *Conformance testing based on formal methods*, Proceedings FORTE '90, North-Holland, 1990.

[9] ISO/IEC, *Information technology - Open Systems Interconnection - Conformance testing methodology and framework - Part 1: General concepts*, ISO/IEC 9646-1, 1994.

[10] ISO/IEC, *Information technology - Open Systems Interconnection - Conformance testing methodology and framework - Part 2: Abstract Test Suite specification*, ISO/IEC 9646-2, 1994.

[11] ISO/IEC, *Information technology - Open Systems Interconnection - Conformance testing methodology and framework - Part 3: The Tree and Tabular Combined Notation (TTCN)*, ISO/IEC 9646-3, 1992.

[12] ISO/IEC, *Information technology - Open Systems Interconnection - Conformance testing methodology and framework - Part 3: The Tree and Tabular Combined Notation (TTCN): Amendment 1: TTCN Extensions*, ISO/IEC 9646-3 DAM 1, 1993.

[13] Kernighan, B., Ritchie, D., *The C Programming Language*, Prentice Hall Software Series, 2nd Edition, 1988.

[14] Levine, J., Mason, T., Brown, D., *lex & yacc*, O'Reilly & Associates, 1992.

[15] de Meer, J., Burmeister, J., Rennoch, A., Schröer, I., *An Approach to a Conformance Testing Methodology and the COAST Test Systems*, Proceedings PTS III, North-Holland, 1991.

[16] Plotkin, G., *A structural approach to operational semantics*, Aarhus University, Computer Science Department, 1981.

[17] Stevens, W. Richard, *UNIX Network Programming*, Prentice Hall Software Series, 1990.

[18] Walter, T., Ellsberger, J., Kristoffersen, F., Merkhof, P.v.d., *A Common Semantics Representation for SDL and TTCN*, Proceedings PSTV XII, North-Holland, 1992.

[19] Walter, T., Plattner, B., *An Operational Semantics for Concurrent TTCN*, Proceedings PTS V, North-Holland, 1992.

[20] Walter, T., *PROSPECT - A Contribution to Protocol Specification, Conformance, and Interoperability*, Dissertation 10419, ETH Zürich, 1993.

[21] Walter, T., Grabowski, J., *Towards the new Test Specification and Implementation Language 'TelCom TSL'*, 5. GI/ITG Workshop *Formal Description Techniques for Distributed Systems*, Kaiserslautern, June 1995.

[22] IUT, *Functional Specification and Description Language SDL*, ITU-T Recommendation, 1992.

PART FIVE

Test Generation 1

12

Test sequence generation for adaptive interoperability testing

Sungwon Kang and Myungchul Kim
Korea Telecom Research Laboratories
Sochogu Umyundong 17, Seoul 137-792, Korea
e-mail: kangsw@sava.kotel.co.kr, mckim@sava.kotel.co.kr

Abstract

When testing communicating systems, nondeterminism makes it a more difficult and evasive process. Adaptive testing is an efficient approach to testing nondeterministic systems. In this paper, we develop an interoperability test generation method for adaptive testing. Also we define a measure of testing cost and compare our method with the conventional approach.

Keywords

Test case generation, adaptive interoperability testing, nondeterminism

1. INTRODUCTION

In order to ensure interoperability of communication networks, it is essential to verify correctness of implementations of communication protocols as well as to verify correctness of their specifications on which implementations are to be based. Specifications of implementations constituting a system should be such that correct implementations interoperate. The activity widely called validation and verification, however, is almost always incomplete for most nontrivial protocol specifications. Verifying implementation correctness is called conformance testing and is in comparing the behavior of implementation under test with respect to the expected behavior according to its specification. It also is incomplete for nontrivial protocols. Such incompleteness leaves the direct examination of system behavior as an essential and integral part of interoperability assurance activity.[1]

Recently as the need for systematic interoperabilty testing increases, there has been significant related work ranging from investigation from practitioner's point of view [Bonnes 90][Vermeer 94] to theoretical study with varying degree of formalism [Arakawa 92][Castanet

[1] Often a specification is given in a parameterized form so that the real specification is obtained by instantiating the parameters with concrete values. In such a case, inability of implementations to interoperate may be due to incompatibility of parameters. Ideally such incompatibility should not be regarded as a major concern of testing but of validation and verification.

94][Kajiwara 94][Luo 94]. Ideally interoperability test sequence generation should be done, as with conformance test sequence generation, in such a way that minimizes the cost of the testing process and maximizes the quality of the test result, which tend to be conflicting goals. In general, approaches from practitioners emphasize the former aspect whereas approaches from theoreticians emphasize the latter aspect.

In this paper, we develop an interoperability test generation method which accommodates these conflicting goals. For the second goal our test generation method is based on state space analysis of the global system. And for the first goal our test generation method is directed to adaptive testing [Kim 94] and global system state space is compressed through the so called slow environment principle [Arakawa 92][Luo 94]. When dealing with a system of communicating entities, nondeterminism is a problem that we cannot get away with. Adaptive testing is an elegant approach to controlling and observing behaviors of nondeterministic systems. The slow environment assumption enables us to abstract from nondeterminism to a significant extent.

This paper is organized as follows. In Section 2, we define a system of communicating finite state machines model and define interoperability with respect to the model. In Section 3, we describe our test generation method for adaptive testing. In Section 4, the test generation method is combined with interoperability test generation to give test sequence generation for adaptive interoperability testing. Also an example of interoperability test generation is given and improvement over the conventional non-adaptive method is measured. In Section 5, we summarize contributions of this paper and suggest further research problems.

2. A SYSTEM OF COMMUNICATING IOSM'S AND INTEROPERABILITY

Definition 2.1 An IOSM M is a 5-tuple $<St, s_0, L_{in}, L_{out}, Tr>$ where:

(1) $St = \{s_0, \ldots, s_{n-1}\}$ is a set of states,

(2) $s_0 \in St$ is the initial state,

(3) $L_{in} = \{v_1, \ldots, v_m\}$ is a set of input symbols,

(4) $L_{out} = \{o_1, \ldots, o_p\}$ is a set of output symbols and

(5) $Tr \subseteq \{s\text{–}v/\mathbf{o}\rightarrow s' \mid s, s' \in St \wedge v \in L_{in} \wedge \mathbf{o} \in (L_{out})^* \}$

In spite of the name, the definition of IOSM here differs from that of [Phalippou 91]. Bold face letters are used to denote a sequence of symbols and the empty sequence is denoted as ε or –. L_{in} and L_{out} are respectively called *input alphabet* and *output alphabet*. Tr is a set of transitions of M. We call $v/\mathbf{o} \in (L_{in}/(L_{out})^*)$ a *label*. If $v_1/\mathbf{o}_1, \ldots, v_k/\mathbf{o}_k \in (L_{in}/(L_{out})^*)$, we denote by $s\text{–}\ v_1/\mathbf{o}_1 \ldots v_k/\mathbf{o}_k \rightarrow s'$ that

$$\exists s_1,\ldots,s_{k-1} \in St\text{: } s\text{–}v_1/\mathbf{o}_1\rightarrow s_1 \wedge s_1\text{–}v_2/\mathbf{o}_2\rightarrow s_2 \wedge \ldots \wedge s_{k-1}\text{–}v_k/\mathbf{o}_k\rightarrow s'$$

In this paper, we adopt the *completeness assumption* which requires that transitions be defined for every input symbol, i.e.

$$\forall s \in St,\ v \in L_{in} : \exists s_1 \in St,\ \mathbf{o} \in (L_{out})^*\text{: } (s\text{–}v/\mathbf{o}\rightarrow s_1) \in Tr.$$

In the set Tr of transitions of a *deterministic* IOSM, there is only one transition for any state with the same input symbol, i.e., Tr satisfies the condition:

$$\forall s, s_1, s_2 \in St,\ v \in L_{in},\ \mathbf{o}_1, \mathbf{o}_2 \in (L_{out})^*\text{: } (s\text{–}v/\mathbf{o}_1\rightarrow s_1 \wedge s\text{–}v/\mathbf{o}_2\rightarrow s_2) \rightarrow \mathbf{o}_1 = \mathbf{o}_2 \wedge s_1 = s_2.$$

Note that Definition 2.1 does not restrict IOSM to be deterministic.

The general class of nondeterministic IOSM's poses challenges to testing since controllability based upon observation of external behavior is very limited. So in this paper we confine our discussion to the class of observable IOSM's defined as follows.

Definition 2.2 (*n*-observable IOSM) Let M be as in Definition 2.1. Then M is *n*-observable if and only if there is $n \geq 1$ such that

$$\forall s, s_i, s_j \in St: \exists \mathbf{x} \in (L_{in}/(L_{out}^*))^n: (s\text{–}\mathbf{x}\rightarrow s_i \wedge s\text{–}\mathbf{x}\rightarrow s_j) \rightarrow s_i = s_j.$$

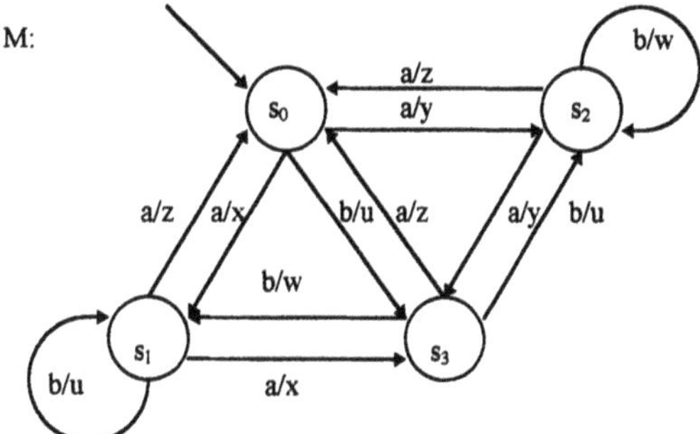

Figure 1 An observable IOSM.

We call 1-observable IOSM *observable IOSM*. Figure 1 depicts an observable IOSM.

Definition 2.3 Let M be as in Definition 2.1. Let v, $v_1, \ldots, v_k \in L_{in}$ and s, s'$\in$St. Then:

(1) $\sigma(s,v) = \{(s', v/\mathbf{o}) \mid s\text{–}v/\mathbf{o}\rightarrow s' \in Tr\}$
(2) $\sigma(s,\mathbf{v}) = \{(s', v_1/\mathbf{o}_1 \ldots v_k/\mathbf{o}_k) \mid s\text{–}v_1/\mathbf{o}_1 \ldots v_k/\mathbf{o}_k \rightarrow s' \in Tr \wedge \mathbf{v} = v_1 \ldots v_k\}$
(3) $s(\mathbf{v}) = \{v_1/\mathbf{o}_1 \ldots v_k/\mathbf{o}_k \mid (s', v_1/\mathbf{o}_1 \ldots v_k/\mathbf{o}_k) \in \sigma(s,\mathbf{v}) \wedge \mathbf{v} = v_1 \ldots v_k\}$
(4) $M(\mathbf{v}) = s_0(\mathbf{v})$

A member of M(**v**) is a sequence of labels and is called a *run* (or *trace*) of M for the input sequence **v**.

Definition 2.4 Let M_1 and M_2 be two IOSM's with the same input alphabet L_{in}. Then M_1 and M_2 are *observationally equivalent* ($M_1 \equiv_o M_2$) if and only if

$$\forall \mathbf{v} \in L_{in}^*: M_1(\mathbf{v}) = M_2(\mathbf{v}).$$

The above definition stipulates so called *trace equivalence*, i.e. two IOSM's are said to be observationally equivalent if and only if, given any input sequence, the set of traces for that sequence are the same. A non-observable IOSM can always be transformed into a trace equivalent observable IOSM [Luo 94]. Various methods were developed to determine conformance by testing for deterministic IOSM's [Sidhu 89]. Trace equivalence was often used as a conformance relation between two IOSM models, one for a specification and the other for an implementation.

IOSM defined as a machine that interacts with environment is not adequate for modeling situations where more than one IOSM's communicate with each other. So we now introduce the notion of communicating IOSM.

Definition 2.5 A *Communicating IOSM*(= CIOSM) M is a 5-tuple $<St, s_0, L_{in}, \{M\}\times N\times L_{out}, Tr>$ where St, s_0, L_{in} and L_{out} are as in Definition 2.1 and $N = N'\cup\{Env\}$ where N′ is the set of identifiers for CIOSM's such that $M\in N'$. The set of transitions is now:

$$Tr \subseteq \{s\text{–}v/\mathbf{u}\rightarrow s' \mid s,s'\in St \wedge v\in L_{in} \wedge \mathbf{u}\in(\{M\}\times N\times L_{out})^*\}$$

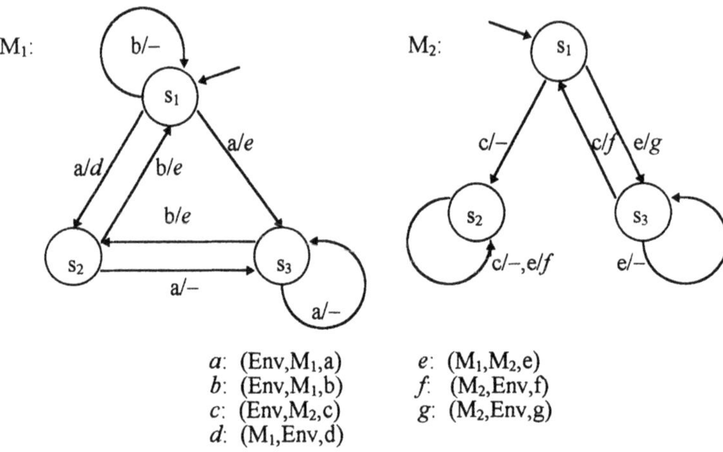

Figure 2 CIOSM's M_1 and M_2.

The difference between IOSM and CIOSM is that CIOSM sends messages of the form (*sender*, *receiver*, *symbol*) to a specific receiver, more than one message in a single transition. Other notions defined for IOSM can be extended for CIOSM in an obvious way.

Definition 2.6 A system Σ of n CIOSM's is $<\{(M_i,Q_i) \mid 1 \le i \le n\}, \{Env\}\times N\times L_{\Sigma,in}, N\times\{Env\}\times L_{\Sigma,out}, s_{\Sigma,0}>$ where:

(1) M_i, $1 \le i \le n$, is a CIOSM $<St_i, s_{i,0}, L_{i,in}, L_{i,out}, Tr_i>$ as in Definition 2.5.

(2) Q_i, $1 \le i \le n$, is an input queue for M_i.

(3) $L_{\Sigma,in}$ is a set of external input symbols.

(4) $L_{\Sigma,out}$ is a set of external output symbols.

(5) The initial state of the system is $s_{\Sigma,0} = <(s_{1,0}, Q_0), \ldots,(s_{n,0}, Q_n)>$ where $Q_i=\varepsilon$ and $s_{i,0}$ is the initial state of M_i, $1\le i \le n$.[2]

The term *external* indicates association with the environment Env. In this model, there is one explicitly defined input queue for each process and implicit communication channels between

[2] Definition 2.6 shows that though we see a system of CIOSM's mostly as a black-box, it is assumed that the names of the individual CIOSM's are known.

each pair of processes. Since a system Σ can contain more that one CIOSM's, an actual input from the environment is of the form (Env,i,v) where i is the identifier of the receiving CIOSM and v is an external input symbol. External outputs have Env as its unique receiver.

A system of CIOSM's Σ is said to be *closed* if no transitions consume inputs from or produce outputs to the environment. We call system state changes caused by consumption of an external input by a CIOSM or of an external output by the environment *observable* (or *visible*) *state changes.* Other system state changes are called *internal* (or *invisible*) *state changes.*

Figure 2 which is a slight modification of an example in [Luo 94] describes a system consisting of two CIOSM's, M_1 and M_2. The input queues are not shown. In the next definition, we make precise the global behavior of a system of CIOSM's.

Definition 2.7. Let Σ be a system of CIOSM's as in Definition 2.6. Let $w \in N \times N \times L_{\Sigma,in}$, $u \in N \times N \times L_{\Sigma,out}$, $\mathbf{w} \in (N \times N \times L_{\Sigma,in})^*$, $\mathbf{u} \in (N \times N \times L_{\Sigma,out})^*$, s be a state of Σ. Then

(1) $\sigma(s,(Env,i,v)) = \{(s',(Env,i,v)/\mathbf{u}) \mid s = <\ldots,(s_{i,j},Q_i.v),\ldots> \wedge$
$\exists i' \in N:(s_{i,j} -(i',i,v)/(i,i_1,o_1)\ldots(i,i_m,o_m) \rightarrow s_{i,j'}) \in Tr_i \wedge$
$s' = \mu(s,\{(s_{i,j'},Q_i)/(s_{i,j},Q_i.v)\} \cup \{(s_{i_k,j_k},Q.o_j)/(s_{i_k,j_k},Q) \mid 1 \le k \le m \wedge i_k \neq Env\})$
$\mathbf{u} = \varphi((i,i_1,o_1)\ldots(i,i_m,o_m))\}$

where $\mu(s,\varnothing) = s$
$\mu(<\ldots,(s,Q),\ldots>, P \cup \{(s',Q')/(s,Q)\}) = \mu(<\ldots,(s',Q'),\ldots>,P)$
$\varphi(\varepsilon) = \varepsilon$
$\varphi((i,Env,o).\mathbf{u}) = (i,Env,o).\varphi(\mathbf{u})$
$\varphi((i,i',o).\mathbf{u}) = \varphi(\mathbf{u}) \qquad \text{if } i' \neq Env$

(2) $\sigma(s,\varepsilon) = \{(s,\varepsilon/\varepsilon)\}$
$\sigma(s_1,w_1\ldots w_{m+1}) = \{(s_{m+2},\mathbf{w}w_{m+1}/\mathbf{u}u_{m+1}) \mid (s_{m+1},\mathbf{w}/\mathbf{u}) \in \sigma(s_1,w_1\ldots w_m) \wedge$
$(s_{m+2},w_{m+1}/u_{m+1}) \in \sigma(s_{m+1},w_{m+1})\}$

(3) $s(\mathbf{w}) = \{\mathbf{w}/\mathbf{u} \mid (s',\mathbf{w}/\mathbf{u}) \in \sigma(s,\mathbf{w})\}$

(4) $\Sigma(\mathbf{w}) = s_{\Sigma,0}(\mathbf{w})$

In (1), σ with a single external input defines a set of new states reached by processing of a single input from the environment followed by a sequence of data movements (including a sequence of internal state changes) out of the receiving CIOSM. The reception of a message and sending messages as a response to it constitutes an atomic action. Thus input from the environment is instantly placed into the input queue of the receiving CIOSM. A system state in which input queues are all empty is called a *stable state.*[3] μ is a function that updates states. φ extracts a sequence of external inputs or outputs from a sequence of inputs or outputs, respectively.

(2) extends σ to a sequence of external inputs. (2)-(4) for systems of CIOSM's are similar to (2)-(4) for IOSM's of Definition 2.2. In a system defined by Definition 2.7, *deadlock* is said to occur when any combination of (external and internal) inputs cannot change system state. *Livelock* is said to occur when system state changes forever and diverges.

[3] This contrasts with the definition of stable state in [1] in which the system waits for input from the environment regardless of whether inputs queues and channels are empty or not.

Definition 2.8 Let Σ_1 and Σ_2 be systems of CIOSM's with the same set of identifiers for CIOSM's N and the same input alphabet L_{in}. Then

Σ_1 and Σ_2 are *observationally equivalent* (denoted $\Sigma_1 \equiv_o \Sigma_2$)
if and only if
$\forall \mathbf{w} \in (\{Env\} \times N \times L_{\Sigma,in})^*: \Sigma_1(\mathbf{w}) = \Sigma_2(\mathbf{w})$.

Definition 2.9 Let Σ_1 be an implementation of a specification Σ_2 of a system of CIOSM's. Then

Σ_1 *interoperate with respect to* Σ_2 (denoted $\underline{interop_{\Sigma 2}}(\Sigma_1)$) if and only if $\Sigma_1 \equiv_o \Sigma_2$

This is a relative definition of interoperability. An absolute definition either should be contained in a given specification explicitly or should be agreed upon in advance by specification providers and implementors. An implication of the definition is that interoperability testing of implementations is a conformance testing of the global state space defined by the whole system. This conformance testing of interoperability is restricted by the observability and controllability that are allowed in a particular system configuration (or architecture) and here we assert that the practical limit on observability and controllability is set by the slowness of the environment which justifies us in focusing only on stable states. Then the additional ingredient interoperability testing provides is to fill the gap left by usual conformance testing. The real work of interoperability assurance should come at the stage of specification development. But the fine points that may have been missed at the time of specification writing (including validation and prototyping) can be augmented through testing.

3. TEST SEQUENCE GENERATION FOR ADAPTIVE TESTING

In this section, we concentrate on test generation for nondeterministic IOSM's. In Section 4, we will apply the test generation method developed here for adaptive testing to interoperability test generation.

Definition 3.1 A *path tree* of a state s_i (denoted T_{Pi}) of IOSM M is a tree such that

(1) The root of T_{Pi} corresponds to the initial state of M
(2) Each node corresponds to a state of M
(3) Each edge of T_{Pi} corresponds to some transition of M with the matching departure and destination states and the matching label.
(4) All leaf nodes of T_{Pi} correspond to s_i.

For the IOSM M in Figure 1, we may select the path trees as in the T_{Pi} column of Table 1. Thus for example, a/x.a/x and a/y.a/y are two branches of {a/x.a/x, a/y.a/y}[4] and traversal of either branch takes M to s_3. A path tree may not be unique. In P_i column are selected paths to the states of M. In this example, a path tree that exhibits the essential difference of the adaptive method from the conventional approach is T_{P3}. In the conventional approach, only one path to a state is selected at the time of test generation. In the adaptive approach, more that one paths to a state are selected and, depending on the responses from the implementation under test, different branches are followed.

[4] We use set notation to represent path trees as well as state identification trees.

	P_i	W_i	T_{P_i}	T_{S_i}
i = 0	ε	{a/x, a/y}	{ε}	{{a/x.a/x, a/y.a/y}}
i = 1	a/x	{b/u.a/x.b/w}	{a/x}	{{b/u.a/x.b/w}}
i = 2	a/y	{b/w.a/y}	{a/y}	{{b/w.a/y}}
i = 3	**a/x.a/x**	**{b/u.a/y}**	**{a/x.a/x, a/y.a/y}**	**{{b/u.a/y, b/w.a/x}}**

Table 1 Path trees and state identification trees for M in Figure 1.

For the purpose of state identification, notions of varying generality such as Distinguishing Sequence, UIO sequence, W set, Wp sets and generalized Wp sets have been used [Chan 89][Chow 78][Sabnani 88][Luo 94]. Depending on whether a single sequence is used or a set of sequences is used for state identification purpose, it may be called *state identification sequence* or *state identification set*. We now introduce a further generalization along the line of state identification set.

Definition 3.2 A *state identification tree* of a state s_i (denoted Ts_i) of IOSM M is a tree such that

(1) The root of Ts_i corresponds to s_i
(2)-(3) are as in Definition 3.1
(4) Each path from the root to a leaf node is a state identification sequence.

The notion of state identification tree can be generalized to the notion of state identification forest for the cases where a single state identification sequence does not exist. A *state identification forest* of a state s_i (denoted Fs_i) of IOSM M is a set of trees such that each tree satisfies (1)-(3) of Definition 3.2 and

$$\{P \mid P \text{ is a path from the root to a leaf node of } T \wedge T \in Fs_i\}$$

is a state identification set.

The notions of *UIO* and *Partial UIO tree* were used in [Kim 94] for a similar purpose. Main differences of our approach are that we do not require that all transitions with the same input symbol should appear in the tree as branches and we use trees not only for state identification but also for tours to states. Thus even with the presence of the status message feature which makes use of input/output pairs of status enquiry/status name, our method can still give efficiency improvement.

For the IOSM M in Figure 1, we may select the state identification trees as in the Ts_i column of Table 1. By selecting W_0 as in Table 1, it takes two sequences of labels to distinguish s_0 from the other states. However, Ts_0 uses a tree each branch of which alone is enough to distinguish s_0 from the other states. But the two sequences were chosen so that the first labels of them constitute all possible transitions that can be taken upon receiving the input a. For state s_3, a sequence of length 2, i.e. b/u.a/y, is a state identification sequence but, in the adaptive approach, two sequences b/u.a/y and b/w.a/x are used. But again they were selected so that the first label has the common input b and u and w are all possible outputs for b.

For testing, the cost of test suite execution is more important than the cost of test suite generation since once generated a test suite is used repeatedly. In order to measure efficiency of a test suite execution, we define a sort of worst-case test case execution function $\xi_w(M,t)$ where M is the name of an IOSM or a system and t is a test case. Since tree in our method a test case is a of which a branch is selected dynamically and a case is finished when a leaf node is reached, in order to apply ξ_w, we make the following assumptions about execution behavior:

(i) Once a path is selected, the same path is not taken again and all other possible paths have been traversed once and

(ii) Once a correct subsequence is taken, then the subsequence of it is always followed (with no need of backtracking) in the next traversal.

(iii) Only after all other possibilities are exhausted obeying the above two restrictions, the responses expected for the test case being executed are obtained.

What would happen in the real world differs from this. However, the model gives some idea about test suite execution efficiency. We illustrate this point with the example in Figure 1.

With the assumptions (i), (ii) and (iii), in order to reach s_3 via P_3 it takes three tours from the initial state, i.e. a/y, a/x.a/z and a/x.a/x in that order. We denote this as

a/y
a/x.a/z
a/x.a/x

or in a linear fashion as (a/y, a/x.a/z, a/x.a/x).[5] In contrast, with T_{P_3} it takes two tours, e.g. (a/x.a/z, a/x.a/x). Then ξ_w(M, a/x.a/x) = (a/y, a/x.a/z, a/x.a/x) and ξ_w(M, {a/x.a/x, a/y.a/y}) = (a/x.a/z, a/x.a/x). And if we let |x| denote a function that counts the number of labels in x, |(a/y, a/x.a/z, a/x.a/x)| = 5 and |(a/x.a/z, a/x.a/x)| = 4. Thus the path tree approach is more efficient for the tour to s_3.[6]

Similar efficiency improvement can be achieved though state identification trees under the same assumptions. So when identifying state s_3 using T_{S_3} instead of a singleton identification set W_3, T_{S_3} requires only 2 trials (b/w.a/z, b/w.a/x) instead of 3 trials with W_3, i.e. (b/w, b/u.a/z, b/u.a/x). When traversal to s_3 is combined with the state identification of s_3, the following comparison can be made:

P_3 combined with W_3	T_{P_3} combined with T_{W_3}
a/y	a/x.a/z
a/x.a/z	a/x.a/x.b/w.a/z
a/x.a/x.b/w	**a/x.a/x.b/w.a/x**
a/x.a/x.b/u.a/z	
a/x.a/x.b/u.a/y	

If we compute test case execution effort in terms ξ_w, then the cost of adaptive testing is 10 whereas that of the conventional testing is 14.

When selecting T_{P_i} and T_{W_i}, however, discretion should be used. A helpful heuristic is to avoid a very long identification sequence while trying to minimize backtracking to the initial state due to a nondeterministic selection of unwanted transitions. Thus as many transitions from a state as possible should become branches of the state identification tree as long as they have the same input symbol. When a node has an incomplete set of children for an input due to absence of a single state identification sequence or efficiency consideration, then the node should be

[5] Return to the initial state can be achieved by appending pre-calculated paths to the initial state or by appending the reset primitives at the end of each test case. But this is an issue that can be treated independently from the subject of this paper.

[6] One might consider that after a/x.a/z we could get a/y. This suggests using only the assumptions (i) and (iii) but then analysis is far more complicated.

marked with the output sequences not in the branches so that those output sequences are not considered as faults.

Some approach uses *tree structured test suite* [Kim 94]. Such an approach is efficient for finding faulty implementations but is rather difficult to give analysis of the cause of failure. Also when an implementation passes a test suite, it is no more efficient than *tabular test suite* approach such as ours in which test cases have corresponding predetermined test purposes.

4. INTEROPERABILITY TEST GENERATION FOR ADAPTIVE TESTING

In this section, we assume the systems of CIOSM's under consideration (both as specifications and implementations) are free of livelocks. A similar assumption was made for the systems of CNFSM's in [Luo 94]. The assumption makes it possible to generate test suites which can detect all interoperation errors in a finite amount of time.

In [Luo 94], a system of communicating finite state machines was assumed to operate in a *slow environment*. In a slow environment, inputs can be sent from the environment to the system only in situations where all the queues are empty. [Arakawa 92] assumes the *single stimulation principle* which requires that a single stimulation must be given at stable states where no change to the system can occur in a stable state without a further stimulation. We assume the slow environment assumption.

Interoperability Test Generation Algorithm

Input: A system Σ of n CIOSM's as defined in Definition 2.6 which has no livelocks.
Output: An interoperability test suite Π.

Step 1: Model Σ as a trace-equivalent nondeterministic IOSM M using the slow environment assumption.
Step 2: Derive a minimal trace-equivalent observable IOSM M′ from M.
Step 3: For each state s_i of M′, construct a path tree T_{Pi} and a state identification tree Ts_i (or a forest Fs_i).
Step 4: Let k be the number of states of M′ and m be the upper bound m ($\geq$ k) on the number of states in the global state space of the system implementation under test. Then a state cover Π_1 and a transition cover Π_2 are constructed as follows:

$$\Pi_1 = \{\mathbf{x}.\mathbf{y}.Ts_j \mid s_0 - \mathbf{x} \rightarrow s_i \wedge \mathbf{x} \in T_{Pi} \wedge s_i - \mathbf{y} \rightarrow s_j \wedge \mathbf{y} \in \bigcup_{0 \leq i \leq m-k} L^i \wedge s_i, s_j \text{ are states of M}'\}$$

$$\Pi_2 = \{\mathbf{x}.\mathbf{z}.Ts_j \mid s_0 - \mathbf{x} \rightarrow s_i \wedge \mathbf{x} \in T_{Pi} \wedge s_i - \mathbf{z} \rightarrow s_j \wedge \mathbf{z} \in L^{m-k+1} \wedge s_i, s_j \text{ are states of M}'\}$$

where $L^0 = \{\varepsilon\}$ and $L^{q+1} = L^q.(L_{in}/(L_{out})^*)$

Step 5: Let Π_Σ be the result of optimizing $\Pi_1 \cup \Pi_2$ by repeatedly applying the rule:
Remove a test case t if
(1) all branches of t are subsequences of some deterministic test case or
(2) a test case is a subtree of another test case.

Without the slow environment assumption, a global state space graph with a potentially large number of nodes could be constructed in Step 1. With the assumption, the number of states in

M is bounded by $|St_1| \times |St_2| \times \ldots \times |St_n|$. In Step 2, by transforming into a minimal observable IOSM, a further reduction in the size of state space may be possible.

"complete-testing assumption" [Luo 94] or "all weather conditions" [Milner 80] is the assumption that all possible paths of reachability graph can be traversed, i.e. with sufficiently many attempts all alternatives of nondeterministic behavior can be selected. With the assumption, observational equivalence of a system of CIOSM's and hence interoperability become properties that can be verified.

Proposition Let Σ_1 be an implementation of a specification Σ_2 of a system of CIOSM's, $t \in ((N \times N \times L_{\Sigma,in})/(N \times N \times L_{\Sigma,out})^*)^*$, $\underline{exec}(\Sigma_1,t)$ the resulting set of traces of executing a test case t sufficiently many times with respect to Σ_1 and $(v_1/\mathbf{w}_1.v_2/\mathbf{w}_2.\ldots.v_q/\mathbf{w}_q)^{in} = v_1v_2\ldots v_q$. Then

$$\underline{interop}_{\Sigma 2}(\Sigma_1) \text{ if and only if } \forall t \in \Pi_{\Sigma 2}: \underline{exec}(\Sigma_1,t) = \Sigma_2(t^{in})$$

Thus the above algorithm generates an interoperabililty test suite that guarantees the interoperation of the system implementation with respect to the system specification but passing interoperabilty test alone does not guarantee that individual components of the system have been correctly implemented.

We illustrate the algorithm with the system of CIOSM's in Figure 2. To keep the example to a manageable size, we let m = k in this example.

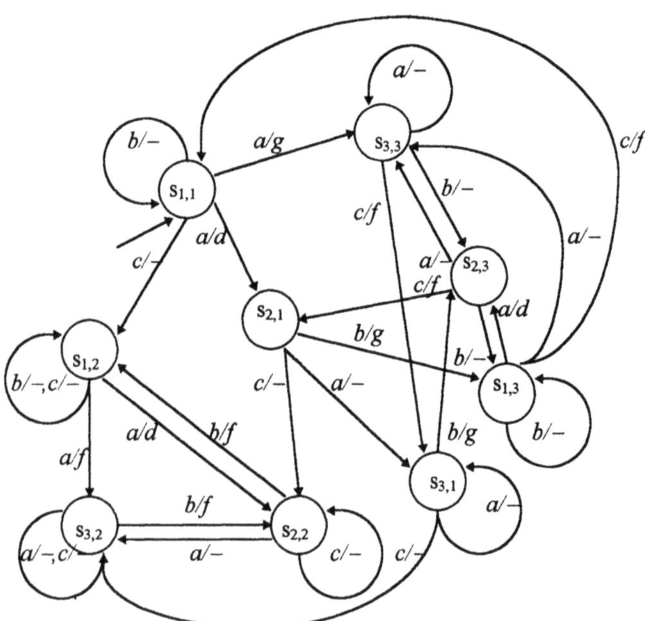

Figure 3 System state space as an observable IOSM.

	Conventional Method		Adaptive Method	
	P_i	W_i	T_{Pi}	T_{Wi}
$s_{1,1}$	ε	**{a/g}**	{ε}	**{{a/d.b/g, a/g}}**
$s_{1,2}$	c/–	**{a/d, a/f}**	{c/–}	**{{a/d.b/f, a/f}}**
$s_{1,3}$	a/d.b/g	**{c/f.a/g}**	{a/d.b/g}	**{{c/f.{a/d.b/g, a/g}}}**
$s_{2,1}$	a/d	**{b/g.c/f.a/g}**	{a/d}	**{{b/g.c/f.{a/d.b/g, a/g}}}**
$s_{2,2}$	c/–.a/d	**{b/f.a/d}**	{c/–.a/d}	**{{b/f.{a/d,a/f}}}**
$s_{2,3}$	a/g.b/–	**{b/–.c/f.a/g}**	{a/g.b/–}	**{{b/–.c/f. {a/d.b/g, a/g}}}**
$s_{3,1}$	**a/d.a/–**	{b/g.c/f.b/g}	**{a/d.a/–, a/g.c/f}**	{{b/g.c/f.b/g}}
$s_{3,2}$	c/–.a/f	{b/f.a/–}	{c/–.a/f }	{{b/f.a/–}}
$s_{3,3}$	a/g	{b/–.c/f.b/g}	{a/g}	{{b/–.c/f.b/g}}

Table 2 Path trees and state identification trees for M_1 and M_2 in Figure **2.**

By Step 1, a global state space graph with a large number of nodes is constructed. In Step 2, by transforming into a minimal observable IOSM and applying the slow environment assumption, the number of states is reduced to 9. Figure 3 is the result of Step 2. Step 3 is where we apply the adaptive method of Section 3 and the results are in the rightmost two columns of Table 2.

In Table 2, we emphasized the differences between the conventional method and the adaptive method by bold face letters. To build the particular table, we observe that the path to $s_{3,1}$ is the only reasonable case for adaptive tour to a state and $s_{1,1}$ and $s_{1,2}$ are definitely the cases for adaptive state verifications. Based upon adaptive state verifications of the two states, verification of many other states can be naturally done in an adaptive way. However, trying to do so for $s_{3,1}$, $s_{3,2}$ or $s_{3,3}$ would result in too lengthy test cases and was avoided here.

By Step 4, state covers and transition covers are constructed. The resulting state covers are shown in Appendix A and contain 10 test cases for the conventional method and 9 for the adaptive method. The transition covers are shown in Appendix B and contain 34 and 30 test cases, thus together with state covers totaling 44 and 39 test cases, respectively. The result of Step 5 is shown in Appendix B with bold face letters. In the final compressed test suite obtained with the adaptive method there are 24 test cases, which is fewer by 3. For the particular example, all the test sequences in the state cover happened to be subsequences of those in the transition covers.

Appendix C contains examples of applying ξ_w to both approaches together with testing costs for all groups of test cases measured by it. Our adaptive method requires significantly less effort (155) compared with the conventional approach (229).

5. CONCLUSION

In this paper, we developed a test sequence generation method which can be applied to IOSM's with observable nondeterminism and applied it to interoperability test generation. Since systems of CIOSM's almost unavoidably show nondeterministic behavior, any useful interoperability test generation approach should cope with it. Since non-observable IOSM's can be transformed into equivalent observable machines, our method works for any systems of CIOSM's as long as the assumption of no livelocks is met.

Also in this paper, we defined a measure of worst case test execution cost and used it to compare the results of applying our approach and the conventional approach to a non-trivial example. For that example, the comparison shows 11% improvement in the number of test cases and 32% improvement of the testing cost in terms of ξ_w. Since an adaptive approach is particularly suitable for dealing with nondeterminism, we believe the improvement will be greater for systems with highly nondeterministic behavior.

In comparison with the conventional approaches, our method tends to require more effort at the test generation step. For the overall cost of testing, however, test generation cost is much less important than the cost of actual testing itself. For once a test suite is generated, it usually is used over and over again.

In relation to interoperabiltiy test generation for adaptive testing, the following are among the further research problems:

(i) Other than the slow environment assumption, what assumption can contribute to reducing the size of global system space without decreasing (or significantly decreasing) controllability and observability which are crucial to test result analysis?

(ii) Random state space analysis to validation and verification is a particularly efficient method for realistic protocols. Since nondeterministic behavior can be considered "random" in some sense, how could the adaptive method be combined with random state space analysis approach to yield a good interoperability test generation method?

REFERENCES

[Arakawa 92] Arakawa, N. and Soneoka, T., "A Test Case Generation Method for Concurrent Programs", *Protocol Test Systems, IV*, J. Kroon, R. J. Heijink and E. Brinksma (Eds.), Elsevier Science Publishers B. V. (North-Holland), 1992.

[Bonnes 90] Bonnes, G., "IBM OSI Interoperability Verification Services", The 3rd IWPTS, 1990.

[Castanet 94] Castanet, R. and Kone, O., "Deriving Coordinated Testers for Interoperability", *Protocol Test Systems , VI(C-19)*, O. Rafiq(Ed.), Elsevier Science B. V. (North-Holland), IFIP, 1994.

[Chan 89] Chan, W. Y. L., Vuong, S. T. and Ito, M. R., "An Improved Protocol Test Generation Procedure Based on UIOs", *SIGCOMM '89 Symposium: Communication Architecture and Protocols in Computer Comm. Review 19(4)*, , pp. 283-294, September 1989.

[Chow 78] Chow, T. S., "Testing Software Design Modeled by Finite-State Machines". *IEEE Trans. on S.E., Vol. SE-4, No. 3*, May 1978.

[Kajiwara 94] Kajiwara, M., Ichikawa, H., Itoh, M. and Yoshida, Y., "Specification and Verification of Switching Software", *IEEE Transactions on Communications, Vol. Com-33, No. 3*, March 1985.

[Kim 94] Kim, B. and Chun, W., "Generating Test Cases for Nondeterministic FSMs Using UIOTrees and PUIOtrees", Technical Report, Comp. Eng., Chungnam National University, 1994.

[Luo 94] Luo, G., Bochmann G. and Petrenko, A., "Test Selection Based on Communicating Nondeterministic Finite-State Machines Using a Generalized Wp-Method", *IEEE Transactions on S.E., Vol 20, No. 2*, February 1994.

[Milner 80] Milner, R., "A Calculus of Communicating Systems," LNCS, Vol. 92, 1980.

[Phalippou 91] Phalippou, M., "The Limited Power of Testing", Proceedings of the 4th International Workshop on Protocol Test Systems, The Hague, October 1991.

[Sabnani 88] Sabnani, K. and Dahbura, A., "A Protocol Test Generation Procedure", Computer Networks and ISDN Systems, Vol. 15, pp. 285-297, 1988.

[Sidhu 89] Sidhu, D. P. and Leung, T. K., "Formal Methods for Protocol Testing: A Detailed Study", *IEEE Trans. on S.E., Vol. 15, No. 4*, April 1989.

[Vermeer 94] Vermeer, G.S. and Blik, H., "Interoperabilty Testing: Basis for the Acceptance of Communication Systems", *Protocol Test Systems, VI(C-19)*, Elsevier Science Publishers B. V. (North-Holland), 1994.

Appendix A. Comparison of state covers

	Conventional Method (10 test cases)	Adaptive Method (9 test cases)
$s_{1,1}$	*a/g*	*{ε}.{a/d.b/g, a/g}*
$s_{1,2}$	*c/–. a/d*	*{c/–}.{a/d.b/f, a/f}*
	c/–. a/f	
$s_{1,3}$	*a/d.b/g. c/f.a/g*	*{a/d.b/g}.{c/f.{a/d.b/g, a/g}}}*
$s_{2,1}$	*a/d. b/g.c/f.a/g*	*{a/d}.{b/g.c/f. {a/d.b/g, a/g}}*
$s_{2,2}$	*c/–.a/d. b/f.a/d*	*{c/–.a/d}.{b/f.{a/d, a/f}}}*
$s_{2,3}$	*a/g.b/–. b/–.c/f.a/g*	*{a/g.b/–}.{b/–.c/f.{a/d.b/g, a/g}}*
$s_{3,1}$	*a/d.a/–. b/g.c/f.b/g*	*{a/d.a/–, a/g.c/f}.{b/g.c/f.b/g}*
$s_{3,2}$	*c/–.a/f. b/f.a/–*	*{c/–.a/f}.{b/f.a/–}*
$s_{3,3}$	*a/g. b/–.c/f.b/g*	*{a/g}.{b/–.c/f.b/g}*

Appendix B. Transition covers and compressed test suites

	Conventional Method (34 (**27**) test cases)	Adaptive Method (30 (**24**) test cases)
$s_{1,1}$	*a/d. b/g.c/f.a/g*	*{ε}. a/d. {b/g.c/f.{a/d.b/g, a/g}}*
	a/g. b/–.c/f.b/g	*{ε}. a/g. {b/–.c/f.b/g}*
	b/–. a/g	*{ε}.* ***b/–. {a/d.b/g, a/g}***
	c/–. a/d	
	c/–. a/f	*{ε}. c/–. {a/d.b/f, a/f}*
$s_{1,2}$	*c/–. a/d. b/f.a/d*	*{c/–}. a/d. {b/f.{a/d, a/f}}*
	c/–. a/f. b/f.a/–	***{c/–}. a/f. {b/f.a/–}***
	c/–. b/–. a/d	
	c/–. b/–. a/f	***{c/–}. b/–. {a/d.b/f, a/f}***
	c/–. c/–. a/d	
	c/–. c/–. a/f	***{c/–}. c/–. {a/d.b/f, a/f}***
$s_{1,3}$	***a/d.b/g. a/d. b/–.c/f.a/g***	***{a/d.b/g}. a/d. {b/–.c/f.{a/d.b/g, a/g}}***
	a/d.b/g. a/–. b/–.c/f.b/g	***{a/d.b/g}. a/–. {b/–.c/f.b/g}***
	a/d.b/g. b/–. c/f.a/g	***{a/d.b/g}. b/–. {c/f.{a/d.b/g, a/g}}***
	a/d.b/g. c/f. a/g	*{a/d.b/g}. c/f. {a/d.b/g, a/g}*
$s_{2,1}$	***a/d. a/–. b/g.c/f.b/g***	***{a/d}. a/–. {b/g.c/f.b/g}***
	a/d. b/g. c/f.a/g	***{a/d}. b/g. {c/f.{a/d.b/g, a/g}}***
	a/d. c/–. b/f.a/d	***{a/d}. c/–. {b/f.{a/d, b/f.a/f}}***
$s_{2,2}$	***c/–.a/d. a/–. b/f.a/–***	***{c/–.a/d}. a/–. {b/f.a/–}***
	c/–.a/d. b/f. a/d	
	c/–.a/d. b/f. a/f	***{c/–.a/d}. b/f. {a/d.b/f, a/f}***
	c/–.a/d. c/–. b/f.a/d	***{c/–.a/d}. c/–. {b/f.a/d, b/f.a/f}***
$s_{2,3}$	***a/g.b/–. a/–. b/–.c/f.b/g***	***{a/g.b/–}. a/–. {b/–.c/f.b/g}***
	a/g.b/–. b/–. c/f.a/g	*{a/g.b/–}. b/–. {c/f.{a/d.b/g, a/g}}*
	a/g.b/–. c/f. b/g.c/f.a/g	***{a/g.b/–}. c/f. {b/g.c/f.{a/d.b/g, a/g}}***
$s_{3,1}$	***a/d.a/–. a/–. b/g.c/f.b/g***	***{a/d.a/–, a/g.c/f}. a/–. {b/g.c/f.b/g}***
	a/d.a/–.b/g. b/–.c/f.a/g	***{a/d.a/–, a/g.c/f}. b/g. {b/–.c/f. {a/d.b/g, a/g}}***
	a/d.a/–. c/–. b/f.a/–	***{a/d.a/–, a/g.c/f}. c/–. {b/f.a/–}***

$s_{3,2}$	*c/–.a/f. a/–. b/f.a/–* *c/–.a/f. b/f. b/f.a/d* *c/–.a/f. c/–. b/f.a/–*	*{c/–.a/f}. a/–. {b/f.a/–}* *{c/–.a/f}. b/f. {b/f.{a/d, a/f}}* *{c/–.a/f}. c/–. {b/f.a/–}*
$s_{3,3}$	*a/g. a/–. b/–.c/f.b/g* *a/g. b/–. b/–.c/f.a/g* *a/g. c/f. b/g.c/f.b/g*	*{a/g}. a/–. {b/–.c/f.b/g}* *{a/g}. b/–. {b/–.c/f.{a/d.b/g, a/g}}* *{a/g}. c/f. {b/g.c/f.b/g}*

Appendix C. Comparison of testing costs in terms of ξ_w

	Conventional Method (ξ_w = 229)	Adaptive Method (ξ_w = 155)
$s_{1,1}$	4	3
$s_{1,2}$	30	14
$s_{1,3}$	37	28
$s_{2,1}$	24	17
$s_{2,2}$	39	20
$s_{2,3}$	20	15
$s_{3,1}$	*a/g* *a/d.a/–. a/–. b/g.c/f.b/g* *a/g* *a/d.a/–.b/g. b/–.c/f.a/d* *a/d.a/–.b/g. b/–.c/f.a/g* *a/g* *a/d.a/–. c/–. b/f.a/–* 26	*a/g.c/f. a/–. b/g.c/f.b/g* *a/g.c/f. b/g. b/–.c/f.a/d.b/g* *a/d.a/–. c/–. b/f.a/–* 18
$s_{3,2}$	26	21
$s_{3,3}$	23	19

13

Fault-tolerant UIO Sequences in Finite State Machines

Kshirasagar Naik

Computer Networks Laboratory, Department of Computer Software, University of Aizu, Aizu-Wakamatsu City, Fukushima, 965-80 JAPAN

Abstract

The central idea in the paper is to define the concept of *strength* of a UIO sequence in verifying a state in a faulty implementation. The strength of a UIO sequence is a quantitative measure of the number of output faults required to render the verification power of the sequence ineffective. Thus, UIO sequences of higher strengths can tolerate more output faults while verifying a state. We present an algorithm to compute all UIO sequences of maximal strength in a state machine.

Key Words: FSM, Protocol Testing, State Verification, UIO Sequence, Fault Detection, Strength of a UIO Sequence, Fault-tolerant UIO Sequence

1 Introduction

Testing of communication protocols remains to be a subject of continued research. The four test sequence generation methods are the transition tour, the D-, W-, and U-method [SL 89]. The last three are called *formal* methods, because they verify the next state of a transition. State verification is the main argument in favor of using formal methods over the transition tour method. However, a verification sequence fails to verify the next state of a transition if the verification sequence *loses its strength* to verify the state due to faults in the implementation. By *losing strength*, we mean the verification sequence in the specification is not a verification sequence in the implementation. Chan, *et. al.* showed how a UIO sequence in a specification no more remains a UIO sequence in a faulty implementation [CVI 89].

Once a UIO sequence loses its property to verify a state, there is the danger of faults being undetected. However, in a faulty implementation, there is every likelihood that some UIO sequences lose their property. Given an FSM, UIO sequences can be considered as static properties of the FSM. Or, we can say that the verification power of UIO sequences are static properties of the FSM. In our opinion, on the other hand, faults in an implementation are dynamic "properties" of the implementation. These are dynamic in the sense that one specification FSM can have multiple faulty implementations with one implementation being different from another. While using UIO sequences for state verification, we are faced with the problem of coping with dynamic properties of implementations using the static properties of a specification FSM. In this paper, we will find answers to the following questions on UIO sequences.

(i) Assuming that an implementation could be arbitrarily faulty, what is the measure of *goodness* of a UIO sequence?

(ii) In the presence of faults, how good are UIO sequences of minimal lengths?

(iii) Can non-minimal UIO sequences with additional properties verify states of a faulty implementation?

(iv) Are two UIO sequences of identical length for the same state equally good?

(v) How to compute the upper bound on the measure of goodness of UIO sequences in a given machine?

The above questions lead to a notion of *fault-tolerance* capability of UIO sequences. That is, to what extent can a UIO sequence tolerate implementation faults and retain its state verification property? We begin our study with a review of why a UIO sequence fails to verify a state in an implementation. Such a review motivates us to define the concept of *strength* of a UIO sequence. The strength of a UIO sequence is a quantitative measure of its fault-tolerant capability.

In Section 2, we state some graph theoretic terms, a basic test sequence, and a fault model. In Section 3, we review how state faults can be masked under various conditions. In Section 4, we present the idea of fault-tolerant UIO sequences and present an algorithm to generate all UIO sequences of maximal strengths. Some concluding remarks are given in Section 5.

2 Protocol Specification and Test Sequence

We organize this section in three parts. In the first part, we present a few FSM and graph-theoretic terms. The general structure of a test sequence is outlined in the second part. Assumptions about an implementation and a fault model are stated in the third part.

2.1 FSM and Graph-theoretic Terms

A finite-state machine M is a tuple: $M = (S, I, O, \delta, \lambda)$, where $S = \{s_1, \ldots, s_n\}$ is a finite set of states; $I = \{a_1, \ldots, a_p\}$ is a finite set of inputs; $O = \{b_1, \ldots, b_r\}$ is a finite set of outputs; $\delta : S \times I \rightarrow S$ is the state transition function; and $\lambda : S \times I \rightarrow O$ is the output function.

For convenience we extend the domain of λ and δ from an input symbol to a string of symbols. For instance, for state s_i and input sequence $x = a_1, \ldots, a_k$, the corresponding output sequence is denoted by $\lambda(s_i, x)$ and the final state is denoted by $\delta(s_i, x)$. Also, we extend the transition and output functions from single states to sets of states. That is, if Q is a set of states and x is an input sequence, then $\delta(Q, x) = \{\delta(s, x) | s \in Q\}$, and $\lambda(Q, x) = \{\lambda(s, x) | s \in Q\}$.

An FSM M is viewed as a directed graph $G = (V, E)$, where the set of vertices $V = \{v_1, \ldots, v_n\}$ represents the set S of states of M, and a set of edges $E = \{(v_j, v_k; a/b) | v_j, v_k \in V\}$ represents all transitions in M. Each edge $e_{jk} = (v_j, v_k; a/b) \in E$ represents a state transition from s_j to s_k with input a and output b. For the sake of brevity, when λ and δ are known, we also denote a transition by a pair (v_j, a). Notation $x@y$ denotes the *concatenation* of strings x and y.

We denote an implementation of M by $M' = (S, I, O', \delta', \lambda')$. Though we do not know the next state function δ' and output function λ' of the implementation, we use them only for explanatory purpose and not to compute any state or output.

2.2 Basic Test Sequence

In FSM-based testing, the basic strategy is to apply the following three steps for all states and inputs. First, move the implementation to some state s_i and apply an input a. Second, verify that the implementation produces the desired output $\lambda(s_i, a)$. Third, verify that the implementation has moved to state $\delta(s_i, a)$ by applying a state verification sequence to the implementation and observing the corresponding response.

Let $TX(s_j, s_i)$ denote an input sequence that transfers machine M from state s_j to s_i. Let $VS(s_k)$ denote an input sequence that verifies that M is in state s_k. We leave out the output parts from these sequences, because they can be readily computed from the specification machine. Let $\Omega(s_k, VS(s_k))$ denote the set of transitions executed by M while $VS(s_k)$ is applied in state s_k.

Assume that implementation M' is in a known state s_j and we want to test the output and next state functions associated with transition (s_i, a). Thus, a basic test sequence for state s_i and input a is defined as an input sequence:
$BTS(s_i, a) = TX(s_j, s_i)@a@VS(s_k)$, where $s_k = \delta(s_i, a)$.

The three commonly used state verification sequences are distinguishing sequence [KOHA 78], Unique Input/Output (UIO) sequence [SD 88], and pair-wise distinguishing sequence (W-sequence) [CHO 78], [FBK 91]. The shortest path consisting of verified transitions from s_j to s_i or a *reset* input is generally selected to be a transfer sequence $TX(s_j, s_i)$.

The subsequence $a@VS(s_k)$ of $BTS(s_i, a)$ is largely known as a *test segment.* Thus, a complete test sequence for a machine is an appropriate concatenation of a collection of test segments and transfer sequences. For the purpose of minimizing the length of the test sequence, various kinds of optimization techniques involving *overlapping* of test segments can be used [ADLU 91], [MP 93], [UZ 93].

2.3 Assumptions and Fault Model

We assume that a protocol behavior is modeled as a completely specified, deterministic, reduced, and strongly connected finite state machine. We make the following assumptions about an implementation.

(i) An implementation has the same number of states as the specification.

(ii) Faults in an implementation are restricted to output (λ) and state (δ) functions.

(iii) All the faults in an implementation are *permanent.*

3 Fault Masking

Assume that an implementation is in state s_i and we want to test the output and next state functions for input a in state s_i. In the UIO method, we need a basic test sequence of the following form:
$BTS(s_i, a) = a@I'(s_k)$,
where $s_k = \delta(s_i, a)$ and $I'/\lambda(s_k, I')$ is a UIO sequence for state s_k.

An output fault can be readily detected if the observed output $\lambda'(s_i, a)$ is not equal to $\lambda(s_i, a)$. However, detecting a state fault is a non-trivial task in spite of using a state verification sequence. This is due to the fact that one fault can mask another [CVI 89], [LS 92], [MP 94]. In the following, we illustrate three kinds of masking:

K1: Output faults in some transitions can mask a state fault in some other transition;

K2: State faults in some transitions can mask a state fault in some other transition.

K3: A state fault masks itself depending on the structure of the FSM.

A. K1 and K2 Type of Fault Masking

Consider the following scenario. Let the implementation of $\delta(s_i, a)$ be faulty, that is, $\delta(s_i, a) = s_k$, $\delta'(s_i, a) = s_j$, and $s_j \neq s_k$. Since we are interested in verifying state s_k, let $I'/\lambda(s_k, I')$ be a UIO sequence for state s_k. Now we have the following four cases.

Case 1: Output functions of all transitions in $\Omega(s_j, I'(s_j))$ are correctly implemented. Thus, $\lambda'(s_j, I') = \lambda(s_j, I')$.

Case 2: Some output functions of transitions in $\Omega(s_j, I'(s_j))$ are faulty such that $\lambda'(s_j, I') = \lambda(s_k, I')$.

Case 3: All state transitions in $\Omega(s_j, I'(s_j))$ are correctly implemented.

Case 4: Some state transitions in $\Omega(s_j, I'(s_j))$ are faulty, but the associated outputs are correctly implemented.

In *Case 1*, $\lambda(s_k, I') \neq \lambda'(s_j, I')$, because $I'/\lambda(s_k, I')$ is a UIO sequence for s_k. Since the observed sequence $\lambda'(s_j, I')$ is different from the expected sequence $\lambda(s_k, I')$, we can deduce that implementation of $\delta(s_i, a)$ is faulty.

However, in *Case 2*, faults in the implementation of output functions of some transitions in $\Omega(s_j, I'(s_j))$ produce an expected output sequence, thereby effectively masking the state fault $\delta'(s_i, a) = s_j$. Thus, *Case 2* results in a $K1$ type of fault masking.

The implication of *Case 3* is similar to that of *Case 1* or *Case 2* depending on the correctness of implementation of the output functions.

Case 4 leads to a $K2$ type of fault masking, because the correct output of the faulty transitions in $\Omega(s_j, I'(s_j))$ produce the expected output, thereby effectively masking the fault under consideration.

Example 1: In Fig. 1, we illustrate the $K1$ and $K2$ types of fault masking. Fig. 1(a) shows a part of a specification. We assume that abc/xyz_1 and abc/xyz_2 are the UIO sequences of states $s2$ and $s3$, respectively. Figs. 1(b) and (c) show two kinds of erroneous implementations. We focus on the implementation of the edge $(s1, s2; d/w)$. In both the implementations, edge $(s1, s2; d/w)$ has been implemented as $(s1, s3; d/w)$. We denote the faulty implementation of this transition by fault

$f1$. Additionally, edge $(s7, s9; c/z2)$ has been implemented as $(s7, s9; c/z1)$ with an output fault in Fig. 1(b). This output fault is denoted as $f2$. Edge $(s5, s7; b/y)$ has been implemented as $(s5, s6; b/y)$ with a state fault in Fig. 1(b). This output fault is denoted as $f3$. It is easy to see that the state fault $f1$ has been masked by the output fault $f2$ and state fault $f3$ in Fig. 1(b) and Fig. 1(c), respectively.

B. K3 Type of Fault

In the $K3$ type of fault [CVI 89], a state fault can mask itself depending on the structure of the machine. Let the transition $(s_i, s_j; a/b)$ be implemented as $(s_i, s_i; a/b)$. If there exists an input sequence I' such that

(i) $\lambda(\delta(s_j, a), I') = \lambda(s_i, I')$ and

(ii) $a@I'/\lambda(s_j, a@I')$ is a UIO sequence for s_j,

the state fault masks itself. Thus, given an FSM and a UIO sequence for each state, it is possible to check whether $K3$ types of faults in an implementation could go undetected. The simplest way to avoid the $K3$ type of fault masking is not to select a UIO sequence satisfying the above properties.

In the following, we discuss an approach to cope with the $K1$ type of fault masking by introducing the concept of *strength* of a UIO sequence.

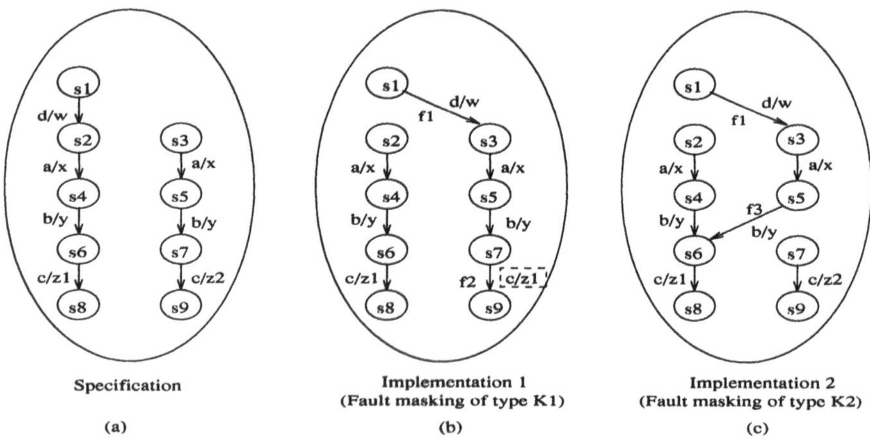

Figure 1: Illustration of Fault Masking.

4 Fault-tolerance Capability of a UIO Sequence

In this section, we define the *strength* of a UIO sequence and give an algorithm to compute all UIO sequences of maximal strength in an FSM.

4.1 Strength of a UIO Sequence

Definition: Given a state s in a machine M and input sequence I', the *transition trace* $TT(s, I')$ is defined as the sequence of transitions traversed in M as a result of applying I' at s.

Definition: Let s_i and s_j be two states and I' be an input sequence in a machine M. Let $O_i = \lambda(s_i, I')$ and $O_j = \lambda(s_j, I')$. The *difference* between the two output sequences O_i and O_j, denoted by $O_i \sim O_j$, is defined as a sequence of 0 and 1, such that the kth. symbol of $O_i \sim O_j$ is a 1 if and only if:

(i) the kth transitions in $TT(s_i, I')$ and $TT(s_j, I')$ have different outputs and

(ii) in case the kth. transition in $TT(s_i, I')$ is identical to any of its predecessors in $TT(s_i, I')$, its identical predecessor has not contributed a 1 to $O_i \sim O_j$, and

(iii) in case the kth. transition in $TT(s_j, I')$ is identical to any of its predecessors in $TT(s_j, I')$, its identical predecessor has not contributed a 1 to $O_i \sim O_j$.

The number of 1s in $O_i \sim O_j$ is called the *degree of difference* between the two output sequences and is denoted by $DoD(O_i, O_j)$.

Remark: In the above definition, the first condition for assigning a 1 to the kth. entry of the difference sequence is obvious. The last two conditions allow us to consider the output fault of a transition exactly once, even though the transition may appear many times in the transition trace.

Definition: Let $M = (S, I, O, \delta, \lambda)$ be an FSM. The *strength* of a UIO sequence $I'/\lambda(s_i, I')$ for state s_i is the *minimum* of $DoD(\lambda(s_i, I'), \lambda(s_j, I')), \forall s_j \in S$ and $s_j \neq s_i$. We denote the strength of a UIO sequence UIO_k for state s_i by $strength(s_i, UIO_k)$.

Remark: Obviously, the minimum strength of any UIO sequence for any state is one. In case a state has many UIO sequences, their strengths could be different.

The motivation for defining the strength of a UIO sequence is as follows. Consider transition $(s_i, s_j; a/b)$ which has been incorrectly implemented as $(s_i, s_k; a/b)$ with a state fault. Assume that $UIO_{s_j} = I'/\lambda(s_j, I')$ is used to verify state s_j. If $strength(s_j, UIO_{s_j}) = 1$, one output fault in the set of transitions $\Omega(s_k, I'(s_k))$ is necessary to mask the state fault. If $strength(s_j, UIO_{s_j}) = 2$, two output faults in the set of transitions $\Omega(s_k, I'(s_k))$ are necessary to mask the state fault. In general, if $strength(s_j, UIO_{s_j}) = m, m \geq 1$, $m-1$ output faults in $\Omega(s_k, I'(s_k))$ can mask the state fault. Thus, higher the strength of a UIO sequence, more output faults are necessary to mask a state fault. Theoretically, in order to detect the state fault, we need a UIO sequence whose strength is more than the number of output faults in an implementation.

Remark: All minimal length UIO sequences have a strength of one. Thus, a single output fault can render them ineffective.

Example 2: Consider machine $M1$ and its implementation in Figs. 2(a) and (b), respectively. Transition $(s_6, s_3; b/1)$ has been incorrectly implemented with a state fault as $(s_6, s_5; b/1)$ and transition $(s_1, s_3; a/1)$ has been incorrectly implemented with an output fault as $(s_1, s_3; a/0)$. We want to test the implementation of state transition $(s_6, s_3; b/1)$. We will consider two cases, such that the first case involves a UIO sequence with a strength of *one* and the second case involves a UIO sequence with a strength of *two*. In both the cases, we assume that the implementation is in state s_6.

Case 1: Consider the minimal length UIO sequence $aaa/000$ for state s_3.
$\lambda(s_1, aaa) = 100,\ \lambda(s_2, aaa) = 010,$
$\lambda(s_3, aaa) = 000,\ \lambda(s_4, aaa) = 110,$
$\lambda(s_5, aaa) = 001,\ \lambda(s_6, aaa) = 101.$
$DoD(\lambda(s_3, aaa) \sim \lambda(s_1, aaa)) = 1,$
$DoD(\lambda(s_3, aaa) \sim \lambda(s_2, aaa)) = 2,$
$DoD(\lambda(s_3, aaa) \sim \lambda(s_4, aaa)) = 2,$
$DoD(\lambda(s_3, aaa) \sim \lambda(s_5, aaa)) = 1,$
$DoD(\lambda(s_3, aaa) \sim \lambda(s_6, aaa)) = 2.$

Therefore, $strength(s_3, aaa/000) = 1$. A basic test sequence to test the implementation of $(s_6, s_3; b/1)$ is:
$BTS(s_3, b) = b@aaa,$
where input b is expected to take the implementation to state s_3 and the aaa sequence is expected to verify that the implementation indeed moves to state s_3. However, input b takes the implementation to state s_5 and the sequence aaa in state s_5 produces the "expected" output 000, because transition $(s_1, s_3; a/1)$ has been implemented as $(s_1, s_3; a/0)$ with an output fault. Thus, the basic test sequence fails to detect the state fault due to a $K1$ type of fault masking.

Case 2: Consider a non-minimal length UIO sequence $aaaba/00010$ for state s_3. From $M1$,
$\lambda(s_1, aaaba) = 10001,\ \lambda(s_2, aaaba) = 01001,$
$\lambda(s_3, aaaba) = 00010,\ \lambda(s_4, aaaba) = 11010,$
$\lambda(s_5, aaaba) = 00111,\ \lambda(s_6, aaaba) = 10111.$
$DoD(\lambda(s_3, aaaba) \sim \lambda(s_1, aaaba)) = 3,$
$DoD(\lambda(s_3, aaaba) \sim \lambda(s_2, aaaba)) = 3,$
$DoD(\lambda(s_3, aaaba) \sim \lambda(s_4, aaaba)) = 2,$
$DoD(\lambda(s_3, aaaba) \sim \lambda(s_5, aaaba)) = 2,$
$DoD(\lambda(s_3, aaaba) \sim \lambda(s_6, aaaba)) = 3.$

Therefore, $strength(s_3, aaaba/00010) = 2$. A basic test sequence to test the implementation of $(s_6, s_3; b/1)$ is:
$BTS(s_3, b) = b@aaaba,$
where the first input b is expected to take the implementation to state s_3 and the $aaaba$ sequence is expected to verify that the implementation indeed moves to state s_3. However, input b takes the implementation to state s_5 and the sequence $aaaba$ in

state s_5 produces the output 00011. Since the output of the implementation 00011 differs from the expected output $\lambda(s_3, aaaba) = 00010$, one can conclude that there is a state fault in the implementation of transition $(s_6, s_3; b/1)$. That is, the basic test sequence using a UIO sequence of strength *two* can tolerate *one* output fault in the implementation of any transition.

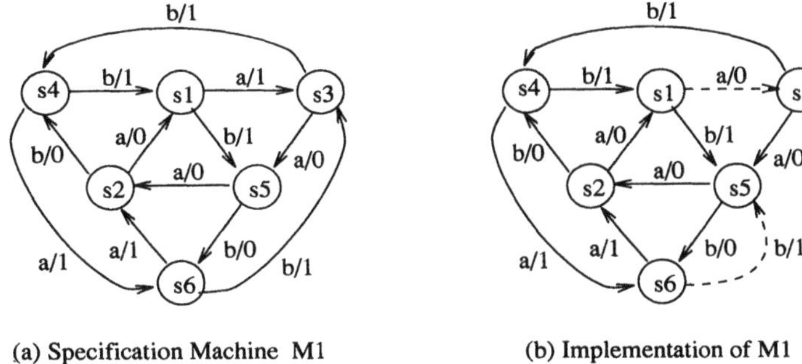

(a) Specification Machine M1 (b) Implementation of M1

Figure 2: A specification [UZ 93] and a faulty implementation.

4.2 Generation of UIO Sequences of Maximal Strength

One can obtain UIO sequences of minimal lengths using the algorithm in [SD 88]. Given a UIO sequence for a state, one can always obtain a UIO sequence of longer length for the same state and check its strength. In order to generate UIO sequence of higher strength, the UIO sequence needs to be made longer. However, it is difficult to define a termination condition for this approach.

Therefore, we will take a different approach to obtaining UIO sequences of higher strength. We will modify another algorithm to compute UIO sequences [NAIK 94] for this purpose. We first introduce a *path vector* and *perturbation* of a path vector. Based on the perturbation idea, next we define a *UIO-tree*. We incorporate three termination conditions into the perturbation process so that the algorithm terminates after generating all UIO sequences of maximal strengths.

Definition: Given an FSM M, a *path vector* (PV) is a collection of state pairs, $(v_1/v_1', \ldots, v_i/v_i', \ldots, v_k/v_k')$, with the following properties:
(1) v_i and v_i' denote the head and tail state, respectively, of a path, where a path is a sequence of state transitions and
(2) an identical sequence of input/output is associated with all the paths in the path vector.

Given a path vector $PV = (v_1/v_1', \ldots, v_i/v_i', \ldots, v_k/v_k')$, the *initial vector (IV)* is the collection of head states of PV, that is $IV(PV) = (v_1, \ldots, v_i, \ldots, v_k)$. Similarly, the *current vector (CV)* is the collection of tail states of PV, that is $CV(PV) = (v_1', \ldots, v_i', \ldots, v_k')$.

A path vector is said to be a *singleton* vector if it contains exactly one state pair. A path vector $PV = (v_1/v'_1, \ldots, v_i/v'_i, \ldots, v_k/v'_k)$ is said be a *homogeneous* vector if all members of $CV(PV)$ are identical. It may be noted that a singleton vector is also a homogeneous vector.

For an n-state machine, we define a unique *initial* path vector: $(s_1/s_1, \ldots, s_i/s_i, \ldots, s_n/s_n)$ such that a *null* path is associated with all state pairs.

Now we introduce the idea of *vector perturbation.* That is, given a path vector PV and an edge label a/b, how to compute a new vector PV' from PV and a/b.

Definition: Given a path vector $PV = (v_1/v'_1, \ldots, v_i/v'_i, \ldots, v_k/v'_k)$ and an edge label a/b, *perturbation* of PV with respect to edge label a/b, denoted by $PV' = pert(PV, a/b)$ is defined as:
$PV' = \{v_i/v''_i | v''_i = \delta(v'_i, a) \wedge \lambda(v'_i, a) = b \wedge v_i/v'_i \in PV\}$.

Given a reduced machine and its initial path vector, we can infinitely perturb all the path vectors for all edge labels. One can imagine the perturbation function $PV' = pert(PV, a/b)$ as an arc from a node PV to a new node PV' with edge label a/b. Also, given a PV and a set of edge labels L, we can arrange the new $|L|$ nodes $pert(PV, a/b)$, $\forall a/b \in L$ on one level. That is, all the path vectors of a given machine can be arranged in the form of a tree with successive levels $1, 2, \ldots\ldots\infty$. Such a tree is called a *UIO-tree.*

Note: In the graphical representation of a path vector $PV = (v_1/v'_1, \ldots, v_i/v'_i, \ldots, v_k/v'_k)$ in a UIO-tree, we represent PV in two rows of states. The top row denotes $IV(PV)$ and the bottom row denotes $CV(PV)$.

Theoretically, a UIO-tree is a tree with infinite levels. However, we need to prune the tree based on some conditions, called *pruning* conditions. After each perturbation $PV' = pert(PV, a/b)$, we check the following pruning conditions:

C1: $CV(PV')$ is a homogeneous vector containing more than one element.

C2: On the path from the initial node to PV, there exists PV'' such that $PV' \subseteq PV''$.

C3: On the path from the initial node to PV, there exists a singleton node and a node PV'' such that $PV' \subseteq PV''$. (Note: the singleton node and PV'' can be the same node.)

If one of the pruning conditions is satisfied, we declare PV' to be a *terminal* node. Now we explain why $C1$, $C2$ and $C3$ are the required termination conditions. Let us consider condition $C1$. If $CV(PV')$ is a homogeneous vector containing more than one element, any further perturbation of the path from the initial node to PV' will not lead to any UIO sequence. Condition $C2$ states that there is no need of perturbing a node if the node or a superset of the node appears before. This is because, if the node or its superset appears before, then it has already been perturbed in all possible ways and perturbing it once again will not lead to new information. Condition $C3$ deals with UIO sequences. If a UIO-tree contains a singleton node, then the sequence of input/output associated with the path from the initial node to the singleton node is a UIO sequence for the *head* state of the node. If we are simply interested in UIO sequences of minimal lengths, we can stop perturbing a singleton node [NAIK 94]. However, we are interested in UIO sequences of maximal strengths. Therefore, we further perturb a singleton node until we know that further perturbation will not add extra information to the UIO sequence. This happens when the singleton node appears before either as itself or as a subset of a larger node. In the rest of the paper, we use the terms *path vector* and *node* in a UIO-tree interchangeably. In the following, we present an algorithm to compute maximal strength UIO sequences.

Algorithm 1: Generating UIO Sequences of Maximal Strength.
Input: $M = (S, I, O, \delta, \lambda)$ and L.
Output: UIO sequences of maximal strengths.
Method: Execute the following steps.
Step 1: Let Ψ be the set of nodes denoting path vectors in the UIO-tree. Initially, Ψ contains the initial vector marked as *non-terminal.*
Step 2: Find a non-terminal member $\psi \in \Psi$ which has not been perturbed. If no such member exists, then go to **Step 5**.
Step 3: Compute $\psi' = pert(\psi, a_i/b_i)$ and add ψ' to Ψ, $\forall a_i/b_i \in L$. Mark ψ to be perturbed. Update the UIO-tree.
Step 4: If $pert(\psi, a_i/b_i)$, computed in Step 3, satisfies condition $C1$, $C2$, or $C3$, then mark $pert(\psi, a_i/b_i)$ as a *terminal* node and go to Step 2.
Step 5: Collect all the UIO sequences for every state and compute their strengths.
Step 6: If a state has at least one UIO sequence, select a UIO sequence of maximal strength. After selecting UIO sequences of maximal strength, terminate the algorithm. **End of Algorithm 1**

Theorem: Given a machine $G = (V, E)$, a state v has a UIO sequence iff the UIO-tree has a singleton node ψ such that $v = IV(\psi)$.

A proof of the above theorem can be found in [NAIK 94]. Obviously, the algorithm has an exponential complexity. Termination of the algorithm is guaranteed by the three conditions $C1$, $C2$, and $C3$.

Example 3: We applied Algorithm 1 to the specification machine of Fig. 2 to compute the UIO-tree. A part of the UIO-tree is shown in Fig. 3. Node s_3/s_1 is a singleton node indicating that the sequence of input/output $aaa/000$ leading from the initial node to s_3/s_1 is a UIO sequence of minimal length for state s_3. However, to obtain UIO sequences of higher strength, we further perturb node s_3/s_1 as shown in Fig. 3 to obtain nodes s_3/s_3 and s_3/s_5. Node s_3/s_3 is a terminal node because condition $C3$ is satisfied. Node s_3/s_5 is further perturbed to obtain nodes s_3/s_2 and s_3/s_6. Node s_3/s_2 is a terminal node because condition $C3$ is satisfied. We continue perturbing node s_3/s_6 to obtain nodes s_3/s_2 and node s_3/s_3, which are both terminal nodes. For machine $M1$, the maximal strength of any UIO sequence for any state is two.

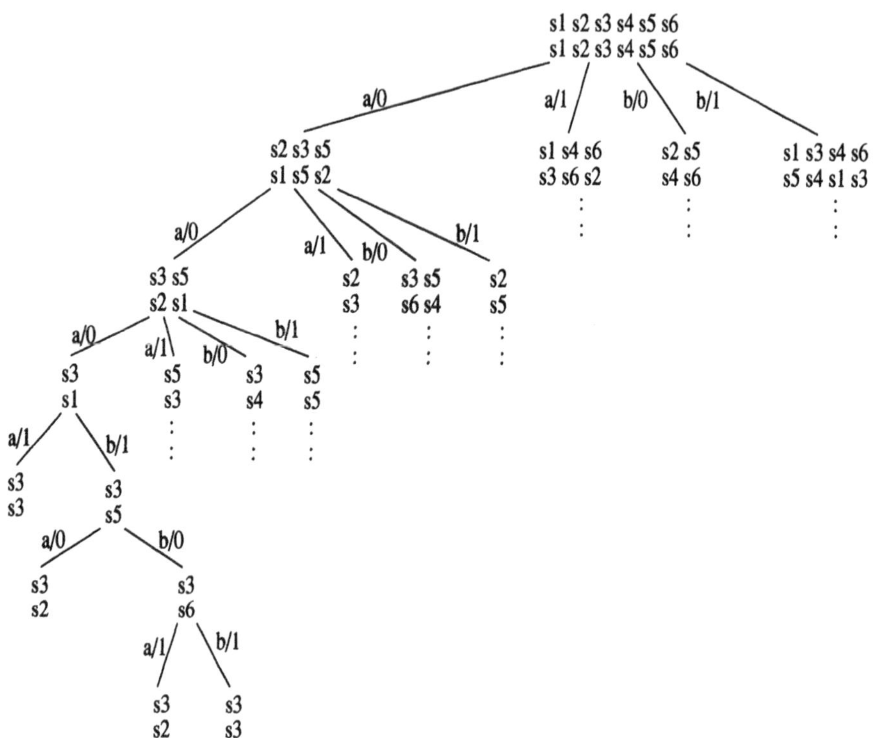

Figure 3: A Partial UIO-tree for specification machine M1.

5 Concluding Remarks

In this paper, we discussed how output and state faults can mask some other state faults by rendering the state verification power of a UIO sequence ineffective. The idea of strength of a UIO sequence was useful in defining the fault-tolerance capability of UIO sequences. UIO sequences of higher strengths are generally longer than the minimal length sequences. Since a minimal length UIO sequence has a strength of one, a single output fault can easily render the UIO sequence useless! Since fault detection, rather than short execution time, is presumed to be the main objective of testing, using the minimal length UIO sequences does not seem to be a good idea. UIO sequences of maximal strengths, though longer than the minimal length UIO sequences, are better than the minimal length UIO sequences.

We believe that our work will lead to new research on test generation involving minimization of test sequences with maximal fault detection for a given FSM specification. Presently, the idea of overlapping of test segments is widely used in minimizing a test sequence. However, our work reveals the requirement for a new way of scheduling test segments.

As an extension of our work, one can apply the fault-tolerance idea to other state verification techniques. In our opinion, this work will provide a quantitative and formal basis for comparing the fault detection capabilities of test sequences generated using the D-, W-, and U-methods. More research is required to be done to generate UIO sequences that can tolerate both output and state faults.

References

[KOHA 78] Z. Kohavi, *Switching and Finite Automata Theory*, Second Ed. New York: McGraw-Hill, 1978.

[CHO 78] T. S. Chow, "Test Design Modeled by Finite State Machine," IEEE Trans. on Software Eng., Vol. SE-4, No. 3, 1978, pp. 178-187.

[SD 88] K. K. Sabnani and A. T. Dahbura, "A Protocol Test Generation Procedure," Computer Networks and ISDN Systems, Vol. 15, No. 4, 1988, pp. 285-297.

[CVI 89] W. Y. L. Chan, S. T. Vuong, and M. R. Ito, "An Improved Protocol Test Generation Procedure Based on UIO's," in Proc. of SIGCOMM'89, 1989, pp. 283-294.

[SL 89] D. P. Sidhu and T. K. Leung, "Formal Methods in Protocol Testing: A Detailed Study," IEEE Trans. on Software Eng., Vol. 15, No. 4, April 1989, pp. 413-426.

[ADLU 91] A. V. Aho, A. T. Dahbura, D. Lee, and M. U. Uyar, "An Optimization Technique for Protocol Conformance Test Generation Based on UIO Sequences and Rural Chinese Postman Tours," IEEE Trans. on Communications, Vol. 39, No. 11, Nov. 1991, pp. 1604-1615.

[FBK 91] S. Fujiwara, G. v. Bochmann, F. Khendek, et. al., "Test Selection Based on Finite State Models", IEEE Trans. on Software Eng., SE-17, No. 6, June 1991, pp. 591-603.

[LS 92] F. Lombardi and Y.-N. Shen, "Evaluation and Improvement of Fault Coverage of Conformance Testing by UIO Sequences," IEEE Trans. on Comm., Vol. 40, No. 8, August 1992, pp. 1288-1293.

[UZ 93] H. Ural and K. Zhu, "Optimal Length Test Sequence Generation Using Distinguishing Sequences," IEEE/ACM Trans. on Networking, Vol. 3, No. 1, June. 1993, pp. 358-371.

[MP 93] R. E. Miller and S. Paul, "On the Generation of Minimal-Length Conformance Tests for Communication Protocols," IEEE/ACM Trans. on Networking, Vol. 1, No. 1, Feb. 1993, pp. 116-129.

[MP 94] R. E. Miller and S. Paul, "Structural Analysis of Protocol Specifications and Generation of Maximal Fault Coverage Conformance Test Sequences," IEEE/ACM Trans. on Networking, Vol. 2, No. 5, Oct. 1994, pp. 457-470.

[NAIK 94] K. Naik, "Efficient Computation of Unique Input/Output Sequences in Finite State Machines," *submitted for publication.*

14

Guaranteeing full fault coverage for UIO-based testing methods

Ricardo Anido[*]
Ana Cavalli
Institut National des Télécommunications
9, rue Charles Fourier, 91011 Evry Cedex, France
ranido@hugo.int-evry.fr, Ana.Cavalli@hugo.int-evry.fr

Abstract

This paper presents an analysis of the fault coverage provided by the UIO-based methods for testing communications protocols. Formal analysis of the fault coverage for the non-optimized method and for some of its optimized versions are presented. A test is said to provide *full coverage* if no erroneous implementation can pass the test. In the case of optimizations based on the Rural Chinese Postman Tour (Aho et al. 1991) it is shown that unless certain conditions are met the method does not guarantee full fault coverage, even when, as suggested in (Chan et al. 1989), the uniqueness of UIO sequences (or Partial UIO sequences) are verified in the implementation. The result of the analysis suggests how the existing methods for generating test sequences should be changed in order to guarantee full fault coverage.

Keywords

Communication protocols, conformance testing, verification, finite state machines, test generation, test coverage.

1.0 INTRODUCTION

The use of a precise set of communication rules, called a *protocol*, is essential for the design and implementation of distributed systems and communication networks. A protocol defines all possible interactions among the components of the system. After the system has been implemented, the protocol implementation must be verified to conform to its specification, to ensure that the system will operate correctly. This procedure is known as conformance testing, and can be accomplished by applying a sequence of inputs to the implementation, by means of an external tester, and verifying if the sequence of outputs is the one specified.

If a test sequence is capable of detecting all erroneous implementations, it is said to provide *full* fault coverage. There are many methods for generating automatically a test sequence to verify a given implementation against a specification. Several of these methods are based on the Unique Input/Output (UIO) technique. A number of papers have studied the fault coverage

---

*. On leave from Departamento de Ciência da Computação, Universidade Estadual de Campinas, Brazil. Work supported by Fundação de Amparo à Pesquisa do Estado de São Paulo (FAPESP), grant 94/3667-3.

of the UIO methods and its optimizations, mainly by *evaluating* the fault coverage through simulation of some test cases. This paper presents a formal analysis of the fault coverage of the UIO methods, showing that, under certain conditions, this evaluation is not needed, since the UIO method and its optimizations can be shown to guarantee full fault coverage. The result of the analysis can be used to improve existing test generation methods so that they guarantee full fault coverage.

The paper is organized as follows. In the rest of this section the basic concepts of test generation are reviewed. The UIO method and its optimizations are well known, but since they are the main subject of this paper they are described in some detail before their fault coverage is analysed in Sections 2 and 3, respectively. In Section 4 the work presented in this paper is related to other approaches in the literature. Finally, Section 5 summarizes the main results.

1.1 Basic concepts

A protocol specification is typically composed by a control portion and a data portion. This paper deals with the control portion only; other approaches are oriented to the analysis of control and data dependencies (Show and Ural, 1993).The control portion of a protocol, which will be referred as protocol specification, can be modelled as a deterministic finite-state machine (FSM) with a finite set of states $S = \{s_1, s_2,..., s_n\}$, a finite set of inputs $I = \{a_1, a_2,..., a_k\}$, and a finite set of outputs $O = \{x_1, x_2,..., x_m\}$. The next state ($\sigma$) and output ($\varphi$) functions are given by a set of mappings $\sigma: S \times I \rightarrow S$ and $\varphi: S \times I \rightarrow O$.

The FSM is usually also represented by a direct graph $G = (V, E)$, where the set $V = \{v_1, v_2,..., v_n\}$ of vertices represents the set of states S, and a directed edge represents a transition from one state to another in the FSM. Each edge in G is labelled by an input a_r and a corresponding output x_q. An edge in E from v_i to v_j which has label a_r/x_q means that the FSM, in state s_i, upon receiving input a_r produces output x_q and moves to state s_j. A triplet $(s_i, s_j, a_r/x_q)$ is used to denote a transition in the text.

The graph representation is useful for describing and reasoning about test generation algorithms. Within this context, some basic definitions from graph theory are briefly reviewed. A graph is said to be strongly connected if for any pair of distinct vertices v_i and v_j there is a walk which starts on v_i and ends on v_j. A *walk* W over a graph is a finite, non-null, sequence of consecutive edges. *Head(W)* and *Tail(W)* denote respectively the vertex where the walk W starts and the vertex where it ends. A *path* is a walk in which each edge of G appears exactly once.

An input/output sequence $U= (a_{r1}/x_{q1}, a_{r2}/x_{q2}, ... a_{rn}/x_{qn})$ is said to be *specified* for state s_i in an specification FSM if there exists a walk W with origin s_i in the graph representation of the FSM such that $W = \{(s_i, s_{j1}, a_{r1}/x_{q1}), (s_{j1}, s_{j2}, a_{r2}/x_{q2}), ... , (s_{jn-1}, s_{jn}, a_{rn}/x_{qn})\}$. A FSM is said to be *fully specified* if from each state it has a transition for every input symbol; otherwise the FSM is said to be *partially specified.* If a FSM is partially specified and a non specified transition is applied, under the *Completeness Assumption* the FSM will either stay in the same state without any output or signal an error. In this paper we consider FSMs under the completeness assumption. The *initial state* of a FSM is the state the FSM enters immediately after power-up. A FSM is said to have the *reset* capability if it can move from any state directly into the initial state with a single transition, denoted "*ri/null*" or simply "*ri*". State s_i is said to be weakly equivalent to state s_j if any specified input/output sequence for s_i is also specified for s_j. If two states are weakly equivalent to each other they are said to be strongly equivalent. It is assumed here that the FSMs are deterministic; that is, for some state $s_i \in S$, with two associated transitions (s_i, s_j, a_r, x_q) and (s_i, s_k, a_w, x_p), $a_r \neq a_w$.

A graph representation of a FSM is depicted in Fig. 1. For the FSM represented, $I=\{a,b,ri\}$ and $O=\{0,1,null\}$. Reset edges are not shown in the figure, but are assumed to be labelled "*ri/null*" and directed towards the initial state v_1, the designated initial state.

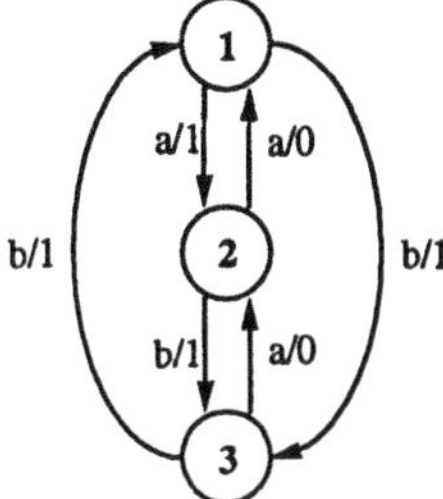

Figure 1 A graph representation of a FSM.

1.2 Test generation techniques

The purpose of test generation is to produce a sequence of inputs, called a test sequence, which can be applied to an implementation to verify that it correctly implements the specification. There is a number of necessary assumptions that must be made in order to make the experiment possible: (1) the specification FSM is strongly connected, so that all states can be visited; (2) the specification FSM does not have strongly equivalent states (it is minimized); (3) there is an upper bound on the number of states in the implementation FSM (otherwise one could always construct a machine which would pass a given test sequence by using as many states as there are transitions in the sequence). In relation to assumption (3) it is usual in the literature to consider that the implementation has no more states than the specification. Note that if the implementation is correct, by this assumption it will have the same number of states than the specification. If the implementation is not correct, however, it may have fewer states than the specification and still pass a test sequence which does not provide full fault coverage. We consider also that the specification FSM is deterministic, and completely specified.

One of the simplest methods for generating test sequences is the *transition tour* method (Naito and Tsunoyama, 1993): a test sequence is generated by simply applying random inputs, constructing a random walk over the graph representing the specification FSM until all transitions have been traversed. Obviously, the sequence generated may contain redundant inputs which in turn generate *loops* in the walk; these redundant inputs may be removed using a reduction procedure (but it is interesting to note that some redundancy may be important to enhance the fault coverage of the test sequence, as will be discussed in Section 4). As an example, the following test sequence is generated using the transition tour method for the automata of Figure 1:

ri/null a/1 b/1 b/1 a/1 b/1 a/0 a/0 a/1 a/0 b/1 *[SEQ 1]*

In general, the fault coverage for tests generated by transition tours is worse than that obtained by other methods considered in this paper. Intuitively, this derives from the fact that verifying whether a transition produces the correct output is not enough to guarantee that the transition is correct: one should also verify that the new state of the implementation FSM after the transition is the one expected. That is what was done in the first methods for generating testing sequences, developed in the 60's.

Kohavi's book gives a good exposition of these earlier results on testing FSMs, motivated mainly by testing of switching circuits (Kohavi, 1978). If the specification FSM has a *distinguishing sequence* (a sequence which produces a different output for each different state), the test procedure presented in (Kohavi, 1978) for testing an implementation FSM is divided into two parts. In the first part, called *state identification*, the implementation is forced to display the response of each state to the distinguishing sequence, while in the second part, called *transition identification*, each transition is verified.

The rationale for the state identification part is that the distinguishing sequence method includes a "hidden" assumption, namely that the distinguishing sequence is valid not only for the specification, but also for the implementation. The transition identification part is carried out by applying an input which causes the desired transition to be exercised and identifying the new state by means of the distinguishing sequence.

The problem with the method presented above is that not all FSMs have distinguishing sequences, and these can be in general very long. In (Sabnani and Dahbura, 1988) it was first presented the idea of using a Unique Input-Output (UIO) sequence as a means of solving these shortcomings. A UIO for state s_i, denoted *UIO(i)*, is an input/output sequence with origin s_i such that there is no $s_j \neq s_i$ for which *UIO(i)* is an specified sequence for starting state s_j. Most FSMs have UIO sequences for every state, and UIOs are never longer than distinguishing sequences, being in practice usually much shorter. An extension to the UIO method can be used when some states have no UIOs, as seen in Section 3.6.

1.3 Conformance

Since the implementation is tested as a black box, the strongest conformance relation that can be tested is *trace equivalence*: two FSMs are trace equivalent if the two cannot be distinguished by any sequence of inputs. That is, both implementation and specification will generate the same outputs ("trace") for all specified input sequences. To prove trace equivalence it suffices to show that (a) there is a set of implementation states $\{p_1, p_2, ...p_n\}$ respectively isomorphic to specification states $\{s_1, s_2, ...s_n\}$, and (b) every transition in the specification has a corresponding isomorphic transition in the implementation.

2.0 FAULT COVERAGE OF THE NON-OPTIMIZED UIO METHOD

The original paper introducing the UIO method proposed to use only the transition identification part as a testing sequence, apparently assuming it would suffice to provide good fault coverage. In (Chan et al. 1989) it was shown, by examples, that if the state identification part is not performed as well, some errors in the implementation may remain undetected. The argument is the same presented above for distinguishing sequences: the validity of the method is based on the fact that the UIO is unique both in the specification and in the implementation.

In this section a formal analysis of the fault coverage for the basic UIO method with the modification proposed in (Chan et al. 1989) is presented. Although the full fault coverage has already been argued in (Chan et al. 1989) (in a less formal manner), the results from this section will be used when analysing the fault coverage of the optimizations of the UIO method.

2.1 State identification

The state identification part of the test for the UIO method is slightly different from the one presented above for distinguishing sequences. To verify that each UIO is unique to a state in the implementation, it must be verified that the UIO for one state is accepted by that state and is *rejected* by all other states.

The procedure for verifying the rejection of *UIO(j)* for state s_i is the following:

1. The implementation is put into state p_i, presumably isomorphic to s_i, by applying a *reset* followed by some path *Preamb(i)* from the initial state s_1 to s_i. For efficiency reasons *Preamb(i)* should be some shortest path from s_1 to s_i but that is not relevant. However, once chosen, *Preamb(i)* must be fixed for the duration of the test.
2. *UIO(j)* is applied to the implementation and the output is checked to verify that the resulting output is not what it would be expected if the *UIO(j)* were applied to s_j.

The procedure for verifying the acceptance of *UIO(i)* for state s_i is similar, with the obvious difference that in step 2 it must verified that the output is indeed the one expected. However, UIO acceptance does not need to be tested in the state identification part, since if an implementation state does not accept its UIO the transition identification part will detect the error. The state identification part for the UIO method therefore consists of verifying, for all pairs i, j $(i \neq j)$, the rejection of *UIO(j)* by state p_i. An implementation state p_i, which rejects *UIO(j)* for all $(i \neq j)$ will be said *UIO-isomorphic* to specification state s_i. Note that the reset feature plays an important role in this part of the test.

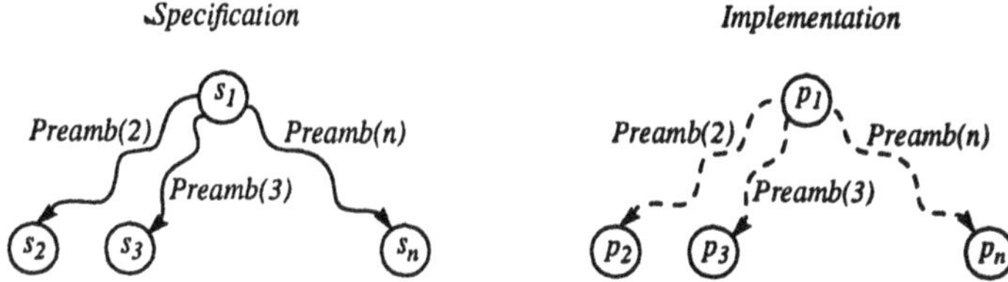

Figure 2 State UIO-isomorphism after the state identification part.

The state identification part determines that the implementation has at least n states p_1, $p_2, \ldots, p_n$ which are accessible using the same preamble as their respective UIO-isomorphic states $s_1, s_2, \ldots, s_n$ in the specification. In other words, after the state identification part the following two properties hold, for all i:

Property 1: There exists one, and only one, implementation state p_i which is UIO-isomorphic to specification state s_i.

Property 2: Implementation state p_i is reachable from initial implementation state p_1 by using *Preamb(i)*.

Note that although several transitions have already been used in *Preamb(i)* and in *UIO(i)*, the state identification part does not guarantee that these transitions are correct in the implementation. The state identification only asserts what the two properties state. Due to multiple faults, *Preamb(i)* may traverse faulty transitions and still take the implementation to the desired state p_i, and UIO may traverse faulty transitions and still give the correct output. This situation is depicted in Fig. 2, where dotted lines are used to emphasise the fact that *Preamb(i)* may include faulty transitions.

2.2 Transition identification

The procedure for testing transition $t = (s_i, s_j, a_r/x_q)$, a transition from state s_i to s_j with input/output a_r/x_q, is the following:

1. The implementation is put into state p_i, known to be UIO-isomorphic to s_i, by applying a reset followed by *Preamb(i)*;
2. Input a_r is applied and the output is checked, to verify that it is x_q as expected;
3. The new state of the implementation is tested to verify it is state p_j as expected, by applying *UIO(j)* and checking that the resulting output is the one expected.

In the transition identification part, all transitions in the specification are tested using the procedure above. A graphical representation of a transition test can be seen in Fig. 3.

2.3 Fault coverage

In this section the UIO method as described is shown to provide full fault coverage, i.e., there are no faulty implementations with at most the same number of states as the specification which can pass a test generated by the method.

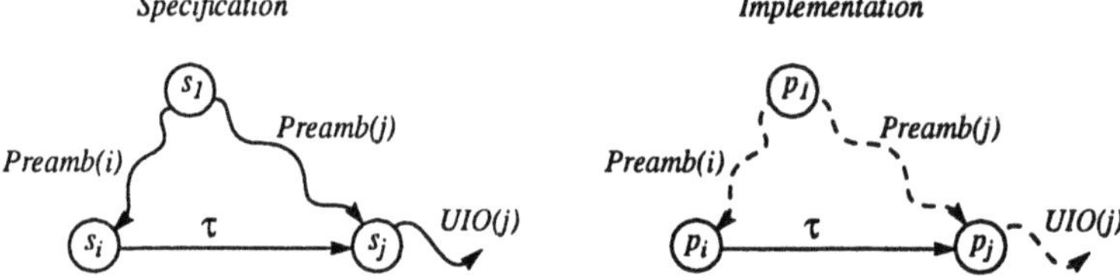

Figure 3 Transition identification part.

Consider a specification FSM with *n* states. If the implementation has fewer than *n* states, it will fail the state identification part, since some specification states will not have a corresponding UIO-isomorphic implementation state. If the implementation has *n* states, the transition identification part will test each transition $\tau = (s_i, s_j, a_r/x_q)$, by the sub-sequence (the symbol @ means concatenation, the transition under test is shown in bold)

$$ri \ @ \ Preamb(i) \ @ \ \mathbf{a_r/x_q} \ @ \ UIO(j)$$

To show the correctness of the test method the two properties defined above are used. Suppose that there is a transition $\tau = (s_i, s_j, a_r/x_q)$ which is wrongly implemented but the error remains undetected after the test. Since all transitions are tested individually, the only way a faulty transition can escape detection is if the transition is not tested at all in the implementation. That is possible only if (a) the implementation is not in state p_i, UIO-isomorphic to s_i, when the transition is tested; (b) p_k, the state the implementation is in when the transition is tested, is UIO-isomorphic to state s_k and there exists a transition $\tau_a = (s_k, s_j, a_r/x_q)$ in the specification. However, (a) cannot happen because by property (2) the implementation is indeed in the corresponding UIO-isomorphic state p_i before the transition τ is applied, i.e., the intended transition will be tested in the implementation. If τ produces the correct output, there is obviously no output fault. And if *Tail(τ)* passes *UIO(j)*, by property (1) *Tail(τ)* must indeed be p_j, UIO-isomorphic to s_j. Therefore, there is no transition which is faulty and is not detected by the test. In other words, the UIO method without any optimization provides full fault coverage.

3.0 FAULT COVERAGE OF OPTIMIZATIONS TO THE UIO METHOD

Several optimizations have been proposed to the UIO method over the years (Aho et al. 1991, Chen et al. 1990, Lombardi and Shen 1992, Miller and Paul 1991, Shen and Lombardi 1992). Interestingly, all optimizations focused on the transition identification part, disregarding completely the state identification part. The general (hidden) assumption seems to be the same one made by (Sabnani and Dahbura, 1988), i.e., the transition identification part would suffice to provide full fault coverage.

All these optimizations, if used as proposed, without the corresponding state identification part, will not provide full fault coverage, since they are all basically the UIO method. The safety of some optimizations proposed is now analysed. The main result is to show that optimizations based on a "global optimization" technique will only provide full fault coverage under certain conditions.

3.1 Rural Chinese Postman Optimization

Let us call a *test segment* the subsequence tr @ $(UIO(Tail(tr))$ used to test a single transition tr. The transition identification part therefore consists of as many test segments as there are transitions in the specification FSM. As seen, in the non-optimized UIO method each test segment is preceded by a *reset* followed by a *Preamble* sequence, with the purpose of bringing the implementation into state $Orig(tr)$ so that tr can be applied and tested. In (Aho et al. 1991)it was first presented the idea of optimizing the cost of connecting the test segments, by using transfer sequences which used not only *reset* and *Preamble* sequences, but could include any specified transition.

The optimization is elegantly presented as a Rural Chinese Postman Tour Problem, which is NP-complete for the general case, but has a low-degree polynomial time solution for weakly connected graphs (they also showed that if a FSM has a reset feature or has a self loop for each state, its corresponding graph is weakly connected). The optimization problem is formulated as follows. The FSM is represented as a graph $G = (V, E)$. Consider the graph $G' = (V', E')$ such that $V' \equiv V$ and $E' = E \cup E_C$, where E_C is the set of all test segments. That is,

$$E_C = \{(s_i, s_k, tr \text{ @ } UIO(j)) \mid (s_i, s_j, tr) \in E \text{ and } Tail(UIO(j)) = s_k\}$$

In G', traversing an edge in E_C corresponds to realizing a test segment; the cost of the traversal is usually taken to be the total length of the test segment. Notice that the edge-induced subgraph $G[E_C] = (V, E_C)$ is a spanning subgraph of G'. Therefore the optimization objective becomes traversing each edge in E_C at least once with a minimum cost tour of G'. Such a tour is a Rural Chinese Postman Tour. Similar to the Chinese Postman Problem, the problem is first reduced to that of finding a symmetric augmentation graph $G^* = (V^*, E^*)$, constructed such that $V^* = V'$ and E^* contains all edges in E_C, and possibly some edges in E. The basic idea is to minimize the number of edges chosen from E and at the same time make the graph G^* symmetric, that is, a graph for which every vertex has an in-degree equal to its out-degree. Finally, an Euler tour can be constructed in linear time from the symmetric, strongly connected graph G^*. The tour is then used as the transition identification part for the test sequence.

In Section 2, a central argument in the proof of full fault coverage of the basic UIO method is that, when testing transition $tr = (s_i, s_j, a_r/x_q)$, it can be guaranteed that the implementation is definitely into the state UIO-isomorphic to s_i before applying tr. That guarantee is given by the state identification part, which uses the same preamble sequence. In what follows it is shown that, due to limited controllability of the implementation, that guarantee may be lost when the RCP optimization is introduced.

Analysis of the fault coverage

Suppose there is one transition τ which is faulty in the implementation, and the fault is not detected by the state identification part. Since a test segment is executed to specifically test each transition, if the implementation passes the test sequence the only possibility is that when executing the test segment τ @ *UIO(Tail(*τ*))* some other transition is traversed instead of τ. And since the error remains undetected, there must be that the transition mistakenly traversed produces the same output (otherwise the error would be detected) and takes the implementation to the same state as τ should (otherwise *UIO(Tail(*τ*))* would fail). That is, if the erroneously implemented transition is specified as $\tau = (s_i, s_j, a_r/x_q)$, there must exist another transition $\tau' = (s_k, s_j, a_r/x_q)$ in the specification.

Therefore, for a faulty transition to remain undetected the only possibility is that the specification includes two edges going to the same state with the same input/output label, and one of these edges represents the faulty transition. Let us call state s_j ***convergent*** if there are edges going from states s_i and s_k into s_j with the same input/output label. Edges going into the same state with the same input/ouput label will be called *converging* edges, or transitions. The reasoning above leads to the following lemma:

Lemma 1: All implementation errors in transitions which are not converging are detected by the UIO method with the Rural Chinese Postman optimization.

Proof: The state identification part guarantees that each state in the specification has a corresponding UIO-isomorphic state in the implementation, i.e., each implementation state p_j will reject the *UIO(i)* for all $i \neq j$. When a test segment is executed and no error is detected, the transition under test must have produced the expected output and must have ended in the correct state. If a faulty transition τ is not converging, its test segment will fail, since either τ will not produce the expected output or *UIO(Tail(*τ*))* will fail. ♦

A first result can then be presented:

Proposition 1: For a FSM which does not include a convergent state the UIO method with the Rural Chinese Postman optimization provides full fault coverage.

Proof: It follows directly from *Lemma 1*.♦

Let us assume the specification FSM includes a convergent state s_j, with converging transitions $\tau = (s_i, s_j, a_r/x_q)$ and $\tau' = (s_k, s_j, a_r/x_q)$. If the test segment for τ succeeds and τ is admittedly faulty, it must be that the test segment was executed when the implementation was in state p_k instead of p_i. That is, when preparing to execute the test segment for τ, a transfer sequence was applied which supposedly should take the implementation into p_i but took it into p_k instead. That is only possible if there is a transition which is faulty and was traversed when executing the transfer to p_i. Therefore, if the error in τ is to remain undetected there must exist a faulty transition which is traversed when preparing to verify τ. The analysis is now divided in three separate cases. The first case is when no UIO used in the test sequence traverses a converging edge; the second case is when UIOs are allowed to use a converging edge; and the third case is when neither the UIO sequences nor transfer paths between test segments include a converging edge.

In the first case, by *Lemma 1*, if the test sequence succeeds all transitions used in all UIOs are correctly implemented. Therefore, when any UIO is executed, it leaves the implementation in the state it is supposed to. In particular, the UIO which is executed immediately before the test segment for the erroneously implemented transition $\tau = (s_i, s_j, a_r/x_q)$ leaves the implementation in the correct state; let p_y denote that state, as depicted in Fig. 4.

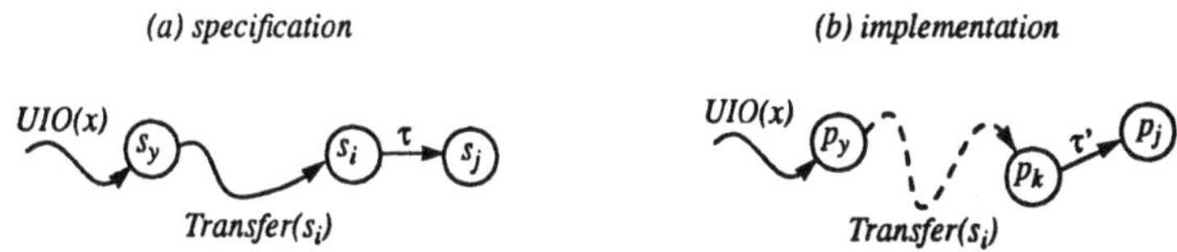

Figure 4 :*Transfer(si)* does not take the implementation into p_i.

Since it was established that there must exist a faulty transition which is traversed when preparing to verify τ, as the transfer path starts from the correct state there must exist another transition τ_1 which is also faulty in the transfer path from s_y to s_i. Note that τ_1 could not be the same transition τ, otherwise the transfer path from s_y to s_i would not include τ_1.

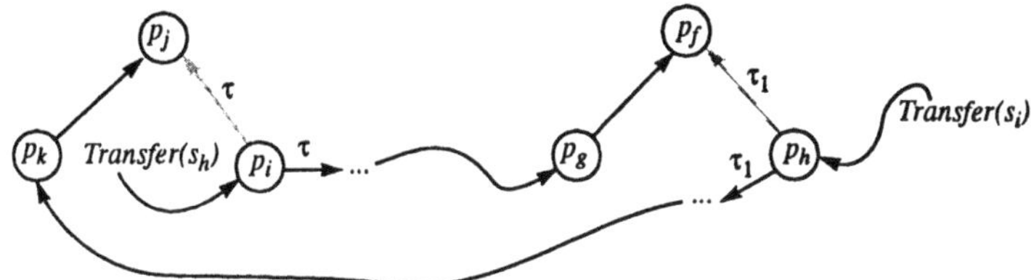

Figure 5 :Interdependence of errors.

Using for τ_1 the same arguments used for τ, one comes to the conclusion that (*i*) τ_1 must be a converging edge; and (*ii*) if the error in τ_1 is to remain undetected there must be a faulty transition which is traversed when preparing to verify τ_1. Unfortunately, as depicted in Fig. 5, the sequence of faulty transitions does not have to continue unlimitedly, i. e., this scenario does not depend on the existence of transitions τ_2, τ_3, ... (in which case it would have been proved that faulty transition τ could not exist). There may exist an interdependence of faults which causes transition τ_1 to be traversed when preparing to verify τ, and transition τ to be traversed when preparing to verity τ_1, such that both errors remain undetected. Note that only some of the transitions and states are shown in Fig. 5. In this example, transitions $\tau = (s_i, s_j, a_r/x_q)$ and $\tau_I = (s_h, s_f, a_w/x_l)$ are erroneously implemented, but τ_1 is traversed when preparing to test τ (resulting that the implementation is erroneously put into state p_k, and τ is not really tested) and τ is traversed when preparing to test τ_1 (and as a result τ_1 is not correctly tested). Therefore the following result can be presented:

Proposition 2: The UIO method with the Rural Chinese Postman optimization provides full fault coverage if none of the UIOs used traverses a converging transition and the specification FSM includes at most one pair of converging transitions.

Proof: The state identification part guarantees that each state in the specification has a corresponding UIO-isomorphic state in the implementation. As discussed in the previous paragraph, if the specification includes only one pair of converging transitions, any transition error would be detected in the transition identification part. ♦

In the second case, some UIOs may include converging edges. Therefore, in this case the state the implementation is after the UIO is not guaranteed to be the one expected. Fig. 6 shows a possible scenario of the situation (again, only some of the specified transitions and states are shown). In this example the faulty transition $\tau = (s_i, s_j, a/0)$ is the last transition of some *UIO(x)*, and is chosen by the optimization to be verified immediately after *UIO(x)* is executed, using the transfer path *b/1* (note that as τ is the last transition in *UIO(x)*, *Tail(UIO(x))* = s_j). That is, the test sequence includes (the transition under test is shown in bold)

... UIO(x) @ b/1 @ **a/0** @ UIO(j) ...

Fig. 6a shows a specification and Fig. 6b shows a possible (wrong) implementation; only part of the specified states and transitions are shown. In this situation, the test segment for τ will succeed despite the erroneous implementation, so that the fault in τ remains undetected.

Figure 6 :Another case of interdependence of errors

Therefore, when UIOs are allowed to use converging edges, even if the specification contains one only pair of converging edges the RCP optimization does not guarantee full fault coverage.

In the third case, when neither UIOs nor transfer paths traverse a converging edge, it is guaranteed that before executing each test segment the implementation is in the state it is supposed to be. That leads to another result:

Proposition 3: The UIO method with the Rural Chinese Postman optimization provides full fault coverage if neither UIOs nor transfer paths traverse a converging transition.

Proof: The state identification part guarantees that each state in the specification has a corresponding UIO-isomorphic state in the implementation. If neither UIOs nor transfer paths use converging edges, by Lemma 1 it is guaranteed that when a test segment is applied, it is applied with the implementation in the correct UIO-isomorphic state. Since the correct transition is tested by each test segment, any error will be detected. ♦

If the specification FSM includes one or more convergent states, Proposition 3 shows how the Rural Chinese Postman optimization should be modified in order to guarantee full fault coverage. The choice of UIOs should be restricted to those which do not traverse converging edges, and converging edges should not be used during the symmetric augmentation of graph $G[E_C]$. The first restriction is not normally difficult to satisfy, since in most FSMs states present a choice of UIOs; the second restriction is also not difficult in general to satisfy, since instead of using a converging edge from s_i to s_j a *virtual edge* (any path from s_i to s_j not including a converging edge) in graph $G[E_C]$ can be used.

If these two restrictions cannot be satisfied, to guarantee full fault coverage all converging transitions should be tested using the non-optimized reset-preamble method, before applying the RCP optimization to test the remaining transitions.

Example

As an example of a FSM which may cause fault masking when using the RCP optimization consider the FSM of Fig. 7.

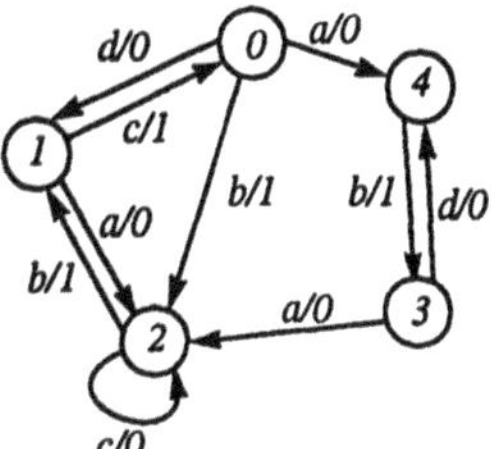

i	*UIO(i)*	*Preamb(i)*
0	*d/0 a/0*	-
1	*c/1*	*d/0*
2	*c/0*	*b/1*
3	*d/0 b/1*	*a/0 b/1*
4	*b/1 d/0*	*a/0*

Figure 7 An example FSM and its UIO and Preamble sequences.

Note that state *2* is convergent, with converging transitions *(1, 2, a/0)* and *(3, 2, a/0)*, and that the first of these transitions was chosen to be used in *UIO(0)* (this represents therefore an example for the second case in the analysis). The state identification part consists of verifying the rejection of subsequences *ri @ Preamb(i) @ UIO(j)* for all pairs *i, j (i ≠ j)*. A RCP optimization produces the following sequence for the transition identification part, where transitions used in transfers are printed in bold:

*ri/null b/1 @ UIO(2) @ **b/1** c/1 @ UIO(0) @ **b/1** a/0 @ UIO(2) @ c/0 @ UIO(2) @ **b/1 c/1** d/0 @ UIO(1) @a/0 UIO(4) @ **b/1** d/0 @ UIO(4) @ b/1 @ UIO(3) @ a/0 @ UIO(2) @ b/1 @ UIO(1)*

Expanded, this sequence produces:

ri/null b/1 c/0 b/1 c/1 d/0 a/0 b/1 a/0 c/0 c/0 c/0 b/1 c/1 d/0
c/1 a/0 b/1 d/0 b/1 d/0 b/1 d/0 b/1 d/0 b/1 a/0 c/0 b/1 c/1

Consider now the implementation FSM depicted in Fig. 8, in which transition *(1, 2, a/0)* is erroneously implemented as *(1, 4, a/0)*. Note that the state identification and the transition identification parts are successfully executed for this implementation. Therefore, the error remains undetected.

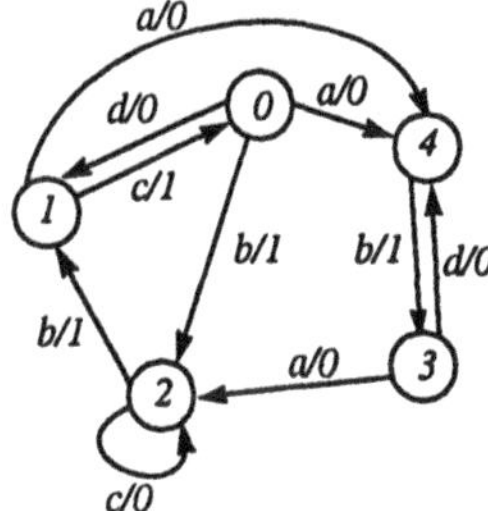

Figure 8 An erroneous implementation which passes the test.

The error would be detected if *UIO(0)* is chosen to be *b/1 c/0*, and care is taken so that no convergent edges are used in transfers. A RCP optimization produces in this case the following sequence for the transition identification part (again, transitions used in transfers are shown in bold):

*ri/null b/1 @ UIO(2) @ **b/1** c/1 @ UIO(0) @ **b/1** a/0 @ UIO(2) @ c/0 @ UIO(2) @ **b/1 c/1** d/0 @ UIO(1) @ a/0 @ UIO(4) @ **b/1** d/0 @ UIO(4) @ b/1 @ UIO(3) @ a/0 @ UIO(2) @ b/1 @ UIO(1)*

Expanded, it results in:

ri/null b/1 c/0 b/1 c/1 b/1 c/0 b/1 a/0 c/0 c/0 c/0 b/1 c/1 d/0
c/1 a/0 b/1 d/0 b/1 d/0 b/1 d/0 b/1 d/0b/1 a/0 c/0 b/1 c/1
which correctly detects the implementation error.

3.2 RCP with multiple UIOs

It was shown in (Shen and Lombardi, 1992) that using different UIOs for identifying a state in different test segments can reduce the total length of the transition part. The basic idea is to obtain a graph $G[E_C]$ which is closer to symmetry, so that fewer edges need to be added to make it symmetric. However, as already noted in (Yao et al. 1993) , the fact that the uniqueness of any UIO-sequence used must be verified, when using multiple UIOs any gain in the transition identification part may be lost by an increase in the state identification part.

In relation to fault-coverage, the fact that multiple UIOs are used does not change any of our previous results, assuming of course their uniqueness is verified in the state identification part.

3.3 RCP with overlaps

To further minimize the transition identification part, overlapping of test segments can be used. If the last part of a test segment T_i coincides with the first part of another test segment T_j, they can be merged so that the overlapping edges would serve to both T_i and T_j. If T_i is completely contained in T_j, T_i does not have to be executed at all, and can be removed from the test sequence. The "full overlap" optimization was in fact proposed in the original UIO paper, (Shen et al. 1992); the general overlap was mentioned as a possible extension in (Aho et al. 1991) but the first solution appeared in (Chen et al. 1990).

In (Chen et al. 1990), *overlap links*, with negative cost, are introduced into the graph $G[E_C]$ to capture the concept of overlaps into the optimization. The optimization problem is then solved as a minimum cost — maximum cardinality matching problem in a bipartite graph.

Rather than presenting the proposed method in more detail, let us certify ourselves that the overlapping of test segments does not introduce any possibility of fault masking. Suppose two test segments $T_i = t_i$ @ $UIO(Tail(t_i))$ and $T_j = t_j$ @ $UIO(Tail(t_j))$ are overlapped, with T_i being executed first. Accordingly to our previous results, it is assumed also that neither UIOs nor transfer paths traverse converging edges. Therefore when test segment T_i starts, it is guaranteed that the implementation is indeed in the UIO-isomorphic state it should be, such that the correct transition t_i is exercised. If the correct transition is exercised, any output or transfer error would be detected, either by the transition failing to produce the correct output or by the $UIO(Tail(t_i))$ failing to produce the correct output. Consider now transition t_j. As the two test segments are made to overlap, t_j must be a component of $UIO(Tail(t_i))$, which means t_j itself is not a a converging edge. However, as it has been seen, if a transition error is to remain undetected, the transition must be a converging edge. Therefore, by *Lemma 1*, any error in t_j will be detected.

3.4 RCP with multiple UIOs and overlaps

In (Miller and Paul 1991) it is shown how multiple UIOs and overlaps can be combined to obtain a further reduction on the transition identification part. It is interesting to note that they proposed different algorithms depending whether the specification FSM includes a convergent state or not. That is, they noticed convergent states are a possible cause of trouble, although they did not pursue the issue. As explained in the paper, their approach can be seen as first finding the test sequence (for the identification part only) and then justifying that all needed test segments are included in the sequence found.

In all their cases, they show there must be a sequence of possibly overlapping test segments embedded into the sequence found, such that all test segments are executed. As we have already examined, multiple UIOs and overlaps do not interfere with the fault coverage. Therefore, in relation to fault coverage our results apply also in this case.

Although interesting, this approach has the same drawback as any method based on multiple UIOs: any gain in the length of the transition identification part due to the use of multiple UIOs incurs an increase in the length on the state identification part.

3.5 Greedy overlap

Another optimization method, presented in (Chen et al. 1990), differs from all previous methods described in that it does not use a global optimization to minimize the transfer sequences between test segments. Rather, it uses a greedy algorithm to construct step by step the test sequence for the transition identification part. As in this method the sequence produced is basically a different concatenation of overlapped test segments (possibly using multiple UIOs), all our results are also valid in this case.

3.6 Partial UIOs

Some FSMs do not possess UIO sequences for every state. Fig. 9 shows a FSM which does not have a UIO for state *1*: if *UIO(1)* starts with input *a* it cannot distinguish state *1* from state *3*; and if the *UIO(1)* starts with *b* it cannot distinguish state *1* from state 2.

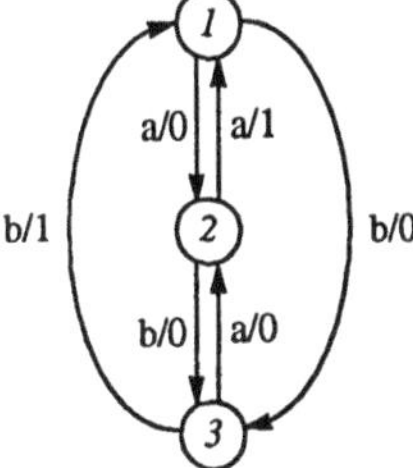

Figure 9 A FSM which does not have UIO for all states.

The approach proposed in (Sabnani and Dahbura 1988) to verify a state s_j that does not have a UIO is to use a *signature*, a sequence which distinguishes s_j from each of the remaining states one at a time. A signature for s_j uses *(n-1)* subsequences *IO(j,m)*, each of which distinguishes s_j from one other state s_m. Before applying each IO subsequence the implementation must be put back into state s_j. Suppose that after the subsequence *IO(j,m)* is applied the specification is in state s_k. The authors propose to use a transfer subsequence *Tr(k,j)*, which is some shortest path from s_k to s_j, in order to bring the implementation back to state s_j. Therefore, the signature for state s_j will be formed by concatenating the subsequences *IO(j,m) @ Tr(Tail(IO(j,m)),j)* for all

$m \neq j$. That is, the test segment for a transition $t = (s_i, s_j, a_r/x_q)$, where s_j is a state which does not have a UIO, is composed by the sequence

$$a_r/x_q \ @ \ IO(j,1) \ @ \ Tr(Tail(IO(j,1)),j) \ @ \ IO(j,2) \ @ \ Tr(Tail(IO(j,2)),j) \ @ \ \ldots$$
$$\ldots \ @ \ Tr(Tail(IO(j,n\text{-}1)),j) \ @ \ IO(j,n) \qquad \textit{[SEQ 2]}$$

In (Chan et al. 1989) and (Chun and Amer, 1992) it was shown this signature method does not work in the general case, and a variation was proposed. The first improvement suggested in (Chan et al. 1990) is noting that a single IO sequence may distinguish state s_j from not only one other state, but from a group of states. An *IO set* for a state is composed by a certain number of IO sequences. Each sequence $IO(i, E_k)$ in an IO set distinguishes state s_i from a subset of states $E_k \subset S$. E_k is called the exclusion set for that IO sequence in relation to state s_i. Therefore IO sets for different states may have different sizes (if the IO set has only one element the IO sequence is in fact a UIO), so that by selecting appropriate IO sequences for an IO set the size of the signature can be reduced. IO sets can be seen as a generalization of the UIO concept: an UIO is fact an IO set with one only element.

The second improvement proposed in (Chan et al. 1990) is not using a transfer function to concatenate IO sequences, since transfer sequences are a possible cause of fault masking when using signatures. In the transition identification part, instead of traversing the transition only once and verifying its final state as in [SEQ 2], the transition is traversed a number of times equals to the size of the IO set for that state, and each time a different IO sequence is used to verify the final state. At each time the reset-preamble sequence is used to bring the implementation to the correct state prior to traversing the transition. That is, transition $tr = (s_i, s_j, a_r/x_q)$ is verified by the sequence

$$ri/null \ @ \ Preamb(i) \ @ \ a_r/x_q \ @ \ IO(j,E_1) \ @$$
$$ri/null \ @ \ Preamb(i) \ @ \ a_r/x_q \ @ \ IO(j,E_2) \ @$$
$$\ldots$$
$$ri/null \ @ \ Preamb(i) \ @ \ a_r/x_q \ @ \ IO(j,E_m)$$

In (Chun and Amer 1992), this style of signature is called Partial UIOs, and an algorithm for generating them is presented. Provided that, as proposed in (Chan et al. 1990), the uniqueness of IO sets is also verified in the state identification part, their use for states which do not possess UIOs does not affect the validity of our results. Note however in this case the reset-preamble technique is fundamental not only when verifying the uniqueness of the IO sets, but also when using these IO sets in the transition identification part. Any tentative of "optimizing" the transition identification part involving the use of a transfer path to join the IO sequences (for example by using all test segments $a_r/x_q \ @ \ IO(j,E_k) \ @ \ Tr(Tail(IO(j,E_k)),i)$ as edges of the graph $G[E_c]$ when doing the RCP optimization) introduces the possibility of fault masking. The proof is similar to the ones presented earlier in the paper: to consider there is a fault transition and to show a scenario where the fault is not detected.

4.0 Related work

Several researchers have addressed the fault-coverage evaluation (as opposed to analysis) of testing methods, and in particular UIO-based methods. Some, however, considered only the transition identification part in their analysis (Lombardi and Shen 1992, Motteler et al. 1993, Zhu and Chanson 1994), while others assumed the optimization itself would not interfere with the fault coverage (Yao et al. 1993). Most of the work on fault coverage evaluation uses the mutation technique: from the specification FSM a certain number of *mutant* (faulty) FSMs are randomly generated and verified with a given test sequence. The number of mutant FSMs which pass the test sequence is used as a measure of the fault coverage.

Another technique to evaluate the fault coverage is to estimate the number of FSM which could pass a given test sequence. This can be done by considering the test sequence as a form of FSM specification, and applying a minimization technique to this 'specification' (Yao et al. 1994). The number of minimized machines not isomorphic to the real specification FSM gives a measure of the test sequence fault coverage. A more or less similar approach was proposed in (Zhu and Chanson, 1994), where a technique is used to reduce the number of possibilities when reconstructing, from the test sequence, all possible FSMs which would pass the given test sequence. The problem of these "exhaustive" approaches is that even with the reduction techniques the task of enumerating all viable solutions may be still too hard to be feasible in the general case.

Our paper shows that for UIO-based methods, under certain conditions, these evaluation techniques are not required, since it can be guaranteed that by applying only safe optimizations the fault coverage of the generated sequence is total. The previous work which more relates to our approach is (Lombardi and Shen, 1992), where the authors also derive a set of rules to guarantee a better fault coverage for a given test sequence. However, their analysis is complicated by the fact that they considered only the transition identification part, and that they divided the possible faults in three different types.

Since the main interest of this paper was to investigated the fault coverage of the UIO-based *methods*, its results always apply to the worst case. But it must be noted that besides the fault coverage provided explicitly by the test generation method, any test sequence carries an *intrinsic* fault coverage. For example, the simple sequence

ri/null a/1 b/1 b/1 a/1 b/1 a/0 a/0 a/1 a/0 b/1 b/1 b/1 a/0 *[SEQ 3]*

which is an extension of [SEQ 1], the "random walk" sequence generated by the transition tour method, possesses an unexpected coverage power. In fact, [SEQ 3] provides full fault coverage, with no need for state identification or reset-preamble sequences, and is much shorter than test sequences generated by UIO based methods. The fault coverage can be verified in this small example by case analysis, or with a tool similar to (Yao et al. 1994) or (Zhu and Chanson, 1994).

This intrinsic fault coverage can be understood by realizing that some information may gained by the simple fact that one specific transition is concatenated after some other transition in the test sequence. For example, if a test sequence includes the subsequence *"a/0 a/2 a/1"*, in order to successfully execute it any deterministic implementation must possess at least three different states. Each transition added to the end of a test sequence will increase or diminish the test intrinsic fault coverage depending on its relation to all previous transitions.

Unfortunately, it seems to be difficult to devise a method which can exploit this intrinsic fault coverage in the general case.

5.0 Conclusions

This paper analysed the fault coverage of the basic UIO method and some of its optimizations. It presented the conditions under which these methods are guaranteed to provide full fault coverage. The main result is to show that optimizations to UIO-based methods are not safe in the general case. If the specification FSM of a protocol is fully specified (or the completeness assumption can be invoked), propositions 1-3 presented in this paper offer the conditions for the full coverage of the test sequence. If a FSM does not include any converging state, the UIO method optimizations analysed were shown to provide full fault coverage. This paper also showed that the optimizations to UIO-based methods offer full fault coverage when no converging edge is used in UIOs or transfer paths. If these restrictions cannot be adhered to, the only way to guarantee full fault coverage is to use the reset-preamble technique to test the converging transitions before applying the optimizations to test the other transitions.

6.0 References

A. V. Aho, A. T. Dahbura, D. Lee, M. U. Uyar (1991), An optimization technique for protocol conformance test generation based on UIO sequences and rural chinese postman tours, *IEEE Transactions on Communications*, vol. 39, no. 11, November.

Bochmann, A. Petrenko, M. Yao (1994), Fault coverage of tests based on finite state models, *Proc. 7th IFIP International Workshop on Protocol Test Systems*, pp. 55-74, Japan, November.

W. Y. Chan, S. T. Vuong, M. R. Ito (1989), An improved protocol test generation procedure based on UIOs, *Proc. SIGCOM89*, pp. 283-294.

M. S. Chen, Y. Choi, A. Kershenbaum (1990), Approaches utilizing segment overlap to minimize test sequences, *Proc. 10th International IFIP Symposium on Protocol Specification, Testing and Verification*, pp. 67-84, Canada, June.

W. Chun, P. D. Amer (1992), Improvements on UIO sequence generation and partial UIO sequences, *Proc. 12th International IFIP Symposium on Protocol Specification, Testing and Verification*, pp. 234-249, Florida, USA, June.

P. Kars (1994), Test coverage estimation by explicit generation of faulty FSMs, *Proc. 7th IFIP International Workshop on Protocol Test Systems*, Japan, November.

Z. Kohavi (1978), Switching and finite automata theory, McGraw-Hill.

F. Lombardi, Y. N. Shen (1992), Evaluation and improvement of fault coverage of conformance testing by UIO sequences, *IEEE Transactions on Communications*, vol. 40, no. 8, pp. 1288-1293, August.

R. E. Miller, S. Paul (1991), Generating minimal length test sequences for conformance testing of communications protocols, *Proc. IEEE INFOCOM 91*, pp. 970-979.

H. Motteler, A. Chung, D. Sidhu (1993), Fault coverage of UIO-based methods for protocol testing, *Proc. 6th IFIP International Workshop on Protocol Test Systems*, pp. 21-33, France.

S. Naito, M. Tsunoyama (1981), Fault detection for sequential machines by transitions tours, *Proc. IEEE Fault Tolerant Computer Systems.*

K. K. Sabnani, A. T. Dahbura (1988), A protocol test generation procedure, *Computer Networks and ISDN Systems*, vol. 15, no. 4, pp. 285-297.

Y. Shen, F. Lombardi, A. T. Dahbura (1992), Protocol conformance testing using multiple UIO sequences, *IEEE Transactions on Communications*, vol. 40, no. 8, pp. 1282-1287, August.

H. Show, H. Ural (1993), Data flow oriented test selection for LOTOS, Technical Report TR 93-12, Department of Computer Science, University of Otawa, Canada.

M. Yao, A. Petrenko, G. Bochmann (1993), Conformance testing of protocol machines without reset, *Proc. 13th Symp. Protocol Specification, Testing and Verification*, Belgium, May.

M. Yao, A. Petrenko, G. Bochmann (1994), A structural analysis approach to the evaluation of fault coverage for protocol conformance testing, *Proc. IFIP Conference on Formal Description Techniques - FORTE 94*, pp. 389-404, Switzerland, October.

J. Zhu, A. T. Chanson (1994), Fault coverage evaluation of protocol test sequences, *Proc. 14th IFIP Symp. Protocol Specification, Testing and Verification*, Canada.

Ricardo Anido received his Ph.D. from Imperial College, London, in 1989. He is on the faculty of the Department of Computer Science at Universidade Estadual de Campinas, Brazil, where he served as chairman from 1992 to 1994. Dr. Anido is currently spending a sabbatical year at Institut National des Telecommunications, France. His current research interests are distributed algorithms, testing and debugging distributed programs, and software fault-tolerance.

Ana Rosa Cavalli, received the Doctorat d'Etat es Mathematics Sciences and Informatics in 1984 from the University of Paris VII, Paris, France. In 1981, she joined the LITP (Laboratoire d'Informatique Theorique et Programmation) of the C.N.R.S., Paris, where she worked on automatic proof methods for temporal logics and their applications to the specification and verification of protocols. From 1985 to 1990, she was a staff research member at the CNET(Centre National d'Etudes des Telecommunications), where she worked on software engineering and formal description techniques. Since 1990, she joined as professor the INT (Institut National des Telecommunications). Currently, A.R. Cavalli is responsible at INT of the research group "Methods and tools for testing". She is a member of the ISO/ITU-T experts group on "Formal Methods in Conformance Testing" and an associate member of the C.N.R.S. She has served as a member of Program Committees for the 7th IFIP International Workshop on Protocol Test Systems (IWPTS) and for the 6th, 7th and 8th IFIP International Conference on Formal Descritpion Techniques (FORTE) and for CFIP95 (Colloque Francophone sur l'Ingenierie des Protocoles). She has published over 50 publications. Her research interest include formal description techniques and verification of protocols, formal methods in conformance testing, methodology of distributed computing, logics and proof methods for distributed systems.

PART SIX

Testability

INVITED LECTURE

2 A TESTABILITY FRAMEWORK

2.1 Communication software testability

Design For Testability (DFT) is understood as the process of introducing some features into a protocol specification or implementation that facilitate the testing process. DFT applied at the implementation level deals with a particular realization on a given platform and it is often called "instrumentation", whereas DFT at the specification level affects all possible implementations regardless of the implementation process.

For a better understanding of communication software testability issues, we give the following rather general definition:
"A software has the property of testability if it includes facilities allowing the easy application of testing methods, the detection or isolation of existing faults".

Observability and controllability of software are widely acknowledged as two important attributes that influence software testability. Although these attributes can be expressed in many different ways, it is often easier to describe them in terms of various criteria [DsFo91], [KDC94].

The set of properties that characterize testable software is not yet well-defined, but software designers have an intuitive idea as to what constitutes testable software. Therefore, this concept corresponds to several desires developed to confront difficulties perceived daily in testing. These difficulties are summarized as follows:

- The selection of the best test suite with minimal length and maximal coverage of faults may be difficult. A test set may be infinite, in which case, total test coverage cannot be achieved unless a fault domain is restricted in advance. This problem is not always due to the test selection techniques in use. The software design or test architecture can also be the cause for the undetected faults during the test campaign.
- The application of a test suite to the implementation under test may be difficult since most test suites are derived from protocol specifications. Such test suites are called abstract test suites. The advantage of abstract test suites is their reusability for testing different implementations of the same specification. The efforts spent in adapting these test suites for specific implementations can diminish the advantage of using such suites.
- It is not easy to identify the design schemes that lead to parts which are untestable or difficult to test.
- The analysis and interpretation of test results (traces) is another difficulty frequently encountered in testing. The problem of finding a matching trace in the specification necessitates the identification of each input and output in both the expected and the observed traces.
- Another important issue in the software development process is the fault diagnostics. When a test case fails, it may be difficult to determine the cause of the failure. The presence of multiple faults in the implementation can make the fault isolation problem even more complex.

2.2 A generic framework

Design for testability can be seen as an iterative process of modifying specifications and/or implementations and measuring the testability of the newly obtained specification. These transformations can be partially automated and integrated in the traditional software development cycle as described in Figure 1. The ultimate goal of these transformations is to produce a more testable specification/ implementation [Dsso91].

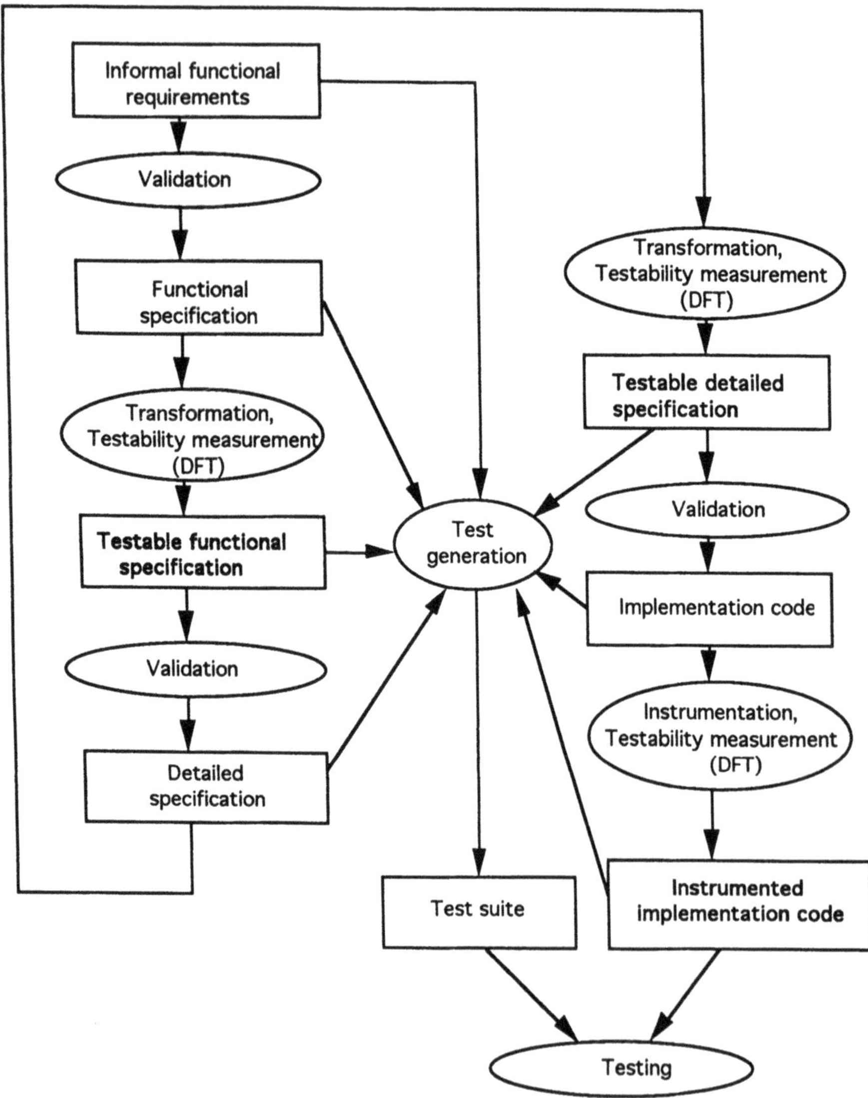

Figure 1 Communication software development cycle

Figure 2 illustrates how the testability of a specification can be improved based on certain testability measures which give an indication on the quality of the end product. The methodology seems simple to apply when testability problems are well identified and for which solutions exist. However for the general case, the Figure 2 raises the following questions:

1. What are the appropriate transformations the protocol designer could apply to the given specification?

2. What are the proper testability measures that the protocol designer should use?
3. When will the process of transformation terminate?

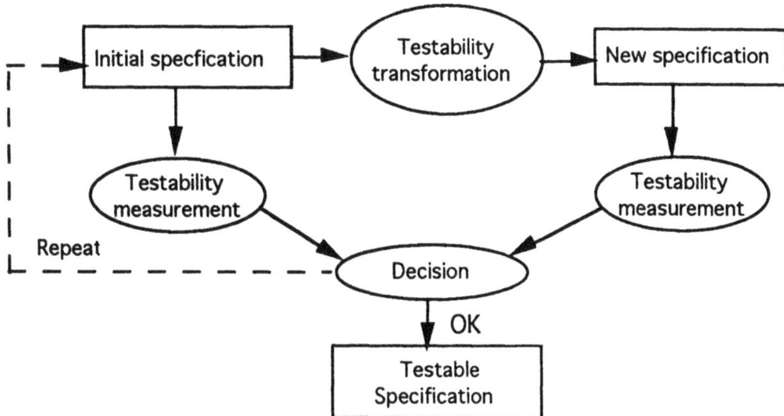

Figure 2 Testability enhancement based on transformation

Generally speaking, a testability transformation should be guided by the identification of what is not easy to test. Designers should try to determine factors that degrade the testability of a particular specification. The list of these factors includes [Dsso91], [VLC93]:

- the formal description language,
- the complexity of the specification,
- nondeterminism due to concurrency,
- the degree of freedom allowed in the specification,
- the level of abstraction that can be used for testing,
- the test architecture,
- the test strategy.

Unfortunately, most existing protocols have been designed and documented without testing requirements in mind [VLC93]. Therefore, it is not realistic to expect that we could alter basic communication functions to improve testability of future protocol implementations. Thus, the design for testability may only attempt to improve the effectiveness of the testing process under certain constraints. It is still possible to automate the procedure of testability transformation for some factors. In particular, consider the partially specified finite state machine (PFSM) model, and a testing strategy based on state identification. If the PFSM has no distinguishing sequence, then the machine is hard to test, and the cost of testing is high. In this case, a systematic approach can be applied: find a minimal augmentation of a specification such that all states becomes distinguishable at a minimal cost. This augmentation can be based on specifying "don't care" transitions that do not generate additional end-to-end exchange; this approach is elaborated in the accompanying paper [YPD95].

For question 2, [Free91], [VMM91] and [PDK93] proposed various testability measures. These measures focus on a particular aspect of testability, and cannot be applied at each step in communication software development process. A model which permits the measurement of testability of software, for the purposes of comparison, would help designers in overcoming many difficulties. Ideally, this model should reflect all aspects of what we call "easily testable" software. However, different testing strategies require different measures to support the management of the design and testing processes in communication software development. The testability measure might be seen rather as a vector in which each element is a measure of a particular factor or aspect of testability [KDC94].

For the last question, the termination of the process of transformation is related to the satisfaction of testability criteria. The desired degree of testability that leads to a confidence in the design should be used as a threshold.

2.3 Instrumentation of implementations

As we said already, the DFT activity can be applied at different steps in the software development cycle. DFT applied to the detailed specification will affect a particular implementation. It offers a way to define various test interfaces depending on designers', and testers' specific needs.

Communication software systems are usually composed of many communicating modules. Each module can also have its own internal structure (Figure 3). Such a multilevel structure may be used as additional information for grey-box testing [PYD94].

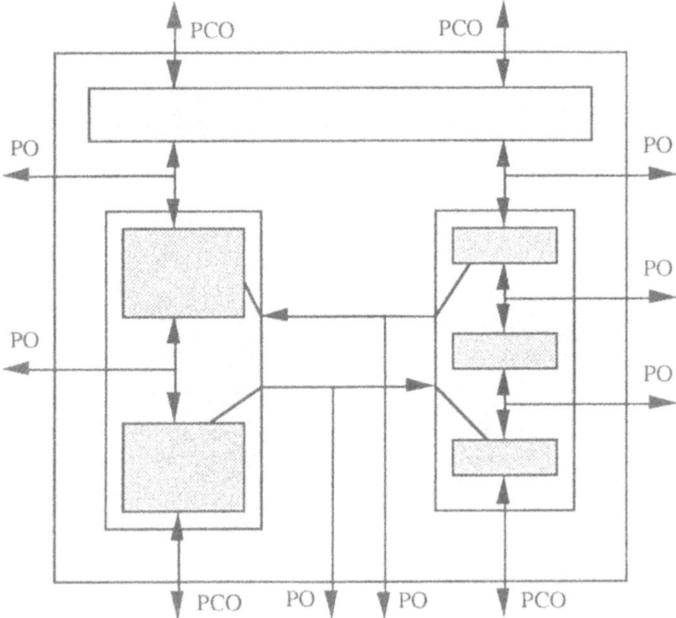

Figure 3 Multilevel structure of communication software

To enhance the testability of such modular systems, it is desirable to make their internal structure visible for testers. This can be done by instrumenting an implementation with internal interaction points accessible by testers. These are points of observation (PO) in addition to external points of control and observation (PCO).

There are several ways of systematic instrumentation such as the addition of primitives, selective broadcasting and the direct call to trace procedures. The instrumentation based on the use of formal description techniques can be often done easily especially when it uses normal constructs of the formal languages.

Addition of primitives
This consists of increasing the functionality of the interaction point by adding a specific primitive whose role is to take the trace of any event happening at this interaction point. This trace may be immediately analyzed by an observer or stocked in memory for later use.

Selective broadcasting
Here each event is selectively broadcasted to its original destination and to the observer. However this method changes substantially the semantics of an interaction point in the usual specification languages (Estelle, LOTOS, SDL).

Direct call to trace procedure
It consists of adding calls to a trace procedure within the code. The choice of the locations where these calls are to be added may correspond to the interaction points of the structure of the software to be implemented, or to a fixed location.

The observation of an implementation may be granular where the degree of granularity depends on what we can control/observe and on what we wish to control/observe. This granularity is based on the inclusion of test points and points of observation. Notice that these points are not necessarily interaction points. We consider the following levels of granularity of instrumentation: structure based instrumentation, transition based instrumentation and state based instrumentation.

Structure based instrumentation
This is the most appropriate level of observation for communication software since it requires an approach based on modular decomposition and communication between modules. It allows the observation and/or control of all specified interaction points (internal or external). It may be associated with a multilevel structure where each module may have its own structure.

Transition based instrumentation
Some software do not allow the application of the modular approach. However if a piece of software is based on transitions (automata), it is possible to include points of observation at each transition thus permitting the recording of their traces. We could also have PCO's since each transition may be executed separately.

State based instrumentation
This type of instrumentation is based on the so-called "read-state" message. If the protocol designer is free to add an additional input ("read-state" message), then new outputs, one per state, are added into the protocol. In this case, the problem of constructing an easily testable implementation becomes trivial. However, such a solution seems expensive, since a protocol entity should support more messages than originally required, so in practice, it has not been accepted as a universal solution.

To conclude our general discussion on testability issues, we notice that formal methods for DFT of protocols have not yet been explored. Note that the existing formal methods and techniques for improving testability which have been developed mainly in the hardware area, see, for example, [ShLe94], [ABF90], [Jose78], cannot be applied in this domain, since they either rely on the structure of an implementation, or they change the set of states, or the sets of input/output events, in a way that is similar to the adding of a read-state message in protocol engineering.

3 DESIGN FOR TESTABILITY FOR THE FSM MODEL

The finite state models in general, and the FSM model in particular, have been extensively used in conformance testing of communication protocols [PBD93], [BoPe94] as well as in hardware and software testing. Testing based on the finite state models has the advantages that the concept of full or complete fault coverage of tests can be formally addressed, and FSM parameters which influence the testability of FSM-based implementations can be identified.

3.1 Basic definitions

An FSM M is a 6-tuple (S, X, Y, h, s_1, D_S), where S is a set of states with s_1 as the initial state; X - a finite set of input symbols; Y - a finite set of output symbols; D_S - a specification domain which is a subset of $S\times X$; h - a behavior function $h: D_S \rightarrow \mathbb{P}(S\times Y)$, and $\mathbb{P}(S\times Y)$ is the powerset of $S\times Y$ [PBD93]. According to the semantics of the input/output behavior all transitions are labeled by a pair of input and output, however, the null (input or output) event can be additionally introduced to implicitly model situations involving spontaneous transitions, for example timeouts [PBD93]. We say that there is a transition from state s to state s' labeled by the pair x/y if $(s',y) \in h(s,x)$, in this case, $s' \in h^1(s,x)$ and $y \in h^2(s,x)$, where h^1 is the first projection (the transfer function) of the behavior function, and h^2 is the second projection (the output function) of the behavior function. The set of states reached after an input/output sequence α/β is accepted by the initial state of M is defined by $h^1{}_\beta(s_1,\alpha)$.

The above definition of the FSM model includes as special cases various classes of FSMs studied in the literature.

Depending on the specification domain, an NFSM can be partially or completely specified. If $D_S = S\times X$ then M is completely specified or *complete*, otherwise it is *partial*.

An FSM is said to be *observable* if starting from any given state the machine can reach only one state in response to any input/output sequence accepted for the given state. In a *deterministic* machine, an output sequence is not required to determine a unique state reached when an input sequence is accepted by the given state. All deterministic machines are observable, however, an observable machine can still be *nondeterministic*. The equivalent transformation of nonobservable machines into observable forms is possible [Star72], [LBP94]. Similar to the case of automata determinization [HoUl79], a nonobservable machine with n states may have up to 2^n states in its observable form. An example of this transformation is given in Section 3.3.

Two states are said to be *distinguishable* if there exists an input sequence accepted by both states which produces different sets of output sequences when it is applied to each of them, otherwise these states are *compatible* states. Compatible states with the same set of acceptable input sequences are called *equivalent*.. Opposed to the compatible states, the equivalent states can always be merged without affecting the specified behavior of the machine. Thus, we exclude the class of FSMs with equivalent states from our discussion. The FSM is *reduced* if all its states are pairwisely distinguishable, otherwise it is *nonreduced*. A deterministic, reduced complete FSM is usually referred to as a *minimal* machine, since every such machine possesses a unique reduced form.

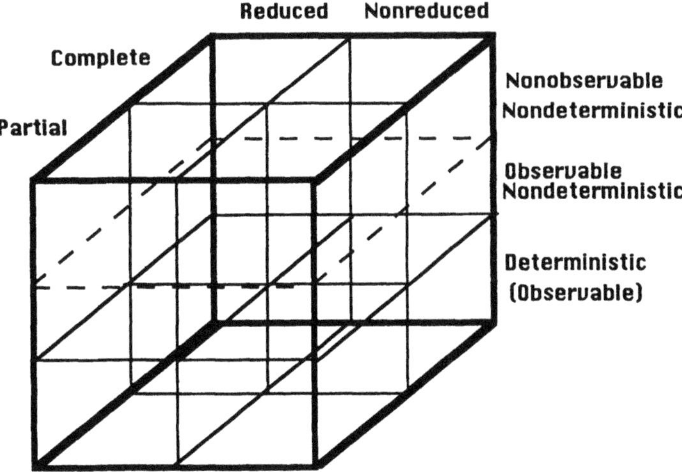

Figure 4 Classification of FSMs

Figure 4 shows various classes of FSMs. Intuitively clear, that the testability of machines from different classes varies. To assess the testability of a given FSM specification one could first find the class to which the FSM belongs. The class indicates the testability issues typical for all machines of this class, as we discuss in the following section. A more precise testability analysis requires the FSM testability parameters such as controllability and observability, formally defined later in the paper.

3.2 Testability and transformation of FSMs

Within the generic testability framework instantiated with the FSM model we first address the following general questions:

Q1 - How can one identify the testability problems in a given FSM specification?
Q2 - How can one estimate the degree of testability of a given FSM?
Q3 - What are the transformations that can be applied by the protocol designer?
Q4 - What freedom has the designer in choosing a specific transformation to apply?

Q1 - The identification of testability problems of an FSM can be viewed as a classification problem of FSMs [PBD93], see Figure 4. It is well known that nondeterministic FSMs are usually less testable than deterministic machines. Within nonderministic FSMs nonobservable machines are less testable than the observable ones. The major factor that affects the testability in this case is the controllability. Nonreduced FSMs are less testable than reduced FSM and the major factor of testability that is important in this case is the observability, more specifically the fuzziness of the machine. Completely specified machines tend to be more testable than partially specified ones. Given two machines defined over the same sets of states and inputs, a partial machine has fewer transitions than a complete one, however, identification of its states may require longer sequences than that for the latter.

Q2 - A particular type of the given machine gives an idea on the FSM testability, however, to assess it in a more accurate way we have to first instantiate the definitions of controllability and observability for the FSM model. We address this issue later in the paper.

Q3 - Designers might have the possibility to modify an FSM by acting on the following elements: states, inputs, outputs, transition and output functions. Transformations may involve one or several elements. Certain transformations are well known, and their objectives are described in the literature. A minimization algorithm brings an FSM from the class of nonreduced FSMs to the class of reduced FSMs can be viewed as a transformation method. The addition of "read-state" message is another transformation. [YPD95] gives a transformation method for a partial FSM which increases the degree of its specifiedness.

The designer should categorize properly the specification, clearly define testability objectives in terms of which class of FSMs he/she would like to obtain after the termination of the transformation process given in Figure 2. Depending on the testability objectives that one can define, some heuristics can be developed in order to find minimal modifications of an FSM that meet the defined objectives.

Q4 - The choice of a specific transformation is related to the freedom that the designer might have. It depends also on the reached step in the development process. In the case where a specification is a standard, and there exist already its implementations, then transformations are restricted to those that do not generate end-to-end exchange (problem of symmetric protocols), in terms of additional PDU's. In the case where a lower layer can absorb all additional PDUs or a peer entity is designed with the ability to ignore them, then a more fundamental transformation might be allowed.

3.3 Controllability of FSMs

Let V be a state cover of M, i.e. for every state s in S there is at least one transfer sequence $v \in V$ such that $s \in h^1(s_1,x)$. If the machine M is not initially connected then there exists at least one state that cannot be reached from the initial state. In this case, we may assume that this state has an infinite transfer sequence. Note that FSMs with unreachable states usually require lengthy test suites [PHK95]. A state cover constitutes a skeleton of test suites with guaranteed or full fault coverage in the sense that every transition from every state specified in the machine should be tested by such a test suite. The machines may possess several state covers, therefore the amount of efforts required to bring the machine to all possible states depends on the state cover chosen for testing. To get an objective characterization of the controllability of the given machine we should rely on a state cover with the minimal amount of these efforts.

In fact, this amount depends on a number of factors. First, the length $L(v)$ of a transfer sequence v is important when we are required to force the machine to reach a certain state. Next, if the machine is nondeterministic then not only one, but several states can be reached when the transfer sequence is applied to the initial state. If, moreover, the FSM is not observable then even observing the output sequence produced by the transfer sequence does not help much in determining the final state of the machine. In fact, only states of its observable form are controllable. These are sometimes called multi-states of the original nonobservable machine [FuBo91]. Finally, the machine may produce in response to the transfer sequence not just a single output sequence but a number of them. It implies than we have to repeatedly apply the same transfer sequence until a desirable output sequence is produced indicating that a proper (multi-) state is reached (cf. the so-called complete testing assumption in [LBP94]). We integrate the above factors into a single formula based on the notion of the weight of a state cover constructed for the observable form of the machine.

Given state cover V of an observable machine, the *weight* $\omega(V)$ of V is defined as follows:

$$\omega(V) = \sum_{v \in V} L(v)\,|h^2(s_1,v)|$$

Then the *controllability* $C(M)$ of the given machine M is defined as $1/\omega_{min}(V)$, where $\omega_{min}(V)$ is the minimal value of the weight among all its possible state covers of its observable form.

Example
Figure 5 (a) shows an example of a nonobservable FSM. Its equivalent observable form is given in Figure 5 (b). Compared to the original machine, there is a new state introduced, viz. the combination of states 2 and 3. The set $V=\{\varepsilon, a, ab\}$ is a minimal state cover. $\omega_{min}(V) = 0 + 1(2) + 2(3) = 8$. Thus, $C(M) = 1/8$.

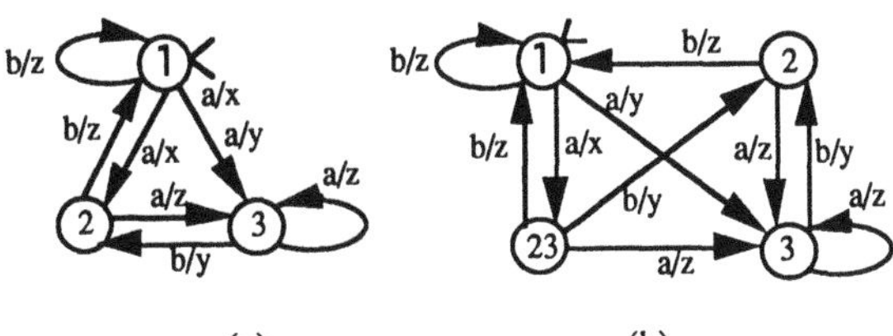

(a) (b)
Figure 5 A nonobservable FSM and its observable form.

Assume that the behavior specified by the given nonobservable FSM can be restricted such that it becomes completely deterministic. This can be achieved if two outgoing transitions labeled with the output x are removed. The modified machine has the same minimal state cover $\{\varepsilon, a, ab\}$. However, the controllability has now increased to 1/3.

Machines not initially connected are the least controllable as there exists a transfer sequence of the infinite length. The controllability of an initially connected deterministic FSM is simply determined by the overall length $L(V_{min})$ of its minimal state cover V_{min}. In the best possible case, for every state except for the initial state, there exists just a single input which takes the machine from the initial state into this state. Then $C(M) = 1/n\text{-}1$, where n is the number of states in M. In the worst case, the maximal length of a transfer sequence may reach n-1, thus the total length of a state cover does not exceed $n(n\text{-}1)/2$. Thus, the controllability of initially connected deterministic FSMs has the following lower and upper bounds:

$2/n(n\text{-}1) \leq C(M) \leq 1/n\text{-}1.$

Next we switch our discussion to the notion of observability of FSMs.

3.4 Observability of FSMs

Testing with complete fault coverage usually requires the equivalence between states to be verified [Vasi73] . At the same time, black-box testing implies that each state has to be identified. If a given machine is not observable then we can never identify the state reached by the machine after it has accepted an input/output sequence. This is why all currently available test derivation methods for FSMs deal only with observable forms whose states are multi-states of the original nondeterministic machine. Certain observable machines are "more observable" than the others if their states require less test events for identification, thus the definition of the observability of observable FSMs should be based on the notion of states' distinguishability. As discussed above, an arbitrary FSM may be either reduced or not. Opposed to reduced observable machines, the nonreduced ones usually require complete test suites of an exponential length, since compatible states can be implemented as a single state. The influence of compatible states on the observability of the given FSM can be characterized in the following way.

Given an arbitrary observable FSM M, we compute a partition Π of the set S of its states into subsets $\Pi = \{S_1, ..., S_f\}$ such that every subset includes only pairwise distinguishable states. The number $f(M)$ of subsets in the minimal partition $\Pi_{\min}$ is called the *fuzziness* of M [YePe89], [LPB94].

If M happens to be complete and minimal, or partial and reduced, then $f(M) = 1$. The larger the number of pairs of compatible states, the higher the fuzziness usually is. For a "fully unreduced" machine $f(M) = n$.

Example

Consider a nonreduced FSM shown in Figure 6.

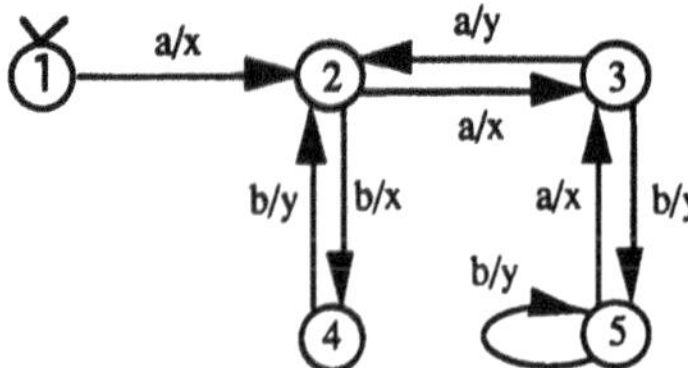

Figure 6 A nonreduced FSM

States 1 and 4 are compatible states. The set of states has the following minimal partition on subsets pairwise distinguishable states: $\Pi_{\min} = \{1,2,3,5;\ 4\}$. Thus, the fuzziness of the machine is two.

The fuzziness characterizes the influence of compatible states on the observability of the given machine, however, distinguishable states require a special consideration.

Assume that the given machine M is reduced, i.e. it has no compatible states. In this case, all states can be uniquely identified (observed) using a so-called *characterization set* W of the given FSM which contains input sequences to tell each pair of states apart [Vasi73], [Chow78], [LeYa94]. The number $|W|$ of sequences in W and their total length $L(W)$ characterize the amount of efforts required to identify states of M. The machine usually possesses several characterization sets. The *weight* $\omega(W)$ of the given W set can be determined as follows:

$$\omega(W) = |W|\, L(W).$$

Then the *distinguishability* $D(M)$ of the given reduced observable machine M is defined as $1/\omega_{\min}(W)$, where $\omega_{\min}(W)$ is the minimal value of the weight among all its possible characterization sets.

The distinguishability of the reduced machine completely characterize its observability. The most observable machines are FSMs which possess a distinguishing sequence of the length one, e.g. the "read-state" input. Thus, $D(M) \leq 1$. There exists also the least upper bounds for the W set of complete reduced machines. In particular, as shown in [TyBa75], $L(W) \leq n(n-1)/2$. In this case, $|W| = n$. Then $D(M) \geq 2/n^2(n-1)$ for complete reduced machines.

Reduced partial machines tend to be less observable than complete machines since parameters of W sets have higher tight bounds. The total length of sequences of a characterization set W has the least upper bound of $[(n^2-n+1)^2-1]/8$ for reduced partial FSMs

[PeBo95]. Then $D(M) \geq 8/n[(n^2-n+1)^2-1]$ for this class of machines.

The observability of a nonreduced partial machine M depends on both parameters, its fuzziness $f(M)$ and distinguishability $D(M)$. The definition of a characterization set is extended to cover nonreduced machines [Petr91]. In particular, such a set should contain separating sequences for every pair of distinguishable states. Then the distinguishability of a nonreduced machine can be evaluated in a similar manner. The observability of such machine is characterized by both, the distinguishability and fuzziness parameters.

Example

Consider again the FSM in Figure 6. As shown above, its fuzziness is two. If now we find the minimal characterization set, such as $W = \{aa, bb\}$ then its distinguishability can be evaluated: $D(M) = 1/\omega_{\min}(W) = 1/8$. Now we compare these parameters of the given machine with those of its reduced form. It is obtained by merging compatible states 1 and 4. The fuzziness of the reduced machine is just one. The set $W' = \{aa, b\}$ is the minimal characterization set, is the distinguishability of the reduced form becomes now 1/6. A test suite that is complete for the given nonreduced machine (Figure 6) in the class of all FSMs with up to four states derived in [Petr91] has forty test events. The length of a corresponding test suite for the reduced form is only 25. Thus, transforming a nonreduced machine into its reduced form may improve its observability, and hence its testability.

4 CONCLUSION

Design for testability integrated in the communication software life cycle attempts to reduce the development costs and at the same time to improve the quality of a final product. To achieve these objectives, formal methods for testability evaluation and enhancement should be developed within an appropriate framework.

In this paper, we have proposed in the first part, a generic framework of design for testability where DFT activity is integrated in the communication software development process. A generic model for testability transformation based on modification of a given specification and testability measurement has been explained. In the second part, we have discussed the notion of testability in terms of controllability and observability. We have considered the FSM model of a protocol machine and proposed a classification method based on the formalized notions of controllability and observability. We have addressed one particular problem of DFT, the problem of analyzing the testability of a given module specified as a finite state machine. Transformations in this paper are not the main objective, but they are used to illustrate how testability can be enhanced by finding an augmentation of the given protocol behavior such that a newly obtained specification is more testable than the original one. A measure of testability of a protocol entity is assumed to be based on the shortest length of a test suite needed to achieve guaranteed coverage of certain faults.

Future work will address testability evaluation and enhancement of protocol specifications modeled by extended FSMs and written in FDTs such as SDL, Estelle, and LOTOS.

5 REFERENCES

[ABF90] M. Abramovici, M. A. Breuer, and A. D. Friedman, Digital Systems Testing and Testable Design, Computer Science Press, Oxford, England, 1990.

[BoPe94] G. v. Bochmann and A. Petrenko, "Protocol Testing: Review of Methods and Relevance for Software Testing", ISSTA'94, ACM International Symposium on Software Testing and Analysis, Seattle, U.S.A., 1994, pp. 109-124.

[BPY94] G. v. Bochmann, A. Petrenko, and M. Yao, "Fault Coverage of Tests Based on

Finite State Models", the Proceedings of IFIP TC6 Seventh International Workshop on Protocol Test Systems, 1994, Japan.

[Chow78] T. S. Chow, "Testing Software Design Modeled by Finite-State Machines", IEEE Transactions on Software Engineering, Vol. SE-4, No.3, 1978, pp.178-187.

[DsFo91] R. Dssouli and R. Fournier, "Communication Software Testability", IFIP Transactions, Protocol Testing Systems III (the Proceedings of IFIP TC6 Third International Workshop on Protocol Test Systems), Ed. by I. Davidson and W. Litwack, North Holland, 1991, pp.45-55.

[Free91] R. S. Freedman, "Testability of Software Components", IEEE Transactions on Software Engineering. Vol. SE-17, No. 6, June 1991.

[FuBo91] S. Fujiwara and G. v. Bochmann, "Testing Non-deterministic State Machines with Fault Coverage", IFIP Transactions, Protocol Test Systems, IV (the Proceedings of IFIP TC6 Fourth International Workshop on Protocol Test Systems, 1991), Ed. by Jan Kroon, Rudolf J. Heijink and Ed Brinksma, 1992, North-Holland, pp. 267-280.

[Jose78] J. Joseph, "On Easily Diagnosable Sequential Machines", IEEE Transactions on Computers, Vol. C-27, February, 1978, pp.159-162.

[HoUl79] J. E. Hopcroft, J. D. Ullman, Introduction to Automata Theory, Languages, and Computation, Addison-Wesley Publishing Company, Inc., 1979, 418p.

[KDC94] K Karoui, R. Dssouli, O. Cherkaoui et A. Khoumsi, " Estimation de la testabilité d'un logiciel modélisé par les relations", publication #921, Département d'Informatique et de Recherche Opérationnelle (DIRO), Université de Montréal.

[LBP94] G. Luo, G. v. Bochmann, A. Petrenko, "Test Selection based on Communicating Nondeterministic Finite State Machines using a Generalized Wp-Method", IEEE Transactions on Software Engineering, Vol. SE-20, No. 2, 1994, pp.149-162.

[LPB94] G. Luo, A. Petrenko, and G. v. Bochmann, "Selecting Test Sequences for Partially-Specified Nondeterministic Finite State Machines", the Proceedings of IFIP TC6 Seventh International Workshop on Protocol Test Systems, 1994, Japan.

[LeYa94] D. Lee and M. Yannakakis, "Testing Finite-State Machines: State Identification and Verification", IEEE Trans. on Computers, Vol. 43, No. 3, 1994, pp. 306-320.

[Petr91] A. Petrenko, "Checking Experiments with Protocol Machines", IFIP Transactions, Protocol Test Systems, IV (the Proceedings of IFIP TC6 Fourth International Workshop on Protocol Test Systems, 1991), Ed. by Jan Kroon, Rudolf J. Heijink and Ed Brinksma, 1992, North-Holland, pp. 83-94.

[PBD93] A. Petrenko, G. v. Bochmann, and R. Dssouli, "Conformance Relations and Test Derivation", IFIP Transactions, Protocol Test Systems, VI, (the Proceedings of IFIP TC6 Fifth International Workshop on Protocol Test Systems, 1993), Ed. by O. Rafiq, 1994, North-Holland, pp.157-178.

[PDK93] A. Petrenko, R. Dssouli, and H. Konig, "On Evaluation of Testability of Protocol Structures", IFIP Transactions, Protocol Test Systems, VI, (the Proceedings of IFIP TC6 Fifth International Workshop on Protocol Test Systems, 1993), Ed. by O. Rafiq, 1994, North-Holland, pp.111-123.

[PeBo95] A. Petrenko, G. v. Bochmann, "On Fault Coverage of Protocol Tests", submitted to Computer Networks and ISDN Systems on Protocol Testing, 1995.

[PHK95] A. Petrenko, T. Higashino, and T. Kaji, "Handling Redundant and Additional States in Protocol Testing", IWPTS'95, France.

[PYD94] A. Petrenko, N. Yevtushenko, R. Dssouli, "Testing Strategies for Communicating FSMs", IWPTS'94, Japan, 1994.

[Star72] P. H. Starke, Abstract Automata, North-Holland/American Elsevier, 1972, 419p.

[ShLe94] M. L. Sheu and C. L. Lee, "Symplifying Sequential Circuit Test Generation", IEEE Design and Test of Computers, Fall 1994, pp. 28-38.

[TyBa75] T. Tylaska and J. D. Bargainer, "An Improved Bound for Checking Experiments that Use Simple Input-Output and Characterizing Sequences", IEEE Transactions on Computers, Vol. C-24, No.6, 1975, pp. 670-673.

[Vasi73] M. P. Vasilevski, "Failure Diagnosis of Automata", Cybernetics, Plenum Publishing Corporation, New York, No.4, 1973, pp.653-665.

[VLC93] S. T. Vuong, A. A. F. Loureiro, and S. T. Chanson, "A Framework for the Design for Testabilitiy of Communication Protocols", IFIP Transactions, Protocol Test Systems, VI, (the Proceedings of IFIP TC6 Fifth International Workshop on Protocol Test Systems, 1993), Ed. by O. Rafiq, 1994, North-Holland, pp.89-108.

[VMM91] J. Voas, L. Morrell and K. Miller, "Predicting Where Faults Can Hide From Testing", IEEE Software, March 1991.

[YePe89] N. Yevtushenko and A. Petrenko, "Fault-Detection Capability of Multiple Experiments", Automatic Control and Computer Sciences, Allerton Press, Inc., New York, Vol.23, No.3, 1989, pp.7-11.

[YPD95] N. Yevtushenko, A. Petrenko, R. Dssouli, K. Karoui, and S. Prokopenko, "On the Design for Testability of Communication Protocols", IWPTS'95, France.

Acknowledgments

This work was partly supported by the HP-NSERC-CITI Industrial Research Chair on Communication Protocols at Université de Montréal, the NSERC Strategic and Individual Research Grant.

6 BIOGRAPHY

Rachida Dssouli received the Doctorat d'université degree in computer science from the Université Paul-Sabatier of Toulouse, France, in 1981, and the Ph.D. degree in computer science in 1987, from the Université de Montréal, Canada. She is currently an Associate professor at the University of Montréal. Her research area is software engineering and her research interests include requirements engineering, software evolution, protocol specification, testing and testability.

Kamel Karoui is a Ph.D. student of the Université de Montréal, Canada.

Alexandre Petrenko received the Dipl. degree in electrical and computer engineering from Riga Polytechnic Institute in 1970 and the Ph.D. in computer science from the Institute of Electronics and Computer Science, Riga, USSR, in 1974. Since 1992, he has been with the Université de Montréal, Canada. His current research interests include communication software engineering, protocol engineering, conformance testing, and testability.

Omar Rafiq received the Doctor ès-Sciences degree (1983) in computer science from the University of Bordeaux-I where he was an assistant professor from 1974 to 1978. He spent 10 years in banking, research and industry before joining the University of Pau in 1987 as professor. He is currently a full professor at the same university. His research interests include computer networks, protocol engineering and distributed processing.

16

Design for testability of protocols based on formal specifications

Myungchul Kim, Samuel T. Chanson and Sangjo Yoo*
Korea Telecom Research Laboratories
Sochogu Umyundong 17, Seoul, Korea 137-792
*E-mail:*mckim@sava.kotel.co.kr
** University of British Columbia*
Department of Computer Science, University of British Columbia,
*Vancouver, B .C. Canada V6T 1Z2, E-mail:*chanson@cs.ubc.ca

Abstract
In this paper, we propose a generic scheme which instruments a formal protocol specification automatically to enhance the testability of the implementation. This approach is a special case of design for testability. It is cost-effective considering the entire cycle of protocol development. The advantage of automatic instrumentation is that the user need not pay special attention to testing problems in the design phase. Unlike most other techniques, our scheme also works with existing protocol specifications since it does not affect the original design. Our models address the problems of controllability and observability, and can handle both sequential and concurrent formal protocol specifications.

Keywords
Design for testability, formal specifications, protocols

1 INTRODUCTION

The world's market for computer products is becoming very competitive. In order to win market share, some companies are changing the entire production cycle from a sequential one (called waterfall model) to a parallel model. The so-called concurrent engineering practice [IEEE 91] has emerged in recent years with the aim of making the entire production cycle as parallel and integrated as possible based on many considerations such as manufacturability, testability, performance, quality, installability, reliability, safety, and serviceability. The concept is based on the premise that if these aspects are taken into consideration in the design phase, the total cost and time of the production cycle will be significantly reduced. So far, most of the work on concurrent engineering has been focussed

---

*currently on leave at the Hong Kong University of Science and Technology.

on hardware products. This paper addresses the design for testability of communication protocols based on formal specifications.

Design for testability on hardware, especially chip testing, is quite mature [Fujiw 85]. In fact, some techniques have been standardized. Chips that are built following the standards can be tested more easily, precisely and in a cost-effective manner using standard techniques through Test Access Ports [Parke 89]. The approach can be classified as gray-box testing, which is in between white-box and black-box testing, as it allows partial access in a controlled fashion to the internals of the system.

The ultimate objectives of both hardware and software testability are similar: avoid or minimize the state space explosion problem in order to reduce the time and cost of testing. However, there are major differences between hardware and software with respect to the testing process [Hoffm 89]. The differences are mainly due to the fact that in hardware testing, we would like to determine whether an implementation (such as a chip) is an accurate copy of the circuit design since the correctness of the circuit design has been established in an earlier verification process. This is because implementations from the proven valid design may still contain serious errors introduced in the manufacturing process. On the other hand, for software, once an implementation is proven correct by testing, copies of the same implementation are always correct. In addition, hardware consisting of distributed circuits or boards often has a single physical global clock which is difficult to provide for software running in a distributed system.

Communication protocols are inherently distributed and concurrent. Understanding and analyzing concurrent programs (or specifications) are much harder than those for sequential ones because the execution order of the sequential programs is fixed (or totally ordered) for a given set of inputs. Since Lamport's pioneering work [Lampo 78], there has been considerable research on the study of concurrency, logical clocks and global states in distributed systems. The logical clock [Matte 89, Fidge 91] is a common technique used to determine whether or not two events are concurrent.

The International Organization for Standardization (ISO) and the International Telecommunication Union - Telecommunication (ITU-T) have developed formal description techniques (FDTs), viz., Estelle, LOTOS, and SDL [ISOb, ISOc, ITU], for specification of communication protocols and services in order to avoid the ambiguity of standard documents written in English. The FDTs can be used in the design phase for providing precise description of the products and for rapid prototyping. In this paper, we assume that the implementations are done according to the design described by an FDT, including the modular structure. In the case of automatic implementation of a protocol from a formal specification, the assumption is usually valid. Even when manually implementing a protocol from a design described by an FDT, the implementation process is easier and more straightforward if the modular structure of the formal specification is followed.

The main contributions of this paper are:

- Automatic instrumentation of formal specifications with respect to design for testability,
- Provision of a practical view of controllability and observability of an implementation under test (IUT) following a formal specification,
- Ability to handle both sequential and concurrent specifications, and
- Provision of a mechanism for error location.

The rest of the paper is organized as follows. Section 2 surveys some related work on design for testability. The generic instrumentation schemes for testability in terms of controllability and observability on sequential specifications, and controllability on concurrent specifications are presented in Section 3. In order to provide observability on concurrent specifications, a formal model is proposed in Section 4. Finally, Section 5 concludes the paper.

2 RELATED WORK

Existing work does not offer a standard mechanism which provides testability for formal specifications. The following is a brief survey of some papers on design for testability of software, and in particular, protocols.

Dssouli and Fournier [Dssou 90] suggested a modification in the software development process to accommodate the notion of testability between the implementation phase and the design phase of the software development cycle. By comparison, our model provides a generic approach to supporting design for testability on formal specification by automatic instrumentation.

After investigating the meaning of software testability formally, Freedman [Freed 91] defined domain testability by applying the concepts of observability and controllability which are used in assessing the testability of hardware components to software. A domain testable program is observable and controllable in that it does not exhibit any test input-output inconsistencies. This concept applies to sequential programs (or specifications) only.

Ellsberger and Kristoffersen [Ellsb 92] identified some aspects of SDL specifications that are difficult to test, i.e., asynchronous communication, the time model, and nondeterminism. They also presented suggestions for improving testability in these aspects of SDL which are difficult to be generalized to other specification techniques.

Petrenko and Dssouli [Petre 93] proposed a testability metric based on Finite State Machines (FSM) under the complete fault coverage assumption. The metric can be applied to multiple FSMs.

In [Vuong 93], a framework was presented with respect to the factors that affect testing and testability of communication protocols in the context of analysis, design, implementation, and testing phases of the protocol engineering cycle. While the framework provides a good basis to reason about protocol design for testability, no detailed solution is given that is directly usable on specifications given in FDTs.

ITU-T and ISO have created a task force on "Formal Methods in Conformance Testing"[FMCT]. Specification styles for testability are proposed. The style is a guideline for increasing the testability of specifications given in FDTs. There is no easy way to automate the recommendations.

3 INSTRUMENTATION SCHEMES TO ENHANCE TESTABILITY

In the ISO conformance testing methodology, an IUT is viewed as a black-box for testing purposes [ISOa]. The protocol specification may consist of a single module (i.e., sequential) or multiple modules (i.e., concurrent). Again, we assume that an FDT is used in the design phase and the modular structure described by the FDT is preserved in the implementation.

The objective of design for testability for protocols is to provide precise and efficient ways of testing the protocol implementation (i.e., the IUT). We believe that testing will be made much easier if some internal behavior of the IUT are exposed to the testers, i.e., treating the IUT as a gray-box for testing purposes. Even though we consider the IUT as a gray-box, there are inherent uncertainties due to nondeterminism. The sources of nondeterminisms are summarized below:

1. Nondeterminism in the specification: In order to make the specification concise and provide flexibility to the implementors, FDTs allow the specification of nondeterministic actions. These intentional nondeterministic actions can be translated into deterministic ones during the implementation phase in a standard way (like the translation from Nondeterministic Finite State Automata to Deterministic Finite State Automata).

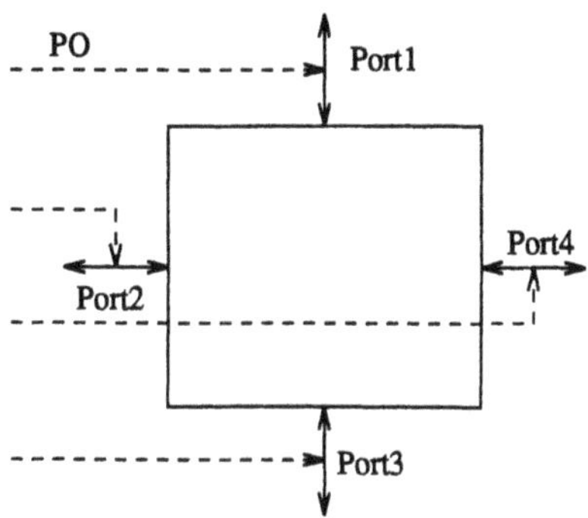

Figure 1 Points of observation.

2. Nondeterminism arising from the concurrent environment: Concurrent behavior in distributed systems is difficult to specify and analyze. In a distributed system, it is generally not possible to determine the total order of events in the system because of the lack of a global clock. Only the partial order of events in the system can be computed, and for concurrent events there is no general way of telling which one has occurred first [Kim 93]. This unintentional nondeterminism is inherent in distributed systems.
3. Nondeterminism due to the black-box approach: The order of messages observed from the Points of Observation (PO) and/or the Points of Control and Observation (PCO) between modules as shown in Figure 1 does not always correspond to the order of event occurrences caused by these messages inside the IUT.

For example, consider two input messages where the input message at port 1 arrives at the IUT earlier than the input message at port 2. However, the messages may be executed in the reverse order of arrival. In order to determine the right order, we may need to enumerate every possible combination of message inputs and outputs beforehand and compare them with the observed behavior. This unintentional nondeterminism may cause state space explosion in trace analysis.

Controllability and observability can be applied at three levels: data, transition and module. Controllability (observability) on data means the ability to assign (print) values to (of) internal variables or parameters of the interaction primitives. However, in the context of conformance testing, it is not valid to modify the internal data. Controllability at the transition/module level is the ability to select specific transitions/modules for testing. Observability at the transition/module level refers to the ability to observe the execution order of the transitions/modules. In this paper, we propose techniques for controllability and observability at the transition and the module levels only. To simplify the discussion, we assume that each module is described in Normal Form Specification (NFS) [Sarik 87]. The NFS is written in Estelle [ISOb] which consists of a single module without procedures, functions, and *while* statements.

3.1 Overall framework

Formal specifications with enhanced testability can be produced by an automatic instrumentation process on the original formal specifications as shown in Figure 2. The instrumentation schemes proposed are generic in the sense that they work for all specifications that satisfy the assumptions given in the previous section. Therefore, the process relieves the protocol designer from having to be concerned with testability issues in the design phase. The instrumented specification can be used to generate the implementation with enhanced controllability and observability automatically (or semi-automatically) using existing tools, or manually.

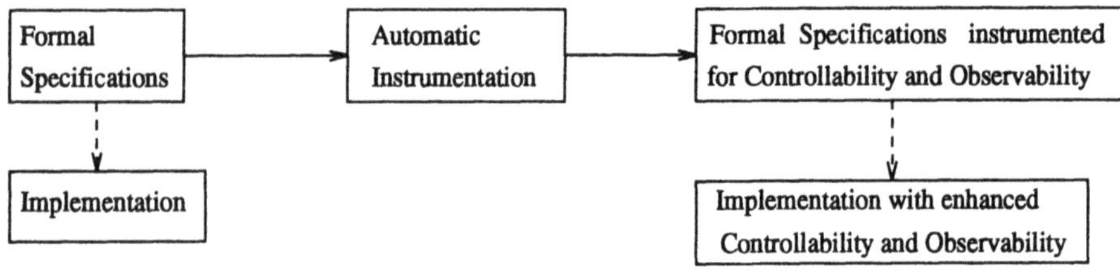

Figure 2 Overall framework.

3.2 Controllability on sequential specification

The issue here is to be able to select the proper transition to be tested. For controllability on a sequential specification, we propose that each transition be instrumented by appending a particular condition just after the PROVIDED clause (if exists). The condition is used to select a particular transition out of possibly more than one candidate transitions due to nondeterminism. The instrumented specification will also work for existing testers not designed to take advantage of the proposed scheme. However, in that case, the selection of nondeterministic transitions cannot be controlled.

The scheme is very simple and is given below. Operations 1, 2 and 3 below are just to specify if the controllability feature should be activated. Operation 4 says that if the existing condition for firing a condition is satisfied and the controllability option is selected, then the transition is fired only if it is the transition specified. Note that the instrumentation does not add illegal behaviour not permitted by the original specification. Its only function is to select a transition out of a set of possible transitions (due to nondeterminism) for testing.

Rules for instrumentation:

1. Interaction primitives "TEST_control" and "TEST_uncontrol" are included in the channel interaction primitives of a sequential specification.
2. The global variable "TEST_var_control = 0" and state set "stateset all = [list_of_all_states]" are defined.
3. The following transitions are added. If the sequential specification or the IUT receives "TEST_control" / "TEST_uncontrol" at whatever state when we need to control / not to control the specification or IUT, set the global variable "TEST_var_control = 1" / "TEST_var_control = 0", respectively.

```
from all
when TEST_control
begin
   TEST_var_control = 1;
end
```

```
from all
when TEST_uncontrol
begin
   TEST_var_control = 0;
end
```

4. A particular condition is appended to the existing condition in the PROVIDED clause as follows:

```
PROVIDED (existing_condition) AND ((!TEST_var_control) OR
     (Input_Primitive.Data = id_of_tr))
```

In order to run an IUT with/without the controllability feature, interaction primitives "TEST_control" / "TEST_uncontrol" are provided. For example, if the feature is to be activated, the interaction primitive "TEST_control" is sent by the tester to the IUT. In that case, the global variable "TEST_var_control" is set to '1' which is used in Operation 4.

By using transition identifier "id_of_tr" in the data field of the input interaction primitive "Input_Primitive.Data", a particular transition can be selected to be fired. Note that no additional PCO is used. If we use an additional PCO for controllability, we will encounter unintentional nondeterminism arising from the concurrent environment as discussed in Section 3. For this reason, the information for controllability is stored in the data field of the interaction primitives. This approach is different from testability in hardware. Hardware testing supports additional PCOs since a global physical clock is usually available.

3.3 Observability on sequential specification

In order to deal with the problems of unintentional nondeterminism due to the black-box approach described in Section 3, an additional output statement with a message identifying the transition that has just been executed is inserted at the end of every transition in the Estelle (NFS) specification. For each module, this output statement is directed to a particular port (PO) created for the purpose of observation where a log of the trace will be recorded. The trace provides the precise execution order of transitions within the module independent of the number of interaction ports in the module. Note that for debugging as well as for testing purposes, it is desirable to have a trace of the execution events even when the controllability feature is used.

The scheme for observability on sequential specification is given below. The instrumentation does not change the semantics of the original specification as only output statements are added. Again, the scheme allows the observability feature to be 'turned off' so that the instrumented specification can work with conventional testers.

1. Interaction primitives "TEST_observe" and "TEST_unobserve" are included in the channel interaction primitives of a sequential specification.
2. Global variable "TEST_var_observe = 0" and state set "stateset all = [list_of_all_states]" are defined.
3. The following transitions are added in the specification. If the sequential specification or the IUT receives "TEST_observe" / "TEST_unobserve" on whatever states, set the global variable "TEST_var_observe = 1" / "TEST_var_observe = 0", respectively.

```
from all                          from all
when TEST_observe                 when TEST_unobserve
begin                             begin
   TEST_var_observe = 1;             TEST_var_observe = 0;
end                               end
```

4. At the end of each transition, the following statement is inserted. The statement outputs to a particular PO ("PO_one" say) an interaction primitive "OBSERVE" whose parameter is the identifier of the transition "id_of_tr".

```
if(TEST_var_observe)   OUTPUT PO_one.OBSERVE(id_of_tr);
```

Interaction primitives "TEST_observe"/ "TEST_unobserve" are provided in order to select/unselect the observability feature. If the feature is needed, the tester sends the interaction primitive "TEST_observe" to the IUT. In that case, the global variable "TEST_var_observe" is set to 1 which is used in Operation 4. If the performance of the IUT is critical or we do not need to observe the IUT, the interaction primitive "TEST_unobserve" is sent instead.

The proposed mechanisms also allow us to control and observe spontaneous transitions as well as transitions containing timers which are difficult problems in conformance testing. Debugging and error location are also made easier since a complete trace of the transition path is maintained.

3.4 Controllability for concurrent specification

Since an NFS specification consists of a single module, we view a concurrent specification as consisting of multiple NFSs. According to our assumption, an IUT derived from a concurrent specification retains the modular structure of the concurrent specification. Also we assume that Estelle, which is the description technique of NFS, supports parameters in the configuration statements (such as init, connect, attach, release, disconnect, and detach) in order to simplify the discussion. Controllability in the concurrent environment means the ability to select a set of specific modules to be tested. The scheme for controllability of concurrent specification is presented below.

1. For those interaction ports which are not exposed to the outside environment, mirrored interaction ports (MIPs) with the same interaction primitives are created in the outermost module as shown in the example in Figure 3.

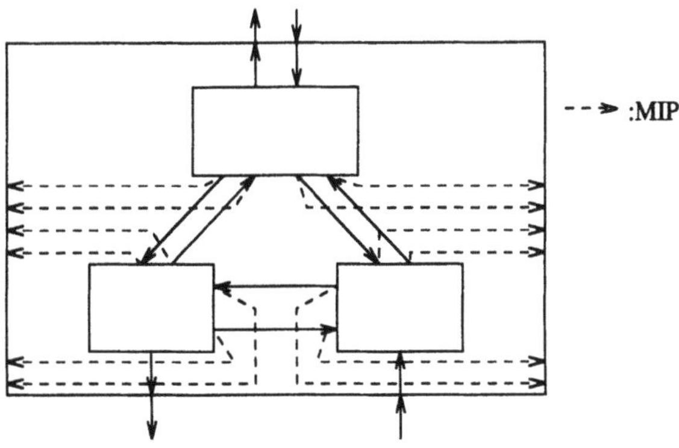

Figure 3 Mirrored interaction ports.

2. Depending on which modules are to be tested, we need to configure the modular structure of the IUT accordingly. Thus a capability for dynamic configuration of the module structure is needed.

A way to do this is to provide the following instrumentation:

```
stateset all = [list_of_all_states];
DATA = ...;
```

```
STATEMENTTYPE = (INIT, CONNECT, ATTACH, RELEASE, DISCONNECT, DETACH)
channel example(user, provider)
by user
    .
    CONFIGURE(ACTION:STATEMENTTYPE; DATA_A;DATA, DATA_B:DATA);
    .


from all
when X.CONFIGURE
provided ACTION = INIT
    begin
       init DATA_A with DATA_B;
     end

from all
when X.CONFIGURE
provided ACTION = CONNECT
    begin
       connect DATA_A to DATA_B;
     end
  .
  .
  .
```

Since all configuration statements of Estelle are provided through the interaction primitive "CONFIGURE" and its parameters "ACTION", "DATA_A" and "DATA_B", it is possible to configure the IUT dynamically at any time. The interaction port "X" can either be a port in the original formal specification or a MIP in the formal specification that has been transformed for testability. For example, if we want to initialize a module in an IUT, we send the interaction primitive "CONFIGURE" to the module with ACTION "INIT", module variables in "DATA_A" and module type in "DATA_B".

Alternately, we may enumerate every possible combination of modules, and then make a set of transitions providing a configuration capability for each combination. This scheme may cause transition explosion if the number of modules is large, but will not need the assumption that parameters are supported in the configuration statements of Estelle.

A formal model is proposed in the following section to deal with the problem of observability based on concurrent specification.

4 FORMAL MODEL

In this section, we shall use the term module to mean an implementation executing as a single process. Let us assume that a concurrent module M_{conc} consists of a set of sequential modules M_1, ..., M_i, ..., M_n; a concurrent formal specification S_{conc} consists of a set of sequential formal specifications S_1, ..., S_i, ..., S_n; and, a concurrent trace T_{conc} from M_{conc}

consists of a set of sequential traces T_1, ..., T_i, ..., T_n from individual modules M_1, ..., M_i, ..., M_n (See Figure 4). The concurrent formal specification S_{conc} is assumed to be error-free and conform to the standards.

Definition 1 A sequential trace T_i of module M_i consists of a chronologically ordered sequence of event vectors t_i^k, k = 1, 2, ... If M_i has p interaction ports, then the event vector $t_i^k = <(in_{i,1}^k, out_{i,1}^k), ..., (in_{i,p}^k, out_{i,p}^k)>$ where $in_{i,j}^k$ and $out_{i,j}^k$ are the input and output messages respectively at interaction port j associated with the k-th event vector in trace T_i. $in_{i,j}^k$ and $out_{i,j}^k$ may be null.

Notice that we have a trace structure equivalent to the one presented in our earlier paper [Kim 91] which dealt with trace analysis based on single modules. An event vector consists of zero or one input, and x output messages where $0 < x < p$.

Definition 2 A sequential module M_i is pointwise conformant to the formal specification S_i on an event vector t_i^k if the behavior of the event vector is permissible (i.e., contained) in the specification. This is denoted as pconf(M_i, S_i, t_i^k).

Note that even if more than one path (or transition) of the specification satisfies the event vector, the pconf relation still holds based on the above definition.

Definition 3 A trace T_i conforms to the sequential module M_i, written as $conf_{seq}(M_i, S_i, T_i)$ iff pconf(M_i, S_i, t_i^k) for all t_i^k in T_i.

The sequential trace T_i from Definition 1 consists of the external behavior of a module M_i with respect to its environment (i.e., the other modules). If $conf_{seq}(M_i, S_i, T_i)$ is valid, the trace T_i conforms to the specification of M_i. In addition, the input/output messages to/from the other modules in T_i are also valid since we assume that S_{conc} is error-free.

Definition 4 A concurrent trace T_{conc} of a concurrent module M_{conc} conforms to the concurrent specification S_{conc} of M_{conc}, written as $conf_{conc}(M_{conc}, S_{conc}, T_{conc})$ iff $conf_{seq}(M_1, S_1, T_1) \wedge \ldots \wedge conf_{seq}(M_i, S_i, T_i) \wedge \ldots \wedge conf_{seq}(M_n, S_n, T_n)$, i.e., $\wedge_{i=1}^{n} conf_{seq}(M_i, S_i, T_i)$.

Note that the trace analysis proposed does not depend on the communication scheme (synchronous or asynchronous). The trace analysis of concurrent modules with respect to concurrent formal specifications provides verdicts based on a set of trace analyses of single modules with sequential formal specifications.

Since there are two types of problems that could arise in a concurrent (or distributed) environment but not in a sequential environment, namely deadlocks and data races, we need to show how they can be detected. A concurrency model was proposed for this purpose based on time-event sequence diagrams [Kim 93]. To detect data races, we note that a set of $conf_{seq}(M_i, S_i, T_i)$, i.e., trace analyses with respect to the single modules in the system, enable us to construct the time-event sequence diagram used by the concurrency model. The concurrency model provides a way to identify the sets of concurrent events which can be checked for data races by determining if there are read/write or write/write conflicts to shared variables. An example in Ada is given in [Kim 93]. The external behavior (i.e., a trace T) of a module M which does not conform to its specification S may cause deadlocks. Deadlock detection depends on the communication scheme used. In the

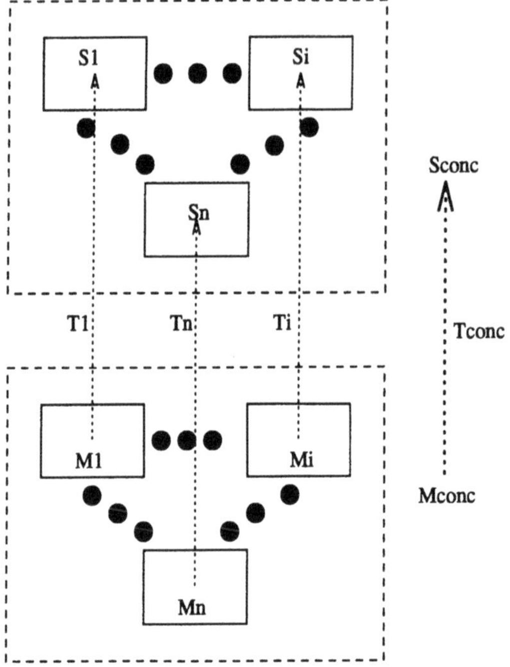

Figure 4 Trace analysis based on concurrent specifications.

synchronous scheme, non-conformance in terms of input or output message may lead to deadlocks. However, asynchronous schemes will not cause deadlocks even though input or output messages from a module M do not conform to the specification S. This is because asynchronous messages do not block and wait, thus removing a necessary condition for deadlock to occur. So far, most of the work in protocol testing has been concerned with the sequential aspects of protocols. The ideas presented in this section is a first step in dealing with testing of concurrent modules which is an extension of our previous work on trace analysis of sequential module [Kim 91]. More research is needed to work out the theory and to build the tools.

5 CONCLUSIONS

The ISO conformance testing methodology treats an IUT as a black-box which may consist of one or more modules. This makes testing difficult since nondeterminism may arise from the tester's point of view because internal actions are not observable. In this paper, we have proposed a generic scheme to provide precise and efficient means of instrumenting the formal specification to enhance controllability and observability in testing protocol implementations consisting of a single module. A framework for testing multiple concurrent modules is also presented. The proposed schemes can work with existing specifications and testers also. To our knowledge, this is the first work that has adopted this approach. Future work includes techniques for data level design for testability (so that we can pinpoint the cause of an error), and the theory and tools to test concurrent modules.

REFERENCES

[Dssou 90] Dssouli R. and Fournier R. (1990) Communication Software Testability. *The 3rd Int'l Workshop on Protocol Test Systems*, McLean, Virginia.

[Ellsb 92] Ellsberger J. and Kristoffersen F. (1992) Testability in the context of SDL. *IFIP Symposium on Protocol Specification, Testing, and Verification, XII*, Lake Buena Vista, Florida.

[Fidge 91] Fidge C. (1991) Logical Time in Distributed Computing Systems. *IEEE Computer*, vol. **24**, no. **8**, 28 – 33.

[FMCT] ISO SC21 P.54 and ITU-T SG10 Q.8 (1994) Formal Methods in Conformance Testing.

[Freed 91] Freedman R. S. (1991) Testability of Software Components. *IEEE Tr. on Software Engineering*, vol. **17**, no. **6**, 553 – 564.

[Fujiw 85] Fujiwara H. (1985) Logic Testing and Design for Testability. MIT Press.

[Hoffm 89] Hoffman D. (1989) Hardware Testing and Software ICs. *The Pacific NW Software Quality Conference*, 234 – 244.

[IEEE 91] IEEE (1991) Special Report on Concurrent Engineering. *IEEE Spectrum* , vol. **28**, no. **7**.

[ISOa] ISO IS 9646 (1991) OSI Conformance Testing Methodology and Framework.

[ISOb] ISO 9074 (1989) ESTELLE - A Formal Description Technique Based on an Extended State Transition Model.

[ISOc] ISO 8807 (1989) LOTOS - A Formal Description Technique Based on the Temporal Ordering of Observational Behavior.

[ITU] ITU-T (1988) Specification and Description Language (SDL) Recommendations Z.100. *ITU-T Blue Book.*

[Kim 91] Kim M., Chanson S. T. and Vuong S. T. (1991) Protocol Trace Analysis based on Formal Specifications. *Fourth Int'l Conference on Formal Description Techniques*, Sydney, Australia, 399 – 414.

[Kim 93] Kim M., Chanson S. T. and Vuong S. T. (1993) Concurrency Model and Its Application to Formal Protocol Specifications. *IEEE INFOCOM*, San Francisco, California, 766 – 773.

[Lampo 78] Lamport L. (1978) Time, Clocks, and the Ordering of Events in a Distributed System. *Comm. ACM*, vol. **21**, no. **7**, 558 – 565.

[Matte 89] Mattern F. (1989) Virtual Time and Global States of Distributed Systems. *Parallel and Distributed Algorithms*, North-Holland, 215–226.

[Parke 89] Parker K. P. (1989) The Impact of Boundary Scan on Board Test. *IEEE Design & Test of Computers*, vol. **6**, no.4, 18 – 30.

[Petre 93] Petrenko A., Dssouli R. and et al. (1993) On Evaluation of Testability of Protocol Structures. *The 6th Int'l Workshop on Protocol Testing Systems*, Pau, France, 115 –

127.

[Sarik 87] Sarikaya B., Bochmann G. and et al (1987) A Test Design Methodology for Protocol Testing. *IEEE Tr. on Software Engineering*, vol. **13**, no. **5**, 518 – 531.

[Vuong 93] Vuong S. T., Loureiro A. A. F. and Chanson S. T. (1993) Toward a Framework for the Design for Testability of Communication Protocols. *The 6th Int'l Workshop on Protocol Testing Systems*, Pau, France, 91 – 111.

17

On the design for testability of communication protocols

N. Yevtushenko[1], A. Petrenko[2], R. Dssouli[2], K. Karoui[2], S. Prokopenko[1]

*1 - Tomsk State University,
36 Lenin str., Tomsk, 634050, RUSSIA.
2 - Université de Montréal,
C.P. 6128, succ. Centre-Ville, Montréal, H3C 3J7, CANADA,
Phone: (514) 343-7535, Fax: (514) 343-5834,
{petrenko, dssouli, karoui}@iro.umontreal.ca*

Abstract

Design For Testability (DFT) is understood as the process of introducing some features into a protocol entity that facilitate the testing process of protocol implementations. DFT at the implementation level deals with a particular realization on a given platform, whereas DFT at the specification level affects all possible implementations regardless of the implementation process. The fact that protocols are usually specified only partially, facilitates DFT at the specification level. In this paper, we address one particular problem of DFT, the problem of finding a minimal augmentation of the given protocol behavior (an FSM) such that a newly obtained specification is more testable than the original one, while maintaining sets of defined states and events. We propose an approach to augmenting a partially specified FSM such that a test suite for the resulting FSM with guaranteed fault coverage is shorter than that for the original FSM.

Keywords

Conformance testing, protocol testability, design for testability, partial FSMs, test derivation

1 INTRODUCTION

Behavior of a protocol entity is usually described for a number of situations. A situation in the entity can be understood as a combination of its state and a current input event from a peer entity or from the user of this entity. A service definition and protocol specification explicitly define situations, sometimes called valid, which can happen during normal or abnormal course of communication. If the protocol behavior is unspecified in a valid situation then the protocol is said to have an error, called an unspecified reception. A well-formed protocol is neither under- nor over-specified. An over-specified protocol has unreachable states or transitions which can never be executed. However, even well-formed protocols leave certain situations undefined. These situations are regarded as invalid, in the sense that some events can never happen in a particular state. There are also certain signals from the user which do not require any communication with a remote entity and are processed locally. Local behavior is usually not standardized and can be implemented in various ways. For this reason, protocols are widely recognized as partially specified systems [BoPe94], [PBD93], [SiLe89]. This feature facilitates the process of implementing a protocol. Undefined situations are utilized for optimizing a protocol implementation on a given platform. Various criteria can be applied for optimization. For complex protocols, testability of their implementations becomes a primary concern of communication software designers.

In general terms, testability of a protocol entity means that it has some features that will facilitate the testing process [DsFo91], [VLC93], [PDK93]. Design for testability (DFT) is understood as the process of introducing these features into the protocol entity. DFT at the implementation level deals with a **particular** realization on a given platform, whereas DFT at the specification level affects **all** possible implementations regardless of the implementation process. Unfortunately, most existing protocols have been designed and documented without testing requirements in mind [VLC93]. To improve the protocol's testability, one should rather rely only on those combinations of states and events for which the protocol behavior is not defined. Assuming that the behavior in such situations can be arbitrary defined, the problem of DFT at the specification level can be formulated as follows.

We must find a minimal augmentation of the given protocol behavior such that a newly obtained specification is more testable than the original one. A measure of testability of a protocol entity is assumed to be inversely proportional to the shortest length of a test suite needed to achieve guaranteed (complete) coverage of certain faults [PDK93]. To the best of our knowledge, formal methods for DFT of protocols have not yet been explored. Note that the existing formal methods and techniques for improving testability which have been developed mainly in the hardware area, see, for example, [ShLe94], [ABF90], [Jose78], cannot be applied in this domain, since they either rely on the structure of an implementation, or they change the set of states, or the sets of input/output events, in a way that is similar to the adding of a read-state message in protocol engineering.

In this paper, we consider a simple FSM model of a protocol machine (at least its control portion) and assume the following scenario of DFT. In the first step, an initial FSM specification is derived from the given requirements. If the FSM is not completely specified then its undefined transitions are "don't care" transitions which model situations where the requirements do not restrict any further protocol behavior. The problem now is how to augment the partially specified FSM by converting "don't care" transitions into defined transitions such that a test suite for the resulting FSM with guaranteed fault coverage is shorter than that for the original FSM. In the last step, the augmented specification of the protocol and its test suite are released for implementation. Even if the scenario seems somewhat idealistic, we believe that such an approach should spare efforts required to produce conforming implementations.

This paper is structured as follows. In Section 2, we present some basic definitions and concepts. In Section 3, we discuss the influence of machine's parameters on the size of a complete test suite and derive several formulae for estimating its length. Based on this discussion, we introduce the basic idea underlying our method for assigning "don't care" transitions presented first for a machine with a single input in Section 4 and then generalized for an arbitrary machine in Section 5. Section 6 contains application examples, including the INRES protocol. We conclude in Section 7 by presenting some open research issues.

2 BASIC NOTIONS AND DEFINITIONS

Throughout this paper we make use of the following definitions. A *partial Finite State Machine* (FSM) A is a 6-tuple $(S, X, Y, \delta, \lambda, D_A)$, where S is a set of n states; X and Y are finite sets of inputs and outputs; δ and λ are transition and output functions; D_A is a set of *defined* transitions of A, that is a subset of $S \times X$. We assume that each input labels at least one defined transition. An initialized machine also has a designated initial state s_0. A becomes a *complete* (completely specified) machine if $D_A = S \times X$. Transitions in $(S \times X) \setminus D_A$ are undefined or *"don't care"* transitions. Here, we assume the so-called "undefined by default" convention for undefined transitions [PBD93], that is, if $(s, x) \in (S \times X) \backslash D_A$ then $\delta(s, x)$ and $\lambda(s, x)$ can be set to (assigned) any s', $s' \in S$ and any y, $y \in Y$, respectively.

A sequence $x_1 \ldots x_k$ of the set X^* of all possible input sequences is called an *acceptable* input sequence for state s if there exist k states $s_1, \ldots, s_k$ from S such that $\delta(s, x_1) = s_1$ and $\delta(s_i, x_{i+1}) = s_{i+1}$, $i = 1, \ldots, k$-1. We use X_i^* to denote a set of all input sequences acceptable for state s_i and X_A^* for the state s_0. Two states s_i and s_j of FSM A are said to be *distinguishable* if there is an input sequence $\alpha \in X_i^* \cap X_j^*$ such that $\lambda(s_i, \alpha) \neq \lambda(s_j, \alpha)$. If any two states of A are distinguishable then A is a *reduced* machine. As usual, states s_i and s_j of A are *equivalent* iff $X_i^* = X_j^*$ and $\lambda(s_i, \alpha) = \lambda(s_j, \alpha)$ for every sequence α.

Let A be a reduced partial FSM. Given a pair of states s_i and s_j, we choose a sequence α_{ij} that distinguishes them, and form a set $W_i = \{\alpha_{ij} \mid s_j \in S \text{ and } j \neq i\}$. The set W_i is called an *identifier* of state s_i. If the W_i has just a single identifying sequence, i.e. $|W_i|=1$ then it is often called a UIO-sequence. If there exists a set W of input sequences such that any sequence from W is acceptable for any state and W is a state identifier of any state then we refer to the set W as a *characterization* set of A. In the case where the W set consists of just a single identifying sequence, we refer to this sequence as a *distinguishing* sequence [Henn64]. Machines with such sequences usually possess very short tests.

We say that the FSM A is *connected* if for any state s there is an input sequence β such that $\delta(s_0, \beta) = s$. A set $V = \{\varepsilon, \beta_1, \ldots, \beta_{n-1}\}$ of input sequences is said to be a *state cover* if for any state s_i of S there is a sequence $\beta_i \in V$ which takes FSM A from its initial state into state s_i, where ε is the empty sequence: $\delta(s_0, \varepsilon) = s_0$, thus $V \ni \varepsilon$.

Let a connected reduced FSM A have a characterization set W. If $D_A = S \times X$ then W always exists. Assume the sets V and W have the following properties:

1) if a sequence β belongs to V then any prefix of it also belongs to V, i.e.

if $\beta x \in V$ then $\beta \in V$; (2.1)

2) if a sequence α belongs to W then any suffix of it also belongs to W, i.e.

if $x\alpha \in W$ then $\alpha \in W$. (2.2)

As an example, a homogeneous distinguishing sequence satisfies (2.2), a sequence is *homogeneous* if it is a sequence of the same symbol, i.e. $x^r = xx^{r-1}$, $r \geq 1$.

It is known that if V and W have the properties (2.1) and (2.2) then the set $TS = TC@W$ is a *complete* test suite for the FSM A in the class $\mathfrak{I}_n$ of all FSMs with up to n states [Vasi73]. Here "@" stands for concatenation operation on sets, and the set TC is a *transition cover* that contains V as well as any sequence β_i from V concatenated with any input that is acceptable for state $s_i = \delta(s_0, \beta_i)$. [Vasi73] gives this version of the so-called W-method [Chow78].

A complete test suite can be shortened if each sequence of the set W is applied after any sequence from V. However, after any sequence from TC that is not in V, only a part of the set W is applied, namely, a corresponding state identifier. This is the main idea of the Wp-method [FBK91]. We use the following notation for such test suites: $TS = V@W \cup TC \otimes W$. Here we consider slightly generalized versions of these methods to cover partial reduced machines. For more discussions on partial FSMs the reader is referred to [Petr91], [PBD93], [BoPe94].

3 LENGTH OF A TEST SUITE AND "DON'T CARE" TRANSITIONS

3.1 Estimating a test size

Generating a complete test suite for a given FSM involves many choices, most of which are left without any guidance in most existing test derivation methods. This makes it extremely difficult to determine length of a complete test suite until the test suite is actually derived. Its length depends on many factors [BPY94]. Among them, properties of the state cover, transition cover and characterization set chosen for the test derivation seem most essential. We note that the known bounds [Vasi73] are derived for the case where state covers and characterization sets satisfy the properties (2.1) and (2.2). These properties facilitate the test length estimation as well. Here we look for other properties of sets V and W that provide shorter test suites. We first consider an example.

Example. The FSM A shown in Figure 3.1 possesses several state covers with the property (2.1). We choose two of them of different lengths: $V_1 = \{\varepsilon, a, aa, b\}$ and $V_2 = \{\varepsilon, a, aba, ab\}$. V_1's length is four, whereas V_2 contains six symbols.

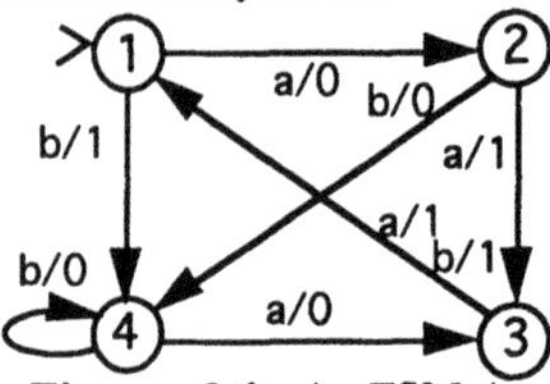

Figure 3.1 An FSM A.

The machine also possesses different characterization sets with the property (2.2): $W = \{aaa\}$, $W' = \{a, b\}$. In the first case, there is a homogeneous distinguishing sequence aaa, from which it is easy to obtain state identifiers as follows: $W_1 = \{aaa\}$, $W_2 = \{aa\}$, $W_3 = \{aa\}$, $W_4 = \{aaa\}$. In the second case, $W'_1 = W'_2 = W'_3 = W'_4 = \{a, b\}$. Total length $L(W)$ of W is three; $L(W') = 2$, but it contains two sequences. Based on these sets, it is possible to construct various test suites complete in the class $\Im_n$, where n=4. The Wp-method gives the following results:

1) $V_1 = \{\varepsilon, a, aa, b\}$, $W = \{aaa\}$. $TS1 = \{aaaaaa, aabaaa, abaaa, baaa, bbaaa\}$. $L(TS1)= 26$.

2) $V_1 = \{\varepsilon, a, aa, b\}$, $W' = \{a, b\}$.
$TS2= \{aaaa, aaab, aaba, aabb, aba, abb, baa, bab, bba, bbb\}$. $L(TS2) = 34$.

3) $V_2 = \{\varepsilon, a, aba, ab\}$, $W=\{aaa\}$. $TS3 =\{aaaa, abaaaaa, ababaaa, abbaaa, baaa\}$. $L(TS3) = 28$.

4) $V_2 = \{\varepsilon, a, aba, ab\}$, $W' = \{a, b\}$.
$TS4 = \{aaa, aab, abaaa, abaab, ababa, ababb, abba, abbb, ba, bb\}$. $L(TS4) = 38$. ❑

As can be seen from this example, a characterization set with fewer sequences yields a shorter test suite even if its total length is not minimal. The number of transfer sequences in a state cover cannot be reduced, however, a state cover of a shorter length usually leads to a shorter test suite. We now present some estimations that confirm these observations in the general case.

We consider a test suite $TS = TC@W$ of an FSM A which has a characterization set W assuming that the sets V and W satisfy properties (2.1) and (2.2), respectively. We have seen from the above example that, all things being equal, length of this test suite is determined by lengths of sets V and W and by the cardinality of W. If A is a partial machine then the size of a test suite depends also on the number of defined transitions. Let p be the number of defined transitions of A, i.e. $p \leq mn$, where m is a number of inputs.

Proposition 3.1. Given an FSM A with n states, a state cover V with the property (2.1), a transition cover TC of p defined transitions, and a characterization set W with the property (2.2), the total length $L(TS)$ of a complete test suite $TS = TC@W$ does not exceed

$$L(TC)|W| + L(W)(p-n+1). \tag{3.1}$$

Proof. In fact, the total length of a set $TC@W$ does not exceed the value of

$$\sum_{v_i \in TC, w_j \in W} [L(v_i) + L(w_j)].$$

Then $L(TC@W) \leq [L(v_1) + L(w_1)] + ... + [L(v_r) + L(w_t)] = [L(v_1) + ... + L(v_r)]\ |W| + L(W)r$, where r is a number of sequences in TC. Thus
$L(TC@W) \leq L(TC)|W| + L(W)|TC|$.

Now it is sufficient to show that the number of sequences in TC is $(p - n + 1)$. Consider the set of input sequences TC and a corresponding successors tree of A with its initial state as a root of this tree. Due to (2.1), the tree has exactly p edges from its internal nodes, therefore this tree has p nodes excluding the root of this tree. Since n nodes are internal, there are exactly $(p - n + 1)$ terminal nodes, i.e. $|TC| = p - n + 1$. ❑

Consider again the FSM A shown in Figure 3.1. In the case of a completely specified FSM, $p = mn$. Assume that V_1 and W are chosen for test derivation. For the test suite $TS1$ we have $L(TC) = 12$, $p\text{-}n\text{+}1 = 5$, $|W| = 1$, $L(W) = 3$, and therefore the length is 27. It is close to the actual length of 26. In the case where V_2 and W' are chosen, we have $L(TC) = 14$, $p\text{-}n\text{+}1 = 5$, $|W'| = 2$, $L(W) = 2$, and $14 \cdot 2 + 5 \cdot 2 = 38$. This is length of the test suite $TS4$.

Since $(p\text{-}n\text{+}1)$ is the number of sequences in the set TC, we usually have $L(TC) >> (p\text{-}n\text{+}1)$. For this reason, **length of a complete test suite primarily depends on the number of sequences in the set W, rather than on its total length.**

It is also worth noting than based on (3.1), it is possible to derive the least upper bound on the length of a complete test suite. Assume for simplicity that the machine is completely specified. Then it is known that the bound is $O(mn^3)$, where m is the number of inputs [Vasi73], [LeYa94]. We need a more precise estimation. [TyBa75] gives $L(W) \leq n(n\text{-}1)/2$, $|W| \leq n\text{-}1$, $L(V) \leq n(n\text{-}1)/2$ provided that the properties (2.1) and (2.2) hold. It is not difficult to check that $L(TC) \leq n(m\text{-}1) + n(n\text{-}1)/2 + 1$. $(p\text{-}n\text{+}1) = nm - n + 1$. Then
$L(TS) \leq [n^2m + 2nm - 2n + 2](n\text{-}1)/2 = L_{max}$.

Thus, $L(TS) \leq L_{max} < mn^3$ for completely specified reduced machines with $n \geq 2$ states and $m \geq 2$ inputs. The least upper bound on tests for partially specified machines remains unknown, as they are still the subject of active research [BPY94], [BoPe94].

Next, we estimate length of a test suite for a partial FSM that possesses homogeneous identifying sequences.

Proposition 3.2. Given an FSM A with n states, a state cover V with the property (2.1), a transition cover TC of p defined transitions, and a characterization set W with k homogeneous identifying sequences of length up to h, there exists a complete test suite of length not more than
$[L(V) + hn]k + L(TC) + h(p\text{-}n\text{+}1)$. (3.2)

Proof. Based on the Wp-method, we construct a complete test suite of the form $TS = V@W \cup TC \otimes W$. The set $V@W$ means that every sequence from V is concatenated by k sequences of length up to h. In the worst case, this part of the test suite is no longer than $L(V)k + hnk$. The set $TC \otimes W$ means that just a single identifying sequence of length up to h is applied to the state reached after any sequence from the transition cover TC. Similar to the previous proof, $|TC| = p\text{-}n\text{+}1$, and the total length of sequences of the set $TC \otimes W$ does not exceed the value $\sum_{v_i \in TC} [L(v_i) + h] = \sum_{v_i \in TC} L(v_i) + h|TC| = L(TC) + h(p\text{-}n\text{+}1)$. ❑

Corollary 3.3. The length of a complete test suite of an FSM A with n states and a transition cover TC of p defined transitions is no less than
$L(TC) + (p\text{-}n\text{+}1) = L_{min}$. (3.3)

(3.1)-(3.3) can be used to determine the expected size of a test suite which is complete for the given machine in the class of all machines with an equal or fewer number of states.

3.2 A criterion for assigning "don't care" transitions

Practice has not yet provided us with protocol machines such that the length of complete test suites approaches the least upper bound L_{max}. However, there exists a class of FSMs for which the length of a complete test suite meets the lower bound L_{min}. These are machines with a distinguishing sequence of length one. Unfortunately, protocol machines seldom fall into this class. It is known that certain machines have neither distinguishing nor state identifying seqences, and there are machines whose states have identifying (UIO) sequences, but only of exponential length [LeYa94]. Their least upper bound remains unknown, especially for partial FSMs. In the general case, n states may require identifiers consisting of up to n-1 sequences, and it is possible to construct a sequence of length up to $n(n-1)/2$ distinguishing two states in a given partial reduced FSM. Thus, in most cases, we deal with partial FSMs which are not easily testable in the sense that they require a complete test suite of length far from the best possible.

If a protocol machine is specified completely then some measures can be taken to improve the protocol testability at the implementation level only. However, if it is specified only partially then its testability can be improved at the specification level ensuring that all implementations derived from the augmented specification are more testable than that derived from the original specification. Given a partial FSM with n states, t inputs, m outputs and p defined transitions, there are $(nt-p)$ "don't care" transitions. Each undefined transition can be either left undefined or transformed into a defined one in nm different ways. Thus, there exist $(nm+1)^{(nt-p)}$ FSMs which are quasi-equivalent to the given machine [Gill62]. We call them *augmented machines* with respect to the original machine. An exhaustive procedure would enumerate all $(nm+1)^{(nt-p)}$ machines, derive a complete test suite for each of them and search for the one with the shortest test suite. We wish to avoid such a brute force search, called also *"perebor"*, and should find a criterion to guide the process of assigning "don't care" transitions.

As follows from the discussion of Section 3.1, a homogeneous distinguishing sequence ensures $|W|=1$ and leads to a short test suite. Since such a sequence may not always exist, the next best case is when each state possesses a homogeneous identifying sequence which might be common for several states. Thus, the transformation of "don't care" transitions that maximizes the number of states possessing homogeneous identifying sequences can be regarded as a successful transformation. The approach we take is based on these heuristics.

However, the numerical characteristics of homogeneous identifying sequences alone such as their number and total length are not sufficient to choose the best assignment of "don't care" transitions. The problem is that their assignment affects length of a test suite in two opposite ways. On the one hand, if newly added transitions create short identifying or even diagnostic sequences of the machine which had no such sequences prior to the assignment then the size of the test suite most probably will be reduced. On the other hand, the number of defined transitions increases, thus, additional test sequences are required to test new transitions. If the increase in length caused by an assignment exceeds the savings gained then the assignment would deteriorate the testability of the given machine. For this reason, the expected or actual length of a complete test suite should eventually be used to estimate the effect of assignments.

We demonstrate later in this paper that if a state has at least one "don't care" transition then it is always possible to construct an augmented machine such that this state has a homogeneous identifying sequence. Moreover, if the necessary and sufficient conditions established below are satisfied then there exists an augmented machine with a homogeneous distinguishing sequence. In the worst case, the length of identifying sequences reaches n, the number of states.

4 AN FSM WITH A SINGLE INPUT

In this section, we assume that a given machine has just one input and propose a method for finding assignments of all "don't care" transitions such that their initial states possess identifying sequences in an augmented machine. We also show that under certain conditions the augmented machine has a distinguishing sequence. By the construction, the obtained sequences are also homogeneous for the original machine which contains the machine with a single input as its submachine.

4.1. Auxiliary notions

Let B be a reduced partial FSM. Consider its submachine A which is obtained by restricting its input set X to an arbitrary input $x \in X$ which labels at least one "don't care" transition. Thus, $A = (S, \{x\}, Y, \delta, \lambda, D_A)$ is an FSM with a single input x.

The state transition graph of the FSM A has a *cycle* (s_1->s_2->...->s_k) if $\delta(s_i, x) = s_{i+1}$ for $i = 1,...,k-1$, $k \geq 1$; and $\delta(s_k, x) = s_1$. For each state of the cycle, any input sequence is acceptable. If length of an input sequence ω is a multiple of k, i.e. $\omega = x^{mk}$, $m \geq 1$, then

$\delta(s_i, x^{mk}) = s_i$ for all $i = 1,...,k$. (4.1)

Given a sequence x^{mk-1} of length mk-1, we also have

$\delta(s_i, x^{mk-1}) = s_{i-1}$ for all $i = 2,...,k$ and $\delta(s_1, x^{mk-1}) = s_k$. (4.2)

State s is said to be a *starting* state of A if there is no transition leading to this state. Consider an arbitrary path s_1->s_2->...->s_k from a starting state s_1. We say that the path *terminates in state* s_k (a terminating path) if (s_k, x) is a "don't care" transition; or the path *cycles* (a cycling path) if $\delta(s_k, x) = s_{i+1}$ for some $k > i \geq 1$, i.e. s_{i+1}->...->s_k is a cycle. The behavior of the FSM A is defined in every state of a cycling path, thereby such path does not traverse any state with a "don't care" transition.

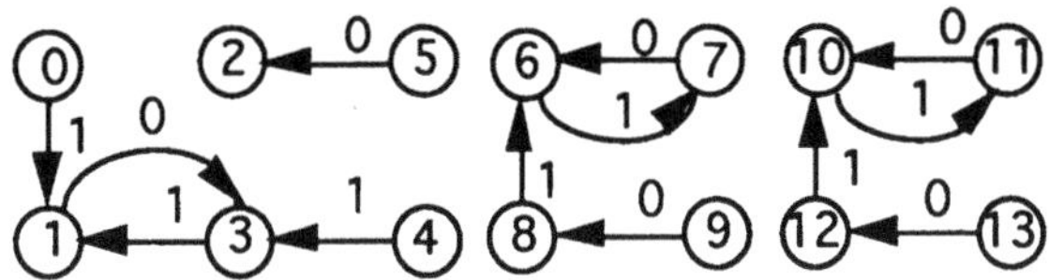

Figure 4.1 The FSM with a single input.

Example. Consider an FSM shown in Figure 4.1. Its transitions are labeled with outputs only, since the machine has just one input. Starting states of A are 0, 4, 5, 9, 13. State 2 has a "don't care" transition. There is only one terminating path from the starting state 5 which terminates in state 2. Four other paths are cycling. All states except 2 and 5 accept all input sequences. The FSM has the following sets of equivalent states: {1, 7, 11}, {0, 3, 6, 10}, {4, 8, 12}, {9, 13}. Notice that starting states 9 and 13 are equivalent, but each of them is not equivalent to any other state of its path. Moreover, they are not equivalent to any other state of this machine. ❑

Based on this observation, we claim a more general property of an FSM with a single input.

Proposition 4.1. Let the FSM A have a cycling path $P = (s_1$->...->$s_k)$ from a starting state s_1 that is not equivalent to any other state of this path. Then A has a cycling path $P' = (s'_1$->...->$s'_k)$ from a starting state s'_1 that is not equivalent to any other state of A which is not a starting state.

Proof. If the starting state s_1 of P is not equivalent to any other state of A then P itself is the path of the proposition. Suppose therefore that there is another state s'_j equivalent to s_1 and s'_j is an intermediate state of a cycle (s'_1->...->s'_t) or of a cycling path (s'_1->...s'_j->...->s'_t), where $j>1$. For any input sequence, the successors of the equivalent states are also equivalent, therefore states $\delta(s_1, x^t)$ and $\delta(s'_j, x^t)$ are equivalent.

If s'_j is an intermediate state of a cycle (s'_1->...->s'_t) then, because of (4.1), $\delta(s'_j, x^t) = s'_j$. In this case, state s_1 and state $\delta(s_1, x^t)$ of the same path are also equivalent. This contradict our assumption that the state s_1 that is not equivalent to any other state of this path.

Assume then s'_j is an intermediate state of a cycling path (s'_1->...s'_j->...->s'_t), where $j>1$. Now, we must show that the starting state s'_1 is not equivalent to any state of $P = (s_1$->...->$s_k)$. In fact, if states s'_1 and s_p, $p \geq 1$ are equivalent, so are states $\delta(s'_1, x^{j-1})$ and $\delta(s_p, x^{j-1})$. $\delta(s'_1, x^{j-1}) = s'_j$ and is equivalent to s_1. Now, states $\delta(s_p, x^{j-1})$ and s_1 are required to be equivalent as well. This is the contradiction. We can exclude the path (s_1->...->s_k) from our consideration. As a result, we could only find another cycling path whose starting state is equivalent to the starting state s_1 of the given path P. ❑

A cycling path whose starting state is not equivalent to any other starting state of cycling paths is termed a *dominant* cycling path of the FSM A. In the above example, among four cycling paths, there are two dominant cycling paths: (9,8,6,7) and (13,12,10,11). As will be shown in next section, dominant cycling paths play an important role for the assignment of "don't care" transitions.

4.2 A single transition

We consider in this section the case where the FSM A with a single input has only one "don't care" transition (s, x) and show that it is always possible to assign this transition such that state s becomes distinguishable from any other state in the newly obtained completely specified FSM.

Given an FSM $A = (S, \{x\}, Y, \delta, \lambda, D_A)$, where $(S \times X) \backslash D_A = \{(s, x)\}$, we consider its state transition graph and determine all its cycles, terminating paths, cycling paths, and dominant cycling paths. Note that a terminating path can only end in state s, but it might be empty if s has no incoming transition in A. There are four possible cases each of which requires a distinct assignment of the "don't care" transition (s, x):

1) There is no cycle in A, so all paths terminate in state s.
2) In A, there are only cycles and paths terminating in state s.
3) A has also cycling paths, but it has no dominant cycling paths.
4) A has a dominant cycling path.

Next we consider how the "don't care" transition (s, x) should be assigned in each of these cases in order to obtain a homogeneous identifying sequence of state s.

Case 1

(s, x) is a "don't care" transition. All paths terminate in s. A has no cycle.
We determine the longest path s_1->s_2->...->s_k->s and define a transition in state s on input x
$\delta(s, x) = s_1$ and $\lambda(s, x) = y$, where output y is such that the sequence $\lambda(s_1, x), ..., \lambda(s_k, x)y$ cannot be represented as any of its proper prefixes repeated several times. If $|Y|>1$ then it is always possible to find such output. Assigning the transition (s, x), we obtain a completely specified FSM A'.

Proposition 4.2. In case 1, state s is distinguishable from any other state of A'.

Proof. In fact, due to the chosen output assignment, state s cannot be equivalent to any other state in the obtained cycle (s_1->s_2->...->s_k->s). Assume therefore, that s is equivalent to state s'_j, $1 \le j \le t$, which belongs to another path s'_1->...->s'_j...->s'_t->s->s_1->...->s_k. If states s and s'_j are equivalent then states $\delta(s'_j, x^{t-j+1}) = s$ and $\delta(s, x^{t-j+1})$ are also equivalent. State $\delta(s, x^{t-j+1})$ belongs to the cycle s->s_1->...->s_k and it is not state s because $t \le k$ and $j \ge 1$. Then state s should be equivalent to another state of the cycle, but this is impossible. ❑

Example 1. Consider the FSM shown in Figure 4.2.

Figure 4.2 The FSM with no cycles.

This machine has a single "don't care" transition in state s_1. There are two paths: s_0->s_1 and s_4->s_3->s_2->s_1 which terminate in state s_1. We choose the longest one and assign $\delta(s_1, x) = s_4$. The output set has two symbols, 0 and 1. If $\lambda(s_1, x)$ is assigned 1 then $\lambda(s_4, x)\lambda(s_3, x)\lambda(s_2, x)\lambda(s_1, x) = 0101$ and it can be represented as its proper prefix 01 repeated two times, i.e. 0101 = (01)(01). In this case, state s_1 would become equivalent to s_3, and so would s_2 and s_4. To distinguish state s_1 from other states we must define $\lambda(s_1, x) = 0$. We obtain a new transition:

s_1-x/0->s_4. Now sequence x distinguishes state s_1 from s_3; sequence xx distinguishes state s_1 from s_4; and xxx distinguishes s_1 from s_0 and s_2. Thus, the sequence xxx is a homogeneous identifying sequence of state s_1 in the augmented FSM. ❑

Case 2

(s, x) is a "don't care" transition. A has cycles, but its paths terminate in s. In this case, we take an arbitrary cycle (s_1->...->s_k) and define a transition from state s to the state s_1, i.e. $\delta(s, x) = s_1$, with the output $\lambda(s, x) = y$ such that $y \neq \lambda(s_k, x)$. By assigning the transition (s, x), we again obtain a new completely specified FSM A'.

Proposition 4.3. In case 2, state s is distinguishable from any other state of A'.

Proof. The obtained FSM A' has some cycles and cycling paths with a cycle (s_1->...->s_k) only. Let s'_1->...->s'_t->s->s_1->...->s_k be such a path. If state s is equivalent to a state in the path then it is also equivalent to another state in the cycle (s_1->...->s_k) since the successors of equivalent states are equivalent for any input sequence. Therefore, we may assume that state s is equivalent to a state s'_1 in a cycle (s'_1->...->s'_t). Consider now an input sequence x^{tk} of length tk. State $\delta(s, x^{tk})$ and state $\delta(s'_1, x^{tk})$ are equivalent states, as they are successors of s and s'_1. By virtue of (4.2) and (4.1), $\delta(s, x^{tk}) = \delta(s_1, x^{tk-1}) = s_k$ and $\delta(s'_1, x^{tk}) = s'_1$. This means s and s_k should be equivalent states, but this is not possible, because $\lambda(s, x) \neq \lambda(s_k, x)$. ❑

Example 2. Consider the FSM with two cycles shown in Figure 4.3.

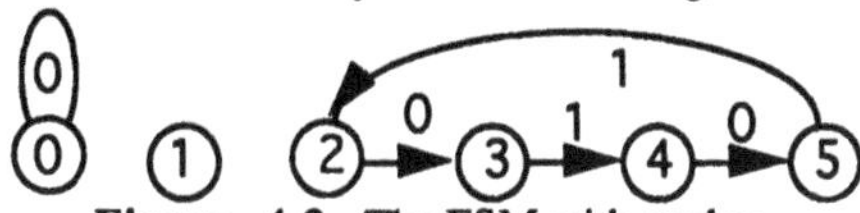

Figure 4.3 The FSM with cycles.

Based on the cycle (s_2, s_3, s_4, s_5), we define $\delta(s_1, x) = s_2$ and $\lambda(s_1, x) = 0$, since $\lambda(s_5, x) = 1$. We obtain a new transition: s_1-x/1->s_2. The machine augmented with this transition has a sequence xxx that distinguishes state 1 from any other state. In fact, $\lambda(s_0, xxx) = 000$, $\lambda(s_1, xxx) = 001$, $\lambda(s_2, xx) = 010$, $\lambda(s_3, xxx) = 101$, $\lambda(s_4, xxx) = 010$, $\lambda(s_5, xxx) = 101$. Alternatively, we could assign $\delta(s_1, x) = s_0$ and $\lambda(s_1, x) = 1$, since $\lambda(s_0, x) = 0$. In either case, the sequence xxx can be used as a homogeneous identifying sequence of s_1 in the completely specified machine. The original machine had no identifying sequence for this state. ❑

Case 3

A has cycling paths, but none of them is dominant. In this case, the starting state of each cycling path is equivalent to another state of this path. To assign the "don't care" transition (s, x) we choose an arbitrary path (s_1->...->s_{i+1}->...->s_k) which ends with the cycle (s_{i+1}->...->s_k), where $i \geq 1$. Since state s_1 is equivalent to a certain state of the path, it is also equivalent to a state s_j of this cycle, $i+1 \leq j \leq k$. Then we assign $\delta(s, x) = s_j$ with the output $\lambda(s, x) = y$ such that $y \neq \lambda(s_{j-1}, x)$. If j=1 then $y \neq \lambda(s_k, x)$. As a result, the augmented machine A' is obtained.

Proposition 4.4. In case 3, state s is distinguishable from any other state of A'.

Proof. We have $\delta(s, x) = s_j$ and $\lambda(s, x) \neq \lambda(s_{j-1}, x)$. Let state s be equivalent in A' to a state s'_p. State s'_p belongs either to a cycle (s'_1->...->s'_t), where $p \leq t$, or to a cycling path s'_1->...->s'_t with a cycle (s'_{r+1}->...->s'_t), where $1 \leq r \leq t$.

In the first case, the equivalence of s and s'_p implies the equivalence of s_j and s'_{p+1} (or s_j and s'_1 if p=t). Moreover, states $\delta(s_j, x^{tk-1})$ and $\delta(s'_{p+1}, x^{tk-1})$ are also equivalent. Because of (4.2), $\delta(s_j, x^{tk-1}) = s_{j-1}$, $\delta(s'_{p+1}, x^{tk-1}) = s'_p$. Thus, if states s_{j-1} and s'_p are equivalent, then states s and s_j should be equivalent as well, but this is impossible, because $\lambda(s, x) \neq \lambda(s_{j-1}, x)$. In the second case, state s is equivalent to state s'_p from the path s'_1->...->s'_t which terminates in cycle (s'_{r+1}->...->s'_t), where $1 \leq r \leq t$. If this path is defined in A then, by the assumption, its starting state is equivalent to another state of the path, and we have the situation considered above. Assume finally that state s is equivalent to state s'_p from the path (s'_1->...->s'_r->...->s->s_1->...->s_k) obtained in A'. Again, state s becomes equivalent to a state of a cycle, and we have exactly the same situation as above. ❑

Example 3. Consider the FSM shown in Figure 4.4.

Figure 4.4 The FSM with equivalent states.

There is a cycling path. Its starting state s_3 is equivalent to state s_1. We add a transition from state s_5 to state s_1 labeled with output $\lambda(s_5, x) = 0 \neq \lambda(s_2, x)$. Then state s_5 becomes distinguishable from any other state by the sequence xxx. ❑

Case 4

A has a dominant cycling path. In this case, there exists a cycling path P = (s_1->...s_{i+1}->...->s_k), $i \geq 1$ such that s_1 is not equivalent to any other state of P, moreover, in accordance with Proposition 4.1, only another starting state of a cycling path might be equivalent to s_1.

We transform the "don't care" transition (s, x) into a defined one in the following way. $\delta(s, x) = s_1$ and $\lambda(s, x)$ is assigned any $y \in Y$. Similarly to the cases considered above, we claim that state s is now distinguishable from any other state of the augmented FSM A'.

Proposition 4.5. In case 4, state s is distinguishable from any other state of A'.

Proof. Assume s is equivalent to a state s'_j. If s'_j is involved in a path (s'_1->...->s'_t) in A then s_1 of P can be equivalent to a starting state of another dominant cycling path, because P is also a dominant path. State s cannot be equivalent to any state of such a path. In this case, s can only be equivalent to some state of a cycling path (s'_1->...->s->s_1->...s_{i+1}->...->s_k). However, s_1 becomes equivalent to some state of the cycle (s_{i+1}->...->s_k). This contradicts our assumption that P is a dominant cycling path. ❑

Example 4. Consider the FSM shown in Figure 4.1. This machine has two dominant cycling paths with states (9, 8, 6, 7) and (13, 12, 10, 11). We can choose the first path and define transition from state 2 to state 9. Regardless of the output of this transition, the sequence xxx becomes an identifying sequence of state 2. ❑

Propositions 4.2 - 4.5 implies the following theorem.

Theorem 4.6. Given an FSM $A = (S, \{x\}, Y, \delta, \lambda, D_A)$, where $(S \times X) \backslash D_A = \{(s, x)\}$, it is always possible to assign its "don't care" transition such that state s of the augmented completely specified FSM is distinguishable from any other state and has a homogeneous state identifying sequence of length not exceeding the number of states. ❑

In certain cases, the augmented FSM has a homogeneous distinguishing sequence, as the following theorem shows. States s_i and s_j are said to be *converging* iff $\delta(s_i, x) = \delta(s_j, x)$ and $\lambda(s_i, x) = \lambda(s_j, x)$.

Theorem 4.7. Given an FSM $A = (S, \{x\}, Y, \delta, \lambda, D_A)$, where $(S \times X) \backslash D_A = \{(s, x)\}$, it is always possible to assign its "don't care" transition such that the augmented completely specified FSM has a homogeneous distinguishing sequence of length not exceeding the number of states iff A has neither converging nor equivalent states.

Proof. If A has at least two converging states then regardless of the transition's assignment, these states would become equivalent in any augmented machine. Assume now that we have assigned the transition (s, x) and state s becomes distinguishable from any other state. Theorem

4.6 states that it is always possible. If two states s_i and s_j are such that any input sequence is acceptable for each of them then they are nonequivalent in A as well as in an augmented FSM. Suppose therefore that for one of these states, say, for state s_i, not all sequences are acceptable. In this case, there is an acceptable sequence ω for s_i of A such that $\delta(s_i, \omega) = s$, since (s,x) is the only "don't care" transition in A. Now state $\delta(s_j, \omega)$ should be equivalent to state s. Then $\delta(s_j, \omega) = s$. The latter is possible only if A has converging states.

The completed FSM has only cycling paths, and the cycles are no longer than n. By construction, every two states are distinguishable by a sequence of length k, where k is the length of a cycle. Thus, $k \leq n$. □

We have examined all configurations possible in a given FSM with a single "don't care" transition in state s and thus, we have devised a technique for converting such a partial FSM into a completely specified FSM where state s possesses an identifying sequence. Under certain conditions, the resulting identifying sequence may also be a distinguishing sequence. Next, this technique will be generalized to cover the case where there exist several "don't care" transitions.

4.3 Several transitions

Given an FSM $A = (S, \{x\}, Y, \delta, \lambda, D_A)$, where $|(S \times X) \setminus D_A| \geq 1$, let the subset S_u contain all states with "don't care" transitions, i.e. $S_u = \{s \mid (s, x) \notin D_A\}$. We also define a subset S_d of states which accept all possible input sequences $\{x\}^*$; these states are involved in cycles or in cycling paths. The set S_d might be empty. For a state $s_i \in S_u$, let S_i denote the set of states from which state s_i is reachable, $s_i \in S_i$. Clearly, $S_i \cap S_j = \emptyset$ for all $i \neq j$ and $S_i \cap S_d = \emptyset$, since A is a deterministic machine. Based on the set $S_d \cup S_i$, we construct a submachine $A_i = (S_d \cup S_i, \{x\}, Y, \delta_i, \lambda_i, D_i)$ of the FSM A by deleting from A all states $S \setminus (S_d \cup S_i)$ along with their transitions. A_i has exactly one "don't care" transition. The technique of Section 4.2 can now be applied.

We present an algorithm for augmenting a given FSM with a single input in order to obtain an identifying sequence.

Algorithm 1.

Input: A partial FSM $A = (S, \{x\}, Y, \delta, \lambda, D_A)$ with a single input and $|S_u| \geq 1$ "don't care" transitions.

Output: An augmented completely specified FSM $A' = (S, \{x\}, Y, \delta', \lambda')$. Each state of S_u has an identifying sequence.

Step 1. Construct the subset S_d for A.

Step 2. Choose a state $s_i \in S_u$ with the maximal $|S_i|$

Construct a submachine $A_i = (S_d \cup S_i, \{x\}, Y, \delta_i, \lambda_i, D_i)$.

Step 3. Call the technique of Section 4.2 to assign (s_i, x) in A_i (and therefore in A).

Step 4. $S_d := S_d \cup S_i$

$S_u := S_u \setminus \{s_i\}$

If $S_u \neq \emptyset$ then GO TO Step 2. □

The resulting machine can be characterized by the following two theorems which are generalized from Theorems 4.6 and 4.7 and are proven in a similar manner.

Theorem 4.8. Suppose that $A = (S, \{x\}, Y, \delta, \lambda, D_A)$, where $|(S \times X) \setminus D_A| \geq 1$ is a given FSM and an FSM A' is the output of Algorithm 1. Then every initial state of "don't care" transitions is distinguishable from any other state in A' and has a homogeneous identifying sequence of length not exceeding the number of states. □

Theorem 4.9. Suppose that $A = (S, \{x\}, Y, \delta, \lambda, D_A)$, where $|(S \times X) \setminus D_A| \geq 1$ is a given FSM and an FSM A' is the output of Algorithm 1. Then the augmented completely specified FSM A' has a homogeneous distinguishing sequence of length not exceeding the number of states iff A has neither converging nor equivalent states. ❑

Example. Consider the FSM A shown in Figure 4.5a. The necessary and sufficient conditions of Theorem 4.9 are satisfied, since the machine has no converging or equivalent states, so Algorithm 1 should augment it in such a way that the resulting machine has a homogeneous distinguishing sequence of length not exceeding the number of states.

Figure 4.5 The FSM and its augmented FSM.

States 2 and 3 have "don't care" transitions, so the set $S_u = \{2, 3\}$. No state of A accepts all input sequences, the set $S_d = \varnothing$. We choose state 2 and find the set $S_2 = \{1, 2\}$. A submachine contains the transition 1->2. It is the case 1 of the technique from Section 4.2. We assign $\delta(2, x) = 1$ and $\lambda(2, x) = 1$. Now $S_d = \{1, 2\}$, $S_u = \{3\}$. $S_3 = \{3, 4\}$. We have the case 2 of Section 4.2. $\delta(3, x) = 1$ and $\lambda(3, x) = 0$. The augmented machine is shown in Figure 4.5b. It has a homogeneous distinguishing sequence xxx. The identifying sequence for state 1 is xx, for 2 - x, and for 3 and 4 - xxx.

5 ASSIGNING "DON'T CARE" TRANSITIONS

We now present an algorithm for augmenting a partial reduced FSM with several inputs. The algorithm uses formulae (3.1) - (3.3) to estimate the expected length of a test suite. If it exceeds the lower bound L_{min} defined by (3.3), the algorithm repeatedly tries all inputs labeling "don't care" transitions and calls Algorithm 1. The Wp-method [FBK91] is used to derive a resulting test suite which is complete in the class of machines with an equal or fewer number of states.

Algorithm 2.
Input: A partial reduced FSM A.
Output: An augmented FSM A' and a complete test suite of length not greater than that of A.
Step 1. Calculate the expected length L_A of a test suite for A.
If $L_A = L_{min}$ then GO TO STEP 3.
Step 2. $C := A$.
Step 2. Let $X_u = \{x_1, ..., x_q\}$ be the set of inputs labeling "don't care" transitions in C.
For each $x_i \in X_u$

Call Algorithm 1 to assign "don't care" transitions labeled with the input x_i.
Add newly defined transitions into the FSM C.
Let the augmented machine be C_i.
Calculate the expected length of a test suite for C_i.

Step 3. Let C^* be an FSM C_i or C with the shortest expected test suite.
If $C^* = C$ then GO TO STEP 4.
$C := C^*$. $X_u = X_u \setminus \{x_i\}$. If $X_u \neq \varnothing$ then GO TO STEP 2.
Step 4. $A' := C^*$. Call the Wp-method to derive a complete test suite for the machine A'. ❑

Remarks on Algorithm 2:
1) We assume that an FSM is given in its reduced form; however, this assumption is indeed not restrictive. The algorithm also accepts FSMs that are not reduced, i.e. that have compatible states [Gill62]. However, under the "undefined by default" convention for "don't care"

transitions, it is recommended first to reduce such a machine by merging compatible states, and then to apply Algorithm 2 to its reduced form. This is because a machine with fewer states usually requires shorter tests. There may exist several reduced forms of a nonreduced partial FSM, unlike the case of complete FSMs; and it is desirable to choose the most testable reduced form in this case. More research is required in this direction.

2) The resulting FSM is not necessarily a completely specified machine, some "don't care" transitions might be left intact. As discussed in Section 3, a shorter state cover usually leads to a shorter test suite. Undefined transitions can be assigned to reduce the total length of a state cover of the machine and eventually that of a complete test suite.

3) The algorithm tries all inputs which label "don't care" transitions. To facilitate its early termination we can arrange inputs such that the overall number of converging and equivalent states for a corresponding input form a non-decreasing sequence. In particular, if there exists an input, such that the necessary and sufficient conditions for the existence of a homogeneous distinguishing sequence are satisfied, then Algorithm 2 assigns the "don't care" transitions labeled with that input.

4) Comparison of possible augmentations with respect to different inputs is based on the expected test suite length. If instead, a test derivation method, such as the Wp-method, is called to derive a test suite whenever its length is required to make a decision, the user can stop the process once a test suite of an acceptable size is obtained. In the worst-case scenario, the method would be called $q(q+1)/2$ times, where q is the number of inputs labeling "don't care" transitions in the original machine.

6 APPLICATION EXAMPLES

Example 6.1

Consider the FSM A shown in Figure 6.1a. It is reduced and partially specified. There are five "don't care" transitions in this machine. Each of the transitions can lead to one of five states with output 0 or 1; alternatively, it can be left intact. Altogether, there exist $11^5 = 161051$ completely and partially specified machines that are augmentations of the given FSM A. A "perebor", i.e. an exhaustive procedure must try all of them, derive a complete test suite for each, and choose a machine with the shortest test suite. Instead, we apply our method.

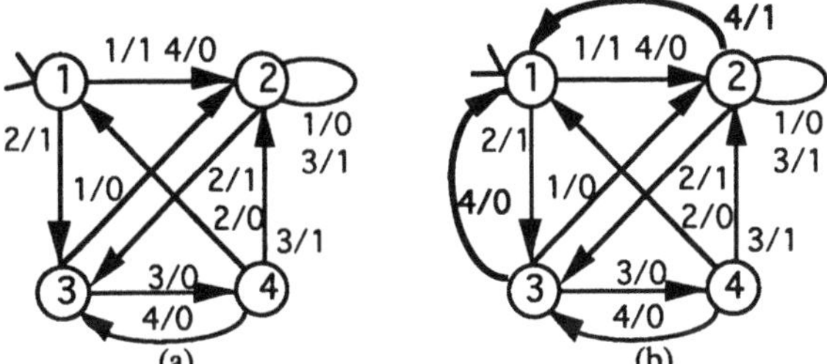

Figure 6.1 The FSM and its augmented FSM.

First we derive a complete test suite for the given machine using the Wp-method. Assuming state 1 as an initial state, the state cover is $V = \{\varepsilon, 1, 2, 23\}$. The transition cover is $TC = \{\varepsilon, 1, 2, 4, 11, 12, 13, 21, 23, 232, 233, 234\}$. The set $W = \{1, 2, 3\}$ is a characterization set of A. The state identifiers are: $W_1 = \{1, 2\}$, $W_2 = \{1, 2, 3\}$, $W_3 = \{1, 3\}$, $W_4 = \{1, 3\}$. The resulting test suite complete in the class of FSMs with up to four states is: {111, 112, 113, 121, 123, 131, 132, 133, 211, 212, 213, 2321, 2322, 2331, 2332, 2333, 2341, 2343, 41, 42, 43}. There are 21 test cases of total length 67. The formula (3.1) gives the expected length of 90. (3.3) returns the lower bound $L_{min} = 30$.

Every input labels at least one "don't care" transition, but only for input 4 are there no converging states. We choose this input and construct a submachine of A. It is, in fact, the machine shown in Figure 4.5a. The corresponding augmented submachine is the one shown in Figure 4.5b. It has a homogeneous distinguishing sequence 444. According to this submachine, two transitions must be added to the original machine, namely 2-4/1->1 and 3-4/0->1. We include them into the FSM *A* and obtain the augmented FSM *A'* shown in Figure 6.1b. The additional transitions are depicted in bold. Notice that three other "don't care" transitions are left intact. *A'* has the following state identifiers (as constructed in Section 3.3): $W'_1 = \{44\}$, $W'_2 = \{4\}$, $W'_3 = \{444\}$, $W'_4 = \{444\}$. Based on the obtained identifiers, we can derive a complete test suite of length 39. As a result, the length is reduced by about 40%. The "don't care" transitions labeled with inputs 1, 2 and 3 remain, since the length cannot be further reduced.

Example 6.2. The INRES protocol

To illustrate the proposed approach to improving the testability of a partially specified protocol machine, we consider the INRES protocol [Hogr91]. The behavior of the responder part of this protocol can be specified by an FSM given in Figure 6.2 (plain lines only).

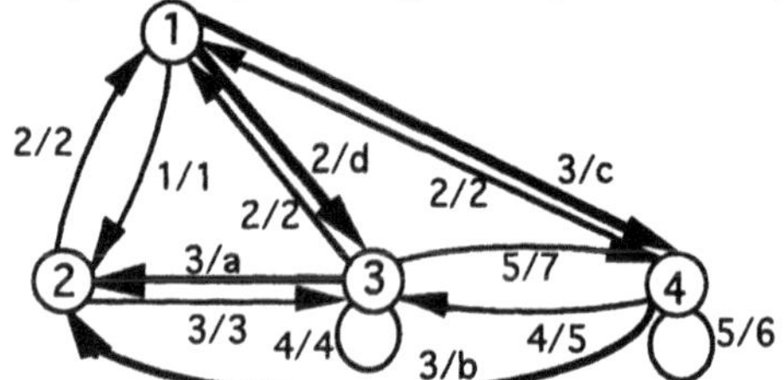

Figure 6.2 The INRES Responder.

The input alphabet is: 1- CR, 2 - IDISr, 3 - ICONrsp, 4 - DT0, 5 - DT1. The output alphabet is 1 - ICONi; 2 - DR; 3 - CC; 4 - ACK0; 5 - ACK0, IDATi; 6 - ACK1; 7 - ACK1, IDATi; 8 - null. The machine is partially specified. The traditional way of augmenting such a machine is based on the completeness assumption [SiLe89], [PBD93]. In particular, all "don't care" transitions are replaced by looping transitions with the null output (not depicted). Following this approach, we first obtain a completely specified FSM and derive a test suite complete for implementations with up to four states, as follows. A state cover is $V = \{\varepsilon, 1, 13, 135\}$. $W = \{41\}$. We apply the Wp-method and obtain a test suite with 19 test cases of total length 76.

Next, we assume that the behavior of the responder for all service primitives from the INRES user can be defined in an arbitrary way, whereas the completeness assumption should still be applied for all incoming PDUs. In particular, the transitions (1, 2), (1, 3), (3, 3), (4, 3) in states 1 and 3 on inputs 2 (IDISr) and 3 (ICONrsp) are "don't care" transitions. This machine requires a complete test suite with 15 test cases of total length 61.

Now we follow the proposed approach to find an augmented machine with a shorter test suite. Applying Algorithm 2 we can obtain the transitions 1-3/*c*->4, 3-3/*a*->2, 4-3/*b*->2 shown in Figure 6.2 as bold lines. Here *a*, *b*, and *c* are different ouputs which can be arbitrary chosen from the set {1, 2, 4, 5, 6, 7, 8}. The obtained machine has a homogeneous distinguishing sequence, that is $W' = \{3\}$ (ICONrsp). There is only one "don't care" transition left in state 1 on input 2. We define a transition 1-2/*d*->3, where *d* is an arbitrary output in order to reduce a state cover. It now has fewer symbols: $V' = \{\varepsilon, 1, 2, 3\}$. Given the sets *W'* and *V'*, we now have a complete test suite (produced by the same method): *TS* = {41, 541, 441, 341, 241, 1241, 13241, 135241, 141, 1541, 1441, 1141, 1344, 1334, 1314, 13544, 13554, 13534, 13514}. It comprises 17 test cases of total length 49. Thus, the obtained version of the INRES responder is more testable than the original one and the version based the completeness assumption. This assumption widely cited in the literature may deteriorate the testability of a protocol, as our example shows.

To assess the effectiveness of the method we have conducted the following experiment. A tool was designed to enumerate all $(1{+}4{\cdot}8)^4 = 1185921$ of the possible augmented machines for the INRES responder, derive a complete test suite for each of them, and find an FSM with the shortest one. A test derivation tool used to generate test suites implements the method developed for partial FSMs in [Petr91]. The experiment shown that all augmented FSMs require no fewer than 49 test events for a complete test suite.

7 CONCLUSION

In this paper, we have addressed one particular problem of design for testability of protocols on the specification level. We have developed an approach to improving testability of the given protocol taking advantage of the fact that a protocol is usually specified only partially and certain state/input combinations can be set in an arbitrary way. The feasibility of the approach was proven on partially specified FSMs with "don't care" transitions. Its effectiveness was demonstrated by conducting an experiment on the INRES protocol.

Though an algorithm given in this paper guarantees that the identifying sequences in the resulting FSM are quite short (their lengths do not exceed the number of states), it does not yet guarantee to produce the shortest possible ones. Thus, our algorithm can be further refined to construct an augmented machine with near-optimal identifying sequences. We continue our research in this direction. The work in progress also concerns the adaptation of the basic ideas underlying the proposed approach to nondeterministic and extended finite state machines.

In this paper, we have also presented some useful estimations of the expected length of complete test suites which are used to guide the process of augmenting partially specified machines. These estimations can also be used to select parameters of transition covers and characterization sets usually left without any guidance by most existing test derivation methods.

We have considered DFT in the context of the test derivation methods that rely on a reset facility, however the presented algorithms can be used in conjunction with other methods which do not use the reset. By augmenting a partial machine, a variety of UIO's or even distinguishing sequences are usually created. A nice property of the resulting machine is that lengths of identifying sequences never exceed the number of states. Therefore, any UIO-based method should yield a short test sequence. The presented approach can also be easily generalized to incorporate additional factors influencing the testability, such as length of transfer sequences (test preambles and postambles), the cost assigned to protocol messages, and others.

Acknowledgments

This work was partly supported by the HP-NSERC-CITI Industrial Research Chair on Communication Protocols at Université de Montréal, the NSERC Individual Research Grant (R. Dssouli) #20629188, and by the Russian Found for Fundamental Research. The authors wish to thank Q. M. Tan who has designed a test derivation tool for his help in experiments and S. A. Ezust for comments.

8 REFERENCES

[ABF90] M. Abramovici, M. A. Breuer, and A. D. Friedman, Digital Systems Testing and Testable Design, Computer Science Press, Oxford, England, 1990.

[BoPe94] G. v. Bochmann and A. Petrenko, "Protocol Testing: Review of Methods and Relevance for Software Testing", ISSTA'94, ACM International Symposium on Software Testing and Analysis, Seattle, U.S.A., 1994, pp. 109-124.

[BPY94] G. v. Bochmann, A. Petrenko, and M. Yao, "Fault Coverage of Tests Based on Finite State Models", the Proceedings of IFIP TC6 Seventh IWPTS'94, Japan.

[Chow78] T. S. Chow, "Testing Software Design Modeled by Finite-State Machines", IEEE Transactions on Software Engineering, Vol. SE-4, No.3, 1978, pp.178-187.

[DsFo91] R. Dssouli and R. Fournier, "Communication Software Testability", IFIP Transactions, Protocol Testing Systems III (the Proceedings of IFIP TC6 Third International Workshop on Protocol Test Systems), Ed. by I. Davidson and W. Litwack, North Holland, 1991, pp.45-55.

[FBK91] S. Fujiwara, G. v. Bochmann, F. Khendek, M. Amalou, A. Ghedamsi, "Test Selection Based on Finite State Models", IEEE Transactions on Software Engineering, Vol. SE-17, No.6, 1991, pp.591-603.
[Gill62] A. Gill, Introduction to the Theory of Finite-State Machines, McGraw-Hill, 1962.
[Henn64] F. C. Hennie, "Fault Detecting Experiments for Sequential Circuits", IEEE 5th Ann. Symp. on Switching Circuits Theory and Logical Design, 1964, pp. 95-110.
[Hogr91] D. Hogrefe, "OSI Formal Specification Case Study: The Inres Protocol and Service", University of Berne, Technical Report IAM-91-012, University of Berne, 1991.
[Jose78] J. Joseph, "On Easily Diagnosable Sequential Machines", IEEE Transactions on Computers, Vol. C-27, February, 1978, pp.159-162.
[LeYa94] D. Lee and M. Yannakakis, "Testing Finite-State Machines: State Identification and Verification", IEEE Trans. on Computers, Vol. 43, No. 3, 1994, pp. 306-320.
[Petr91] A. Petrenko, "Checking Experiments with Protocol Machines", IFIP Transactions, Protocol Test Systems, IV (the Proceedings of IFIP TC6 Fourth International Workshop on Protocol Test Systems, 1991), Ed. by Jan Kroon, Rudolf J. Heijink and Ed Brinksma, 1992, North-Holland, pp. 83-94.
[PBD93] A. Petrenko, G. v. Bochmann, and R. Dssouli, "Conformance Relations and Test Derivation", IFIP Transactions, Protocol Test Systems, VI, (the Proceedings of IFIP TC6 Fifth International Workshop on Protocol Test Systems, 1993), Ed. by O. Rafiq, 1994, North-Holland, pp.157-178.
[PDK93] A. Petrenko, R. Dssouli, and H. Konig, "On Evaluation of Testability of Protocol Structures", IFIP Transactions, Protocol Test Systems, VI, (the Proceedings of IFIP TC6 Fifth International Workshop on Protocol Test Systems, 1993), Ed. by O. Rafiq, 1994, North-Holland, pp.111-123.
[ShLe94] M. L. Sheu and C. L. Lee, "Symplifying Sequential Circuit Test Generation", IEEE Design and Test of Computers, Fall 1994, pp. 28-38.
[SiLe89] D. P. Sidhu and T. K. Leung, "Formal Methods for Protocol Testing: A Detailed Study", IEEE Trans. on Software Engineering, Vol. SE-15, No.4, 1989, pp.413-426.
[TyBa75] T. Tylaska and J. D. Bargainer, "An Improved Bound for Checking Experiments that Use Simple Input-Output and Characterizing Sequences", IEEE Transactions on Computers, Vol. C-24, No.6, 1975, pp. 670-673.
[Vasi73] M. P. Vasilevski, "Failure Diagnosis of Automata", Cybernetics, Plenum Publishing Corporation, New York, No.4, 1973, pp.653-665.
[VLC93] S. T. Vuong, A. A. F. Loureiro, and S. T. Chanson, "A Framework for the Design for Testabilitiy of Communication Protocols", in the Proceedings of IFIP TC6 Fifth IWPTS'93, Ed. by O. Rafiq, 1994, North-Holland, pp.89-108.

9 BIOGRAPHY

Nina Yevtushenko received the Dipl. degree in radio-physics in 1971 and Ph. D. in computer science in 1983, both from the Tomsk State University, Russia. She is now a Professor at that University. Her research interests include the automata and FSM theory and testing problems.
Alexandre Petrenko received the Dipl. degree in electrical and computer engineering from Riga Polytechnic Institute in 1970 and the Ph.D. in computer science from the Institute of Electronics and Computer Science, Riga, USSR, in 1974. Since 1992, he has been with the Université de Montréal, Canada. His current research interests include communication software engineering, protocol engineering, conformance testing, and testability.
Rachida Dssouli received the Doctorat d'université degree in computer science from the Université Paul-Sabatier of Toulouse, France, in 1981, and the Ph.D. degree in computer science in 1987, from the Université de Montréal, Canada. She is currently an Associate professor at the University of Montréal. Her research area is software engineering and her research interests include software specification and testability, protocol testing and observation.
Kamel Karoui is a Ph.D. student of the Université de Montréal, Canada.
Svetlana Prokopenko is a Ph.D. student of the Tomsk State University, Russia.

PART SEVEN

Test Generation 2

18

A Unified Test Case Generation Method for the EFSM Model Using Context Independent Unique Sequences[1]

T. Ramalingom[a] Anindya Das[b] and K.Thulasiraman[c]

[a]Bell-Northern Research Ltd., Ottawa, Canada K1Y 4H7 Tel: (613) 765-5377
E-mail: ramaling@bnr.ca Fax: (613) 763-5782
[b]D.I.R.O., University of Montreal, Montreal, Canada H3C 3J7
[c]School of Computer Science, University of Oklahoma, Norman, OK 73019, U.S.A. On leave from Dept. of Electrical Engineering, Concordia University, Montreal, Canada

A unified method for generating test cases for both control flow and data flow aspects of a protocol represented as an Extended Finite State Machine (EFSM) is presented. Unlike most of the existing methods, the proposed method considers the feasibility of the test cases during their generation itself. In order to reduce the complexity of the feasibility problem without compromising the control flow coverage, a new type of state identification sequence, namely, the Context Independent Unique Sequence (CIUS) is defined. The trans-CIUS-set criterion used in the control flow test case generation is superior to the existing control flow coverage criteria for the EFSM. In order to provide observability, the "all-uses" data flow coverage criterion is extended to what is called the def-use-ob criterion. A two-phase breadth-first search algorithm is designed for generating a set of executable test tours for covering the selected criteria. The approach is also illustrated on an EFSM module of a transport protocol.

Automatic test case generation from protocol standards is a means of selecting high quality test cases efficiently. Recently, International Organization for Standards (ISO) has established a working group for studying the application of Formal Methods in Conformance Testing (FMCT) [5]. One of the primary aims of this group is to enable computer-aided test case generation from protocol standards specified in Formal Description Techniques (FDT) such as Estelle [2], SDL [3], and LOTOS [4]. In this paper, we present a new method for automatically generating test cases for both control flow and data flow aspects of a protocol which is represented as an Extended Finite State Machine (EFSM) as defined in [21].

In order to have better fault coverage [7], some of the test sequence generation methods proposed recently [11, 13, 14] for the EFSM model apply state identification sequences for confirming the states. However, the state identification sequences defined for the FSM model are inadequate for the EFSM model. In this paper, we define a general Unique Input Sequence (UIS) for an EFSM state. We then consider a special type of UIS, called Context Independent Unique Sequence (CIUS) in order to reduce the complexity of the well known feasibility problem associated with the EFSM model that arises during the application of UISs for confirming states.

The test case generation method proposed in this paper addresses both control and data flow aspects of an EFSM. It is known from Finite State Machine (FSM) testing methods that those which use state identification sequences for confirming the tail state of a transition under test have better fault coverage [16, 10, 8]. In particular, the Uv-method has the capability of detecting both label faults and tail state faults in transitions [8]. The control flow fault coverage criterion established in this paper is called **trans-CIUS-set criterion** (defined later) and it is based on the Uv-method. For the data flow coverage, we extend the "all-uses" criterion [17]

[1]This work was done at Concordia University, Montreal, prior to T. Ramalingom joining Bell-Northern Research Ltd., and represents the views of the authors and not necessarily those of BNR Ltd.

to what is called a **def-use-ob criterion**. We shall see that this new criterion is required due to the so called black-box approach of protocol testing and it enhances the observability of the def-use associations. Thus our aim is to generate a set of feasible test cases for the trans-CIUS-set criterion and the def-use-ob criterion. Each test case in the proposed approach corresponds to a test tour which starts and ends at the initial state of the protocol. In the worst case, the cardinality of the set of tours generated is only quadratic in terms of the number of transitions in the protocol.

Most of the existing methods first generate a set of test tours which satisfy the coverage criteria and then check if the generated test tours are feasible [12, 21, 11, 20]. This strategy results in discarding infeasible tours, which in turn affects the coverage criteria. Therefore, an important requirement of our method is to consider the feasibility of the tours during their generation itself. We present a two-phase breadth-first search algorithm which generates a set of feasible test tours which adequately covers the required control flow and data flow criteria. The combined testing method by Miller and Paul [14] addresses the feasibility problem while selecting the test tours. This method does not however handle the feasibility issue effectively while joining different types of test subsequences into a single feasible sequence. Moreover, the trans-CIUS-set criterion and the def-use-ob criterion established in this paper are superior to the respective criterion in [14].

1 The EFSM Model

The EFSM model presented in this paper is inspired from [21]. An EFSM M is a 6-tuple $M = (S, s_1, I, O, T, V)$, where S, I, O, T, V are a nonempty set of states, a nonempty set of input interactions, a nonempty set of formal output interactions, a nonempty set of transitions, and a set of variables, respectively. Let $S = \{s_j \mid 1 \leq j \leq n\}$; s_1 is called the **initial state** of the EFSM. Each member of I is expressed as *ip?i(parlist)*, where *ip* denotes an interaction point where the interaction of type i occurs with a list of input interaction parameters *parlist*, which is disjoint from V. Each member of O is expressed as *ip!o(outlist)*, where *ip* denotes an interaction point where the interaction of type o occurs with a formal list of parameters, *outlist*. Each parameter in *outlist* can be replaced by a suitable variable from V, an input interaction parameter, or a constant. The interaction thus obtained from a formal output interaction is referred to as an **output interaction** or an **output statement**. We will assume that the variables in V and the input interaction parameters can be of types integer, real, boolean, character, and character string only. Each element $t \in T$ is a 5-tuple $t = (source, dest, input, pred, compute\_block)$. Here, *source* and *dest* are the states in S representing the starting state and the tail state of t, respectively. *input* is either an input interaction from I or empty. *pred* is a Pascal-like predicate expressed in terms of the variables in V, the parameters of the input interaction *input* and some constants. The *compute_block* is a computation block which consists of Pascal-like assignment statements and output statements.

A component of a transition can also be represented by postfixing the transition with a period followed by the name of the component. For example *t.pred* represents the predicate component of the transition t. Note that, unlike a variable, the scope of a parameter in an input interaction of a transition is restricted to the transition only. Let m denote the number of transitions in M. We will assume that $m \geq n$. A closed walk which starts and ends at the initial state is referred to as a **tour**. A transition in M with empty input interaction is called a **spontaneous transition**.

A **context** of M is the set $\{(var, val) \mid var \in V$ and val is a value of var from its domain$\}$. A **valid context** of a state in M is a context which is established when M's execution proceeds along a walk from the initial state to the given state.

Let t be a non-spontaneous transition in M. t is said to be **executable** if (i) M is in the state *t.source*, (ii) there is an input interaction of type i at the interaction point *ip*, where *t.input* = *ip?i(parlist)*, and (iii) the valid context of the state and the values of the input interaction parameters in *parlist* are such that the predicate *t.pred* evaluates to *true*. A spontaneous transition t is **executable** if (i) M is in the state *t.source* and (ii) the valid context of the state is such that *t.pred* evaluates to *true*. When a transition is executed, all the statements in its computation block get executed sequentially and the machine goes to the destination state of the transition.

A walk W in M is said to be **executable** if all the transitions in W are executable sequentially, starting from the beginning of the walk. A walk W in M can be **interpreted symbolically** by assuming distinct symbolic values for the local variables at the beginning of W as well as distinct symbolic values for the input interaction parameters along W. Let W be a symbolically interpreted walk. Clearly the conjunction of the predicates along W is also interpreted and is expressed in terms of the initial symbolic values for the local variables and the symbolic values for the input interaction parameters. W is said to be **satisfiable** if the conjunction of the interpreted predicates is satisfiable. Note that a walk which is executable is always satisfiable. However, its converse is not true. This is because none of the possible values for the variables which made W satisfiable may be a valid context at the starting state of the walk. That is, these values are not 'settable' by any of the executable walks from the initial state to the starting state of W.

An EFSM is **deterministic** if for a given valid context of any state in the EFSM, there exists at most one executable outgoing transition from that state.

An EFSM M is said to be **completely specified** if it always accepts any input interaction defined for the EFSM. An arbitrary EFSM M can be transformed into a completely specified one using what is called a **completeness transformation** described next. Given a valid context of a state and an instantiated input interaction, suppose that M does not have an executable outgoing non-spontaneous transition at the state for the given valid context and the input interaction, and that M does not have an outgoing spontaneous transition at the state such that it is executable for the given valid context, then a self-loop transition is added at the state such that it is executable for the given context and the input interaction. The newly added transitions are called **non-core transitions** and they do not have computation blocks.

We assume that the EFSM representation of the specification is deterministic and completely specified. It is assumed that for every transition in the EFSM, it has at least one executable walk from the initial state to the starting state of the transition such that the transition is executable for the resulting valid context. Similarly, we assume that the initial state is always reachable from any state with a given valid context.

1.1 An Example

As an example of an EFSM , let us consider a major module (*AP-module* in [6]) of a simplified version of a class 2 transport protocol [1]. This module participates in connection establishment, data transfer, end-to-end flow control, and segmentation. It has the interaction point labeled U connected to the transport service access point and another interaction point labeled N connected to a mapping module. Here, we represent the EFSM by (S, s_1, I, O, T, V). We would like to note that the EFSM is obtained from the *AP-module* by eliminating a few non-determinisms in certain transitions starting from the data transfer state. Let $S = \{s_1, s_2, s_3, s_4, s_5, s_6\}$. The set of input interactions and the set of output interactions are given below.

I = {U?TCONreq(dest_add, prop_opt), U?TCONresp(accpt_opt),
U?TDISreq, U?TDATreq(Udata, EoSDU), U?U_READY(cr),
N?TrCR(peer_add, opt_ind, cr), N?TrCC(opt_ind, cr),
N?TrDR(disc_reason, switch), N?TrDT(send_sq, Ndata, EoTSDU),
N?TrAK(XpSsq, cr), N?ready, N?terminated, N?TrDC }

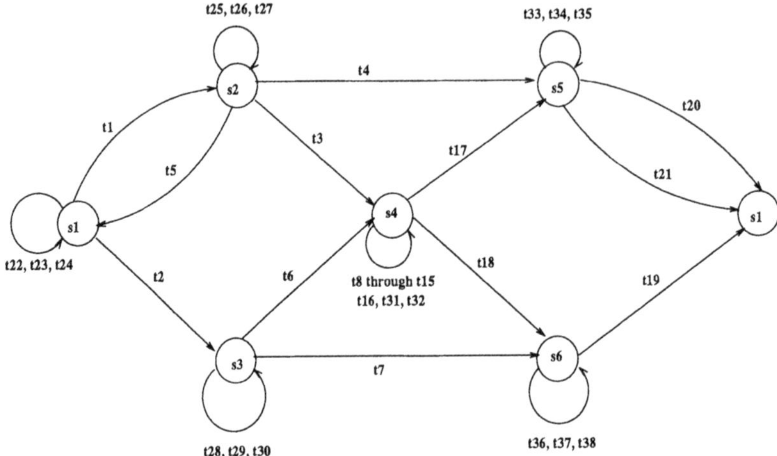

Figure 1: An EFSM for the AP-module in the Class 2 transport protocol

O = {U!TCONconf(opt), U!TCONind(peer_add, opt), U!TDISind(msg), U!TDATAind(data, EoTSDU), U!error, U!READY, U!TDISconf, N!TrCR(dest_add, opt, credit), N!TrDR(reason, switch), N!terminated, N!TrCC(opt, credit), N!TrDT(sq_no, data, EoSDU), N!TrAK(sq_no, credit), N!error, N!TrDC}

V={*opt, R_credit, S_credit, TRsq, TSsq* }. All the variables in V are of integer type. The transitions as described in Table 1 and Table 2 are shown in Figure 1. The state s_1 is repeated in the figure merely for convenience.

1.2 Unique Input Sequence

An input sequence, a sequence of input interactions, is said to be **instantiated** if all the parameters in the sequence are properly instantiated with values. Given an instantiated input sequence X, a state s_i and a valid context C at s_i, $Ewalk(i, X, C)$ denotes the unique walk traversed when X is applied to the EFSM which is currently at s_i with the context C.

A **test sequence** is a sequence of input and output interactions. A sequence of zero or more output interactions between two successive input interactions in a test sequence is the sequence to be observed after applying the preceding input interaction to an EFSM and before applying the succeeding one.

The sequence of input and output interactions along a satisfiable walk W is denoted as **Trace(W)**, known as the **trace of the walk** W. The sequence of input (output) interactions along a walk W is denoted by **Inseq(W) (Outseq(W))**. $Trace(W)$ and $Outseq(W)$ are actually obtained by symbolically interpreting W. Suppose that the actual value of a symbol is known, then the corresponding sequences can be obtained from the above sequences by replacing the symbol by the value throughout the sequences.

Two input interactions are said to be **distinguishable** if: (i) they occur at two different interaction points or (ii) their interaction types are different. We say that two output interactions are **distinguishable** if at least one of the following is true: (i) they occur at two

Tr.	Input	Predicate	Compute-block
t1	U?TCONreq(dst_add, *prop_opt*)		opt:= prop_opt; R_credit := 0; N!TrCR(dst_add,opt,R_credit)
t2	N?TrCR(peer_add, *opt_ind*, *cr*)		opt := opt_ind; S_credit := cr; R_credit := 0; U!TCONind(peer_add, opt)
t3	N?TrCC(opt_ind,cr)	opt_ind ≤ opt	TRsq:=0;TSsq:=0; opt := opt_ind; S_credit := cr; U!TCONconf(opt)
t4	N?TrCC(opt_ind, cr)	opt_ind > opt	U!TDISind(' procedure error'); N!TrDR('procedure error', false)
t5	N?TrDR(disc_reason, *switch*)		U!TDISind(disc_reason); N!terminated
t6	U?TCONresp(accpt_opt)	accpt_opt ≤ opt	opt := accpt_opt; TRsq := 0; TSsq := 0; N!TrCC(opt, R_credit)
t7	U?TDISreq		N!TrDR('User initiated' , true)
t8	U?TDATreq(Udata, *EoSDU*)	S_credit > 0	S_credit := S_credit−1; N!TrDT(TSsq, Udata, EoSDU); TSsq := $(TSsq+1) mod 128$;
t9	N?TrDT(send_sq, Ndata, *EoTSDU*)	R_credit ≠ 0 ∧ send_sq = TRsq	TRsq := $(TRsq+1) mod\ 128$; R_credit := R_credit − 1; U!TDATAind(Ndata, EoTSDU); N!TrAK(TRsq, R_credit)
t10	N?TrDT(send_sq, Ndata, *EoTSDU*)	R_credit = 0 ∨ *send_sq* ≠ *TRsq*	N!error; U!error
t11	U?U_READY(cr)		R_credit := R_credit+cr; N!TrAK(TRsq, R_credit)
t12	N?TrAK(XpSsq, cr)	TSsq≥XpSsq ∧ $cr + XpSsq - TSsq \geq 0$ ∧ $cr + XpSsq - TSsq \leq 15$	S_credit := $cr + XpSsq - TSsq$
t13	N?TrAK(XpSsq, cr)	TSsq≥XpSsq ∧ $(cr + XpSsq - TSsq < 0$ ∨ $cr + XpSsq - TSsq > 15)$	U!error; N!error
t14	N?TrAK(XpSsq, cr)	TSsq<XpSsq ∧ $cr + XpSsq - TSsq - 128 \geq 0$ ∧ $cr + XpSsq - TSsq - 128 \leq 15$	S_credit := $cr + XpSsq - TSsq - 128$
t15	N?TrAK(XpSsq, cr)	TSsq<XpSsq ∧ $(cr + XpSsq - TSsq - 128 < 0$ ∨ $cr + XpSsq - TSsq - 128 > 15)$	U!error; N!error
t16	N?ready	S_credit > 0	U!READY
t17	U?TDISreq		N!TrDR('User initiated', *false*)
t18	N?TrDR(disc_reason, *switch*)		U!TDISind(disc_reason); N!TrDC
t19	N?terminated		U!TDISconf
t20	N?TrDC		N!terminated; U!TDISconf
t21	N?TrDR(disc_reason, *switch*)		N!terminated

Table 1: Core transitions in the transport protocol

Transitions	Input
t25, t28, t31, t33, t36	U?TCONreq(dest_add, prop_opt)
t23, t26, t34, t38	U?TDISreq
t22, t29, t37	N?TrDR(disc_reason, switch)
t24, t27, t30, t32, t35	N?terminated

Table 2: Non-core transitions in the transport protocol

different interaction points, (ii) their interaction types are different, and (iii) if the parameters in a given position in both interactions are constants then they are different.

For example, the output interactions N!TrDR('procedure error', false) and N!TrDR('procedure error', true) are distinguishable. However, N!TrDT(TSsq, Udata, EoSDU) and N!TrDT(TRsq, Udata, EoSDU) are not distinguishable.

An input interaction is obviously distinguishable from an output interaction. The total number of input and output interactions - each occurrence of an interaction is counted - in a sequence is called the **length** of the sequence. Let S_1 and S_2 be two sequences of input and/or output interactions. Assume that they are of the same length. In order to check for distinguishability of the two sequences, starting from the first position the interactions in S_1 and S_2 are checked position-wise. S_1 and S_2 are said to be **distinguishable** if the interactions in at least one position in S_1 and S_2 are distinguishable. Otherwise, they are said to be **indistinguishable**. Two sequences of different lengths are always distinguishable.

Let W be an executable walk at s_j. Let U be an instantiation of $Inseq(W)$. We define U as a **Unique Input Sequence (UIS)** of s_j if $Trace(W)$ is distinguishable from $Trace(W')$, for any satisfiable walk W' at state s_k, for $1 \leq k \leq n, k \neq j$. In this case, W is called an **UIS walk** for U.

2 Test Case Selection Criteria

2.1 Control Flow Coverage Criterion

We would like to apply an UIS of every state at the tail state of the transition under test. As indicated in [13], automatic test case generation for an EFSM is difficult when a general UIS is used. For example, let U be an UIS for s_j, and let W be the UIS walk of U. Let t be an incoming transition at s_j and s_i be the starting state of t. In order to test t, one needs to compute an executable preamble walk P_t from s_1 to s_i and associate values for the input interaction parameters along P_t and t such that $P_t\ t\ W$ is executable. For a given W, it is in general difficult to find a P_t so that the walk $P_t\ t\ W$ is executable. Moreover, if the general UISs are considered, then multiple UISs may be required for a state in order to test all the incoming transitions at that state. Hence a careful selection of the UISs is required.

A walk from a state is said to be **context independent** if the predicate of every transition along the walk, duly interpreted symbolically, is independent of the symbolic values of the local variables at the starting of the walk. Observe that every context independent satisfiable walk is executable.

We introduce a special type of UIS, called **Context Independent Unique Sequence (CIUS)**. Let U_i be an instantiated UIS of s_i and let $U(i)$ be the corresponding UIS walk at s_i. U_i is said to be a **CIUS** of s_i if $U(i)$ is context independent and executable.

Note that all the local variables used in the predicate of each transition in $U(i)$ are defined within $U(i)$ prior to their use. In other words, the predicates along $U(i)$ are independent of any valid context at s_i. Therefore, $U(i)$ can be postfixed to any executable walk from the initial state to s_i and the resulting walk is also executable. This property is very useful in computing

State	CIUS	Transition Seq.
s_1	U?TCONreq(dst_add, prop_opt)	t1
s_2	N?TrDR(disc_reason, switch)	t5
s_3	U?TDISreq	t7
s_4	U?TDISreq	t17
s_5	N?TrDR(disc_reason, switch)	t21
s_6	N?terminated	t19

Table 3: CIUSs for the states in the EFSM of Figure 1

feasible test cases for the control flow coverage. Also, one CIUS of a state is sufficient for testing all the incoming transitions at that state.

In [15], we have developed an algorithm for computing a CIUS for a given state. Table 3 shows the CIUSs for all the states of the EFSM of Figure 1 computed using the algorithm. Note that the parameters in the CIUSs have to be instantiated with certain valid values. We have also found that a few other protocols such as a class 0 transport protocol as specified in [21] and the abracadabra protocol [19] have a CIUS for every state. The maximum length of the CIUSs computed for these protocols is only 2. It should also be noted that there are protocols which may not have a CIUS for every state. For example, the initiator module of the INRES protocol as modeled in [9] does not have a CIUS for one state.

Let U_i be a CIUS for the state s_i, $1 \leq i \leq n$. Let $\mathcal{U} = \{U_i \mid 1 \leq i \leq n\}$. We call $\mathcal{U}$ as a **CIUS set**. Our control flow coverage criterion, namely, the **trans-CIUS-set criterion** is to select a set $\mathcal{T}$ of executable tours such that for each transition t in the EFSM and for each $U_i \in \mathcal{U}$, $\mathcal{T}$ has a tour which traverses t followed by U_i. An executable walk from the initial state to the starting state of a transition t is called a **preamble walk for** t if Wt is also executable. Due to the requirement of applying the entire UIS set at the tail state of a transition under test, the trans-CIUS-set criterion is superior to the existing control flow coverage criteria for the EFSM.

2.2 Data Flow Coverage Criterion

A hierarchy of data flow coverage criteria has been proposed in [17]. It is interesting to know that the "all-uses" is the best criterion among those which can be satisfied by a set of test cases with polynomial order cardinality [17]. Ural and Williams [20] have recently used the all-uses criterion for generating test cases for protocols specified in SDL. Due to the black-box approach of protocol testing, the set of test cases which satisfy the all-uses criterion may not be observable. Therefore, we extend the all-uses criterion to what is called a **def-use-ob criterion**. This criterion facilitates the tester to observe every def-use association in the protocol.

We introduce some definitions before presenting the def-use-ob criterion. A parameter v occurring in the input interaction of a transition t is referred to as a **def** and is denoted by $t.I.v$. Similarly, a variable v in the left side of an assignment statement at the location c in the computation block of a transition t is also said to be a **def** and it is denoted by $t.c.v$. The use of a variable or input interaction parameter v in the predicate of a transition t is called a **p-use** and is denoted by $t.P.v$. The variable/input interaction parameter v used on the right side of an assignment statement at the location $c1$ in the computation block of a transition t is referred to as a **c-use** and is denoted by $t.c1.v$. Similarly, the variable/input interaction parameter v appearing as a parameter in the output interaction at the location $c2$ in the computation block of a transition t is referred to as a **o-use** and it is denoted by $t.c2.v$. By an **use**, we refer to a p-use, a c-use or a o-use.

A **def-use pair** D with respect to a variable/parameter v is an ordered pair of def and use of v such that there exists a walk in the EFSM which satisfies the following: (i) the first transition in the walk is the one where v is defined and the last transition of the walk is the one where v is used and (ii) v is not redefined in the walk between the location where it is originally defined and the location where it is used. Such a walk is called a **def-clear** walk for D. Note that a def-clear walk could be a single transition. A def-use pair is said to be **feasible** if the EFSM has at least one executable tour which contains a def-clear walk for this pair. The def-use pairs can be classified into five types as follows.

type 1: An input parameter v is defined in the input interaction of a transition t_1 and is used in the predicate of the same transition. Such a pair is denoted by $(t_1.I, t_1.P)v$.

type 2: An input parameter v is defined in the input interaction of a transition t_1 and is used in an output statement c_2 in the computation block of the same transition. Such a pair is denoted by $(t_1.I, t_1.c_2)v$.

type 3: An input parameter v is defined in the input interaction of a transition t_1 and is used in an assignment statement c_3 in the computation block of the same transition. Such a pair is denoted by $(t_1.I, t_1.c_3)v$.

type 4: A variable v is defined in an assignment statement c_1 in the computation block of a transition t_1 and is used in the predicate of another transition t_2. Such a pair is denoted by $(t_1.c_1, t_2.P)v$.

type 5: A variable v is defined in statement c_1 in the computation block of a transition t_1 and is used in statement c_2 in the computation block of a transition t_2. Such a pair is denoted by $(t_1.c_1, t_2.c_2)v$.

Let l (l') be a location in transition t (t') where a variable/parameter v (v') is defined (used). Suppose that $X = D_1 D_2 \dots D_k$, where $k \geq 1$, is a sequence of def-use pairs such that (i) D_i is a def-use pair for variable v_i, $i = 1, 2, \dots, k$, (ii) $v_1 = v$ and $v_k = v'$ and the source of D_1 is $t.l$ and the destination of D_k is $t'.l'$, (iii) the use part of D_i is for defining v_{i+1}, where $i = 1, 2, \dots, k-1$ and (iv) if $k = 1$, then $v = v'$. Then, X is called an **information flow chain** from the definition of v at the location l of transition t to the use of v' at the location l' of transition t'. Further, if a walk W has a subwalk W' with t and t' as the first and the last transition such that W' can be expressed as $W' = W_1@W_2@\dots@W_k$, where W_i is a def-clear walk for D_i, for $i = 1, 2, \dots, k$, then, we say that X is an **information flow chain along** W. In this case, we also say that W has an information flow chain from the definition of v at the location l of transition t to the use of v' at the location l' of transition t'. We would like to note that the information flow chain is somewhat similar to the IO-def-chain proposed in [21].

Let $\mathcal{D}$ be the set of all def-use pairs for all the variables and input interaction parameters in the EFSM. A minor modification of the algorithm presented in [9] would suffice to obtain $\mathcal{D}$. This modification is to consider the def-use pairs within a transition. Our **def-use-ob criterion** requires the selection of a set of executable tours such that for each feasible def-use pair $D \in \mathcal{D}$, the set has at least one tour, say T, satisfying the following conditions.

(a) If the use part in D is an o-use, then T contains a def-clear walk for D.

(b) If the use part in D is a p-use, then T contains a def-clear walk $W1$ for D followed by the CIUS walk $U(j)$, where s_j is the tail state of $W1$.

(c) If the use part in D is a c-use, then T contains a walk $W2$ followed by a walk $W3$ such that $W2$ is a def-clear walk for D and $W3$ has an information flow chain from the variable which is defined at the location where the variable for D is c-used to a location where a variable is either o-used or p-used. Moreover, if the information flow chain terminates in a p-use variable, then, in T, $W3$ is followed by the CIUS walk $U(p)$, where s_p is the tail state of $W3$.

Condition (a) takes care of the def-use association for all the def-use pairs in which the use part is an o-use. If the use part of D is a p-use, then apart from meeting the def-use association, by applying the CIUS of s_j, condition (b) enables the tester to check if the predicate of the transition where the p-use occurs evaluates to **true** as expected. On the other hand, if the use part of D is a c-use, then condition (c) enables the tester to observe the effect of the value computed. Actually, this value flows through other intermediate variables along T until it is used in an output statement or in a predicate of a transition. In addition, the correct evaluation of the predicate is ensured by T as in condition (b).

An executable walk W starting from the initial state is called a **preamble walk for** D if it satisfies conditions (a), (b) and (c) where T is replaced by W.

We know that, as per the trans-CIUS-set criterion, each transition followed by the CIUS of the tail state of the transition will be covered by at least one tour. Clearly, this tour also covers all the def-use pairs of types 1 and 2 for the def-use-ob criterion. Henceforth, we assume that $\mathcal{D}$ consists of types 3 4 and 5 only.

We define a new type of Data Flow Graph (DFG) to represent the data flow information on a particular executable walk starting from the initial state. This graph is useful in computing the subset of $\mathcal{D}$, for which this walk is a preamble walk, except possibly for the CIUS walk extension. The data flow graph has four types of nodes: i-node, c-node, p-node and o-node.

- An **i-node** is labeled as (t, I, v) and it corresponds to the definition of the parameter v in the input interaction of the transition t.
- A **c-node** is labeled as (t, c, v) and it corresponds to the definition of the variable v in the assignment statement c of the transition t.
- A **p-node** is labeled as (t, P) and it indicates that the node corresponds to the predicate of the transition t.
- A **o-node** is labeled as (t, c) and it simply denotes that it corresponds to the output statement c in the computation block of the transition t.

The **data flow graph** for the transition t with respect to the walk W which contains t is denoted by DFG$[t, W]$. It contains the data flow information along W for all the input interaction parameters and local variables defined in t. It has one connected directed subgraph, say G, for each definition of a variable or an input interaction parameter, say v, in t. G has a designated node, called the **root node** which identifies the definition of v. A given node in G is considered to be in one of three different levels. The root node is the unique node in the first level. Nodes in level 2 correspond to the direct use of v in statements/predicates in W and W contains a def-clear walk for every def-use pair consisting of the root node and a node in level 2. The root node is connected to all the nodes of level 2. A node is in level 3 if there exists a data flow along W from at least one assignment statement which corresponds to a c-node in level 2 to a predicate, assignment statement, or an output statement corresponding to this level 3 node. A c-node in level 2 is connected to a level 3 node if there exists an information flow chain along W from the level 2 node to the level 3 node.

Figure 2 shows the data flow graph DFG$[t3, t1t3t8]$, for the transition $t3$ in the walk $t1t3t8$ of the EFSM given in Figure 1. In Figure 2, rectangles represent i-nodes as well as o-nodes, whereas the circles and diamonds represent c-nodes and p-nodes, respectively. The second subgraph in this data flow graph, for instance, corresponds to the definition of the input interaction parameter cr. Observe that the edges from $(t3, c4, S\_credit)$ to the level 3 nodes $(t8, P)$ and $(t8, c1, S\_credit)$ indicate that the variable S_credit defined in $t3.c4$ is p-used at the predicate of transition $t8$ and c-used in the definition of S_credit at the first statement in the computation block of $t8$, respectively.

The size of the test cases required for satisfying the coverage criteria is summarized in the following theorem.

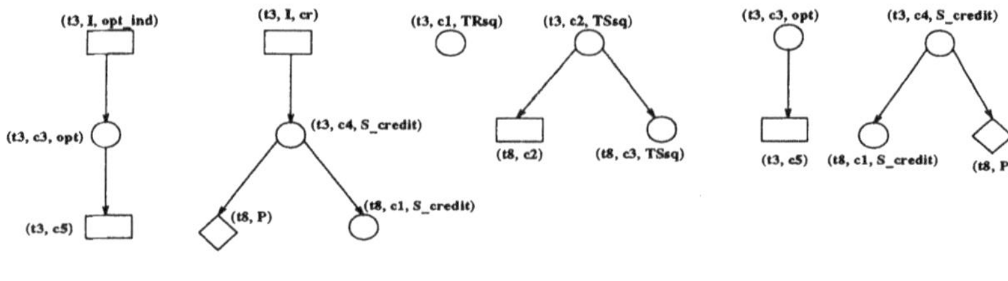

Figure 2: A data flow graph for $t3$ with respect to the walk $t1t3t8$

Theorem 1 *The order of the set of test tours required to satisfy the trans-CIUS-set and the def-use-ob criteria together is quadratic in the number of transitions in the EFSM.*

3 Data Flow Graph Manipulation

In this section, we briefly describe the procedures for constructing and manipulating the data flow graph DFG$[t, W]$ for a given transition t which is a part of a given executable walk W starting from the initial state of an EFSM. These procedures are used in our test case generation algorithm for checking if a walk is a preamble walk for some def-use pairs.

Our first procedure *PredExtendGraph* is for processing a predicate in a given transition. The procedure accepts a walk $W2$, a transition $t2$, where $t2$ is the last transition in $W2$, and a partial subgraph G of DFG$[t3, W2]$, for some transition $t3$ in $W2$. Let G correspond to a variable/parameter u defined at $t3$. G is partial since it does not have the data flow information corresponding to the transitive use of u in $t2$. As described below, *PredExtendGraph* extends the graph G if the value of u is eventually used in the predicate of $t2$. The variable $inlevel2$ ($inlevel3$) is used to ensure that the p-node $(t2, P)$ is created atmost once in level 2 (level 3) of G. This procedure also checks if $W2$ is a preamble walk for a def-use pair along $W2$ where the definition corresponds to the root node of G. For notational convenience, we denote a node at a given level by attaching the level number as a subscript to the label of the node. For example, a c-node (t, c, v) at level 3 is also denoted by $(t, c, v)_3$. Comments are enclosed in braces.

```
procedure PredExtendGraph(G:graph; t2:transition; W2:walk);
begin
   inlevel2 := false; inlevel3 := false;
   Let (t1,x1,u) be the root node of G; { x1 = 'I' or assignment stmt. no. }
   for each variable v used in t2.pred do begin
      Let (t, c) = W2.recentdef(v); {Recent definition of v in W2 is at t.c}
      if ((t,c,v) is the root node of G) then begin { (t,c,v)= t1,x1,u) }
         if (not inlevel2) then begin
            Create a p-node (t2,P) at level 2 in G; inlevel2 := true;
         end;
         Add an edge from (t1,x1,u)_1 to (t2,P)_2 in G;
         if (D = (t1.x1, t2.P)(u)∈ 𝒟 is not yet covered) then begin
            Mark D as covered;
            Obtain a preamble walk for D by appending U(j) to W2,
               where s_j= t2.dest & U(j) is the CIUS walk for U_j;
         end
      end;
      if ((t,c,v) is a node at level 2 in G) then begin
         if (not inlevel3) then begin
```

```
                Create a p-node (t2,P) at level 3 in G; inlevel3 := true;
            end;
            Add an edge from (t,c,v)_2 to (t2,P)_3 in G;
            if (D = (t1.x1, t.c)(u)∈ 𝒟 is not yet covered) then begin
                Mark D as covered;
                Obtain a preamble walk for D by appending U(j) to W2,
                    where s_j= t2.dest & U(j) is the CIUS walk for U_j;
            end
        end;
        if ((t,c,v) is a node at level 3 in G) then begin
            if (not inlevel3) then begin
                Create a p-node (t2,P) at level 3 in G; inlevel3 := true;
            end;
            for each incoming edge e to (t, c, v) do begin
                Let (t', c', v')_2 be the starting node of e;
                Add an edge from (t', c', v')_2 to (t2,P)_3 in G;
                if (D = (t1.x1, t'.c')(u)∈ 𝒟 is not yet covered) then begin
                    Mark D as covered;
                    Obtain a preamble walk for D by appending U(j) to W2,
                    where s_j= t2.dest & U(j) is the CIUS walk for U_j;
                end
            end
        end
    end { for each variable v }
end { PredExtendGraph }
```

StmtExtendGraph and *OutputExtendGraph* are the other two procedures for extending a subgraph of a data flow graph with respect to an assignment statement and an output statement, respectively. They are similar to *PredExtendGraph* [15].

We shall now describe procedure *ExtendDFG*. This procedure accepts a walk $W1$, a transition $t1$ in $W1$, and a transition $t2$ which starts from the tail state of $W1$ and it computes DFG$[t1, W1\ t2]$, the data flow graph for $t1$ with respect to the walk $W1\ t2$. *ExtendDFG* achieves this by extending the already known data flow graph DFG$[t1, W1]$ as per the data flows along $W1\ t2$ from the variables/parameters defined in $t1$ to the variables used in the predicates and the statements in $t2$. Let $W2 = W1\ t2$. Let us assume that the set of def-use pairs in $\mathcal{D}$ which are yet to be covered for the def-use-ob criterion is known at the starting of the procedure. After copying DFG$[t1, W1]$ into DFG$[t1, W2]$, it manipulates each subgraph in DFG$[t1, W2]$ with respect to the variables used in the predicate of $t2$. It calls the procedure *PredExtendGraph* for this purpose. It then sequentially selects every statement in the computation block of $t2$, and updates every subgraph in DFG$[t1, W2]$ by considering all the variables/parameters used in the statement. If it is an assignment statement, then *ExtendDFG* calls the procedure *StmtExtendGraph*; otherwise it invokes *OutputExtendGraph* for updating a given subgraph. The formal description is given below.

```
procedure ExtendDFG(t1:transition;W1:walk;t2:transition);
begin
    Let W2 be the walk obtained by appending t2 to the walk W1;
    DFG[t1,W1] := DFG[t1,W2];
    for each subgraph G in DFG[t1,W2] do
        PredExtendGraph(G, t2, W2);
    { Sequentially process the statements in the compute-block of t2 }
    for each statement c2 in the compute-block of t2 do
        for each subgraph G in DFG[t1,W2] do
            if (c2 is an assignment statement) then
                StmtExtendGraph(G, t2, c2, W2)
            else OutputExtendGraph(G, t2, c2, W2);
end; { ExtendDFG }
```

Our final procedure for DFG manipulation is *ConstructDFG* for constructing DFG$[t, t]$ for every transition t in an EFSM. It is very similar to *ExtendDFG* but for the fact that it starts with an empty data flow graph. It is easy to see that the data flow graph DFG$[t, W]$ of a transition t with respect to a walk W which contains t can be constructed using *ConstructDFG* and *ExtendDFG*.

4 Automatic Test Case Generation

4.1 The Two-Phase Algorithm

We have already established the trans-CIUS-set criterion for the control flow testing and the def-use-ob criterion for data flow testing. The next step is to generate a set of test cases satisfying these criteria. The algorithm presented in this section systematically generates a set of executable test tours for covering the above criteria. It has two phases and it traverses the EFSM in a breadth-first fashion in both phases. The first phase constructs a preamble walk for every transition in the EFSM and for the feasible def-use pairs in $\mathcal{D}$. In the second phase, all preambles computed in the first phase are completed into a set of executable tours.

The step-wise description of the first phase of the algorithm is given below. The salient points in the algorithm are then discussed. For ease of understanding, each step is embedded with comments.

Phase I

Input: An EFSM, CIUS-set $\mathcal{U} = \{U_j \mid 1 \leq j \leq n\}$, Def-use pairs set $\mathcal{D}$. A positive integer K_1.

Output: UFset: set of preamble walks for the coverage criteria.

Step 0 { Data flow graphs initialization }

(i) Construct the data flow graph of each transition with respect to itself.

Step 1 { null walk initialization }

(i) Let P be a null walk at s_1; Let $\mathcal{P} = \{P\}$.

Step 2 { ith iteration of this step computes the set of all executable walks of length i starting from s_1. They are computed from the executable walks of length $i-1$ computed in the previous iteration. This step marks all transitions & def-use pairs covered by the new walks.}

(i) Let $\mathcal{T} = \emptyset$.

(ii) Do Step 2.1 for each $P \in \mathcal{P}$ and for each outgoing transition t from the tail state of P.

(iii) If all the transitions in the EFSM are covered for control flow and all the def-use pairs in $\mathcal{D}$ are covered for data flow or the number of iterations of Step 2 exceeds K_1, a fixed positive integer, then proceed to Step 3.

(iv) Consider $\mathcal{T}$ as $\mathcal{P}$ and repeat Step 2.

Step 3 { For every transition t, and for every CIUS, postfix t followed by the walk along the CIUS to the preamble walk. Also collect the resulting walks for the transitions as well as the preamble walks for the def-use pairs into $UFset$.}

(i) Let both *CFset* and *DFset* to be the empty set.

(ii) For each transition t covered by Step 2 and for each CIUS $U_k, 1 \leq k \leq n$, add $W@t@$ $Ewalk(j, U_k, C)$ to CFset, where W is the preamble walk computed for t, s_j is the tail state of t and C is the context after executing $W@t$.

(iii) For each def-use pair $D \in \mathcal{D}$ covered by Step 2, add the preamble walk for D computed in Step 2 to *DFset*.

(iv) Let *UFset* = *CFset* $\cup$ *DFset*. Delete each walk $W \in$ *UFset* such that W is a prefix of some other walk in *UFset*.

(v) Stop.

Step 2.1

(i) Let $Q = P\ t$. If Q is executable and t is not yet covered for control flow then mark t as covered and take P as the preamble walk for t.

(ii) If Q is executable and either t is not a self-loop or t has at least one assignment statement in its computation block then add Q to $\mathcal{T}$.

(iii) If Q is executable then do Step 2.1.1.

Step 2.1.1

(i) For each $t' \in P$, (a) Compute DFG[t', Q] from DFG[t', P], (b) Mark all the def-use pairs covered by Q, and (c) Construct an appropriate preamble walk for each such pair.

(ii) Consider DFG[t, t] to be DFG[t, Q].

Observe that the first phase starts by constructing DFG[t, t], for every transition t in the given EFSM. This can be done using the procedure *ConstructDFG*. Starting from the initial state, Step 2 traverses the EFSM in a breadth-first fashion, in order to compute the preambles for each transition and for each feasible def-use pair in $\mathcal{D}$. At the starting of the kth iteration of Step 2, $k \geq 1$, $\mathcal{P}$ consists of the set of all executable walks of length $k-1$ which start from the initial state. The kth iteration of this step computes the set of all executable walks of length k by extending the walks in $\mathcal{P}$ by single transitions. The executability of the extended walk is checked only with respect to the last transition since the rest of the walk is known to be executable at this point. This reduces the complexity of the feasibility problem to a great extent.

For each walk $P \in \mathcal{P}$ and for each transition t from the tail state of P, Step 2.1 checks if the walk Q obtained by postfixing t to P is executable. When Q is executable, Step 2.1 uses Step 2.1.1 for computing the data flow graphs pertaining to Q, for determining the def-use pairs in $\mathcal{D}$ covered by Q, and for selecting a preamble walk for every def-use pair covered by Q. Step 2.1.1 can be achieved using the procedure *ExtendDFG* which extends DFG[t', P] to DFG[t', Q], for all t' in P.

Step 2 is repeated until the preambles for all the transitions are computed and all def-use pairs in $\mathcal{D}$ are covered or the number of iterations of Step 2 exceeds a fixed positive integer K_1. K_1 depends on the given EFSM. It has to be chosen in such a way that the preambles for all the transitions are computed in K_1 iterations of Step 2. Recall that, for every transition, the EFSM is assumed to have at least one feasible walk from the initial state such that the transition is executable for the resulting context. Therefore, the preambles for all the transitions are computable in a finite number of iterations of Step 2. Observe that some of the def-use pairs in $\mathcal{D}$ may not be feasible. Also, the problem of finding whether a given pair is feasible or not is undecidable. If $\mathcal{D}$ has some infeasible pairs, then this phase terminates after K_1 iterations of Step 2.

Phase II described below is essentially for completing each walk in *UFset*, computed in Phase I, into an executable tour. These tours are in fact the ones required for the trans-CIUS-set and the def-use-ob criteria. The algorithm is self-explanatory and further description is omitted.

Phase II

Input: The EFSM considered in Phase I and the UFset returned by Phase I

Output: UFTourset, a set of tours for the selection criteria

Step 1 { Initialization }

(i) Let P be a null walk at s_1; Let $\mathcal{P} = \{P\}$.

(ii) Let *UFTourset* be the empty set.

Step 2 { ith iteration of this step computes the set $\mathcal{T}$ of all satisfiable walks of length i ending at s_1. The set of all preambles in *UFset*, which are executable in conjunction with a walk in $\mathcal{T}$ which starts at the tail state of the preambles, are declared to be covered by the tour obtained by prefixing the preamble to the walk. }

(i) Let $\mathcal{T}$ be the empty set.

(ii) Do Step 2.1 for each $P \in \mathcal{P}$ and for each transition t starting from a state other than s_1 and ending at the starting state of P.

(iii) If all the walks in *UFset* are covered, then stop.

(iv) Consider $\mathcal{T}$ as $\mathcal{P}$ and repeat Step 2.

Step 2.1

(i) Let $Q = t\ P$. If Q is satisfiable, then add Q to $\mathcal{T}$.

(ii) Do Step 2.1.1 for each walk W in *UFset* such that $W\ Q$ is a tour provided Q is satisfiable.

Step 2.1.1

(i) If $W\ Q$ is executable then Add $W\ Q$ to *UFTourset* and mark W as covered.

The time and space complexities and correctness of the algorithm are summarized below. The proof of the theorem and a detailed refinement of the above algorithm is presented in [15].

Theorem 2 *Let K_2 (K_1) be the number of times (maximum number of times) Step 2 of Phase II (Phase I) is executed. The time complexity of the algorithm is $O((d^{out}_{max})^{K_1+1} + (d^{in}_{max})^{K_2+1})$ steps, where d^{in}_{max} (d^{out}_{max}) denotes the maximum number of incoming (outgoing) transitions including the self-loops at any state in the EFSM. The algorithm also requires $O((d^{out}_{max})^{K_1} + (d^{in}_{max})^{K_2})$ units of memory. It successfully computes an executable tour for those transitions which have at least one preamble walk of length at most K_1. The algorithm computes an executable tour for every feasible def-use pair in $\mathcal{D}$ which have at least one preamble walk of length at most K_1 excluding their CIUS subwalk extension.*

□

Corollary 1 *For a suitable value of $K_1, 1 \leq K_1 < \infty$, the algorithm successfully computes a set of tours such that (i) the set satisfies the trans-CIUS-set criterion, and (ii) the set satisfies the def-use-ob criterion if $\mathcal{D}$ has only feasible def-use pairs.*

4.2 Fault Coverage

Let us assume that the Implementation Under Test (IUT) is represented as a deterministic, completely specified EFSM having the same set of input interactions and states as the specification EFSM. It is known that some of the FSM-based test sequence generation methods achieve complete fault coverage capability by including the verification of the state identification sequences in the IUT [7, 10, 8]. In the EFSM model, in order to establish that an input sequence is an UIS of a state in the IUT, one has to show that for any valid context of the IUT at that state, the output sequence produced by the IUT while applying the input sequence is different from the output sequence obtained by applying the input sequence at any other state with every valid context. Due to the black-box approach of testing, it is, in general, difficult to achieve this UIS verification requirement. For each incoming transition at a state s_i, our test case generation method generates one feasible tour for applying the CIUS U_i at s_i to see if it provides the expected output, and a tour for applying the CIUS U_j of the

Def-Use Pair	Preamble	Tour
(t3.c4, t8.c1)S_credit	t1 t3 t8 t8 **t17**	t1 t3 t8 t8t17t20
(t6.c2, t9.P)TRsq	infeasible	
(t6.c2, t9.c1)TRsq	infeasible	
(t6.c3, t12.P)TSsq	t2 t6 t12 **t17**	t2 t6 t12t17t20
(t6.c3, t12.c1)TSsq	t2 t6 t12 t8 **t17**	t2 t6 t12 t8t17t20
(t6.c3, t13.P)TSsq	t2 t6 t13 **t17**	t2 t6 t13t17t20

Table 4: Sample data flow test tours for EFSM given in Figure 1

Transition	Preamble	Set of walks	Tour
t6	t2	t2t6t17	t2t6t17t20
		t2t6t31	t2t6t31t17t20
		t2t6t32	t2t6t32t17t20
		t2t6t18	t2t6t18t19
t7	t2	t2t7t19	t2t7t19
		t2t7t36	t2t7t36t19
		t2t7t37	t2t7t37t19
		t2t7t38	t2t7t38t19

Table 5: Sample control flow test tours for the EFSM given in Figure 1

state s_j, $j = 1, 2, \ldots, n, j \neq i$ at s_i to check if it produces the output different from the one obtained when U_j is applied at s_j. Further, these tours can be exercised for different data in their feasible domain. Thus our method establishes the CIUS verification requirement at least partially, while the existing EFSM based test generation methods do not consider this issue. In addition, the test tours selected are all feasible and for a suitable value for K_1, they satisfy the control flow criterion. Therefore, the control flow fault coverage of this method is the same or better than those guaranteed by the existing EFSM based test sequence generation methods.

5 Transport Protocol Test Case Generation

In [15] we have illustrated our test case generation algorithm on the transport protocol given in Figure 1. We shall summarize the results here. Only core transitions are considered for the coverage criteria. There are 80 def-use pairs satisfying the all-uses criterion. Among them 7 are infeasible. Some of the def-use pairs are shown in the first column of Table 4. Phase I computes the preamble walks for all the transitions by the fourth iteration of Step 2. The preamble walks selected for some of the transitions are shown in the second column in Table 5. Note that the walks in the third columns in this table are obtained by appending the preamble walk with the transition followed by a CIUS walk. By the fifth iteration pramble walks for all the feasible def-use pairs have been computed. The second column in Table 4 shows the preamble walks for the selected def-use pairs. Observe that the bold faced transition appended to a walk in the table is for confirming the tail state of the last transition whose predicate transitively uses the value of the variable in the corresponding def-use pair. After deleting the duplicate walks, Phase I produces 128 walks. Phase II for completing these walks in to feasible tours is fairly straight forward for the EFSM in Figure 1. For instance, since none of the incoming transitions ($t5, t19, t20$ and $t21$) at state s_1 has predicate, in the first iteration, all the walks output by Phase I which terminate at the starting states (s_2, s_5 and s_6) of these transitions are completed into executable tours by concatenating the appropriate transitions from $\{t5, t19, t20, t21\}$. With in two iterations of Step 2, Phase II successfully finds a set of executable tours for all the walks selected in the first phase. The last columns of Table 4 and Table 5 show some of the selected tours. This set of tours satisfies both the trans-CIUS-set and the def-use-ob criteria.

Let us examine the fault detection capability of the generated test tours through examples. Suppose that an IUT has a simple control flow fault at the transition $t6$, which originally ends at s_4. Let the tail state of this transition in the IUT be s_2. While applying a test data along the tour $t2t6t17t20$ which is one of the tours for covering the trans-CIUS-criterion for $t6$ (refer to Table 5), it shows an output mismatch. Therefore the fault is detected.

Suppose that the IUT has a variable definition fault at $t3.c4$ where the variable *S_credit* is defined. That is , in $t3.c4$, *S_credit* is replaced by some other variable, say *R_credit*. Let us assume that the default value for all the integer variables is zero. Take the def-use pair $D = (t3.c4, t8.c1)S\_credit$. From Table 4, we see that $T = t1t3t8t8t17t20$ is the required tour for covering D with respect to the def-use-ob criterion. Observe that for any feasible test data for T, the expected sequence along the tour is different from the one observed in the IUT. Thus, the presence of the fault is detected.

6 Conclusion

The Context Independent Unique Sequence defined in this paper is very useful in generating executable test cases for both control and data flow in an EFSM. The trans-CIUS-set criterion is superior to the existing control flow coverage criteria for the EFSM. In order to provide observability, the "all-uses" data flow coverage criterion is extended to what is called the def-use-ob criterion. Finally, a two-phase breadth-first search algorithm is designed for generating a set of executable test tours for covering the selected criteria.

In order to generate the control flow test cases for EFSM model with only integer variables, Li *et al* have recently defined an Extended UIO-sequence (EUIO-sequence, in short)[13]. We observe that if an UIO-sequence is also an EUIO-sequence, then the input part of this sequence becomes a CIUS. While a number of EUIO-sequences are required to test all the incoming transitions at a given state one CIUS is sufficient for this purpose. Also, there is no algorithm presently available for computing EUIO-sequences.

The problem of finding a set of test data for executing each tour selected by a test case generation algorithm such that the data-oriented faults are detected is certainly an interesting research problem. We believe that the set of tours generated by our approach is a good candidate for the test data selection problem, since (i) all the tours generated are executable and (ii) it provides observability of the data flow. The fault based techniques as described in [18] would be helpful to gain more insight on this problem.

Since the EFSM model considered in this paper is similar to a module in Estelle or SDL, an interesting area for future study is to integrate our test case generation method with the existing tools for these FDTs. Such an integrated tool will be useful to automatically generate test cases for real-life protocols specified in Estelle and SDL.

Extending our work to EFSMs which may not have CIUSs for certain states is another direction for further research.

References

[1] ISO TC97/SC6 8073: Information Processing Systems - Open Systems Interconnection - Connection Oriented Transport Protocol Specification.

[2] ISO/IEC 9074: Information Processing Systems - Open Systems Interconnection - Estelle - A Formal Description Technique Based on an Extended State Transition Model, 1987.

[3] CCITT/SGx/WP3-1, Specification and Description Language, SDL. CCITT Recommendations Z.100, 1988.

[4] ISO/IEC 8807: Information Processing Systems - Open Systems Interconnection - LOTOS - a Formal Description Technique Based on the Temporal Ordering of Observational Behavior, June 1988.

[5] ISO SC21 WG1 P54: Information Processing Systems - Open Systems Interconnection - Formal Methods in Conformance Testing, Working Document, June 1993.

[6] G. v. Bochmann. Specifications of a simplified transport protocol using different formal description techniques. *Computer Networks and ISDN systems*, 18:335–377, 1989/1990.

[7] G. v. Bochmann, A. Petrenko, and M. Yao. Fault coverage of tests based on finite state models. In *7th International Workshop on Protocol Test Systems, Tokyo , Japan*, November 1994.

[8] W. Y. L. Chan, S. T. Vuong, and M. R. Ito. An improved protocol test generation procedure based on UIOs. In *ACM SIGCOMM*, pages 283–294, 1989.

[9] S. T. Chanson and J. Zhu. A unified approach to protocol test sequence generation. In *Proc. IEEE INFOCOM*, pages 106–114, 1993.

[10] T. S. Chow. Testing software design modeled by finite state machine. *IEEE Tr. Soft. Engg.*, SE-4(3):178–187, March 1978.

[11] W. Chun and P. D. Amer. Test case generation for protocols specified in Estelle. In J. Quemada, J. Manas, and E. Vazquez, editors, *Formal Description Techniques, III*, pages 191–206. Elsevier Science Publishers B. V. (North-Holland), 1991.

[12] B. Forghani and B. Sarikaya. Semi-automatic test suite generation from Estelle. *IEE/BCS Software Engineering Journal*, 7(4):295–307, July 1992.

[13] X. Li, T. Higashino, M. Higuchi, and K. Taniguchi. Automatic generation of extended UIO sequences for communication protocols in an EFSM model. In *7th International Workshop on Protocol Test Systems, Tokyo, Japan*, November 1994.

[14] R. E. Miller and S. Paul. Generating conformance test sequences for combined control and data flow of communication protocols. In *Proc. 12th International Symposium of Protocol Specification, Testing and Verification*, 1992.

[15] T Ramalingam. *Test case generation and fault diagnosis methods for communication protocols based on FSM and EFSM models.* PhD thesis, Concordia University, Montreal, Canada, 1994.

[16] T. Ramalingam, A. Das, and K. Thulasiraman. Fault detection and diagnosis capabilities of test sequence selection methods based on the FSM model. *Computer Communications*, 18(2):113–122, February 1995.

[17] S. Rapps and E. J. Weyuker. Selecting software test data using data flow information. *IEEE Tr. Soft. Engg.*, SE-11(4):367–375, April 1985.

[18] M. C. Thompson, D. J. Richardson, and L. A. Clarke. An information flow model of fault detection. In *Proc. International Symposium on Software Testing and Analysis*, pages 182–192, Cambridge, USA, June 1993. ACM press.

[19] K. J. Turner, editor. *Using formal description techniques.* John Wiley & Sons, Chichester, England, 1993.

[20] H. Ural and A. Williams. Test generation by exposing control and data dependencies within system specifications in SDL. In *Proc. FORTE'93*, October 1993.

[21] H. Ural and B. Yang. A test sequence selection method for protocol testing. *IEEE Tr. Comm.*, 39(4):514–523, April 1991.

Handling redundant and additional states in protocol testing

A. Petrenko[1], T. Higashino[2], and T. Kaji[2]

*1 - Université de Montréal,
C.P. 6128, succ. Centre-Ville, Montréal, H3C 3J7, CANADA,
Phone: (514) 343-7535, Fax: (514) 343-5834,
petrenko@iro.umontreal.ca*

*2 - Osaka University,
Toyonaka, Osaka 560, JAPAN
Phone: +81-6-850-6607, Fax: +81-6-850-6609
higashino@ics.es.osaka-u.ac.jp; t-kaji@ics.es.osaka-u.ac.jp*

Abstract

This paper addresses the problem of conformance testing of protocols modeled by FSMs with redundant states. Redundant states appear in an FSM which may be nonminimal or nonconnected. The existing test derivation methods usually are not directly applicable to these machines. In this paper, we show that they can be adjusted to cover this class of FSMs and that the traditional assumption on the minimality of machines is not necessary. Another problem with redundant states is that they can cause the appearance of additional states in protocol implementations whose guaranteed detection requires tests of an exponential length. This paper proposes techniques for deriving tests for FSMs with redundant or additional states such that a high fault coverage is achieved while maintaining an acceptable test suite length. The effectiveness of the proposed methods has been evaluated in an experimental way using a benchmark protocol.

Keywords

Conformance testing, FSMs, redundant and additional states, test derivation, fault coverage

1 INTRODUCTION

Conformance testing of a protocol is typically a black-box testing, i.e. it is based on its specification. Formal methods for deriving conformance tests are widely recognized as being capable of producing tests with a high fault coverage [BPY94]. To apply such a method, one usually abstracts a relevant formal model from the available specification of the protocol. For our purposes, we will consider here the FSM-based test derivation methods. They require a single FSM that models the behavior of a given protocol and satisfies certain conditions for their applicability. In particular, the classical model of completely specified, strongly connected, deterministic and minimal FSMs has been regarded in the testing literature (see, e.g. [SiLe89], [LeYa94], [Ural92]) as the most tractable model for test derivation. A complex protocol is rarely given as a single pure FSM with the properties requested by these methods. The protocol may be given, for example, in the form of a single extended FSM or several communicating (pure or extended) machines. FDT's, such as ESTELLE, LOTOS and SDL, produce modular specifications from which an FSM can be abstracted for each module. To test a system of such modules, it is necessary to specify the behavior of the system as one FSM. Thus, a global FSM is constructed from the given system of component FSMs or EFSMs, using a method such as reachability analysis or unfolding. The resulting FSM may not yet be tractable for test derivation, as it may contain equivalent states and states unreachable from the designated initial state. Even if a single FSM is directly derived from a semi-formal description of the protocol, it may still have redundant states. Quite often, equivalent states are intentionally introduced in order to increase the readability of the specification. If we construct an FSM to model a certain functionality of a complex protocol, then states which were distinguishable in the system as a whole may become equivalent in the obtained FSM. Moreover, states reachable in the system might not be reachable from the chosen initial state. However, connected minimal machines with no redundant states are typically accepted for test derivation. Therefore the original FSM with redundant states is customarily replaced by its connected minimal form.

We observe that the traditional approach explained above leads to the loss of structural information about the behavior of redundant states which are an apparent source of additional states in implementations. This information can be used to elaborate alternative techniques for deriving tests directly from an FSM containing redundant states. To the best of our knowledge, there is no systematic procedure for directly treating this class of machines, except by using an exhaustive procedure.

In Section 2, we introduce some basic definitions and concepts. In Section 3, we consider the class of non-reduced connected FSMs and show that the commonly used assumption on the minimality of the specification machine is no longer necessary and the existing test derivation methods can be generalized to cover this class of FSMs. Based on the structural information, we also develop an alternative technique for FSMs with redundant reachable states which may become additional states in the implementations. The case where redundant states are not reachable in the given machine is analyzed in Section 4, and a proper technique for test derivation is proposed. The techniques elaborated in Sections 3 and 4 are applied in Section 5 to a simple protocol, called INRES. Experimental measurements of the fault coverage of various test suites for this protocol are reported in this section as well. Section 6 discusses how the ideas presented in this paper can be incorporated into existing test derivation methods to ameliorate the fault coverage of conformance tests when the possibility of additional states is allowed for. We conclude by presenting some open research issues.

2 PRELIMINARIES

Let $A = (S, X, Y, \delta, \lambda, s_0)$ be an initialized FSM with m states from which we are required to derive a test suite. Here S, X, and Y are finite and nonempty sets of states, inputs and outputs, respectively; δ: $S \times X \rightarrow S$ is the transition function; λ: $S \times X \rightarrow Y$ is the output function; s_0 is the initial state. We thus consider here only completely specified, i.e. complete, deterministic machines. The usual extensions of the two functions from input symbols to strings (sequences) will be used to specify the behavior of these machines. The two properties of a machine, namely, its minimality and connectedness are important for test derivation.

A finite state machine is *reduced* or *minimal* if no two of its states are equivalent. Every complete deterministic FSM possesses a unique (up to the isomorphism) minimal form. Minimality of the machine provides the possibility of identifying states during testing. In particular, a so-called *characterization set* W is a set of input sequences of the minimal machine which tells every two states apart [Koha78].

A machine is *initially connected* if all of its states are reachable from the initial state, i.e. there exists a transfer sequence α such that $\delta(s_0,\alpha) = s$ for every $s \in S$; $\delta(s_0,\varepsilon) = s_0$, where ε is the empty sequence. It is *strongly connected* if every state is reachable from any other state. Initially connected FSMs are usually considered for test derivation in the case where the reliable reset is available in the implementations, whereas the strongly connected FSMs are used in other cases. Initially or strongly connected machines will simply be referred to as *connected* machines. A connected machine possesses a so-called *state cover V* which is a set of transfer sequences, usually one shortest sequence per state. A state cover is used to check all the transitions from every state of the machine.

An implementation FSM *I* is assumed to be a complete initialized machine with the input alphabet *X* of the specification FSM *A*. One must check the equivalence of the two machines *I* and *A* by testing *I* as a black-box. The equivalence of machines is defined as the equivalence of their initial states. *I* is viewed as a black box, and any test suite *TS* which is a set of input sequences, can neither distinguish equivalent states in the FSM *I*, nor bring *I* into a state unreachable from its initial state. Therefore, it makes sense to assume that any implementation is a complete connected minimal machine $I = (T, X, Y, \Delta, \Lambda, t_0)$.

Let $\mathfrak{I}$ be a certain set of FSMs with the input alphabet *X* of the FSM *A*. A test suite *TS* is said to be *complete for A in the class* $\mathfrak{I}$ if for every machine *I* from this set which is not equivalent to *A*, there is an input sequence α in *TS* such that the corresponding output sequences of *A* and *I* are different, i.e. $\lambda(s_0,\alpha) \neq \Lambda(t_0,\alpha)$ for any $I \neq A$, $I \in \mathfrak{I}$. In this case, we also say that the test suite guarantees the complete fault coverage in the class $\mathfrak{I}$. There are several ways in which this class can be specified [PBD93]. As an example, consider the set of FSMs which differ from the FSM *A* in their output functions only. Any test suite covering all of the transitions of *A* (a transition tour) is complete in this class. Another example constitutes the universal set $\mathfrak{I}_m$ of all FSMs with at most *m* states. The test suite complete in the class $\mathfrak{I}_m$ is simply called *m-complete* [BPY94]. *M*-complete test suites were first introduced as checking experiments [Moor56], [Henn64]. Since then much of the research in this field has focused on deriving *m*-complete test suites from connected minimal FSMs (for a recent survey, see [BoPe94]). Note that other classes of completely specified deterministic FSMs have not been studied systematically in the testing theory. As a result, there is no systematic procedure for test derivation which is directly applicable to a machine with unreachable and/or equivalent states.

If a specification machine happens to be nonconnected or nonminimal, then the machine should be transformed into its equivalent form, connected and minimal, before one can apply the currently existing methods of test derivation. Unreachable states are hence deleted and the machine is minimized. We say that the specification FSM has *redundant* states if it does not coincide (up to the isomorphism) with its connected minimal form. A redundant state can be either reachable or unreachable from the initial state.

The problem with the redundant states in the specification is that they could be a source of additional faults. In particular, a redundant state, reachable in the specification FSM and equivalent to another state, can become a new distinct state in an implementation machine. A state unreachable in the specification FSM can become reachable in the implementation machine. Due to these faults, additional states emerge in implementations. An implementation FSM has *additional* states if its connected minimal form has more states than the connected minimal form of the corresponding specification. Keeping in mind that two connected minimal FSMs, if equivalent, have the same number of states, we can easily prove that an implementation FSM with additional states is not equivalent to the specification FSM. Faulty implementations with additional states may escape from detection by tests when the original specification FSM is replaced by its connected minimal form. The replacement results in the loss of the structural information in the specification about the behavior of redundant states which are an apparent source of additional states in implementations. This structural information could offer new ways for deriving test suites with a high fault coverage at reasonable cost. In next sections, we first examine a commonly used assumption on the minimality of the specification machine and then propose several techniques for deriving tests from FSMs with redundant states based on this structural information. The experimental results reported later in this paper confirm the viability of our approach.

3 FSMS WITH REDUNDANT REACHABLE STATES

3.1 M-complete test suites

Assume that an FSM A is connected, but not minimal (non-reduced). As mentioned above, all currently existing formal methods for test derivation require that the complete FSM is minimal. This limitation implies that before tests are derived, the given non-reduced machine should undergo the state minimization procedure. The latter is a classical problem, and there is a suitable algorithm [Hopc71] with complexity O(pm log$m$), where p is the number of inputs and m is the number of states. While the procedure is relatively simple, the question still arises as to the necessity of this step and thus on the necessity of the assumption of minimality itself. We first recall the basic techniques for deriving a q-complete test suite, where q is not less than the number of states in the minimal form.

Suppose that a minimal form of the FSM A, i.e. a reduced FSM B with n states ($n<m$), is obtained. It now becomes possible to systematically derive a q-complete test suite by applying one of the existing methods to the FSM B. We may use, for example, a method such as the W-method [Vasi73], [Chow78]. Like all the others, this method requires the specification machine to be minimal. It also implies that the user previously estimates an upper bound q on the number of states in the minimal form of the implementation machines.

The most widely used assumption suggests that this bound coincides with the number of states in the minimal machine from which tests are produced, i.e. $q = n$. To derive from B an n-complete test suite based on this assumption, we find its state cover V_B and characterization set W. A state cover contains exactly n transfer sequences, one per state. Let X^k be a set of all input sequences which have length up to k, in particular, $X^1 = X \cup \{\varepsilon\}$. By concatenating sequences of the three sets, V_B, X^1 and W, a test suite $TS_B = V_B X^1 W$ is produced. TS_B is proven to be n-complete for the FSM B, i.e. it is complete in the class $\mathfrak{I}_n$ [Vasi73]. We shall demonstrate that a similar result can be obtained directly from the original specification FSM A with a slightly modified version of the W-method. In other words, we show that the state minimization is not an obligatory step of the test derivation.

To this end, we extend the classical notion of a characterization set to cover non-reduced FSMs. A set W of input sequences is said to be a *characterization set* of the FSM A if it distinguishes any two non-equivalent states. The W set induces a partition $\Pi(W)$ of the set S of states into the equivalence classes. $|\Pi(W)| = n$. $\Pi(W)$ is an equivalence partition of A [Gill62]. If A is reduced, then we have the traditional notion of the characterization set, as used for example, in [Koha78], [Vasi73], [Chow78], [FBK91].

Next we define a *class cover* V_c of A as a set of transfer sequences leading A to every class of $\Pi(W)$ from its initial state. $|V_c| \geq |\Pi(W)|$. If A is reduced, then the V_c set is its state cover in the usual sense. It is always possible to choose exactly $|\Pi(W)|$ transfer sequences as a class cover.

Similarly to the original W-method, we concatenate the three sets: V_c, X^1, W, to obtain a test suite $V_c X^1 W$ for the FSM A. There exists a one-to-one mapping between the equivalence classes of $\Pi(W)$ and the states of the minimal form B. Then the class cover V_c is a state cover of B. The characterization sets of A and B coincide, since the equivalent states cannot be distinguished by any input sequence. Thus, we have the following proposition.

Proposition 3.1. The test suite $V_c X^1 W$ is n-complete for the FSM A with m states, where n

is the number of states in the minimal form of A.

We illustrate the idea of deriving tests directly from a non-reduced machine using an FSM A shown in Figure 1 as an example. State 1 is the initial state of the FSM A.

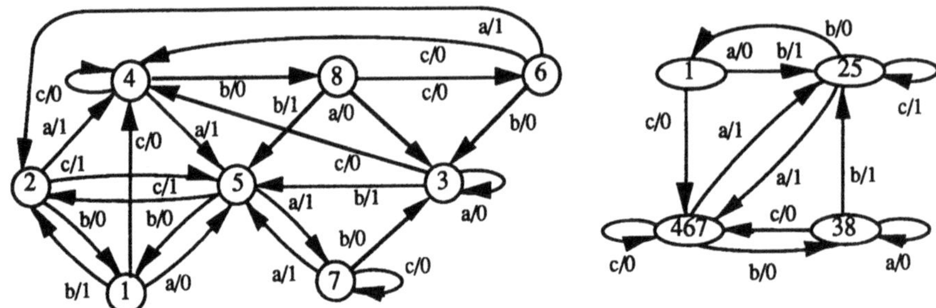

Figure 1 The FSM A.

Figure 2 The minimal form B of A.

First, we find a characterization set W of A by constructing a distinguishing tree, as explained in textbooks [Koha78], [Gill62]. It is sufficient to consider only input sequences of length up to m-1 = 7. This is because two states not distinguished by all input sequences of this length are equivalent [Gill62]. We obtain just a single sequence ac, i.e. W={ac}. It induces the equivalence partition $\Pi(W) = \{1; 2,5; 3,8; 4,6,7\}$. This partition implies the minimal form of A with four states shown in Figure 2. The sequence ac is a distinguishing sequence for the FSM B. However, to derive a test suite complete in the class $\Im_4$, we make no use of the FSM B, in contrast to the W-method in its original form.

Next, we construct a class cover of the FSM A. The minimal set $V_C = \{\varepsilon, a, c, cb\}$ can be chosen. It is easy to see that this set is also a state cover of B.

The final test suite is:

$V_cX^1W = \{\varepsilon, a, c, cb\}\{\varepsilon, a, b, c\}\{ac\} =$
$\{aaac, aac, abac, acac, bac, caac, cac, cbaac, cbac, cbbac, cbcac, ccac\}$.

This very test suite might also be produced from the FSM B by applying the W-method to this machine. If there is only single minimal state cover of B then the two methods always yield the same result.

Next, we note that a test suite complete w.r.t. the number of states in the minimal form of the given specification A might not be the best possible solution from the viewpoint of fault coverage. The problem is that an implementation under test may have been derived from the original (non-reduced) specification and not from its minimal form used for test derivation. Since the original specification has more states than its reduced form, the implementation may also have up to m distinct states, assuming that no state was split into several states during the implementation process. In this case, a test suite V_cX^1W no longer guarantees the detection of faulty implementations with additional states. In fact, this test suite does not even cover all of the transitions in the original machine. Consider the above example. By executing the test suite V_cX^1W against the FSM A we find that five transitions are not covered. It means that even a trivial output fault in any uncovered transition goes undetected. The conclusion is that if the implementation is known to be derived from the original specification, then a test suite should foresee the possible appearance of additional states in it. The problem of deriving complete tests in the case where the number of extra states in implementations is unknown is usually considered undecidable [Gill62]. The original FSM provides at least the upper bound m for the number of states in implementations under the assumption that no state of the specification was split into several states during the implementation process. However, this information is usually lost once a non-reduced FSM is substituted by its minimal form for further test derivation.

Under the above assumption, an m-complete test suite for the minimal FSM with n states has to be derived. It takes the following form, according to the W-method:

$$TS_m = V_B X^{m-n+1} W,$$

where V_B is a state cover, and W is a characterization set of the FSM B. As already discussed, the method in its original form, explicitly calls for the state minimization process.

However, an m-complete test suite can be directly obtained from the given non-reduced FSM. In fact, a statement similar to the Proposition 3.1 can be proven for the class $\mathfrak{I}_q$ with an arbitrary q, $q \geq n$, as well. The structure of such test suite is again based on a characterization set and a class cover of A, but not on that of its minimal form. The missing parameter n is easily deduced from the partition $\Pi(W)$, as this partition is constructed whenever a W set is being derived from the given machine.

In our example, $m = 8$, $W = \{ac\}$, $n = |\Pi(W)| = 4$, m-n+1 = 5, $V_C = \{\varepsilon, a, c, cb\}$ give the test suite

$$TS_8 = \{\varepsilon, a, c, cb\} X^5 \{ac\}.$$

Thus, there is no need to transform the original specification into its minimal form if tests are to guarantee complete fault coverage with respect to either the number of states in the minimal form of the specification or any higher limit, including the number of states in the original specification. In fact, the construction of both the minimal machine and the characterization set rely on a common technique for deciding state distinguishability. It is sufficient to apply the technique only once, thus avoiding an unnecessary preprocessing of the specification FSM.

The presented approach yields test suites with guaranteed coverage for the same classes of implementations as the traditional approach based on the W-method. The proposed adjustment of the W-method extends its applicability to the class of complete non-reduced FSMs. Both minimal and non-reduced machines are then treated in a unified way.

3.2 Test suites complete in a special class of FSMs

If the implementations have up to m-n additional states, then a resulting test suite of the structure $VX^{m-n+1}W$ suffers from exponential growth with respect to m-n. The sequences in the set X^{m-n} provide an exhaustive search for potential additional states in an implementation at the penalty of a sharp increase of test size. Regardless of a method used for test derivation, the universal traversal sequences X^{m-n} have to be incorporated into m-complete test suites [Vasi73], [YaLe91].

In our example, $m = 8$, $n = 4$, and $X^{m-n+1} = X^5$. The number of sequences in X^5 is equal to 364. As a result, the number of test cases in the test suite TS_m for the FSM A (Figure 1) is about eight hundred, and the total length of this test suite is well into the thousands. Being theoretically correct, this solution would hardly be accepted in practice. The exponential growth in lengths of the resulting tests precludes the use of such methods on most real protocols. Typically protocols have dozens of various protocol data units and abstract service primitives. Thus the number of inputs $|X|$ in the corresponding FSM model would immediately trigger the explosion of any test incorporating the universal traversal sequences X^{m-n}. This observation motivates us to look for an alternative solution.

Test suites derived to cover additional states rely on a worst-case assumption typical of automata theory. This assumption uses the minimal form, but not the given specification machine itself. In fact, just a single parameter, the maximal number of states, is used by this assumption. The structural information contained in the original machine is completely ignored. This information indicates reachability of the equivalent states in the machine which are more likely to be implemented as additional states in a faulty implementation. It is natural to expect that an alternative assumption based on this structural information could yield shorter tests with a fault coverage which is acceptable from a practical standpoint. In this section, we try to elaborate such an assumption on reachability of additional states in implementations, and to find a test derivation method for the non-reduced FSMs based on this assumption.

Let V be an arbitrary state cover of the given possibly non-reduced connected FSM A with m states, $|V|\geq m$. By $\Im_V$ we denote a set of all FSMs for which the set V is a state cover. In other words, we assume that every state of a machine in this class is reachable from its initial state with a proper sequence in the set V. It is clear that the number of states in any machine of this class is not more than $|V|$. $|V|\geq n$, where n is the number of states in the minimal form of A, thus $\Im_V \supseteq \Im_n$.

Proposition 3.2. A test suite $TS_A = VX^1W$ is complete for the FSM A with the state cover V and the characterization set W in the class $\Im_V$.

Proof. Consider an FSM I of the class $\Im_V$. If it fails the test suite TS_A, then it is not equivalent to A. Assume therefore, that it passes this test. We must show that I is equivalent to A.

The machine I belongs to the class $\Im_V$. Then by the assumption, all of its states are reachable from the initial state with certain sequences of V. Let the set of its states be T. We have $|T|\leq|V|$. The sequences of the W set are applied to each of the states of I. Then the W set induces a partition of the set T of states in I into the subsets of states which produce the same output reaction in response to W. We denote this partition by $\Pi_I(W)$. It is clear that $|\Pi_I(W)| = |\Pi_A(W)|$, where $\Pi_A(W)$ is the equivalence partition of A. We have $VX^1W = VW \cup VXW$. I passes the test suite, then the FSMs A and I produce the same output sequences in response to the set of input sequences VW. To prove the equivalence of A and I, it remains to demonstrate that $\Pi_I(W)$ is an equivalence partition of I. In this case, the minimal forms of A and I coincide up to the isomorphism [Gill62]. We claim the following lemma.

Lemma 3.3. Let $I = (T, X, Y, \Delta, \Lambda, t_0)$ be a completely specified initially connected deterministic FSM with q states, $q\leq|V|$. If $\Pi_I(W\cup XW) = \Pi_I(W)$ then any two states of the same class of $\Pi_I(W)$ are equivalent.

Proof. We claim that the partition $\Pi_I(W\cup XW)$ is a refinement of the partition $\Pi_I(W)$ of the state set T. Proven by contradiction, if this were not the case, there would be states that are not distinguishable by the set $(W\cup XW)$, but distinguishable by the set W. Suppose that a class of $\Pi_I(W)$ contains two distinguishable states. We choose a class C of $\Pi_I(W)$ and its states t_1 and t_2, such that the length of a sequence β which distinguishes these states is minimal. Thus $\Lambda(t_1,\beta) \neq \Lambda(t_2,\beta)$. β can not be a single input symbol, because in this case, $\Pi_I(W\cup XW) \neq \Pi_I(W)$, a contradiction. Now let the length of β be more than one, i.e. $\beta = x\alpha$. Consider the states $\Delta(t_1,x)$ and $\Delta(t_2,x)$. $\Lambda(\Delta(t_1,x),\alpha) \neq \Lambda(\Delta(t_2,x),\alpha)$. Since the sequence β has the minimal length, the states $\Delta(t_1,x)$ and $\Delta(t_2,x)$ belong to different classes C_1 and C_2 of $\Pi_I(W)$. This means there exists a sequence γ in the W set which distinguishes them. In this case, the sequence $x\gamma$ also distinguishes the states t_1 and t_2, and it belongs to the set XW. This contradicts our assumption that $\Pi_I(W \cup XW) = \Pi_I(W)$. Therefore, any two states of the same class of $\Pi_I(W)$ are equivalent. This completes the proof of the Lemma, and thus completes the proof of Proposition 3.2.

The fault coverage of the test suite VX^1W is proven to be complete in the class $\Im_V$. The size of this class depends on the particular state cover V chosen for test derivation. At one extreme, a state cover V_B of the minimal form B (or equivalently, a class cover) may serve as the set V for the test suite. The corresponding test suite V_BX^1W is n-complete, if the V_B set has no less than n sequences. At the other extreme, we have the set V_BX^{m-n}, which ensures that all states are reached in any implementation with up to m reachable states. The corresponding test suite $V_BX^{m-n+1}W$ is m-complete; moreover, it is complete in the class $\Im_{V_BX^{m-n}} \supset \Im_m$. Consider an arbitrary state cover V, $V_B \subset V \subset V_BX^{m-n}$. It yields a test suite VX^1W complete in the class $\Im_V \supset \Im_n$. The relationships between the classes of implementations covered by the considered test suites are illustrated in Figure 3.

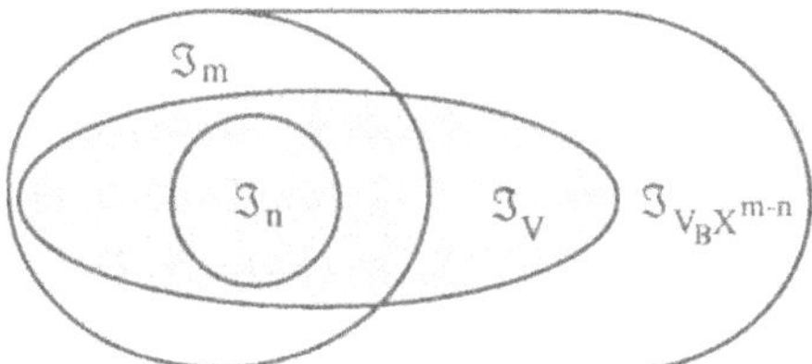

Figure 3 Classes of implementations.

The test suite VX^1W is shorter than the test suite $V_BX^{m-n+1}W$ if $V \subset V_BX^{m-n}$, however, its fault coverage is worse, as it may not cover all the FSMs in the universal set $\Im_m$. Uncovered machines have states which are not reachable when transfer sequences of the chosen set V are applied to their initial states. The test suite based on the selected state cover V reflects a tradeoff between the fault coverage and the size of a test suite.

At this point, different heuristics can be applied to choose a state cover for the given non-reduced machine such that a reasonable tradeoff is reached. As an example, based on the specifics of the given protocol, it might be possible to select a limited number of transfer sequences which could most probably lead to additional potentially non-equivalent states.

Here we present one particular assumption on reachability of these states, namely, **if an additional state exists in a faulty implementation, then it is reachable from the initial state with the shortest transfer sequence of the specification.** Under this assumption, the maximal number of possible states in the implementation detected by the test suite becomes a derivative of the number of the transfer sequences given in the specification. A natural intention to avoid infinite or exhaustive tests is behind this particular assumption.

This assumption suggests that one should construct a test suite based on a state cover of the given machine that contains all possible shortest transfer sequences for every state. In particular, a *canonical* state cover of A is a state cover V such that if α is a transfer sequence of the minimal length taking A from the initial state into a certain state, then $\alpha \in V$. Since the initial state is a very special state in the initialized FSMs, the set V includes only the empty transfer sequence for this state. It is clear that $|V| \geq m$, the number of states in A, and in the case that some state has several minimal transfer sequences, $|V| > m$. Every FSM possesses a unique canonical state cover. Therefore such a state cover characterizes the structure of the given non-reduced FSM. Note that substituting the original non-reduced machine by its minimal form, as required by the existing test derivation methods, results in the loss of these characteristics. As a consequence, m-complete test suites tend to explode in size.

Example. We illustrate the idea using the FSM A (Figure 1). The canonical state cover is

$V = \{\varepsilon, b, aab, cba, c, a, cbc, aa, cb\}$.

There is only one state, namely, state 3, which has two transfer sequences of the same minimal length, aab and cba. All the remaining states have single minimal transfer sequences. The above assumption suggests that in an implementation, up to 9 distinct states can be reached with this state cover. Following the approach suggested by Proposition 2.2, we can derive a test suite

$VX^1W = \{\varepsilon, b, aab, cba, c, a, cbc, aa, cb\}\{\varepsilon, a, b, c\}\{ac\}$.

It is much shorter that the test suite V_BX^5W.

As shown in Figure 3, the test suite VX^1W does not guarantee the detection of all faults within the bound m. However, it is an n-complete test suite, so its fault detection power is no less than any other n-complete test derived from the minimal form. The domain in the class $\mathfrak{I}_m$ uncovered by the test suite based on the canonical state cover represents a class of FSMs with up to m states whose detection is not guaranteed. Intuitively, this class contains those faulty implementations of A where a certain state of A is split during the implementation process into several non-equivalent states. Faults uncovered by VX^1W seem least realistic, in the sense that implementing an FSM through state splitting can be viewed as a rather unusual way of deriving an FSM implementation.

Thus, if we are required to derive a test suite from an FSM, and extra states in its implementations are not excluded, we may now choose between the two following assumptions:

- The number of the additional states in the implementation does not exceed a certain limit. It is a typical worst-case assumption.

- If these states exist then they are reachable from the initial state with shortest sequences defined by the given machine. It is a more realistic assumption, as it relies on the structural information available in the specification.

In general, any particular bound m is not easy to justify. We can only hope that the actual number of states does not exceed m. Selecting a suitable value for the bound m without knowledge of the class of implementations under test and their interior structure is very difficult, perhaps even requiring guesswork, although any faulty machine within this limit will definitely be detected by an m-complete test suite. The penalty comes from the test explosion effect. Every extra guessed state increases the size of the test suite at least by the factor $O(|X|)$.

In the second case, tests do not suffer the explosion. The maximal number of covered distinct states naturally follows from the properties of the given specification machine with redundant states. The penalty is a possibly incomplete coverage of the universe of machines with the number of states defined by the cardinality of the canonical state cover.

The two alternative assumptions usually yield different test suites. The effectiveness of the implied test derivation strategies can be evaluated in an experimental way. In fact, we have conducted such an experiment for a simple protocol (see Section 5). Before we report it, we first consider the case of specification FSMs with unreachable redundant states (until now the specifications were assumed to be at least initially connected).

4 FSMS WITH REDUNDANT UNREACHABLE STATES

A specification machine with redundant states may not necessarily be connected. Here, we focus our attention on the problems arising from the presence of redundant unreachable states in the given specification machine.

Consider an initialized minimal FSM A with some states unreachable from the initial state. The existing test derivation methods cannot be directly applied to such a machine. To apply an existing method for test derivation, we should substitute the original specification by its connected submachine A^*. Suppose that A has m states, whereas A^* has n states, $n<m$. A number of methods can now be applied to derive a test suite under the assumption that the implementation machines have up to n states.

Consider the following example. The FSM A in Figure 4 has state 4 unreachable from the initial state 1.

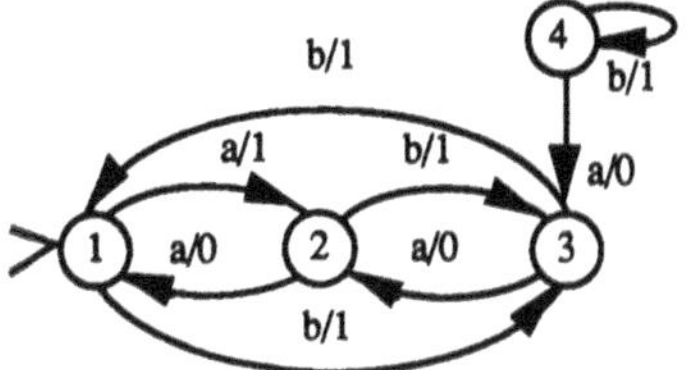

Figure 4 FSM A with an unreachable state.

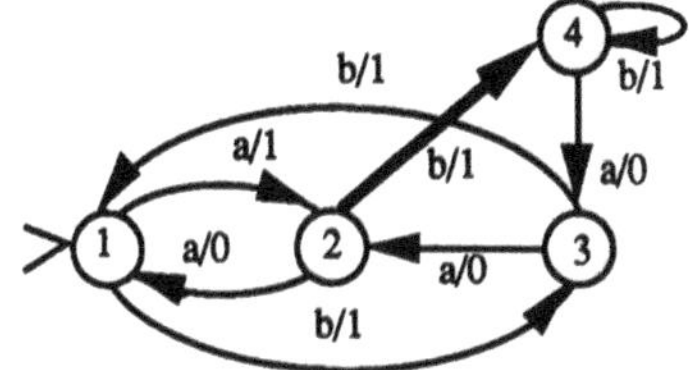

Figure 5 Faulty implementation.

If state 4 with its outgoing transitions is deleted from the original machine, then, for example, the W-method can be employed. A state cover of the connected submachine A^* is $\{\varepsilon, a, b\}$. There is a distinguishing sequence aa, that is $W = \{aa\}$. The test suite $\{\varepsilon, a, b\}\{\varepsilon, a, b\}\{aa\}$ is complete in the class of FSMs with up to three states (3-complete test suite). However, if the transition 2-b/1->3 is implemented as the transition 2-b/1->4, as shown in Figure 5, then this test suite can not reveal the fault. The only known remedy for such a situation is to derive a test suite which is complete w.r.t. four possible states:

$\{\varepsilon, a, b\}X^2\{aa\}$.

As already discussed in Section 3.2, this approach results in a sharp increase in the length of tests. At the same time, it neglects to explore the structural information provided by the original specification, as in the case of the equivalent states. In the presence of unreachable states, this information can be used to increase the distinguishing power of a W set required for state identification, thus increasing the fault coverage of a resulting test suite. An adjustment to the W-method implied by the above suggestion is presented below.

Given an FSM A, a set of input sequences is said to be a *characterization set* W_A of A if it distinguishes every reachable state of A from any other non-equivalent state including unreachable ones.

The new key feature of our notion of the W set is that unreachable states are not just ignored, as they are in the traditional notion of characterization sets. At the same time, it is a further generalization of the notion introduced in Section 3.1. The traditional techniques for constructing characterization sets are applicable to the full extent in this case, since state reachability is not required for deciding state distinguishability. Note that the diagnostic power of a W set could be increased further if we add as a requirement, the pairwise distinguishability of unreachable states. This option might be useful for deriving tests for fault localization. In most cases, however, they become more lengthy than tests for fault detection only.

In the above example, the sequence aa is not a characterization set of the FSM A (states 3 and 4 produce the same output sequence 00). The extended sequence aaa may serve as a characterization set W_A, as it yields different output reactions in the given four states.

A test suite preserves the traditional structure: VX^1W_A, where V is a state cover for all reachable states, i.e. the one of the connected submachine of A.

In our example, the test suite is

$$VX^1W_A = \{\varepsilon, a, b\}\{\varepsilon, a, b\}\{aaa\} = \{aaaaa, abaaa, baaaa, bbaaa\}.$$

The faulty implementation shown in Figure 5 fails this test. Compared to the first test suite, it has four additional test events; however, it is half the size of the test suite

$$VX^{4\text{-}3+1}W = \{\varepsilon, a, b\}X^2\{aa\}$$

produced by the original W-method.

The proposed adjustment to the W-method results in an increase in the length of tests linear with respect to m-n, the number of redundant unreachable states. The exhaustive solution offered by the original W-method is exponential with respect to m-n.

The test suites derived by the proposed technique are n-complete but usually not m-complete. They are complete in a certain class $\mathfrak{I}_W$ of FSMs. The set $\mathfrak{I}_W$ is a superset of $\mathfrak{I}_n$; machines in the set $\mathfrak{I}_W \backslash \mathfrak{I}_n$ can be characterized as follows. An FSM $I \in \mathfrak{I}_W \backslash \mathfrak{I}_n$ if it has an additional state whose reaction to the W set is different from the reaction of any reachable state of A. Note that the W set can now identify not only all of the states of the connected minimal submachine as in the original method, but also redundant states once they become reachable in a faulty implementation.

In the case where the given FSM has both reachable and unreachable redundant states, the presented technique has to be used in conjunction with the technique of Section 3.2.

5 EXPERIMENTAL RESULTS

To illustrate the proposed techniques for test derivation from specifications with redundant states, we consider the INRES protocol [Hogr92]. This simple protocol has been widely used in a number of publications, so we omit here its detailed description. The behavior of the responder part of this protocol can be specified by an EFSM given in Figure 6.

This EFSM has three control states and an internal Boolean variable v. First, we unfold it into a pure FSM. The input alphabet is: 1- CR, 2 - IDISr, 3 - ICONrsp, 4 - DT0, 5 - DT1. The output alphabet is 1 - ICONi; 2 - DR; 3 - CC; 4 - ACK0; 5 - ACK0, IDATi; 6 - ACK1; 7 - ACK1, IDATi; 8 - null. The three control states combined with the two possible values of v give six states as shown in Figure 7.

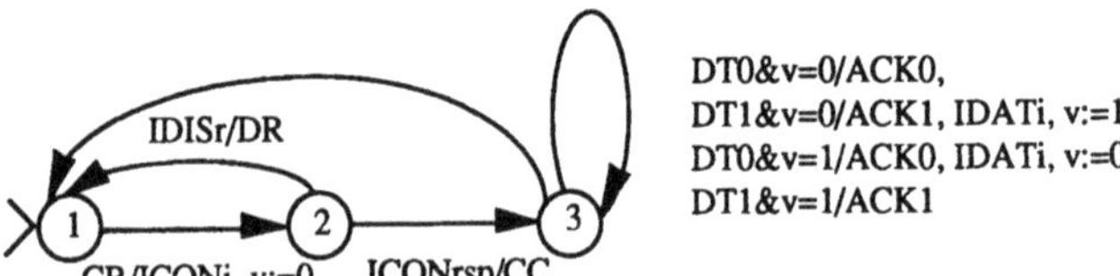

Figure 6 The EFSM of the INRES responder.

The FSM A is completely specified under the completeness assumption [PBD93], the transitions implied by this assumption (looping transitions with the output 8) are not depicted. State 21 is not reachable from the initial state. There are also two equivalent states, namely, 10 and 11. Thus FSM A has both types of redundant states. Its minimal connected form B is given in Figure 8. We use this example to compare several test suites derived by different strategies discussed in the paper.

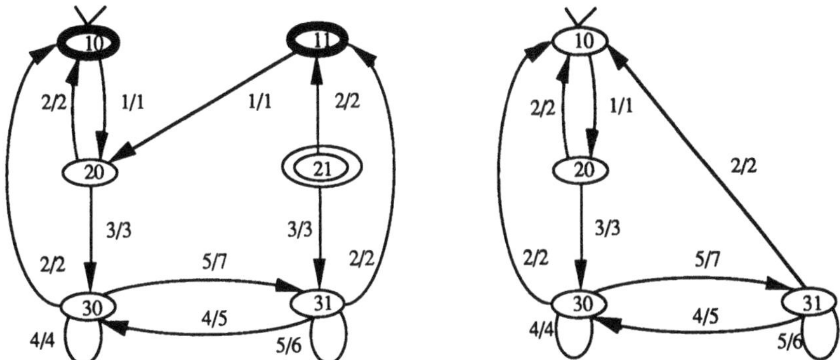

Figure 7 The FSM A unfolded from the EFSM. **Figure 8** The minimal form B of A.

Solution 1

Assuming $m = n = 4$, i.e. that the number of states in implementations never exceeds that of the minimal form B, we apply the method from Section 3.1 to FSM A. Deriving a characterization set $W = \{3, 4\}$ based on reachable states, we find that state 11 is equivalent to 10, i.e. $\Pi(W) = \{10,11; 20; 30; 31\}$. Finally, we determine a class cover $V_C = \{\varepsilon, 1, 13, 135\}$ and obtain a test suite complete in the class $\mathfrak{I}_4$:

$TS1 = \{\varepsilon, 1, 13, 135\}X^1\{3, 4\} = \{\varepsilon, 1, 13, 135\}\{\varepsilon, 1, 2, 3, 4, 5\}\{3, 4\} = \{$ 113, 114, 123, 124, 1313, 1314, 1323, 1324, 1333, 1334, 1343, 1344, 13513, 13514, 13523, 13524, 13533, 13534, 13543, 13544, 13553, 13554, 143, 144, 153, 154, 23, 24, 33, 34, 43, 44, 53, 54 $\}$.

There are 34 test cases of total length 122.

It is easy to see that if the assignment v:=0 of the transition from the first control state of the EFSM is not implemented by mistake, then state 11 is no longer equivalent to state 10 and state 21 becomes reachable. It is intuitively clear that the above test suite hardly has the power to detect such an error.

Solution 2

Now we assume $m = 6$ and derive a test suite complete in the class $\mathfrak{I}_6$ following the same method in the Section 3.1. The class state cover and characterization set are as in Solution 1. The test suite is

$TS2 = \{\varepsilon, 1, 13, 135\}X^3\{3, 4\}$.

This test suite has 850 test cases of total length 4750.

It detects any fault within six states at a very high cost; its length is about 40 times more than that of the $TS1$.

Solution 3

Next we devise a test suite according to the techniques presented in Sections 3.2 and 4. The canonical state cover of the non-reduced FSM A is $\{\varepsilon, 1, 13, 135, 1352\}$. Note that compared to the above solutions, there is an additional transfer sequence 1352 for a reachable redundant state 11. For state identification we use the following characterization set $W = \{34, 4\}$. Note that with this W, it now becomes possible to distinguish all reachable states from the unreachable state 21.

$TS3$ = {ε, 1, 13, 135, 1352}{ε, 1, 2, 3, 4, 5}{34, 4} = {1134, 114, 1234, 124, 13134, 1314, 13234, 1324, 13334, 1334, 13434, 1344, 135134, 13514, 1352134, 135214, 1352234, 135224, 1352334, 135234, 1352434, 135244, 1352534, 135254, 135334, 13534, 135434, 13544, 135534, 13554, 1434, 144, 1534, 154, 234, 24, 334, 34, 434, 44, 534, 54}.

Here are 42 test cases of total length 193. This test suite is only about 60% longer than the $TS1$.

Solution 4

One may view the test suites $TS1$ and $TS2$ as two extreme alternatives. We make a compromising assumption that the number of states is not more than 5 and derive a test suite complete in the class $\mathfrak{I}_5$.

$TS4 = \{\varepsilon, 1, 13, 135\}X^2\{3, 4\}$.

This test suite has 170 test cases of total length 780.

All the above obtained tests provide complete fault coverage in the class $\mathfrak{I}_4$. The test suite $TS2$ also guarantees complete fault coverage in the class $\mathfrak{I}_6$, but $TS1$, $TS3$ and $TS4$ do not. Ideally, we should compare their fault coverage in the class $\mathfrak{I}_6 \backslash \mathfrak{I}_4$. To this end, we have to enumerate all the machines of this set and compare the numbers of FSMs which fail these test suites. The total amount of $(6x8)^{(6x5)} - (4x8)^{(4x5)}$ machines should be checked against the three test suites. This task is beyond our current computing power. Instead, we have decided to perform a restricted, controlled mutation of transitions to estimate the fault coverage. To keep the number of mutants reasonable, output faults are excluded, and only certain transfer faults are simulated for the following transitions of the FSM A (Figure 7):

10-1/1->20 {21}; 20-2/2->10 {11}; 30-2/2->10 {11}; 11-1/1->20 {10, 11, 21, 30, 31}; 11-3/8->11 {20, 21, 30, 31}; 11-4/8->11 {20, 21, 30, 31}; 21-1/8->21 {10, 11, 20, 30, 31}; 21-2/2->11 {20, 21, 30, 31}; 21-3/3->31 {10, 11, 20, 21, 30}; 21-4/8->21 {10, 11, 20, 30, 31}.

This fault model describes $2^3 \times 5^3 \times 6^4$ = 1,296,000 FSMs. The idea behind the chosen mutations is to generate mainly those mutants which have more than four distinct states.

We have designed a tool which constructs a mutant machine according to the user defined fault model and checks whether or not it passes a given test suite. If the mutant passes the test suite then the tool verifies its equivalence to a specification FSM. Among 1,296,000 FSMs, there are 4368 mutants equivalent to the original FSM A (Figure 7). The fault coverage of a test suite is a percentage of non-equivalent (faulty) mutants that fail it [BPY94]. The results of the experiment are reported in Table 1. Times are measured in minutes for a UNIX machine SONY NWS 3470 (17 MIPS, 16 MB).

Note that the test suite $TS2$ was also executed against the generated mutants mainly to check the performance of the tool in the case of lengthy tests. As expected, CPU time of calculating the fault coverage is mainly determined by the total length of a test suite. On the other hand, a test suite with a low fault coverage, such as $TS1$, also consumes much time since each test case must be tried against about 50% of mutants, as opposed to the test suite $TS3$ which can recognize a faulty mutant much faster.

Table 1 Experimental results

Test suites	*# of test cases*	*Total length*	*Fault coverage (%)*	*CPU time (min)*
*TS*1	34	122	49.983277	382.67
*TS*2	850	4750	100.000000	806.27
***TS*3**	**42**	**193**	**99.945495**	**41.48**
*TS*4	170	780	99.666778	517.99

An additional experiment was conducted to estimate more precisely the fault coverage of the test suite *TS*3. The above given list of mutated transitions was extended by including the following:

11-2/8->11 {20, 21, 30, 31}; 11-5/8->11 {20, 21, 30, 31}; 21-5/8->21 {10, 11, 20, 30, 31}.

The extended fault model describes 194,400,000 FSMs, among them 26,016 machines are found to be equivalent to the specification FSM *A*. It took 2571.67 min. of CPU time to determine that *TS*3 has even a higher fault coverage of 99.997794% in this class of faults.

The results of the experiments indicate that the test suite *TS*3 for the INRES protocol derived by the proposed techniques has a high fault coverage with a reasonable length compared to other considered solutions.

6 CONCLUSION

In this paper, we have addressed the problem of test derivation from an FSM specification with redundant states which may create additional states in implementations. We have first demonstrated that the state minimization required by the existing methods for completely specified FSMs is not an obligatory step of the test derivation process. The methods can be easily extended to deal with even non-reduced machines. In other words, our first result is that the traditional assumption on the minimality is not a necessary one. Next, based on an observation that the existing approaches, which allow for the possibility of additional states in implementations, yield tests of exponential length, we have proposed two techniques used in combination. The first one is based on the idea of extending a state cover with an intention to reach additional states in implementations obtained from equivalent states reachable in the specification. The second technique suggests an extension of a characterization set with an intention to identify those additional states obtained from unreachable states. We have shown how these ideas are incorporated into the classical W-method. Our experimental results demonstrate that the proposed approach offers a reasonable compromise between the fault coverage and length of tests.

The two basic ideas of extending state covers and characterization sets can be similarly incorporated into other methods for deterministic machines, such as the UIOv-method [VCI89], the Wp-method [FBK91] and the methods based on harmonized state identifiers [Petr91], [LPB94b], which rely on the reset in the implementations. These ideas can also be used to improve the fault coverage of tests produced by a number of UIO-based methods [SiLe89], [YPB93] which yield a single test sequence. The adjustments concern covering all of the transitions (even from redundant reachable states) and extending the UIO-sequences (or any other state identifiers) in such a way that reachable states are distinguished from unreachable states.

The ideas involved in the proposed techniques seem useful not only for the deterministic case, but also for nondeterministic FSMs. Regardless of the relation used for testing, be it equivalence or reduction, the existing methods, such as the GWp-method [LBP94a], the HSI-method [LPB94b], and the SC-method [PYL93], [PYB94] can be applied only to observable and connected NFSMs. If a given machine is not observable then it is always possible to transform it into an observable form. A classical algorithm [HoUl79] for automata determinization may well produce an observable machine with both equivalent and unreachable states. However, these states deserve a special treatment, as they are, in fact, only certain subsets of states of the original non-observable machine from which an implementation is usually derived. Another open issue of the test derivation from nondeterministic machines arises due to peculiarities of the structures of complete tests for the equivalence and the reduction relations [PYB94] used in conformance testing.

Acknowledgments
This work was partly supported by the HP-NSERC-CITI Industrial Research Chair on Communication Protocols at the Université de Montréal and the Telecommunication Advancement Foundation of Japan. The authors would like to thank S. A. Ezust for comments to a previous version of this paper.

7 REFERENCES

[BoPe94] G. v. Bochmann and A. Petrenko, "Protocol Testing: Review of Methods and Relevance for Software Testing", ISSTA'94, ACM International Symposium on Software Testing and Analysis, Seattle, U.S.A., 1994, pp. 109-124.

[BPY94] G. v. Bochmann, A. Petrenko, and M. Yao, "Fault Coverage of Tests Based on Finite State Models", the Proceedings of IFIP TC6 Seventh International Workshop on Protocol Test Systems, 1994, Japan.

[Chow78] T. S. Chow, "Test Design Modeled by Finite-State Machines", IEEE Trans., SE-4, No. 3, 1978, pp. 178-187.

[FBK91] S. Fujiwara, G. v. Bochmann, F. Khendek, M. Amalou, and A. Ghedamsi, "Test Selection Based on Finite State Models", IEEE Trans., SE-17, No. 6, 1991, pp. 591-603.

[Gill62] A. Gill, Introduction to the Theory of Finite-State Machines, NY, McGraw-Hill, 1962, 207p.

[Henn64] F. C. Hennie, "Fault Detecting Experiments for Sequential Circuits", IEEE 5th Ann. Symp. on Switching Circuits Theory and Logical Design, 1964, pp. 95-110.

[Hogr91] D. Hogrefe, "OSI Formal Specification Case Study: The Inres Protocol and Service", University of Berne, Technical Report, 1991.

[Hopc71] J. E. Hopcroft, "An n log n Algorithm for Minimizing States in a Finite Automaton", Theory of Machines and Computations, NY, Academic Press, 1971, pp. 189-196.

[HoUl79] J. E. Hopcroft, J. D. Ullman, Introduction to Automata Theory, Languages, and Computation, Addison-Wesley, 1979, 418p.

[Koha78] Z. Kohavi, Switching and Finite Automata Theory, NY, McGraw-Hill, 1978.

[LeYa94] D. Lee and M. Yannakakis, "Testing Finite-State Machines: State Identification and Verification", IEEE Trans. on Computers, Vol. 43, No. 3, 1994, pp. 306-320.

[LPB94a] G. Luo, A. Petrenko, and G. v. Bochmann, "Test Selection based on Communicating Nondeterministic Finite State Machines using a Generalized Wp-Method", IEEE Trans., Vol. SE-20, No. 2, 1994, pp. 149-162.

[LPB94b] G. Luo, A. Petrenko, and G. v. Bochmann, "Selecting Test Sequences for Partially-Specified Nondeterministic Finite State Machines", the Proceedings of IFIP TC6 Seventh International Workshop on Protocol Test Systems, 1994, Japan.

[Moor56] E. F. Moore, "Gedanken-Experiments on Sequential Machines", Automata Studies, Princeton University Press, Princeton, NJ, 1956.

[PBD93] A. Petrenko, G. v. Bochmann, and R. Dssouli, "Conformance Relations and Test Derivation", IFIP Transactions, Protocol Test Systems, VI, (the Proceedings of IFIP TC6 Fifth International Workshop on Protocol Test Systems, 1993), Ed. by O. Rafiq, 1994, North-Holland, pp. 157-178.

[Petr91] A. Petrenko, "Checking Experiments with Protocol Machines", IFIP Transactions,

Protocol Test Systems, IV (the Proceedings of IFIP TC6 Fourth International Workshop on Protocol Test Systems, 1991), Ed. by Jan Kroon, Rudolf J. Heijink and Ed Brinksma, 1992, North-Holland, pp. 83-94.

[PYB94] A. Petrenko, N. Yevtushenko, and G. v. Bochmann, "Experiments on Nondeterministic Systems for the Reduction Relation", Universite de Montreal, DIRO, Technical Report #932, 1994, 23p (submitted for publication).

[PYL93] A. Petrenko, N. Yevtushenko, A. Lebedev, and A. Das, "Nondeterministic State Machines in Protocol Conformance Testing", IFIP Transactions, Protocol Test Systems, VI, (the Proceedings of IFIP TC6 Fifth International Workshop on Protocol Test Systems, 1993), Ed. by O. Rafiq, 1994, North-Holland, pp. 363-378.

[SiLe89] D. P. Sidhu and T. K. Leung, "Formal Methods for Protocol Testing: A Detailed Study", IEEE Trans. Vol. SE-15, No. 4, 1989, pp. 413-425.

[Ural92] H. Ural, "Formal Methods for Test Sequence Generation", Computer Comm., Vol. 15, No. 5, 1992, pp. 311-325.

[Vasi73] M. P. Vasilevski, "Failure Diagnosis of Automata", Cybernetics, Plenum Publishing Corporation, NY, No. 4, 1973, pp. 653-665.

[VCI89] S. T. Vuong, W. W. L. Chan, and M. R. Ito, "The UIOv-method for Protocol Test Sequence Generation", Proceedings of IFIP TC6 Second International Workshop on Protocol Test Systems, 1989, Ed. by J. de Meer, L. Machert and W. Effelsberg, North-Holland, pp. 161-175.

[YaLe91] M. Yannakakis, D. Lee, "Testing Finite State Machines", Proceedings of the 23d Annual ACM Symposium on Theory of Computing, Louisiana, 1991, pp. 476-485.

[YPB93] M. Yao, A. Petrenko and G. v. Bochmann, "Conformance Testing of Protocol Machines without Reset", IFIP Transactions, Proceedings of the IFIP 13th Symposium on Protocol Specification, Testing and Verification, Ed. by A. Danthine, G. Leduc and P. Wolper, 1993, North-Holland, pp. 241 - 253.

8 BIOGRAPHY

Alexandre Petrenko received the Dipl. degree in electrical and computer engineering from Riga Polytechnic Institute in 1970 and the Ph.D. in computer science from the Institute of Electronics and Computer Science, Riga, USSR, in 1974. Since 1992, he has been with the Université de Montréal, Canada. His current research interests include communication software engineering, protocol engineering, conformance testing, and testability.

Teruo Higashino received the B.E., M.E., and Ph.D. degrees in Information and Computer Sciences from Osaka University, Osaka, Japan, in 1979, 1981 and 1984, respectively. Currently, he is an Associate Professor in the Department of Information and Computer Sciences at Osaka University. In 1990 and 1994, he was a Visiting Researcher of Dept. I.R.O. at the Université de Montréal, Canada. His current research interests include design and analysis of distributed systems and communication protocols.

Tadashi Kaji is a Master course student in the Department of Information and Computer Sciences at the graduate school of Osaka University. His current research interests include testing and verification of communication protocols.

PART EIGHT

Industrial Applications

Experiences with the Design of B-ISDN Integrated Test System(BITS)

Ki Young Kim, Weon Soon Kim, Beom Kee Hong
Electronics and Telecommunications Research Institute(ETRI)
161 Kajong-Dong, Yusong-Gu, TAEJON, 305-350, KOREA
E-mail : kykim@winky.etri.re.kr
TEL : 82-42-860-4882 FAX : 82-42-860-5440

Abstract

To ensure interoperability between network components, a test system is understood as a useful facility to test and validate the developed system. In this paper, we deal with design and implementation of the B-ISDN Integrated Test System(BITS). The BITS is composed of functions for conformance and interoperability testing, call simulation and protocol monitoring. The BITS is a flexible test platform for B-ISDN protocol testing. The test configuration of the BITS can be easily and efficiently prepared for testing B-ISDN protocol implementations in a real networking world. This paper presents experiences with the design of BITS, especially the tester architecture, the development of test suites for the B-ISDN signalling protocol, and its application to B-NT system developed by ETRI.

Keywords

B-ISDN, ATM, Protocol Tester, Conformance Test

1 INTRODUCTION

In order to verify implementations during the development stage, it is important that the protocol testing is performed in a proper way using appropriate tools. The framework for conformance testing was standardized in ISO/IEC JTC1 to verify the product's compliance with the specifications.

In this paper, we introduce "B-ISDN Integrated Test System(BITS)", which is a useful tool for the validation of the prototype during ATM technology development as well as for the performance evaluation of B-ISDN products. The BITS is composed of functions for conformance and interoperability testing, call simulation, and protocol monitoring. The BITS is designed so that test configurations are easily and efficiently prepared for testing B-ISDN User Network Interface implementations in a real networking world. The development of BITS has been carried out by ETRI, as a part of the Network Test Bed(NTB) project, which is one of the nation-wide governmental project named "HAN/B-ISDN".

The HAN/B-ISDN project is composed of four research areas : network technology, switching technology, transmission technology, and terminal technology. The entire research project is divided into 11 different research project units and is being carried out by several research organizations, universities and manufacturers. The main goal of the NTB project unit is to test the performance of various communication systems that will be developed by several project teams and to evaluate conformance, and interoperability. Protocols might be implemented incorrectly because of the wrong interpretation of standards and misuses of many options. Therefore, a clear verification process is necessary to ensure the interoperability among network components. It seems to be efficient that the development of the testing methodology, the test suites on various kinds of protocols, and the testers are made in early development stage of each network component.

2 DESIGN OF BITS

The design concept of this system is based on a modular structure, with loosely coupled interconnection between the test control part and the test connection part. Thus, the establishment of one's own development and the test environment are allowed. Modular structure introduces overload for a simple system, but upgrades and changes the system functions effectively. Loosely coupled interconnection between the test control part and the test connection part is appropriately applied to the configuration of various test networks. We constructed our own development and test environment. The test control part is named as the "Test Host System(THS)", which is implemented on the same general UNIX workstation as one is used for the development environment in order to make both environments alike. Because of several ATM connection interfaces, it is possible to test the THS itself without another Implementation Under Test(IUT). The test connection part is named as "Test Satellite(TS)", consists of one MVME 147, two ATM Boards, and one PHY with two STM-1 ports. In the near future, multiparty testing will be available.

2.1 System Architecture

The BITS is composed of a Test Host System(THS) and several Test Satellite(TS)s as shown in Figure 1. The THS communicates with the TSs via Ethernet.

2.2 System Functions

The THS has higher layers than the AAL layer on general workstation and the TS has three layers(STM-1 Physical, ATM, AAL 5). For logical connection, the Test Satellite Function(TSF) of the TS uses interprocess communication, frame control, and distribution function of two STM-1 ports. The Message Distribution Function(MDF) in THS executes the message distribution function for the interprocess communication and other application processes. Among the application processes of THS, process for the call simulation and the conformance tests of Q.2931 and Q.SAAL will be used for various test purposes and applied to various types of configuration. Figure 2 shows the functional architecture of BITS.

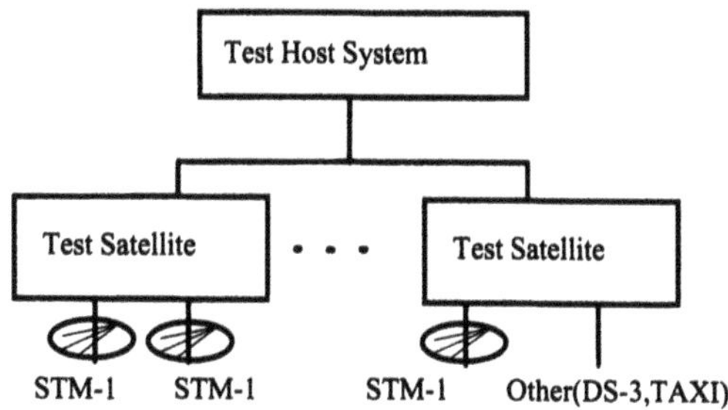

Figure 1 System Configuration Diagram of BITS.

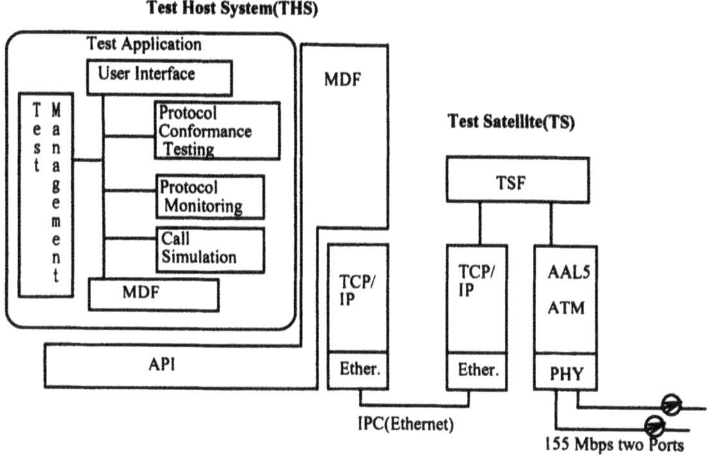

API : Application Programming Interface
TSF : Test Satellite Function
TS : Test Satellite
MDF : Message Distribution Function
THS :Test Host System

Figure 2 Functional Architecture of BITS.

The BITS has the following functions :

- Protocol Monitoring

One basic function of BITS is protocol monitoring implemented by intercepting signals between the BITS and the System Under Test(SUT) or between two SUTs.

- Protocol Conformance Testing
  The conformance testing consists of verifying processes that B-ISDN User-Network Interface Products comply with B-ISDN specifications, performed by a tester. In addition the protocol and the test suite for Q.2931 and Q.SAAL Layer are implemented in BITS.
- Call Simulation
  In order to test the call processing capability of an IUT, we generate the call establishment for testing and assure that the data transfer is completed via the ATM connection. The BITS call simulation function consists of file transfer/receive, frame generation/analysis, and call connection/release.
- Test Management
  The protocol functions of BITS is the realization of the B-ISDN User Network Interface Protocol, the connection handling and the traffic management. The test application process of THS performs conformance testing, call simulation, and protocol monitoring of Q.2931 and Q.SAAL. Simulation procedures and scenario handler are controlled and managed by user control commands.

3 SYSTEM IMPLEMENTATION

3.1 Hardware Implementation

As mentioned before BITS is composed of a THS and several TSs. In this section, we describe the hardware configuration of the TS. Figure 3 shows the hardware block diagram of the TSs. Several hardware block units are divided into two units, Test Control Unit and Test Circuit Unit. The functions of hardware block units are as follows.

The Test Control Unit consists of the Ethernet Interface Unit for Ethernet interface with THS and the Central Processing Unit for test function processing of TS.

In the Test Circuit Unit, most of the processing units were implemented using Field Programmable Gate Array(FPGA). The Physical Interface Unit was designed for different physical interfaces. In order to capture the physical layer data stream without disturbing the communication between BITS and the SUT, Physical Interface Unit receives/transfers the optical signals from/to the SUT and converts them into/from the electrical signals. The traffic Processing Unit generates cells by software. Therefore, the traffic rate is restricted to 1Mbps, and it will be constructed by hardware for supporting 155Mbps rate later on. The debugging for the hardware development is done by a logic analyzer and a digital signal analyzer. Most of the block diagram was designed using Workview of Viewlogic company. The system has its own diagnostic function as a loopback test at the board and system level.

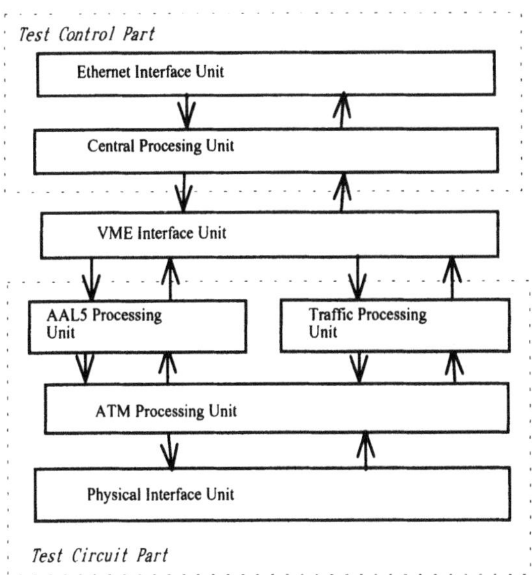

Figure 3 Hardware Block Diagram of a Test Satellite.

3.2 Software Implementation

The software of the BITS is divided into two softwares, one in the THS and the other in the TS. System interface uses TCP/IP for LANs. Software in the THS covers protocol conformance, event generation and analysis for the protocol conformance and interoperability testing in a UNIX system environment. Software for TS executes the test traffic generation and control. Functions for ATM interface control are implemented by using C language and are based on a real time operating system, VRTX. The operation method of the system, the interface type and the number of data for construction management will be determined when the TS downloads a program.

4 TEST EXPERIENCES WITH THE DESIGN OF B-ISDN SIGNALLING PROTOCOL

We have applied BITS to testing of Q.2931 and Q.SAAL. In this section, we explain the test suite of Q.2931 and Q.SAAL conformance testing and X/Motif based on the user interface of BITS. The purpose of testing is to check whether the functions of a System Under Test(SUT) operate normally or not. In order to apply BITS to the various test purposes, the system function and the size of BITS are made to be extendable.

4.1 Conformance Test Methods

Conformance testing, a stage of protocol engineering, includes determining whether a given protocol IUT conforms to the specifications. For this purpose, ISO is developing the OSI conformance testing methodology and framework, describing the various test architectures for different environments, test assessment and Abstract Test Suite(ATS) notation. The test methods designed in ISO/IEC 9646 are divided into two categories, the local and distributed test. The local test method has Upper Tester(UT) and Lower Tester(LT) which are required to access directly the upper boundary and the lower boundary of the IUT, respectively. The distributed test is classified into three kinds of test methods, which are Distributed, Coordinated, and Remote Single layer test method according to the number and the position of the PCO(Point of Control and Observation) and the Test Coordination Procedure(TCP). The Remote Single layer(RS) test method has only LT that is required to achieve the control and the observation of specified interactions.

In this paper, we have chosen the RS test method because it does not put a burden on the implementation, even though the test coverage of RS test method is relatively low. The RS test method is illustrated in Figure 4, the dotted lines indicating that only the desired effects of the TCP are described in the ATS and this test method is applicable to the B-ISDN signalling protocol.

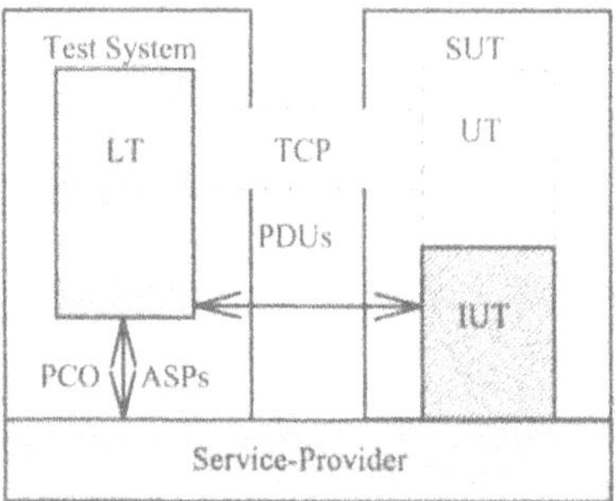

Figure 4 The Remote Single layer test method.

4.2 Development of Abstract Test Suites(ATS) for Q.SAAL and Q.2931

We developed an abstract test suite for the protocol conformance testing, using the following procedures. First of all, we extract the test cases and the test purposes by analyzing the basic standards. Then we generated an Extended Finite State Machine(EFSM.) describing the control procedures of the protocol specifications. The test tool, Conformance Kit, developed by Netherlands PTT was used to get the raw level of the dynamic behavior part of the Abstract Test Suite(ATS). Finally we got the ATS, which was written in Tree and Tabular Combined Notation(TTCN) language, which is a language standardized by ISO for the specification of tests for communication. TTCN has been developed within the framework for ISO's test methodology ISO/IEC 9646. The Test Suite comprises different test cases specified in TTCN which comes together with all the different necessary definitions and declarations. The above procedure was performed manually by considering nondeterministic problems, timer operation and data flow within the raw test suite. Table 1 shows the dynamic behavior of a sample test case for Q.2931 protocol of B-ISDN signalling protocol.

According to the above procedure, we obtained 161 test cases for Q.SAAL-UNI and 397 test cases for Q.2931 User-Side protocol in the Remote Single layer(RS) test method. We applied the same procedure to Q.2931 Network-Side protocol and obtained 420 test cases for the protocol based on the parallel test architecture.

Table 1 An Example of a Test Case Dynamic Behavior

Test Case Dynamic Behaviour					
Test Case Name :	TC00001				
Test Group :	q2931_protocol/pt/N00/				
Objective :	NULL STATE TESTS. VALID TEST EVENTS. Ensure that on receipt of a valid SETUP message the IUT responds with a CALL PROCEEDING message and moves to the state 3.				
Default :	general_default(0)				
Comments :					
Nr	**Label**	**Behaviour Description**	**Constraints reference**	**Verdict**	**Com-ments**
1		+ pr_U00			
2		L ! SETUP **START** T303MAX	SU1(0)		
3	L1	L ? CL_PRr CANCEL T303MAX	CPr1(1)	(PASS)	
4		+ cs_N03(0)			
5		+ po_U00(0)			
6		+ um(1)			
7		GOTO L1			
8		? **TIMEOUT** T303MAX		(FAIL)	
9		+ po_U00(0)			
10					

4.3 Development of an Excutable Test Suite(ETS) for Q.SAAL and Q.2931

In the case of protocol conformance testing, an ATS is not executed as it is in BITS. Note that the procedure of ATS to ETS translation is needed. For protocol conformance testing, the tester covers every transition of the protocol that is represented by an EFSM. We describe the construction of ETS in BITS. In order to achieve the test purpose, every test case of BITS is composed of the following parts : a preamble that drives IUT state into initial state, a test body that accomplishes the test purpose and checks the test results, and a postamble that drives the IUT state into the initial state after the protocol state transition. In the case of testing every transition, test cases will be classified into a valid, invalid, or inopportune test cases. According to the above procedure, the test result turns out to be "pass" or "fail".

Conformance test suite of Q.SAAL

Q.SAAL is the peer-to-peer protocol for the transfer of information and control by selective retransmission and flow control. This protocol is composed of a common part and a service specific part. The service specific part is called SSCS, which is divided into SSCOP and SSCF.

There are two state of IUT in the Q.SAAL conformance testing, one is the active IUT, which connects/releases the Q.SAAL layer, and the other is the passive IUT, in which the protocol tester connects/releases the Q.SAAL layer.

Conformance Test Suite of Q.2931

Q.2931 is the procedure for establishing, maintaining, and clearing of network connections at the B-ISDN User-Network interface. Q.2931 conformance testing verifies that a B-ISDN product complies with the B-ISDN specification. This protocol is composed of Q.2931 User-Side protocol and Q.2931 Network-Side protocol. Each side of the protocol has two states of IUT, Active IUT and Passive IUT. The former connects/releases the Q.2931 layer and for the latter case, the protocol tester connects/releases the Q.2931 layer.

4.4 Configuration of the Test System Interfaces

The test system Interfaces of BITS have been implemented as shown in Figure 5. A TS has two STM-1 interfaces for an IUT and one THS has several TSs with the dynamic allocation. In the near future, various types of physical interfaces(e.g. DS3 interface) will be available.

For the conformance testing and the call simulation of BITS, configured function of each application part establishes a specific application system. The User Interface supports a convenient interface for the functions of the system. The test application system of BITS has protocol monitoring and call simulator, and a conformance tester for basic interconnection and interoperability testing.

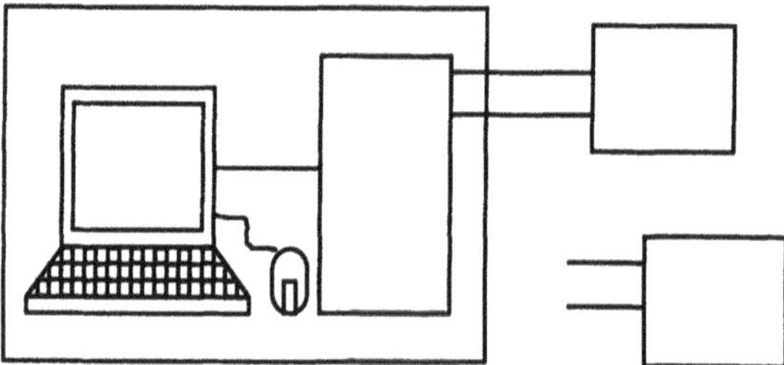

Figure 5 System Interfaces of BITS.

Figure 6 shows a simplified logical display of the BITS user interface. The entire window is composed of 3 frames and each frame is as follows.

Under the Test Environment Setup frame, the user determines the environment that will be executed.. In the Mode Setup button, the test environment and protocol layer of protocol tester and IUT are determined. If the simulation environment is chosen, then testing will be performed inside the THS. If the test environment is the target environment, then the test will be performed through Ethernet to TS. In the Env. Reset button, entire application of BITS is terminated. The Q.2931 Service Request button determines service request environment for Q.2931. The Q.SAAL Service Request button determines service request environment for Q.SAAL.

Under the Conformance Test frame, each protocol and test case are selected, in order to drive test software of Q.2931 layer and Q.SAAL layer and the selected ETS(Executable Test Suite) will be executed. The Test Suite selection button determines the Q.2931 and Q.SAAL test case number and performs Q.2931 ETSs and Q.SAAL ETSs.

Under the Call Simulation, the frame provides a call simulation menu for selecting each file/frame/call function, and the selected function is performed.

Figure 6 Display of the BITS User Interface.

5 CONCLUSIONS

In this paper, we described the design and implementation of a B-ISDN protocol tester to ensure interoperability between network components. The BITS is composed of functions for conformance testing, interoperability testing, call simulation, and protocol monitoring. The design concept of the BITS is that the test configurations can be easily and efficiently prepared for testing the B-ISDN User Network Interface implementations in a real networking world. This paper presented experiences with the design of BITS, especially its tester architecture and the development of test suite for the B-ISDN signalling protocol.

This system was used to test the Centralized Access Node System(CANS), which is one of the products of the HAN/B-ISDN project, and had been exhibited in ION'94. In the near future, we are going to test ATM switch, ATM-MSS(Metropolitan Area Network Switching System), Ethernet TA, Video TA and B-TE as components of Network Test Bed. So far, we have implemented the basic ETSs and full functions are being developed. We also have a plan to develop testing functions for UNI/NNI signalling release II, III for multi-party and multi-connection.

6 REFERENCES

Erik Kwast, Harma Wilts, Hans Kloosterman and Jan Kroon.(1991) User Manual of the Conformance Kit.

ISO/IEC, Information Technology IS-9646.(1993) Open Systems Interconnection - Conformance Testing Methodology and Framework,

ITU-T, Recommendation. I.432(1994) B-ISDN User-Network Interface-Physical Layer Specification

ITU-T, Recommendation.I.361(1994) ATM Layer Specification for B-ISDN

ITU-T, Recommendation. I.363(1994) ATM Adaptation Layer Functional Specification for B-ISDN

ITU-T, Recommendation. Q.2110(1994) Service Specific Connection Oriented Protocol

ITU-T, Recommendation. Q.2130(1994) Service Specific Coordination Function for Signalling at the UNI-SSCF at UNI

ITU-T, Recommendation. Q.2931(1994) B-ISDN Digital Subscriber Signalling No.2(DSS2) User Network Interface Layer 3 Specification for Basic Call/Connection Control

T.Kang, M.Kim, M.Choi.(1994) Design and Implementation of A-monitor for ATM Protocol Monitoring and Analysis, ATNAC '94, Melbourne Australia

About the Authors:

Ki Young Kim is a member of engineering staff at the ETRI since 1988. She studied computer science at the Chonnam National University, Korea, and received a MS from the Chonnam National University in 1993. Current research interests include protocol conformance test, graphical user interface design.

Weon Soon Kim is a senior technical staff at the ETRI since 1983. He studied electronic engineering at the Seoul City University, Korea, and received a MS from the Chungnam National University in 1994. Current research interests include protocol conformance test, traffic of ATM network.

Beom Kee Hong is a senior technical staff at the ETRI since 1982 and a leader about protocol test project of B-ISDN protocol. He studied computer science at the Hongik University, Korea, and received a MS from the Hongik University in 1985. Current research interests include protocol conformance test, ATM network testbed.

21

The Testing of BT's Intelligent Peripheral using abstract test suites from ETSI

N.Webster
Network Intelligence Engineering Centre, BT plc.
BT Laboratories, Martlesham Heath, IPSWICH, IP5 7RE, ENGLAND
Tel. 01473 645910; Email: nick@chaplin.bt.co.uk

Abstract

This paper describes the practical use of ETSI test suites in the testing of an Intelligent Peripheral, which is the Intelligent Network node responsible for in-band speech processing - playing announcements and processing any in-band responses.

Standardised test suites, which are increasingly available from ETSI, offer the potential to cut costs. This is achieved by reducing the need for expensive in-house test development whilst maintaining the testing quality through a set of approved tests which are independent of development.

This paper uses the experiences of testing the Intelligent Peripheral to examine whether these advantages are realised in practice. It describes the test suites, their test coverage, the modifications needed for test execution and the test results and metrics for the testing cycle.

Keywords

Tree and tabular combined notation, abstract test suites, intelligent peripheral.

1 INTRODUCTION

The testing of network elements to ensure successful network operation is an expensive but necessary activity. The complexity of the specifications is continually increasing and with it the costs of testing. Improved specifications have allowed the development of standardised test suites which can help reduce testing costs and provide a powerful set of protocol testing criteria which are independent of development. This helps to ensure the quality of the network elements and their successful interworking in the network.

Increasingly, the documentation package for a protocol or network element includes both a functional specification and a test specification coupled to it. A number of conformance test specifications are available which are based on the ISO 9646 [1] methodology and are written in the Tree and Tabular Combined Notation (TTCN). The availability of commercial tools for editing, compiling and executing TTCN provided the opportunity for standardised test suites to be used in testing the Intelligent Peripheral (IP).

The IP in its network configuration is shown in Figure 1. The role of the IP is to perform in-band speech processing for the Intelligent Network (IN). The IP is controlled from the IN Service Control Point (SCP) using the INAP protocol. The SCP manages the service provided to the caller by first requesting the IP to play selected announcements and then checking the IPs response. The latter contains the speech processed in-band response from the caller encapsulated in an INAP message. The IN Signalling Switching Point (SSP) connects the caller to the SCP and IP using the C7 National User Part (NUP) and INAP respectively. The IP processes the speech channels of the C7 NUP.

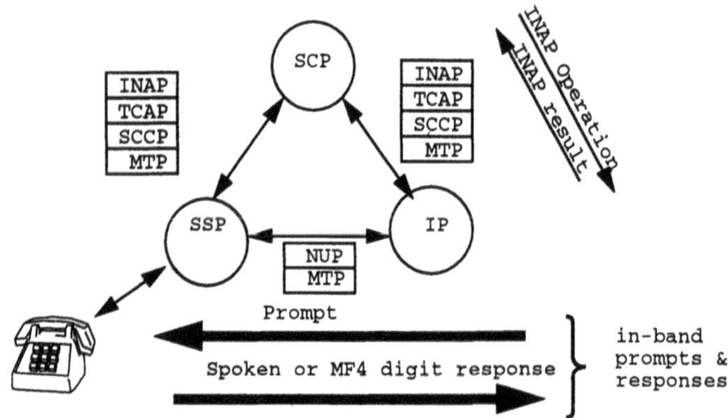

Figure 1: Overview of the Intelligent Peripheral

The project was split into development and test teams. The role of the development team was to add the Intelligent Network (IN) functionality of:-

- ITU blue book SCCP,
- ITU white book TCAP [8] and
- Core ETSI INAP [2]

to an existing speech application platform. The INAP software was developed in-house whilst the SCCP and TCAP software was procured from an external supplier.

The speech applications processing, C7 National User Part (NUP) and Message Transfer Part (MTP) had already been tested with proven in-house tests that will not be described here. The SCCP testing is still in progress and so is not described.

The challenge for the test team was to test the new protocols of the IP, so that the IP was suitable for integration testing with a real SCP and SSP. The use of test case generation tools as described in [9] was considered but not adopted for the following reasons:-

- The tools did not handle large systems,
- The tools were focused towards SDL coverage not data - a vital part of TCAP/INAP,
- "Formal" specifications in SDL were not available

Unfortunately, the tight timescales for the IP project gave little time to develop in-house test suites, so there was no other option but to use the ETSI Abstract Test Suites (ATS). These were at different stages of development:-

- SCCP ATS - stable and mature,
- TCAP ATS [6] - being extensively modified,
- INAP ATS [7] - stable and close to maturity.

An assessment was made of the available TTCN tools based on the criteria of cost, tool support, testing capability and the ability to support the protocols of the IP. We chose Telelogics ITEX editor, and the Siemen's K1197 tester [5]. Together these comprised:-

- protocol encoders and decoders for MTP,NUP,SCCP,TCAP and INAP,
- a TTCN IS compiler,
- a menu driven message building system and
- a proven test environment.

The processes of development and testing were kept independent within the project, so that the tests did not re-use any development software. Therefore, any assumptions made in the development did not percolate into the tests thus weakening them.

2 TESTING METHODOLOGY

This section describes the methodology for the creation of test suites by standardisation bodies and the correction and verification of the test suites for IP testing.

2.1 Test Suite Creation

A necessary pre-cursor to the writing of a protocol test suite is the specification of the protocol in textual and graphical notation. The specification should be complete and correct (although not necessarily formal) to enable developers and network managers to implement the protocol in real equipment.

The specification of the tests then follows, and may take several man-years to complete. The first and most important activity is the creation of a set of test purposes which explain the proposed test method, architecture and give a precise definition of the purpose of each test. The definition is in english text and cross references the protocol specification. The test purpose document is then used as the high-level test design document during the creation of tests in TTCN. Each test purpose is expanded into detailed TTCN test behaviour in the test suite. The writing of the test suite is a skilled manual activity. The test suite is usually the last document to be written and is often several hundred pages long.

2.2 Review and Transformation

Before the test suite could be used in the IP project it was analysed using ITEX and then reviewed for accuracy and completeness. Enhancements to the test suite to make it suitable for IP testing were identified.

The restrictions of the TTCN compiler supplied with the tester were compared against the range of TTCN constructs used in the test suite and the protocol layers supported on the tester. The un-supported constructs were assessed in order to define a strategy for the transformation of the test suite into a form which would execute on the target tester.

2.3 Test Verification

The transformed test suite was compiled on the K1197 for a more rigorous semantic analysis. Errors identified at this stage were removed before recompilation.

The suite was verified against both:-

- a simulated system under test (SUT) and
- the IP.

The simulated SUT was run on one C7 port of the tester and sent protocol messages to the C7 port running the TTCN, via an external looped-back connection. This approach conveniently allowed TTCN events to be checked. It would have been possible to have verified the tests against more comprehensive simulations on the K1197, but the timescales did not permit their development. The test suite was further exercised by running it against the IP implementation. At this stage, internal and external message traces were available to be used as verification tools.

The tester conveniently displayed execution traces of the test steps and constraints which was invaluable in test verification. The test suite parameters could be modified after compilation to configure the tests to match the configuration of the implementation under test.

The running test suite (in this case TCAP) was connected by an internal software interface to the run-time operating system. The interface was comprised of a user-defined *test manager* and *special file*. The *test manager* connected the automatic MTP layers to the *special file*, and could filter or respond to specific MTP messages. The *special file* was the interface between the TCAP test suite and the *test manager*, and converted TTCN events into protocol messages for the *test manager* and visi versa. The *special file* could be easily modified to add new protocol primitives to the test suite.

With the architecture as shown in Figure 2 the *test manager* could provide automatic SCCP, TCAP or INAP layers if so desired. However, since the IP only used connectionless SCCP, an automatic SCCP was not required and therefore the sent and received Protocol Data Units (PDUs) were defined explicitly in the test suite.

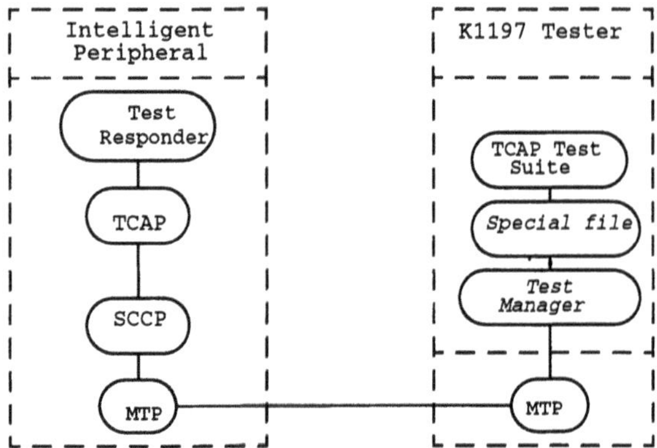

Figure 2: TCAP Test Architecture

3 TESTING THE TCAP LAYER

ETSI supplied early versions of their test suite [6], which comprised 140 test cases and proposed the use of a test responder as the upper tester, replacing the TCAP user during testing. This was a remote coordinated test method [1]. The purpose of the test responder was to send and receive primitives from TCAP, under the control of the test suite. Test management messages for the responder were encapsulated in normal TCAP messages.

3.1 Test Architecture

The TCAP test architecture is shown in Figure 2. Test management messages to the responder were defined as TTCN implicit sends, but this approach was under review by ETSI, and it was not clear precisely how the responder commands would be defined. Thus, owing to the demanding timescales for the IP development, it was decided to design our own responder and responder interface in SDL [3].

We decided to encapsulate the responder commands and responses in the user-defined data within TCAP un-structured dialogues, since the IP only made use of TCAP structured dialogues.This conveniently separated the facilities of TCAP under test from those used in testing. The main assumption made in testing TCAP was that the operation of structured and unstructured dialogues were independent.

The specific responder test sequences were defined for each test, encoded on a SUN workstation and then added to the ATS in TTCN test suite parameters.

Figure 3 shows a typical TTCN test case. Test steps PRE_ID and SEND_BEGIN_RI2 cause test management messages to be sent to the responder to reset it and then send TC-INVOKE and TC-BEGIN primitives to TCAP. This causes TCAP to send a BEGIN message, which is verified by the receive constraint BEGIN_R_I to give a test pass. The message sequence for the test including both TCAP and responder messages is shown in Figure 4.

Line	Behaviour	Constraint	Verdict	Comments
1	+PRE_ID			Reset Test Responder
2	+SEND_BEGIN_RI2			Send Test Responder command
3	START T_WAIT			Initialise Test Timer
4	?BEGIN	BEGIN	PASS	Receive BEGIN - test PASS
5	+POST_TEST			request responder test log
6	TIMEOUT		FAIL	Timer expiry: test FAIL
7	POST_TEST			request responder test log
8	?OTHERWISE		FAIL	Other receive event: test FAIL
9	POST_TEST			request responder test log

Figure 3: TCAP test case

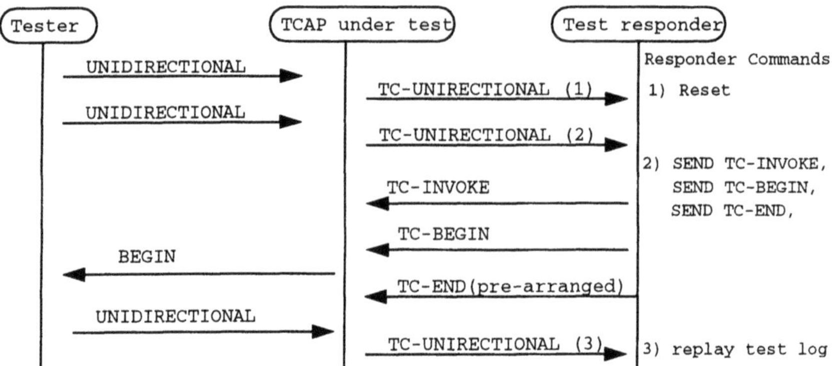

Figure 4: Message Sequence Chart for TCAP test case

The responder was enhanced to extract and store a TCAP dialogue identity from a received primitive and subsequently use it in primitive requests. The responder was also enhanced to improve the verification of the primitive interface to TCAP by recording the sequence of incoming and outgoing TCAP primitives. The recorded sequence could then be replayed at the request of the tester which is the function of test step POST_TEST in Figure 3.

3.2 TCAP ATS Enhancements

Naming Conventions

There was no easily comprehensible naming convention in the ATS. Several overlapping conventions were present which together with the scale of the ATS made it difficult to understand.

Problems

There were a number of unresolved issues within the ATS:-

- whether to use structured or unstructured dialogues for responder communication,
- the responder sequences were not all defined,
- where to place the test responder command within a TCAP message,
- some tests did not assign verdicts,
- some tests did not shut down active dialogues if the test failed,
- sometimes the active dialogues were not returned to the idle state,
- the ATS did not test white book facilities.

An important issue was the tests only tested the blue book facilities supported by white book TCAP. Thus we made sure that the tests we ran tested white book facilities of white book TCAP. We ensured that test verdicts were assigned for each test and the correct state of any dialogue was verified at the end of each test. The tests were made robust, so that if the tests failed, active dialogues were terminated.

The ATS used many different datatypes, which were transformed into three types: integer, boolean and octet string.

ASN.1 on the K1197

The K1197 supported the majority of the ISO 9646 IS constructs but did not support TTCN Abstract Syntax Notation 1 (ASN.1) [4] Protocol Data Unit (PDU) and constraint definitions. Unfortunately ASN.1 was used extensively throughout the test suite. Possible solutions were:-

- TTCN structured constraints,
- octet string constants,
- TTCN test suite operations.

Structured constraints were not feasible for the variable length ASN.1 fields which were present in TCAP. Structured constraints could have been used if the fields had been of fixed length.

Manually encoded octet string constants were viable but would have been of fixed length and thus would not support variable length fields. Additionally software maintenance would have been expensive.

TTCN test suite operations were selected. The approach needed manual coding of the test suite operations, but enabled software re-use in other constraints and allowed easy adaptation to other ASN.1 encoding formats.

3.3 TCAP test results

After running 140 tests, the problems discovered with the TCAP implementation were:-

- large TCAP PDUs and primitives were truncated,
- abnormal TCAP primitive sequences were incorrectly handled,
- the Dialogue Portion specification was easy to misinterpret,
- transactions remained active after termination.

These are described in detail below.

The development of the responder messages was initially slow since the first two problems prevented the transfer of responder messages through TCAP. TCAP was monitored internally to pinpoint the error and to help identify the problems we created protocol messages using the K1197 message definition facility.

When the responder log was replayed to the K1197 at the end of a test, a loss of messages occurred. The reason was that a TCAP invoke state machine was run for each unstructured dialogue and was still active when the same dialogue was re-used, subsequently blocking the sending of a unidirectional PDU. This highlighted how practical issues can impact unexpectedly on the ATS.

The specification of the TCAP dialogue portion was complex and took some time to understand before the TCAP dialogues could be successfully initiated. Once these problems were resolved the test suite was executed and in only a few days it was possible to run all the tests.

Overall the identified faults gave a clear perspective on the suppliers development process and showed that the lower interface of TCAP was well checked for normal and abnormal sequences. However the TCAP primitive interface had been checked only for a limited set of normal sequences.

The metrics for TCAP testing are shown in Table 1.

	TCAP	INAP	additional INAP
ETSI ATS used	yes	yes	no
Status of ETSI ATS	draft (not mature)	draft (mature)	-
TTCN compiler	yes	yes	yes
Test Suite execution	auto	semi-auto	semi-auto
Number of Tests	140 tests	50 tests	20 tests
ATS enhancement	60 days (40%)	15 days (60%)	5 days (70%)
Coding of test suite operations	20 days (13%)	5 days (20%)	2 days (10%)
Responder/Harness specification	10 days (6%)	-	-
Responder/Harness coding	15 days (10%)	-	-
Test Execution including fault removal	30 days (20%)	10 days	5 days
Final test execution	3 days	0.5 day	0.5 day
TOTAL man power	140 days	30 days	12 days
Man power per test	1.0 days	0.6days	0.6 days

Table 1: Metrics for Testing

4 TESTING THE INAP-SRF

The ETSI INAP test suite [7] contained over 50 IP tests and made extensive use of ASN.1. The ETSI test suite used the multi-party [1] remote test method. The INAP test suite also contained tests for the IN SSP which were not used as they were not relevant to the IP testing.

4.1 Test Architecture

The ATS was written to use the services of an automatic TCAP layer running under the ATS, which unfortunately was not available on the K1197. The two solutions were:-

- write a TCAP emulation with an Abstract Service Primitive (ASP) interface,
- transform the TCAP TTCN ASPs to TTCN PDUs (as in the TCAP ATS).

Siemens offered to provide a TCAP emulation but it was decided to transform the ATS since this approach would focus all of the test control within the test suite.

The K1197 emulated the NUP and INAP protocols and a digital telephone was connected to the NUP link to listen to in-band recorded announcements and to make in-band voice or keyed digit responses. The test architecture is shown in Figure 5.

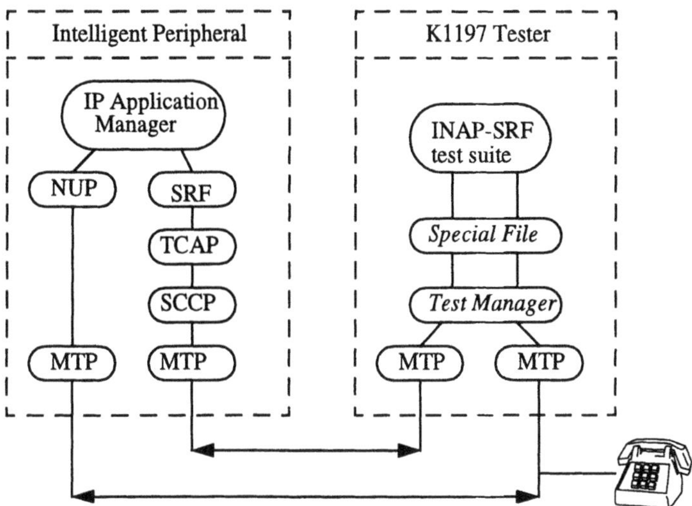

Figure 5: Test Architecture for the SRF

A typical message sequence for an INAP-SRF test is shown in Figure 6. The test begins with a NUP Initial and Final Address Message (IFAM) to the IP, which then sends an INAP Assist Request Instruction (ARI) in a TCAP BEGIN. The INAP Play Announcement is then sent in a TCAP CONTINUE, which replies with an INAP specialised resource report in a TCAP CONTINUE. The IP also sends the Address Complete Message (ACM) and Answer message (ANS) and the call clears with a release (REL).

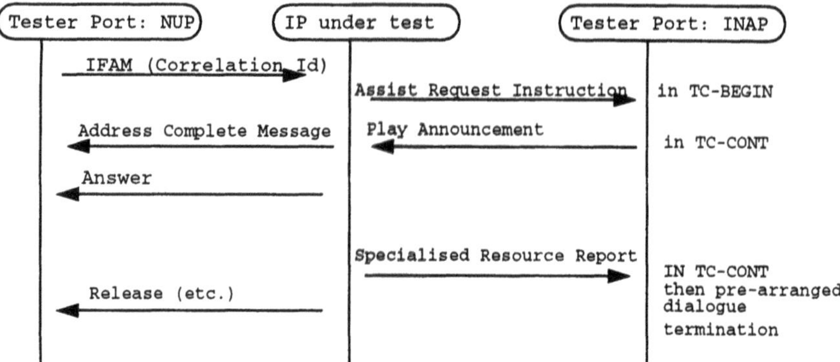

Figure 6: Message Flow for INAP test

4.2 INAP-SRF ATS enhancement

The INAP test suite used a generic bearer call control signalling interface. Thus the ATS was enhanced for the specific SSP-IP NUP interface of the IP by the addition of the associated TTCN PDUs and constraints.

The review of the test suite found a number of errors:-

- verdicts were not always assigned,
- some of the message flows were incorrect,
- call disconnection in some tests was incorrect,
- the use of *ElementarymessageID* was wrong in some tests.

The transformation of the ASP constraints to PDU constraints was straight forward, and the TCAP ATS PDU definitions were re-used in the INAP ATS. The TTCN ASP fields which defined the INAP operations in ASN.1 were each replaced by a test suite operation. It was written so as to use any test suite parameters, constants or variables in precisely the same form as in the ASP definition. Thus the operation of the transformed and original ATS were identical.

The TTCN datatypes were converted into integer, boolean or octet strings.

4.3 Supplementary INAP-SRF Tests

20 new tests were written in TTCN to test:-

- BT's IN extensions and
- call disconnection procedures.

The metrics for these tests are shown in Table 1.

4.4 INAP-SRF Test Results

The metrics for INAP testing are shown in Table 1.

The testing showed the benefits of test suite parameters, which could be modified interactively to match the values supported by the IP. The response of the IP system to un-supported values was not tested explicitly within the ETSI ATS, but was an important feature which was easily checked.

5 DISCUSSION OF THE USE OF ABSTRACT TEST SUITES

Table 1 shows a summary of the testing for the different protocols. It shows the time taken in man-days for the different stages of the testing. The percentage time for each activity is shown in parenthesis.

The INAP ATS has a metric of 0.6 days per test. The 20 supplementary INAP tests have a similar metric of 0.6 days, but this includes the creation of new tests which is more efficient. The TCAP metric of 1.0 days was twice that of INAP, because of the time to correct the draft test suite (60 days) and to clear faults in the IP (30 days). If the ATS had been more mature, less time would have been taken and the overall metric would have improved. The metric could have been further improved if the tester had supported TTCN ASN.1.

TCAP has been successfully tested using an in-house version of the ETSI test suite. A large number of tests have been run and problems highlighted. An equivalent in-house ATS would probably have taken several man-years to develop. The main limitation with the test suite was that only normal sequences were sent by the TCAP test responder. The use of abnormal sequences and large ASPs (including maximum length) needs to be considered to enhance the test suite. This would only be possible during TCAP testing and not during testing of the TCAP user.

The ability to modify the test suite to match the implementation under test is important. This includes the selection of test suite parameters and changes to the TTCN.

5.1 K1197 Tester

The protocol encoders and decoders were invaluable in decoding protocol messages and in the creation of messages during test suite verification. Test suite parameters could be changed interactively and the tests re-run without recompilation.

The support and in-depth technical help from the supplier was good and if any problems were encountered a quick and accurate response was forthcoming. The importance of good support in using new technology cannot be overstated.

5.2 TTCN

The notation of TTCN is powerful and comprehensive, however without the use of naming conventions and design information, the ATS' are difficult to understand and the impact of proposed changes unnecessarily time-consuming to assess. Large TTCN test suites are rather intimidating.

Test engineers would adopt TTCN more readily if there was a simple guide to its use. TTCN is very difficult to review by those who are not fluent in it. Training in TTCN is essential.

The use of a TTCN editor was invaluable to modify and check the TTCN. However because of TTCNs spread-out nature it was sometimes easier to modify the TTCN using a text editor.

6 CONCLUSION

Abstract test suites from ETSI are being used successfully to test an Intelligent Peripheral. They have provided large numbers of effective tests, without the need for expensive in-house test development and have helped to reduce the costs of testing. The use of standardised tests ensured independence from the development process and so BTs Network Interconnection Responsible Officer was able to approve the tests.

The ETSI protocol documentation, encompassing functional and test specifications, was a sound basis for protocol testing. Unfortunately, for state-of-the-art protocols the availability of the test specification was somewhat delayed behind the specifications. Thus draft test suites were used for TCAP and INAP. It was time-consuming to resolve the TCAP problems. Some practical issues had not been considered, which we felt necessary to improve, such as the emphasis of the white book TCAP tests being to test blue book facilities of white book TCAP. It was important that the functionality to bc used by the IP was actually tested. The INAP ATS was much easier to update.

The abstract test suites were modified to correct errors and updated to support the specific implementation and architecture under test. The TTCN tools were not as well developed as equivalent specification tools and need enhancement to make them more usable, so that large and complex test suites can be handled effectively.

The execution of the abstract test suite was achieved successfully, using practical techniques made possible by the flexibility and power of the K1197. The absence of ASN.1 support in the TTCN compiler caused some difficulties for the TCAP and INAP testing but was overcome by the use of TTCN test suite operations.

The testing of the IP using ETSI test suites was successful and was probably more efficient than equivalent in-house test suite developments. However at the moment, standardised test suites are not the solution for testing all network elements. For each network element, a fresh assessment must be made of the protocols and architecture to be tested, the maturity and applicability of the test suites and the facilities of the testers which will run the test suites.

In the future, the standardised test suites are likely to be more widely adopted as the number of available test suites increase and the quality of the TTCN support tools improves.

7 REFERENCES

[1] ISO/IEC: *Conformance Testing Methodology and Framework- Part 3: The Tree and Tabular Combined Notation.* ISO 9646, ISO/IEC, 1992.

[2] ETSI: *Core Intelligent Network Application Protocol Specification.* PrETS 300, 374-1,1994.

[3] CCITT: *Specification and Description Language (SDL).* CCITT Recommendations Z.100, CCITT/ITU 1988.

[4] CCITT: *Abstract Syntax Notation No 1 (ASN.1) Encoding Rules.* Recommendations X.209 CCITT/ITU 1989.

[5] Siemens AG: *Manuals for the K1197* Tester.

[6] ETSI: *TCAP Abstract Test Suite.* Draft version from ETSI SPS2, January 1994.

[7] ETSI: *INAP Abstract Test Suite.* Draft version from ETSI SPS3, July 1994.

[8] CCITT: *Transaction Capabilities.* CCITT Recommendations Q.771-Q.775, CCITT/ITU 1992.

[9] L. Boullier et Al.: Evaluation of some test generation tools on a real protocol example. IWPTS VII - IFIP WG6.1 Proceedings Nov. 1994.

8 ACKNOWLEDGEMENTS

The author wishes to thank his colleagues Mr. P.McGuinness and Mr. M.Duffell for their help with TCAP, and also Hans Neuendorf of Siemens for guidance and technical help in the use of the K1197.

9 BIOGRAPHY

Nick Webster graduated in physics from the University of Bristol in 1973 and has spent most of his working life with British Telecom. He began his career in semi-conductor research, then changed his work-area, and for the last 12 years has been responsible for the design and testing of protocol systems including System X, ISDN and IN. He has concentrated on the application of modern testing technology, such as TTCN, to telecommunications systems.

22

Design of Intelligent OSI Protocol Monitor

Tomohiko Ogishi, Akira Idoue, Toshihiko Kato and Kenji Suzuki
KDD R&D Laboratories
2-1-15, Ohara, Kamifukuoka-shi, Saitama 356, Japan
Telephone : +81-492-66-7370 Facsimile : +81-492-66-7510
E-mail : {ogishi, idoue, kato, suzuki}@hsc.lab.kdd.co.jp

Abstract

As the OSI protocols come to be widely adopted in various communication systems, the testing of OSI protocol implementations becomes important. In this testing, the interoperability testing is performed to examine the interconnectability which cannot be checked in the conformance testing. In this paper, we propose the Intelligent OSI Protocol Monitor which observes PDUs (Protocol Data Units) exchanged between OSI system and analyzes the protocol behaviors as well as the PDU format according to the protocols of OSI 7 layers. This monitor is effectively used to observe the actual communication for a long period and to detect protocol errors which cannot be detected by the conformance testing. This paper describes the detailed design of the Intelligent OSI Protocol Monitor and shows how it works for actual OSI protocols.

Keywords

Interoperability testing, protocol monitor, OSI

1 INTRODUCTION

As the standardization of OSI (Open Systems Interconnection) progresses, OSI protocols come to be widely adopted in various communication systems. As a result, the testing of OSI protocol implementations becomes important in order to realize the interconnection between OSI based communication systems. When a communication system is developed, it will be tested whether the system implements the relevant OSI protocols correctly. This testing is called the conformance testing and is useful to increase the possibility of the interoperability of OSI systems. However, the conformance testing focuses only on the conformity of an OSI system to the standard protocols, and therefore there are some possibilities that two OSI systems which have passed the conformance testing independently cannot communicate with each other, for example, because of some incompatibility of the usage of optional parameters. Also, there are some possibilities that the conformance testing fails to detect some errors included in the systems.

The interoperability testing will be performed to resolve these problems of the conformance testing. In this testing, OSI systems are connected through networks and the interoperability will be examined by making them communicate with each other. Currently, there are some studies on the interoperability testing and they are categorized into two types of methods. One method introduces additional testing programs, such as lower tester and upper tester, in the communicating OSI systems [7, 8, 9]. This method assumes that the additional testing programs apply test sequences and observe responses for them. These test sequences can be generated based on the test case generation method for the conformance testing. The other method only monitors PDUs (Protocol Data Units) exchanged between the communicating OSI systems without using controlled test sequences. In this type of testing, protocol monitors are used to analyze exchanged PDUs [5, 10].

However, these two types of interoperability testing have the following problems:

• As for the interoperability testing using additional testing programs, it is possible that the additional programs may change the original behavior of the communication systems.
• Since the test sequences are generated based on the method adopted by the conformance testing, it may be possible that the errors which cannot be detected in the conformance testing cannot be detected again.
• As for the testing by monitoring exchanged PDUs, the currently available protocol monitors have only the functions to analyze the format of PDUs and exchanged data sequence, and protocol errors should be detected manually by test operators.

In order to resolve these problems, we have adopted the following approach. We select the testing method by monitoring exchanged data for avoiding the effects of additional testing programs. We introduce in a protocol monitor the functionality to analyze the protocol behaviors and to detect protocol errors in the communicating systems according to protocols of OSI 7 layers. We call this protocol monitor as Intelligent OSI Protocol Monitor. The Intelligent OSI Protocol Monitor observes the actual communication between OSI systems, checks both the PDU formats and the protocol behavior, and finds errors if the PDU formats and the behavior does not conform to the standard protocols.

This paper describes the design of the Intelligent OSI Protocol Monitor. The next section and section 3 describe the design principle and the detailed design of the Intelligent OSI Protocol Monitor, respectively. Section 4 gives an example on how this monitor works for the OSI Transport and Session protocols [2, 3]. Section 5 gives some discusses on this monitor and section 6 makes the conclusion on our researches.

2 DESIGN PRINCIPLES

We have adopted the following principles on the design of Intelligent OSI Protocol Monitor.

1. The Intelligent OSI Protocol Monitor observes PDUs exchanged between communicating

OSI systems, and emulates the behavior of individual OSI systems separately according to observed PDUs. In the configuration where OSI systems A and B are communicating, an observed PDU in the direction from A to B will be handled as a sent PDU by the emulation of system A and as a received PDU by the emulation of system B.

2. The Intelligent OSI Protocol Monitor emulates the behavior of an OSI system based on the layered structure model, that is, it emulates the behavior of individual layers in one OSI system. The primitives exchanged between the layers are estimated by the monitor.

3. The emulation of individual layers is invoked by the observation of PDUs based on the following procedure.
• When a received PDU of a layer is observed, the monitor emulates the behavior of the layer in the case that the PDU is applied. If the behavior includes sending out of a PDU or a primitive to the lower layer, the monitor expects the sending, and confirms it when corresponding PDU is observed.
• When a sent PDU of a layer is observed in the case that some PDUs are expected to be sent, the monitor searches for the input (primitives from the higher layer or timeouts) which generates the PDU, and it considers that the input is applied to the layer. If it is a primitive from the higher layer, it is reported to the higher layer as an issued primitive.

4. During the emulation, the monitor checks the following protocol errors.
• PDU format error, such that a PDU does not have the mandatory parameters or the order of parameters is wrong.
• PDU parameter value error, such that parameter values are out of range defined by the protocol.
• PDU mapping error, such that a PDU is included in a wrong type of lower layer's PDU.
• State transition error, such that an inopportune PDU or a PDU including invalid parameter values for the current state is sent out.

If an invalid PDU is sent out, the monitor decides that the system which sends the PDU has protocol errors.

5. The Intelligent OSI Protocol Monitor maintains the state of each layer to emulate the behavior of each layer. If the state of a communicating OSI system is known when the monitoring starts, then the state can be given by test operators. If not, the monitor estimates the state by using IO (Input/Output) sequences which determines the state before or after they are observed.

3 DETAILED DESIGN

3.1 Overview

Figure 1 shows the structure of the Intelligent OSI Protocol Monitor. The monitor consists of the modules depicted in Figure 1. The Frame Capturing Module captures frames transmitted in both directions through the transmission line. As described in section 2, the System Emulating Modules which emulate the behavior of OSI systems are provided separately for the individual systems (systems A and B in Figure 1), the System Emulating Module is decomposed into a set of

Emulating Modules for individual layers. The (N) Emulating Module is provided with sent or received (N)-PDUs or issued or received (N-1)-primitives, which are estimated in the (N-1) Emulating Module. It emulates the (N) layer protocol and generates (N+1)-PDUs or (N)-primitives to be reported to the (N+1) Emulating Module. The monitored PDUs and the errors are shown by the Displaying Module.

Figure 2 shows the internal structure of the (N) Emulating Module. It consists of the (N) PDU Analyzing Module, the (N) State Identification Module and the (N) Behavior Control Module. The (N) PDU Analyzing Module decodes (N)-PDUs along with checking the format of (N)-PDUs. The decoded (N)-PDUs and (N-1)-primitives are given to the (N) Behavior Control Module.

The (N) Behavior Control Module emulates the behavior of the (N) protocol based on the State Transition Table of (N) layer. This emulation is performed for the related PDUs, such as PDUs over the same connection of the connection oriented protocol and PDUs with the same

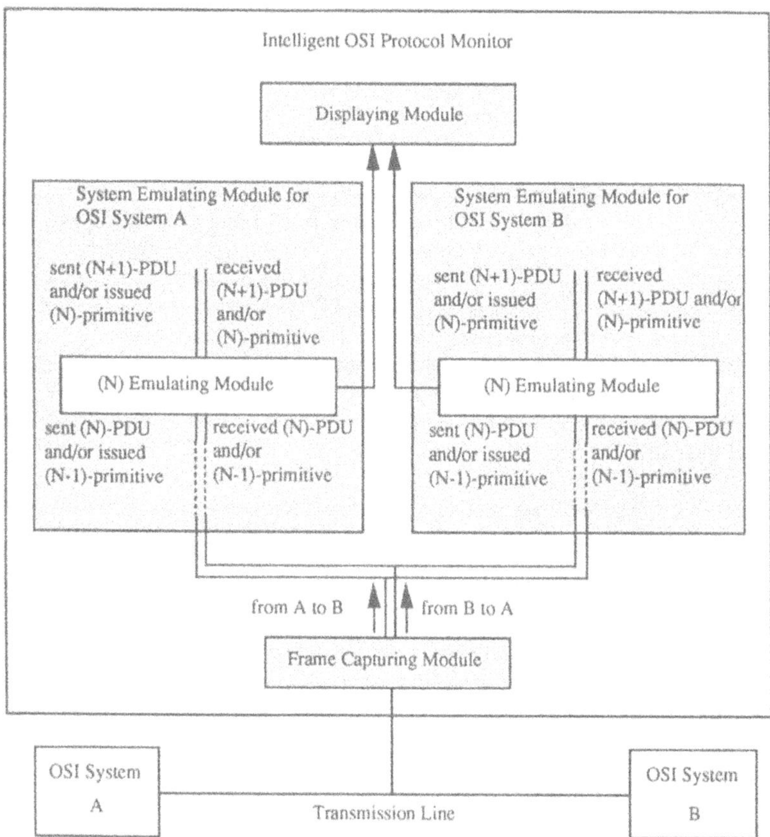

Figure 1 Configuration of Intelligent OSI Protocol Monitor.

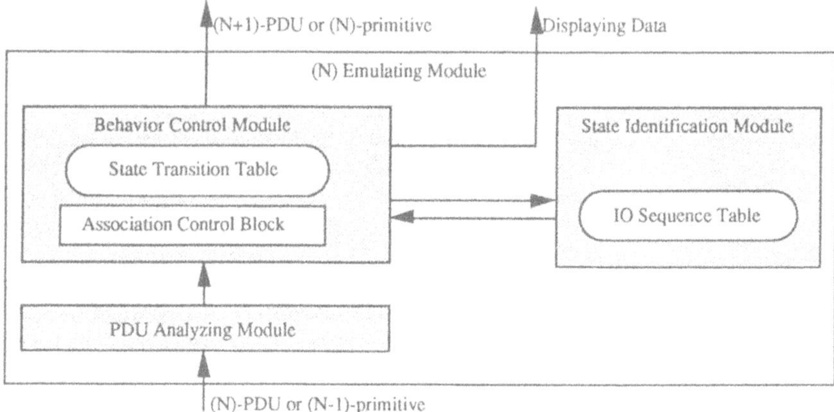

Figure 2 Configuration of (N) Emulating Module.

Data Unit Identifier which CLNP (Connectionless Network Protocol) uses for the segmentation. Therefore, the (N) Behavior Control Module manages the Association Control Block for maintaining the association, such as the connection, between the peer (N) layers. This block includes the emulating-status listing up the candidates of states which the (N) layer may possibly take. In the case that the (N) Behavior Control Module does not know the state of (N) layer, it stores the decoded PDUs and primitives, provides the (N) State Identification Module with the PDUs and/or primitives, and asks for identifying the state of (N) layer. The (N) State Identification Module maintains the IO Sequence Table which specifies the sequences of (N)-PDUs and (N-1)-primitives which can identify the state of (N) layer before or after the sequence is observed. By use of this table, the State Identification Module estimates the state and report it to the Behavior Control Module.

In the case that the (N) Behavior Control Module has identified the state, it emulates the behavior of (N) layer using the State Transition Table with the following approaches.

1. The (N) Behavior Control Module maintains all possible states of (N) layer in the emulating-status of the Association Control Block. When received (N)-PDUs and (N-1)-primitives or sent (N)-PDUs and issued (N-1)-primitives are provided, the transition from each states in the emulating-status is examined. If all transitions from a state are inconsistent with provided (N)-PDUs and (N-1)-primitives, then the state will be deleted from the emulating-status. If all states are deleted from the emulating-status, the (N) Behavior Control Module decides that some protocol errors occurred.

2. As for received (N)-PDUs and (N-1)-primitives, it checks this transition in the State Transition Table. If the transition sends out any (N)-PDUs and/or (N-1)-primitives, this module defers the emulation of this transition until the expected outputs are sent out or until some period of time passes. This is performed by use of the processing-flag in the emulating-status and the processing-

timer. If the transition does not send out any (N)-PDUs and (N-1)-primitives, it is emulated at this stage.

3. As for sent (N)-PDUs and issued (N-1)-primitives, there are two cases of handling by the (N) Behavior Control Module. In the case that the processing-flag is set, the module checks whether the transition focused on sends out the sent (N)-PDU and/or issued (N-1)-primitive. If so, this transition is emulated and the state is changed to the new state for the transition. If not, the (N) Behavior Control Module deletes the state from the emulating-status.

4. In the case that the processing-flag is not set, the (N) Behavior Control Module checks whether the sent out PDUs and/or primitives can be sent in the current state, by referring to the State Transition Table. If so, the module emulates that the event which sends out the PDUs and/or primitives has occurred, and if the event is an (N)-primitive, it is reported as an issued (N)-primitive to the (N+1) Emulating Module. If not, the state is deleted from the emulating-status.

3.2 PDU Analyzing Module

The (N) PDU Analyzing Module is provided with (N)-PDUs together with (N-1)-primitives in which the PDUs are contained. It checks the following items for the PDUs:

- the format of PDUs
- the constraint for PDU parameters defined by the protocol
- the mapping between the (N)-PDU and the (N-1)-primitives.

If there are no errors, the PDUs are handled as valid PDUs sent or received, and the decoded PDUs are provided for the (N) Behavior Control Module together with the (N-1)-primitives. If there are any errors detected, the PDUs are handled as invalid ones and are reported with error information. The PDUs are also provided for the Displaying Module.

3.3 State Identification Module

The (N) State Identification Module checks whether the sequence of (N)-PDUs and (N-1)-primitives provided by the (N) Behavior Control Module match any IO sequence maintained in the IO Sequence Table. If the matched IO sequence is found, the module identifies the state before or after the sequence is observed depending on the type of the IO sequence, and reports the identified state to the (N) Behavior Control Module.

3.4 Behavior Control Module

3.4.1 Data Structure

As described in section 3.1, the (N) Behavior Control Module uses the Association Control Block. In the case of the connection oriented protocol, it includes the following elements:

• the emulating-status which is a list including the state, processing-flag, input-id, transition-id and output-id
• the identifiers for the (N) connection and (N-1) connection
• parameter variables for the connection defined by the (N) protocol, such as reference of the Transport Protocol and the functional unit of the Session Protocol
• the buffer for (N)-PDUs and (N-1)-primitives while the state is not identified
• (N)-SDU buffer for reassembling and resequencing
• the buffer for received (N)-PDUs and (N-1)-primitives.

The emulating-status is used for two purposes. One is to maintain all the possible states which the (N) layer may take and the other is to maintain all the possible transitions caused by received (N)-PDUs and (N-1)-primitives and with (N)-PDUs and/or (N-1)-primitives sent out. The processing-flag shows that a transition is during processing and the observation of outgoing (N)-PDUs and/or (N-1)-primitives is being waited for. The input-id and the transition-id specify the current input and transition in the corresponding entry of the State Transition Table. The output-id indicates which outgoing PDUs and primitives are waited for in the case that the corresponding entry has more than one outputs. The buffer for received (N)-PDUs and (N-1)-primitives is used for deferring the handling of the PDUs and/or primitives as described below.

The State Transition Table used for emulation is different from the table standardized. All possibilities of outputs are written in the action field as alternatives. For example, if an output of a PDU is an option, the alternatives are the output of the PDU and no output.

3.4.2 Emulating Algorithm When State is Identified

When the state of (N) layer is identified, the (N) Behavior Control Module will emulate the behavior of (N) layer based on the following algorithms.

1. Handling of received (N)-PDU and (N-1)-primitive

• The module searches for the entry of the Association Control Block corresponding to the PDU and/or primitive.
• If there is at least one state whose corresponding processing-flag is set in the emulating-status, the handling of the PDU and primitive will be deferred until the handling of the expected sent PDUs and/or issued primitives is completed. The received PDU and/or primitive is stored in the buffer for received (N)-PDUs and (N-1)-primitives in the Association Control Block.
• The received (N)-PDU and/or (N-1)-primitive values are checked with the parameter variables for the connection.
• For each state in the emulating-status, the following procedure is performed for all the transitions specified in the entry for the state and the PDU and/or primitive:

  • If the transition does not send out any (N)-PDUs and (N-1)-primitives, this transition will be emulated. This emulation includes performing the corresponding actions in the transition, such as reassembling and resequencing using (N)-SDU buffers, updating parameter variables corresponding to the parameter values in the PDU and/or primitive, and reporting (N+1)-PDUs and/or (N)-primitives to the (N+1) Emulating Module if they are generated. The current state is changed to the new state transferred by the transition.

• If the transition sends out any (N)-PDUs and/or (N-1)-primitives, the emulation of this transition is deferred by using the following steps:

• setting the processing-flag of the emulating-status,
• storing the identifier of the received (N)-PDU and/or (N-1)-primitive in the input-id of the emulating-status,
• storing the identifier of the transition focused on in the transition-id of the emulating-status,
• storing the identifier of the outgoing (N)-PDU or (N-1)-primitive expected to be observed in the output-id of the emulating-status, and
• start the processing-timer.

The handling of this transition is performed when a sent (N)-PDU and/or an issued (N-1)-primitive is observed as described below. In the case that the processing-timer expires, this transition is also handled in the same way as the case that (N)-PDU and/or (N-1)-primitive observed.

2. Handling of sent (N)-PDUs and issued (N-1)-primitives
• The module searches for the entry of the Association Control Block corresponding to the PDU and/or primitive.
• The sent (N)-PDU and/or issued (N-1)-primitive values are checked with the parameter variables for the connection.
• For each state in the emulating-status, the following procedure is performed:

• If the processing-flag associated with the state is set, the (N) Behavior Control Module checks whether the output specified by the output-id in the transition specified in the State Transition Table which is identified by the state, the input-id and the transition-id is the sent (N)-PDU and/or (N-1)-primitive focussed on. If so, the output-id is set to the next output in the transition if it exists, or the transition is emulated as described above and the next state is stored in the state of the emulating-status with the processing-flag cleared. After the processing-flag is cleared, the buffer for received (N)-PDUs and (N-1)-primitives is checked and the stored PDUs and/or primitives are processed as described above, if they exist.

If the output is not the same as the PDU and/or primitive focussed on, the state will be deleted from the emulating-status.

• If the processing-flag is not set, the (N) Behavior Control Module searches for the transition in the state which sends out the PDU and/or primitive focussed on. This is performed by looking up all of the entries for the state in the State Transition Table. If such a transition exists, the (N) Behavior Control Module emulates it. This emulation includes performing the corresponding actions in the transition, such as reassembling and resequencing using (N)-SDU buffers, and updating parameter variables corresponding to the parameter values in the PDU and/or primitive. The current state is changed to the new state transferred by the transition. If the transition is generated by (N)-primitive given by the higher layer, the (N) Behavior Control Module reports it to the (N+1) Emulating Module.

If the transition sending out the PDU and/or primitive focussed on does not exist, the state will be deleted from the emulating-status.

• If all the states are deleted from the emulating-status as the result of the above procedure, the (N) Behavior Control Module decides that the PDU and/or primitive will be a protocol error.

3.4.3 Handling When State is not Identified

When the state of (N) layer is not identified, the (N) Behavior Control Module stores the (N)-PDUs and (N-1)-primitives in the buffer for (N)-PDUs and (N-1)-primitives in the Association Control Block. The module also provides the identifier of (N)-PDUs and (N-1)-primitives for the (N) State Identification Module. If the state is identified by the (N) State Identification Module, it returns the state before or after a specific (N)-PDU and/or (N-1)-primitive is observed. The (N) Behavior Control Module starts the emulation from the point where the state is identified using the stored (N)-PDUs and (N-1)-primitives, following to the algorithm described in section 3.4.2.

4 EXAMPLE ON HOW INTELLIGENT OSI PROTOCOL MONITOR WORKS

This section demonstrates how the Intelligent OSI Protocol Monitor works by taking as an example OSI Transport Protocol class 4 over CLNS (connectionless network service) and OSI Session Protocol.

4.1 Examples of Tables and Variables

Tables 1 and 2 show a part of the State Transition Tables maintained in the Transport and Session Behavior Control Modules of the Intelligent OSI Protocol Monitor, respectively. An entry of these tables specifies the transition for an input is received in a state. It includes the output both to the higher layer and to the lower layer, and the next state. If there are more than one alternatives for the transition, they are specified in the entry with the identifier such as (trans1).

The Behavior Control Module also maintains the parameter variables which the protocol defines. The Transport Behavior Control Module maintains the following variables corresponding to the sequence number of TPDU:

- V(S) : the sequence number to be sent next
- V(LS) : the lower window edge for sending
- V(US) : the upper window edge for sending
- V(R) : the sequence number to be received next
- V(LR) : the lower window edge for receiving
- V(UR) : the upper window edge for receiving.

Table 3 shows an example of IO Sequence Table maintained in Transport and Session State Identification Module. The estimated state and before/after flag are described together with each IO sequence. '-' for inputs means that the inputs cannot be observed.

Table 1 State Transition Table for Transport Protocol class 4 (partially)

	CLOSED	*WFCC*	*OPEN*	*CLOSING*	*REFWAIT*
TCONreq	CR WFCC				
CC	DR CLOSED	(trans1) TCONconf, AK OPEN (trans2) TDISind,DR CLOSING	AK OPEN	CLOSING	DR REFWAIT
TDTreq			DT OPEN		
DT	CLOSED	DR CLOSING	(trans1) AK OPEN (trans2) OPEN	CLOSING	REFWAIT
DT (eot)	CLOSED	DR CLOSING	(trans1) TDTind, AK OPEN (trans2) TDTind OPEN	CLOSING	REFWAIT
AK	CLOSED	DR CLOSING	OPEN	CLOSING	REFWAIT
timeout	CLOSED	CR WFCC	DT OPEN	DR CLOSING	CLOSED

eot : End of TSDU
DT : DT TPDU with eot=0
DT (eot) : DT TPDU with eot=1

Table 2 State Transition Table for Session Protocol (partially)

	STA01	*STA01B*	*STA02A*	*STA713*
SCONreq	TCONreq STA01B			
TCONconf		CN STA02A		
AC		(trans1) S-P-ABORTind, AB(nr) STA16 (trans2) S-P-ABORTind,AB(r) STA01A	SCONconf STA713	(trans1) S-P-ABORTind, AB(nr) STA16 (trans2) S-P-ABORTind,AB(r) STA01A (trans3) S-P-EXEPTION-REPORTind,ER STA20

AB(r) : AB SPDU with reuse of transport connection
AB(nr) : AB SPDU with no reuse of transport connection

4.2 Examples of Emulation

We show the emulation performed by the Intelligent OSI Protocol Monitor when the PDU sequence depicted in Figure 3 is observed. This sequence represents a normal connection establishment phase in Transport and Session Protocols.

Table 3 An example of IO Sequence Table

IO sequence	*state*	*before/after*
- / CR	WFCC	after
CC / AK	OPEN	after
DT / AK	OPEN	after
- / AK	OPEN	after
CC / DT	OPEN	after
DT / DT	OPEN	after
- / DT	OPEN	after

(a) Transport

IO sequence	*state*	*before/after*
- / TCONreq	STA01	before
TCONind / TCONresp	STA01	before
TCONconf / CN	STA01B	before

"-" means input cannot be observed

(b) Session

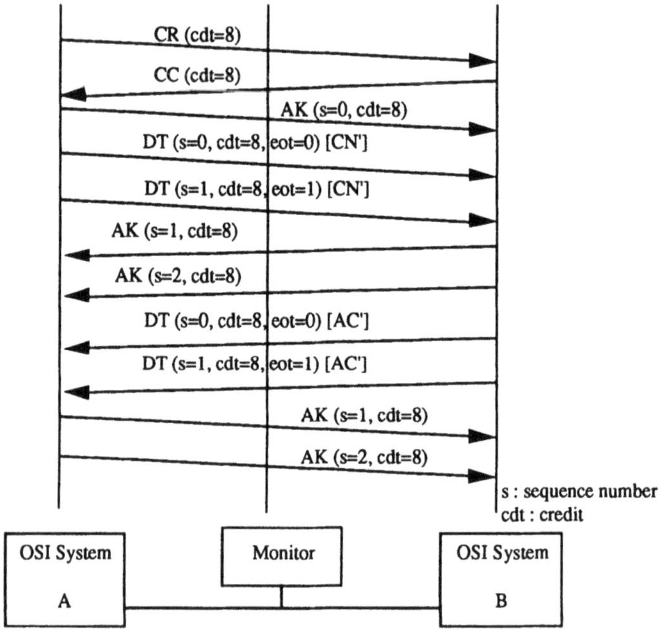

Figure 3 An example of protocol sequences at Transport and Session Protocol

Figure 4 shows how the monitor emulates the behavior of system A, the initiator side. At first, the state is not identified. When CR TPDU is observed, the monitor handles it as a sent PDU. Since the sequence '- / CR' is an IO sequence after observation for state WFCC, the state of Transport layer is identified as WFCC.

Then, the monitor observes the received CC TPDU for system A. Since there are two alternatives for this input, both of which have a sent out TPDU, the following two elements are registered in the emulating-status. They consist of state, processing-flag, input-id, transition-id and output-id.

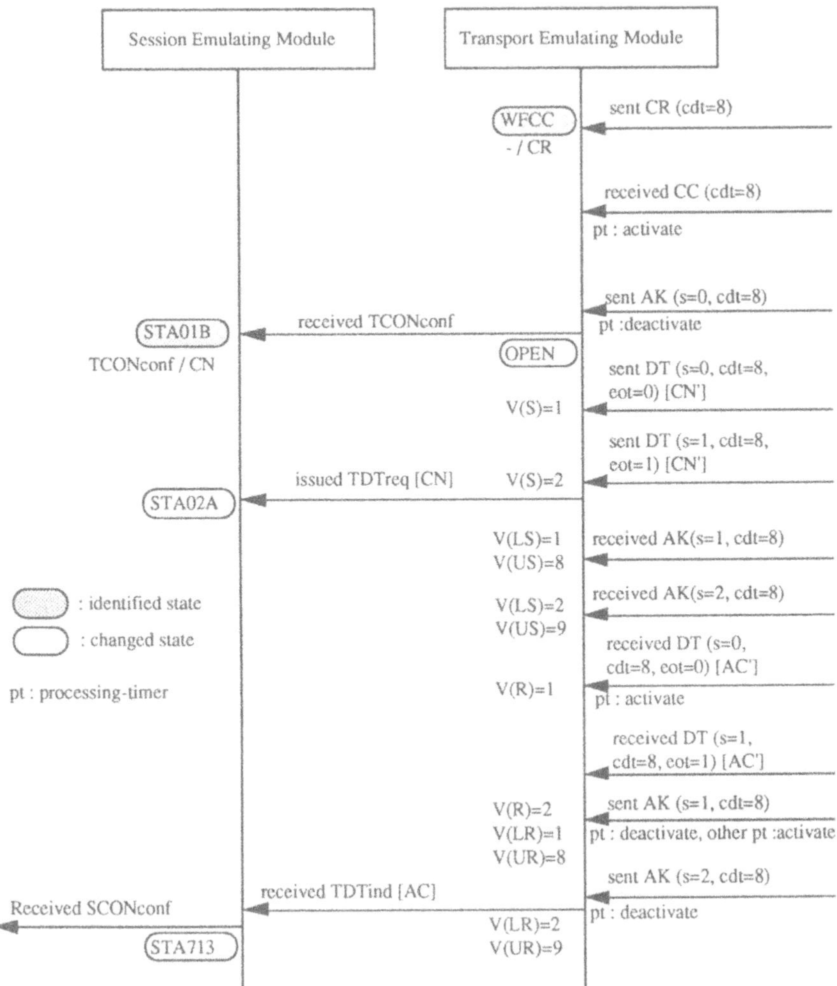

Figure 4 An example of emulation for Transport and Session Protocol

{WFCC, 1, CC, trans1, AK} and {WFCC, 1, CC, trans2, DR}

At this stage, the processing-timer is started in order to wait for the sending of any TPDU.

In the Figure 4, a sent AK TPDU is observed before the processing-timer expires and the monitor emulates that the first element of the emulating-status is selected. A received TCONconf is reported to the Session Emulating Module and the emulating-status is changed to

{OPEN, 0, null, null, null}.

At this time, the state in the Session layer is not identified and TCONconf is not an IO sequence.

Therefore, TCONconf is stored in the buffer.

Then, two DT TPDUs which contain a segmented CN SPDU are observed and V(S) is set to 1 and then to 2. Since the eot parameter of the second DT TPDU is set to 1, the user data is reassembled. At this stage, the Transport Emulating Module will search for the transition which sends DT TPDU in state OPEN and find that the input is TDTreq. TDTreq with CN SPDU is reported to the Session Emulating Module as an issued primitive.

Then, the Session Emulating Module observes the received TCONconf and the sent of TDTreq with CN SPDU and finds that this sequence corresponds to an IO sequence before the observation for state STA01B. The Session Emulating Module determines the state before the received TCONconf is STA01B and performs the emulation for received TCONconf and sent CN. As a result, the state is set to STA02A.

Then, the Transport Emulating Module observes two received AK TPDUs and the V(LS) is updated according to the parameter values. After that, it observes the received DT TPDU and sets V(R) to 1. In this case, two elements of emulated-status,

{OPEN, 1, DT, trans1, AK} and {OPEN, 0, DT, trans2, null},

are registered. Since the next received DT TPDU is observed before the processing-timer expires and next received TPDU is observed, the PDU is stored in the buffer and the handling is deferred. When the sent AK TPDU is observed, the Transport Emulating Module determines the first transition has been selected and handles the DT TPDU stored in the buffer.

Again, the corresponding entry for the DT TPDU has a transition sending out a TPDU, two elements of emulated-status,

{OPEN, 1, DT, trans1, AK} and {OPEN, 0, DT, trans2, null},

are registered. When the last sent AK TPDU is observed, the user data are reassembled and TDTind with is reported to the Session Emulating Module and the variables are updated.

The reported TDTind with AC SPDU is handled by the Session Emulating Module and the received SCONconf is reported to the Presentation Emulating Module and the state is changed.

5 DISCUSSIONS

1. It is considered that our Intelligent OSI Protocol Monitor is applied effectively to the testing of the communication systems which passed the conformance testing. Our monitor can observe the systems' behaviors for a long period and emulate them according to the protocol reference it has by the form of state transition table. If the systems have any errors which are difficult to find, it takes a long period for the errors to appear. Such errors are difficult to find by the conformance testing and our monitor is more appropriate.

2. As described in section 1, some research activities introduce the additional testing programs into the interoperability testing and generate the test sequence based on the generation method of the conformance testing. The test sequence generation method of the conformance testing can be categorized into the transition tour method [6], the UIO (Unique Input Output) sequence method [1] and the state identification method [4]. It is considered that the error detecting capability is the smallest for the transition tour method and the largest for the state identification method. On the other hand, the Intelligent OSI Protocol Monitor can be considered as a testing system which uses uncontrollable sequence as a test sequence. The emulating algorithm adopted

by our monitor trace the behavior of OSI systems according to the state transition table and therefore our algorithm corresponds to the transition tour method. We think that this is reasonable for the monitoring based testing method because of the following reasons:

- Since the sequence is not controllable, the UIO sequence and the distinguishing sequence cannot be applied.
- Our monitor can detect errors by emulating the behavior of a system for a long period and by finding the difference between the system and the reference.

3. There is a delay between the time system A sends a PDU and the time system B receives the PDU, and the monitor captures the PDU in the different timing from them. Therefore, the order of PDUs in which a system actually sends or receives might be different from the order in which the monitor detects.

We handle the PDU crossing depicted in Figure 5 in the following way. In case (a), the Transport Emulating Module of system B understands that CC TPDU is sent out by received CR TPDU in one transition and stores crossed DR TPDU in the buffer. However, in case (b), the Transport Emulating Module of system A detects CC TPDU at WFCC and it differs from the behavior of system A. Then, the Transport Emulating Module would estimate the emulating-status, {WFCC, 1, CC, trans1, AK} and {WFCC, 1, CC, trans2, DR}, and, since the DR TPDU observed next has different reason parameter value from expected DR TPDU, it would be considered as a protocol error. This might be resolved by reordering the emulation for received CC TPDU and sent DR TPDU. In the actual environment, more than one PDUs might cross as shown in case (c). In this case the Transport Emulating Module must consider the possibility of many cases. In order to cope with such a situation, it might be necessary to consider the propagation delay between the monitor and systems and to reordering the emulation.

4. When a protocol error has occurred, the state and variables will be reset or will not be reset according to how severe it is. If the protocol error has been caused by sending an invalid parameter value, only the corresponding variable might be reset, but if it has been caused by sending an invalid type of PDU, the state and all variables would be reset. When a protocol error has occurred in (N) layer, it need to be determined to report the error to (N-1) layer and/or (N+1) layer also for individual errors.

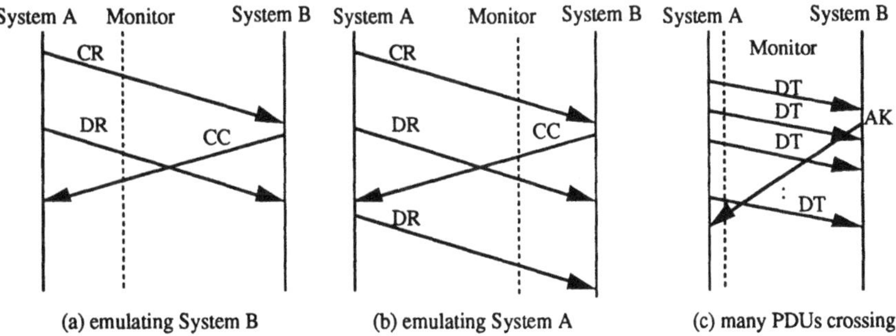

Figure 5 An example of PDU crossing

6 CONCLUSION

In this paper, we have described Intelligent OSI Protocol Monitor which observes PDUs exchanged between OSI system and analyzes the protocol behaviors as well as the PDU format according to the protocols of OSI 7 layers. This monitor would be very useful in the interoperability testing since it can detect protocol errors which cannot be detected by the conformance testing by monitoring the actual communication for a long period.

This monitor emulates the behavior of an OSI system based on the layered structure model. It has the Emulating Module for each layer which emulates the behavior of the protocol in the layer based on the state and the state transition table. This emulation is invoked by the observation of PDUs and all the functions defined in the protocol are emulated. In order to cope with the case when the state of the OSI system is not identified, the monitor provide state identification mechanism by use of IO sequence.

This paper have described the detailed design of the Intelligent OSI Protocol Monitor including the structure and the emulating algorithm. It has also demonstrated how the monitor works for the actual OSI protocols by taking OSI Transport Protocol class 4 over CLNS (connectionless network service) and OSI Session Protocol as examples.

7 ACKNOWLEDGEMENT

The authors wish to thank Dr. Y. Urano, Director of KDD R&D Laboratories, for his continuous encouragement of this study.

8 REFERENCES

[1] Bosik, B.S. and Uyar, M.U. (1991) Finite state machine based formal methods in protocol conformance testing: from theory to implementation, Computer Networks and ISDN Systems, vol 22.

[2] CCITT (1988) Recommendation X.225 - Session Protocol Specification for Open Systems Interconnection for CCITT Applications.

[3] ITU-T (1993) Recommendation X.224 - Protocol for Providing the OSI Connection-Mode Transport Service.

[4] Hennie, F.C. (1964) Fault-detecting experiments for sequential circuits, Proc. of 5th Ann. Symp. on Switching Circuit Theory and Logical Design, 95-110.

[5] Kato, T. and Suzuki, K. (1993) Development of OSI 7 Layer Link Monitor, Proc. of the 46th Annual Convention IPS Japan, vol.1, 211-212.

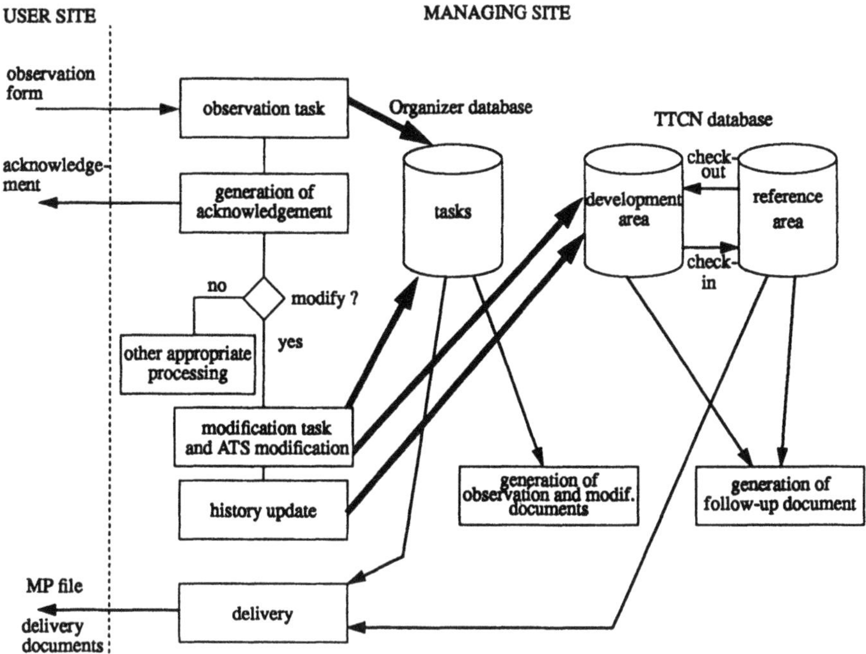

Figure 1 The Numerix solution

3.1 Observation tasks

The information contained in a user site observation form is reported manually within Numerix by the means of an observation task.

The TTCN tables mentioned by the observation form are referenced in the task by hypertext links set by the user. Their contents are thus directly accessible from the task.

The managing site specifies in this task how the observation will be processed, for example:
- do the modification,
- first do a complementary study,
- forward to concerned bodies,
- no processing.

Several observations issued from different user sites may concern the same tables on the same subject. Such observations are called **associated** and lead to a unique processing. These associations are implemented by hypertext links in the *associated observations* field.

The Numerix tool can be used to produce a document from an observation task. This document, including the answer elements, can be sent to the issuing site as an acknowledgement.

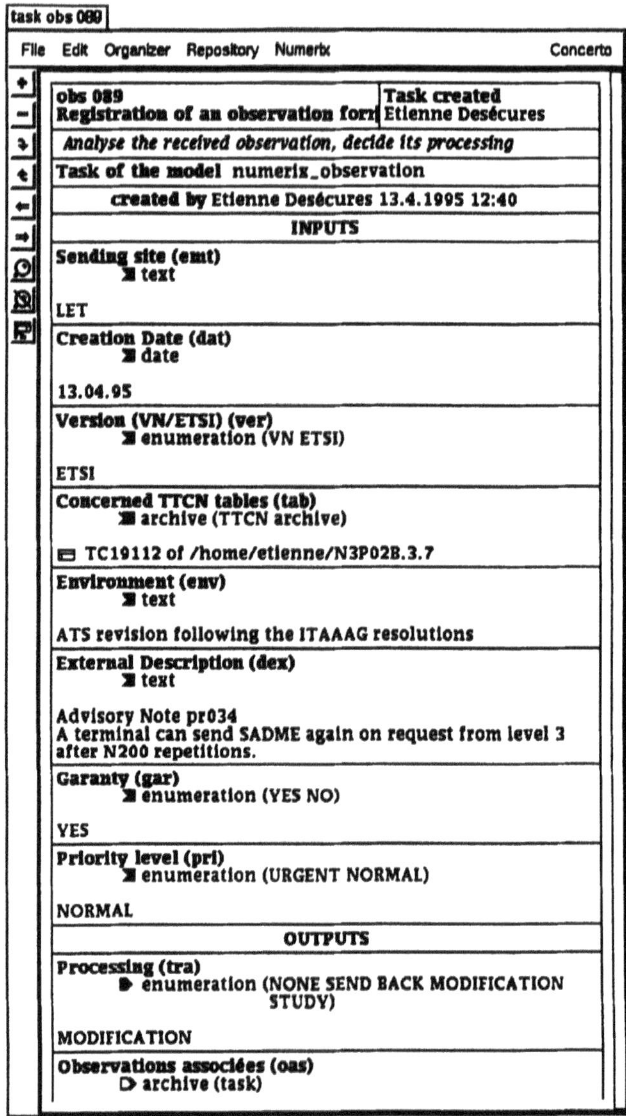

task obs 089

File Edit Organizer Repository Numerix Concerto

obs 089 Registration of an observation form	Task created Etienne Desécures

Analyse the received observation, decide its processing

Task of the model numerix_observation

created by Etienne Desécures 13.4.1995 12:40

INPUTS

Sending site (emt)
text

LET

Creation Date (dat)
date

13.04.95

Version (VN/ETSI) (ver)
enumeration (VN ETSI)

ETSI

Concerned TTCN tables (tab)
archive (TTCN archive)

TC19112 of /home/etienne/N3P02B.3.7

Environment (env)
text

ATS revision following the ITAAAG resolutions

External Description (dex)
text

Advisory Note pr034
A terminal can send SADME again on request from level 3 after N200 repetitions.

Garanty (gar)
enumeration (YES NO)

YES

Priority level (pri)
enumeration (URGENT NORMAL)

NORMAL

OUTPUTS

Processing (tra)
enumeration (NONE SEND BACK MODIFICATION STUDY)

MODIFICATION

Observations associées (oas)
archive (task)

Figure 2 Example of an Observation task description form

3.2 Modification tasks

Every modification undertaken on an ATS is declared in a **modification task**. The various fields contain:

• hypertext links to the tables to be corrected and also to the tables effectively corrected,
• hypertext links to the observation task which initiates the processing, and its associated observation tasks,
• a complete description of the modifications carried out.

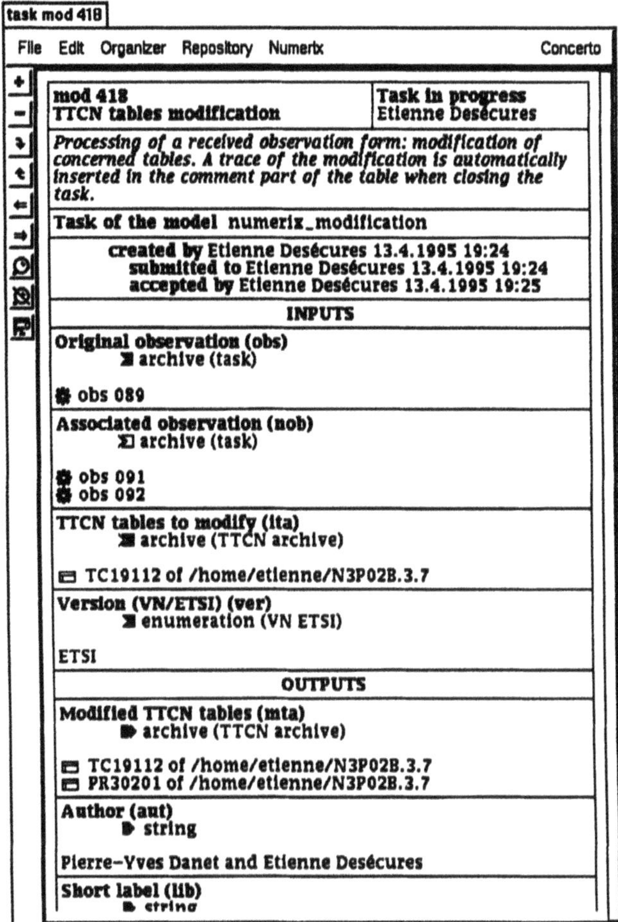

Figure 3 Example of a Modification task description form

The "create modification" command is used to create a modification task from an observation task. The elements of the observation task are automatically reported in the input fields of the modification task.

The Numerix tool can be used to create a document from a modification task.

3.3 History lists

The complete history list of a given table is automatically built by Numerix:
every completion of a modification task inserts automatically a history statement in the *Detailed Comments* field of the modified TTCN table taken from the modification task.

<table>
<tr><th colspan="6">Test Case Dynamic Behaviour</th></tr>
<tr><td colspan="6">Test Case Name: TC19114
Group: /ISDN/Basic_call/Successful/Speech/
Purpose: GLOBAL CALL REF. - STATE R1 - PASSIVE IUT BEHAVIOUR SYNTACTICALLY INVALID TEST EVENTS
Configuration:
Default: DF69902
Comments:</td></tr>
<tr><th>Nr</th><th>Label</th><th>Behaviour Description</th><th>Constraints Ref</th><th>Verdict</th><th>Comments</th></tr>
<tr><td>1</td><td></td><td>+ PR39003</td><td></td><td></td><td>preamble to R0</td></tr>
<tr><td>2</td><td></td><td>L!ERROR START TAC</td><td>ERR225</td><td></td><td>PDU RESTART ACK</td></tr>
<tr><td>3</td><td>L1</td><td>L?STATUSr [(STATUS.CST.CSTV = 61) AND (STATUS.CAU.CVAL = 96)] CAN-
CEL TAC</td><td>ST7 (0)</td><td>(P)</td><td>valid STATUS</td></tr>
<tr><td>4</td><td></td><td>+ CS59003 (61 , 1)</td><td></td><td></td><td></td></tr>
<tr><td>5</td><td></td><td>+ UM59902</td><td></td><td></td><td>unexpected message</td></tr>
<tr><td>6</td><td></td><td>GOTO L1</td><td></td><td></td><td></td></tr>
<tr><td colspan="6">Detailed Comments:
The test step CS59003 is used for checking the IUT state R1.
References to Recommendations
ETS 300 102 5.8.7.1

*HV | 28.2.94 | 1.0 | mod 087 | obs 047 | N3P02 | Preamble to R0 is now PR39003
*HV | 12.4.95 | 1.1 | mod 293 | obs 113 | N3P02 N3B02 | Purpose modified
*HV | 13.4.95 | 1.1 | mod 297 | obs 119 | N3P02 N3B02 | 61 intead of 71</td></tr>
</table>

Figure 4 Example of a TTCN table with a history list which automatically results from modification form processing

Each statement is composed of the modification task main elements: date, version, observation and modification task numbers, embedding suites, short description of the modification. The history list which appears in each modified table can be immediately consulted in every representation of the table (MP, screen, paper). Obviously, these statements do not garble actual comments, which remain modifiable.

The Numerix tool can be used to create a document containing the history list in a more readable way from a given TTCN table.

3.4 Delivery processing

With a single command the Numerix tool allows the production of:

- a delivery document
- a Unix file containing the suite MP.

A modification delivery document contains the set of modification forms that initiated the delivery. A full ATS delivery document contains the GR format of the whole ATS delivered, followed by the list of modification forms. In both cases, modifications are described starting at a chosen version of the ATS ; it may be the last one delivered or a previous one, so that specific deliveries can be done for particular user sites.

Generally the delivery of a suite makes official a new version of that suite. It is important to freeze this new version. Then it is exported to a specific Concerto base and can be consulted through the network by the using sites.

3.5 Version management and ATS composition

The TTCN database is split in two separate areas:
• the "Reference Area",
• the "Development Area".

The reference area contains the stable versions of the ATSs while the ATSs under development or modification are stored in the development area. ATSs in the reference area are frozen: they are write-protected. A modification to a frozen suite implies first the creation of an unfrost copy placed in the development area with a new version number. This operation is called "check-out". The "check-in" operation consists in moving the modified ATS from the development area to the reference area and to freeze it. The link between an ATS version and the previous version is kept. This mechanism, supported by the Configurator environment, is widespread in Concerto and is also applicable to shared fragments.

ATSs can be composed of fragments shared by other suites. These fragments are called ATS components. The current practice at CNET is to use components at the table level, but it could be done at any other level, e.g. Test Groups. Modifications to tables usually lead to the creation of one or several versions of the tables. Before delivering an ATS it is necessary to update it, so that it refers the appropriate, usually the newer, version of every component. This operation is assisted by a special command allowing to choose in a matrix which version of each component is to be taken. The updated ATS can then be frozen and constitutes a stable version which can be delivered and exported to other sites.

4 CONCLUSION

The Numerix solution installation is in progress at CNET. It is too early to make definite conclusions but it can be said that:
• The saving of all the significant events in the organizer database and the facilities to query this database is appreciated by the users,
• The sharing of tables between ATSs allows the reduction of the number of managed tables from 60 000 to 20 000.
• The use of simple and familiar concepts (observation, modification, etc.) enables the users to be operational very quickly,
• Numerix has been designed to be, as far as possible, not too much linked to the method and parameterizable. It could be easily adapted to support another method.

5 REFERENCES

ISO/9646-1 : 1991, *Information technology - Open Systems Interconnection - Conformance testing methodology and framework - Part 1: General concepts*

ISO/9646-2 : 1991, *Information technology - Open Systems Interconnection - Conformance testing methodology and framework - Part 2: Abstract test suite specification*

Pierre-Yves Danet : *Management and maintenance of TTCN Abstract Test Suites ;* OSTC - 9 March 1995

Concerto/Numerix User Manual v1.0

PART NINE

Distributed Testing and Performance

24

Port-synchronizable test sequences for communication protocols

K. C. Tai [*] and Y. C. Young [*#]
Department of Computer Science, North Carolina State University
Raleigh, North Carolina, 27695-8206, USA
Tel: (919) 515-7146, Fax: (919) 515-7896
e-mail: kct@csc.ncsu.edu, yyoung@vnet.ibm.com

Abstract

In conformance testing of a communication protocol, the synchronization between inputs from different testers for the protocol becomes a problem. A **synchronizable test sequence** of a finite state machine (FSM) is a test sequence for which the synchronization problem either does not exist or can be solved by communication between testers. In this paper, for a multi-port FSM with one tester for each port, we define a necessary and sufficient condition under which a test sequence of the FSM does not have the synchronization problem. Such a test sequence is called a **port-synchronizable test sequence**. Our empirical results show that an FSM may contain many port-synchronizable test sequences that are not synchronizable according to a previous definition of a synchronizable test sequence.

Keywords

Protocol testing, synchronizable test sequences, finite state machines

* This work was supported in part by the US National Science Foundation under grant CCR-9309043.

The author is also with IBM, Raleigh, North Carolina.

1 INTRODUCTION

The finite state machine (FSM) model is commonly used for specifying communication protocols. The problem of generating test sequences based on an FSM has been studied for about two decades (Tarnay, 1991) (Sarikaya, 1993). When an implementation of an FSM is tested for conformance, test sequences are derived from the FSM, and testers (or test drivers) for the implementation are constructed according to these test sequences. With the use of multiple testers for an FSM, the synchronization between inputs from different testers becomes a problem. A **synchronizable test sequence** of an FSM is a test sequence for which the synchronization problem either does not exist or can be solved by communication between testers. In (Sarikaya and Bochmann, 1984) the synchronization problem for an FSM with two testers was discussed, and a type of synchronizable test sequence that does not have the synchronization problem was defined. In recent years, several issues on synchronizable test sequences have been studied.

Due to the existence of distributed database systems and communication networks, FSMs with multiple ports are needed to specify protocols (Luo et al., 1993). Also, the use of multi-port FSMs makes the design of communication protocols flexible. Protocol specification languages such as LOTOS, Estelle, and SDL (Turner, 1993) allow the use of multiple ports. In this paper we define a necessary and sufficient condition under which a test sequence of an FSM with multiple ports does not have the synchronization problem. Such a test sequence is called a **port-synchronizable test sequence**. Based on our new definition, more test sequences of an FSM become synchronizable.

This paper is organized as follows. Section 2 provides basic definitions. Section 3 summarizes previous work on synchronizable test sequences of an FSM. Section 4 gives our motivation for extending the definition of a synchronizable test sequence in (Sarikaya and Bochmann, 1984). Section 5 shows how to construct a set of testers from a test sequence, based on the assumption of **port-based testing**, which does not allow different testers for an FSM to communicate with each other. Also, section 5 defines the port-based synchronization problem. Section 6 gives the definition of a port-synchronizable test sequence of an FSM and shows that this definition provides a necessary and sufficient condition under which a test sequence of an FSM does not have the port-based synchronization problem. Section 7 discusses the generation of port-synchronizable test sequences of an FSM. Section 8 shows the results of our empirical studies on synchronizable test sequences. Section 9 concludes this paper.

2 PRELIMINARIES

Below we provide a formal definition of a multi-port FSM, which is different from that in (Luo et al., 1993).

Definition. A **finite state machine** (FSM) M with multiple ports is defined as a 6-tuple M=(S, I, O, T, U, s0), where

- S is the set of states of M

- I is the set of input symbols of M. Each input symbol is of the form P:A, where P denotes a port and A an input message.
- O is the set of output symbols of M. Each output symbol is of the form Q:B, where Q denotes a port and B an output message.
- T is the transition function of M, which maps from D to S, where $D \subseteq S \times I$. (Thus, M is **deterministic**.)
- U is the output function of M, which maps from D to $(O1 \times O2 \times ... \times Ov) \cup \{ \varepsilon \}$, where each Oi, 1<=i<=v, is an output symbol, and ε stands for the empty output. (The output of a transition may contain two or more output symbols with the same port name.)
- s0 is the initial state of M.

In the above definition, each input or output symbol contains two elements: port name and message. The separation of port name and message makes it easier to discuss synchronizable test sequences. If the transition function of an FSM allows a state to have a transition on an input, then this input is said to be **valid** for the state. The output function of an FSM allows zero, one, or more output symbols to be associated with a transition. An FSM communicates with its environment by receiving input symbols and sending output symbols via ports. Each port of an FSM has two unbounded FIFO queues: the input queue, which keeps input messages of the FSM, and the output queue, which keeps output messages of the FSM. It is assumed that the delivery of messages from an FSM to any of its port is FIFO, i.e., messages sent from M to the same port are received in the order sent. For an FSM with two ports, its two testers are commonly referred to as the lower and upper testers. In this paper, the lower and upper testers are referred to as the **L-tester** and the **U-tester**, respectively and the two ports connected to the L- and U-testers are referred to as ports $\mathbf{P_L}$ and $\mathbf{P_U}$, respectively.

Assume that during an execution of an FSM and its environment, the current state of the FSM is S. For a transition T of S, if the first input message at the input port of S is valid for S, then T is said to be **eligible** for S. If S has two or more eligible transitions, then one of them is chosen at random for execution. If S does not have any eligible transitions, then the FSM waits until at least one eligible transition for S becomes available.

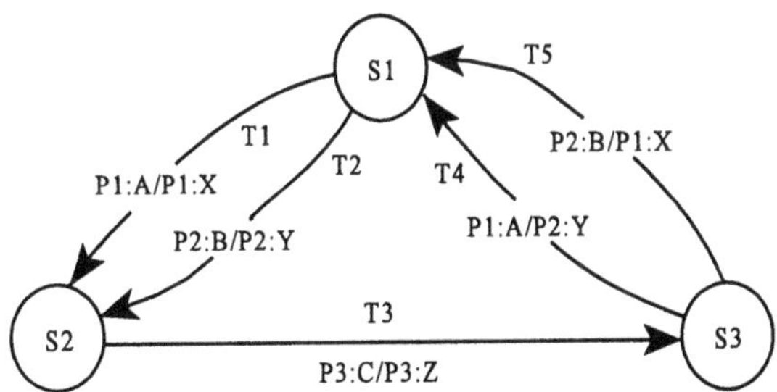

Figure. 1. FSM M1 with Ports P1, P2, and P3

A **transition** T from state Si to state Sj that has P:A as the input symbol and (Q1:B1,Q2:B2,...,Qv:Bv) as the sequence of output symbols is denoted as (Si, Sj, P:A/(Q1:B1,Q2:B2,...,Qv:Bv), and Si is referred to as the **head state** of T or **head**(T), Sj the **tail state** of T or **tail**(T), P the input port of T, A the input message of T, Qi , 1<=i<=v, an output port of T, and Bi an output message of T. P:A/(Q1:B1,Q2:B2,...,Qv:Bv) is referred to as the **label** of T or **label**(T). For the sake of simplicity, if v = 1, P:A/(Q1:B1) is referred to as P:A/Q1:B1. If T has no output symbols, it is denoted as (Si, Sj, P:A/ε). A port is said to be **involved** in a transition T if this port is either the input port or an output port of T.

For an FSM M, let $D_M = (V_M, E_M)$ denote the **digraph** of M, where V_M is a set of vertices, each representing a state of M, and E_M is a set of directed edges, each representing a transition of M. Figure 1 shows the digraph of FSM M1, which has three ports P1, P2, and P3, and three states S1, S2 and S3, with S1 being the initial state. A, B and C are input messages of ports P1, P2 and P3, respectively, and X, Y and Z are output messages of ports P1, P2 and P3, respectively. A tabular representation of M1 is given below, which has inputs as rows and states as columns.

	S1	S2	S3
P1:A	P1:X, S2 (T1)		P2:Y, S1 (T4)
P2:B	P2:Y, S2 (T2)		P1:X, S1 (T5)
P3:C		P3:Z, S3 (T3)	

Each non-empty entry in the above table defines a transition for the corresponding state and input symbol; it contains the output symbol and tail state of this transition, followed by a pair of parentheses enclosing the name of this transition.

Let M denote an FSM. A **transition sequence** of M is a sequence of consecutive transitions in M, and a **test sequence** of M is a transition sequence of M starting from the initial state of M. In this paper, a transition or test sequence is often denoted as a sequence of transition names connected by ".". For example, T1.T3.T5 and T4.T2.T3 are transition sequences of M1 in Figure 1, and the former is a test sequence of M1, but the latter is not. For a transition sequence, its **i/o sequence** refers to the sequence of labels associated with the transitions in the transition sequence. Since a test sequence of M starts with the initial state of M, it does not need to keep information about the head and tail states of transitions in the test sequence. Thus, a test sequence of M can be denoted by its i/o-sequence. For example, T1.T2.T3 of M can be denoted by its i/o sequence is (P1:A/P1:X, P3:C/P3:Z, P2:B/P1:X). Similarly, a transition sequence of M that starts from a given state can be denoted by its i/o sequence. For a transition sequence E of an FSM, if a transition T occurs (immediately) before a transition T' in E, then T is said to be a (the immediate) **predecessor** of T' in E and T' a (the immediate) **successor** of T in E.

A digraph D = (V,E) is **strongly connected** if for every pair of vertices Vi and Vj in V, there exists a path from Vi to Vj. An FSM is said to be strongly connected if its digraph is strongly connected. An FSM is said to be **completely specified** if each state of this FSM has a transition for every possible input symbol. For a state S of an FSM, let IO(S) be the set of i/o sequences that start from S. An FSM is said to be **minimal** if for any two states S and S' of the FSM, IO(S) ≠ IO(S'). In this paper, we assume that each FSM is deterministic, strongly connected, minimal, and possibly incompletely specified, unless otherwise specified.

A **postman tour** (or **transition tour**) of a digraph D = (V,E) is a path in D that starts and ends at the same vertex and covers each edge in E at least once. A postman tour of an FSM is a postman tour in the digraph of this FSM. A digraph D has a postman tour if and only if D is strongly connected.

For a state S in an FSM M, a **unique input/output** (UIO) sequence E is a transition sequence starting from S such that if the sequence of input symbols of E is applied to a state in M other than S, the sequence of corresponding output symbols is different from the sequence of output symbols of E. In other words, an UIO sequence for S can distinguish S from other states in M. The UIO- or U-method for testing an FSM verifies the tail state of a transition of the FSM by using an UIO sequence for the tail state (Sabnani and Dahbura, 1988).

3 PREVIOUS WORK ON SYNCHRONIZABLE TEST SEQUENCES OF AN FSM

(Sarikaya and Bochmann, 1984) considered the synchronization problem for an FSM with two ports (for the L- and U-testers, which do not communicate with each other). Two consecutive transitions are said to have the **synchronization problem** if the input port of the second transition is not involved in the first transition (i.e., the input port of the second transition is neither the input port nor an output port of the first transition.) A test sequence is said to be **synchronizable** if no two consecutive transitions of the test sequence have the synchronization problem. (Sarikaya and Bochmann, 1984) also discussed how to extend existing test sequence generation methods in order to generate synchronizable test sequences. In order to distinguish the type of synchronizable test sequence defined in (Sarikaya and Bochmann, 1984) from other types of synchronizable test sequences, the former is referred to as a **pair-synchronizable test sequence** in the remainder of this paper.

(Chen et al., 1990) defined a **tightly synchronizable test sequence** of a two-port FSM as a test sequence such that for any two consecutive transitions, the input port of the second transition is the same as the output port of the first transition. (Each transition was assumed to have at most one output symbol.) They showed how to construct a graph, called the **duplex digraph**, of a two-port FSM such that a test sequence of the duplex graph is a tightly synchronizable test sequence of the FSM.

(Boyd and Ural, 1991) investigated complexity issues related to pair-synchronizable test sequences of a two-port FSM. They presented a necessary and sufficient condition for the existence of a pair-synchronizable postman tour of an FSM, and this condition can be determined in polynomial time. Also, they showed that the problem of finding a minimum-length pair-synchronizable postman tour of an FSM is NP-complete.

(Ural and Wang, 1993) considered two-port FSMs satisfying a number of conditions, including the necessary and sufficient condition in (Boyd and Ural, 1991) and the condition that each state in the FSM possess two UIO sequences, one for the L-tester and the other for the U-tester. By modifying the algorithm in (Chen et al., 1990) for the construction of a duplex digraph and by providing additional algorithms, they showed how to find a pair-synchronizable UIO-based test sequence of an FSM in polynomial time. (Guyot and Ural , 1995) extended the work by showing the construction of a digraph of an FSM M such that all paths in the digraph are pair-synchronizable test sequences of M and by showing that under certain conditions, a pair-synchronizable test sequence of M can distinguish M from any FSM not isomorphic to M.

(Chen and Ural, 1995) allowed the L- and U-testers for a two-port FSM to communicate with each other, and they considered the cost of such communication in test sequence generation. They showed how to convert a non-pair-synchronizable test sequence into a synchronizable test sequence by adding communication statements between the L- and U-testers. Such synchronizable test sequences are referred to as **LU-synchronizable test sequences** in this paper. Based on the duplexU digraph of an FSM, a minimum-cost LU-synchronizable test sequence using multiple UIO sequences can be generated. Since the generation of such a test sequence is an NP-complete problem, they proposed a heuristic algorithm that yields an LU-synchronizable UIO-based test sequence with its cost within a bound of the minimum cost.

(Luo et al., 1993) discussed the need for FSMs with multiple ports and extended the definition of a pair-synchronizable test sequence for a multi-port FSM. They also investigated the issue of fault coverage by pair-synchronizable test sequences.

The use of synchronizable test sequences is necessary for conformance testing of an implementation of an FSM, since the implementation is treated as a black box. When the implementation is tested by its developers, it can be modified to perform deterministic testing and thus make every test sequence synchronizable. Details on deterministic testing and debugging of concurrent programs can be found in (Tai and Ahuja, 1987) (Carver and Tai, 1991) (Tai et al., 1991) (Tai and Carver, 1995).

4 MOTIVATION FOR EXTENDING THE DEFINITION OF A PAIR-SYNCHRONIZABLE TEST SEQUENCE

In this section, we show that the definition of a pair-synchronizable test sequence is more restrictive than necessary and that some non-pair-synchronizable test sequences are actually synchronizable. Below we first give a general definition of a synchronizable test sequence.

Definition. A test sequence for an FSM M is said to be **synchronizable** if any execution of M and the testers generated according to the test sequence is deterministic (i.e., at any time during an execution of M and these testers, the current state of M has at most one eligible transition).

Notes:

- In this paper, the issue of fault detection is not discussed. Therefore, we consider executions of M and its testers, not executions of an implementation of M and its testers.
- Whether a test sequence of M is synchronizable depends on the test sequence, M, and the construction of testers for the test sequence.

According to (Sarikaya and Bochmann, 1984), two consecutive transitions are said to have the

synchronization problem if the input port of the second transition is not involved in the first transition. In the remainder of this paper, this synchronization problem is referred to as the **pair-based synchronization problem**. To illustrate this problem for a two-port FSM, we first show how the L- and U-testers work for two consecutive transitions T1 and T2. Assume that the input port of T2 is port P_L. The following are three possible relationships between port P_L and T1:

(a) Port P_L is an output port of T1. In this case, the L-tester sends the input message of T2 to port P_L immediately after receiving the output message of T1 from port P . No synchronization problem exists.

(b) Port P_L is not an output port of T1, but is the input port of T1. In this case, the L-tester sends the input message of T2 to port P_L immediately after sending the input message of T1 to port P_L. Thus, the input message of T2 will be received after the completion of T1. No synchronization problem exists.

(c) Port P_L is neither the input port nor an output port of T1 (i.e., only port P_U is involved in T1). In this case, the L-tester sends the input message of T2 to port P_L without any dependency on T1. According to (Sarikaya and Bochmann, 1984), this case creates the pair-based synchronization problem.

Now we examine the effect of case (c). Let S1 and S2 be the head and tail states of T1, respectively. Assume that S1 is the current state, the input message of T2 is the first input message at port P_L, and the input message of T1 is the first input message at port P_U. Also, we assume that S1 contains a transition, say T3, that has the same input port and message as T2 (see below).

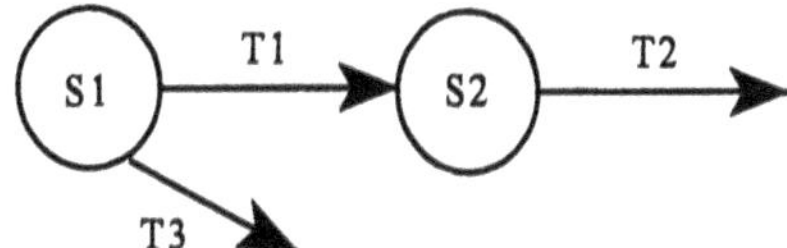

Since the first input messages at both ports P_L and P_U are valid for S1, both transitions T1 and T3 are eligible for S1. If the first input message at port P_L is accepted (i.e., transition T3 is executed), then a synchronization problem occurs, since T1, not T3, is expected to be executed. To prevent this problem to occur, the definition of a pair-synchronizable test sequence in (Sarikaya and Bochmann, 1984) does not allow case (c). As a result, at any time during an execution of a two-port FSM by using a pair-synchronizable test sequence, at most one of ports P_L and P_U contains input messages.

Now we assume that transition T3 does not exist (i.e., state S1 does not contain any transition with the same input symbol as T2). When S1 is the current state, the arrival of the input message of T2 at port P_L does not create the pair-based synchronization problem since the input symbol of T2 is invalid for S1. Therefore, we can allow both ports P_L and P_U to contain input messages as long as only one eligible transition exists for the current state. Following the above discussion, the definition of a pair-synchronizable test sequence is more restrictive than necessary, and some non-pair-synchronizable test sequences are actually synchronizable. Note that the definition of a pair-synchronizable test sequence is based solely on the test sequence. Our new concept of a synchronizable test sequence of an FSM is based on not only the test sequence, but also the transitions of the states in the FSM that are passed through by the test sequence.

Since the multi-port FSM model becomes more commonly used and is more general than the two-port FSM model, we define our new synchronizable test sequences for the multi-port FSM model. In the remainder of this paper, an FSM is assumed to have multiple ports, unless otherwise specified. There are two testing strategies for an FSM: **port-based testing**, which does not allow testers for the FSM to communicate with each other, and **group-based testing**, which divides the ports of the FSM into groups and allows the testers for ports in the same group to communicate with each other. In this paper, we focus on port-based testing.

5 PORT-BASED TESTERS AND SYNCHRONIZATION PROBLEM

In this section, we first show the generation of a set of testers from a test sequence, according to port-based testing. Such testers are referred to as **port-based testers**. We then define the port-based synchronization problem, according to port-based testers..

The port-based tester for port P of an FSM M according to a test sequence E of M, denoted as Tester(P,E,M), contains the following two types of statements:

- send A to P;
- receive B from P;

where A is an input message of M, B is an output message of M, a send operation is non-blocking, and a receive operation is blocking. For the sake of simplicity, Tester(P,E,M) is referred to as Tester(P) if E and M are either implied or immaterial, and as Tester(P,E) if M is implied. When "receive B from P" is executed by Tester(P), if the first output message of M at port P is not B, then Tester(P) has an abnormal termination. It is assumed that the delivery of messages from Tester(P) to port P is FIFO, i.e., messages sent from Tester(P) to port P are received in the order sent. In section 2, the FSM model assumes that each transition is associated with an input symbol and zero, one, or more output symbols. Thus, a transition in an FSM can be viewed as a receive statement, followed by zero, one, or more send statement. For the sake of simplicity, a tester is defined as a sequence of send and receive statements, not as a sequence of transitions. If necessary, a sequence of send and receive statements can be converted into a sequence of transitions.

Algorithm Port_Tester_Gen

Input: a test sequence E of an FSM M

Output: Tester(E), which is {Tester(P,E) | P is a port involved in E}

(1) For each port P involved in E, let Tester(P,E) be empty.

(2) For each transition T in E (starting from the first transition in E),

  (a) if the input symbol of T is P:A, then add "send A to P" to the end of Tester(P,E).

  (b) if the list of output symbols of T is (Q1:B1,Q2:B2,...,Qv:Bv), then

for each i from 1 to v, add "receive Bi from Qi" to the end of Tester(Qi,E).
end of algorithm.

Consider test sequence R1 = T1.T3.T4 of M1 , which is shown in Figure 1. By applying the above algorithm to T1.T3.T4, Tester(R1) contains the following testers for ports P1, P2, and P3:

Tester(P1,R1)	Tester(P2,R1)	Tester(P3,R1)
(T1) send A to P1	(T4) receive Y from P2	(T3) send C to P3
(T1) receive X from P1		(T3) receive Z from P3
(T4) send A to P1		

(To improve readability, each send or receive statement in a tester is preceded with the name of the transition from which the statement is derived.) R1 is not pair-synchronizable since the input port of T3, which is P3, is not involved in T1 (i.e., P3 is neither the input port nor an output port of T1), and the input port of T4, which is P2, is not involved in T3. When an execution of M1 and the above testers starts, state S1 is the current state. When a transition of S1 is to be chosen, it is possible that messages A and C are available at ports P1 and P3, respectively. However, only P1:A is valid for S1 and thus only transition T1 is eligible for S1. After the execution of T1, state S2 becomes the current state of M1. When a transition of S2 is to be chosen, it is possible that messages A and C are available at ports P1 and P3, respectively. However, only P3:C is valid for S2 and thus only transition T3 is eligible for S2. Although R1 is not pair-synchronizable, it does not have the problem of synchronizing inputs form different testers. Below we formally define the synchronization problem based on the use of port-based testers. In the next section, we show how to solve this synchronization problem.

Definition. Let E be a test sequence of an FSM M and let Tester(E) be the set of testers generated by algorithm Port_Tester_Gen according to E. During an execution of M and Tester(E), the **port-based synchronization problem** occurs when a state of M has two or more eligible transitions. (As a result of this problem, the execution of M and Tester(E) is nondeterministic and may result in an abnormal termination.)

6 PORT-SYNCHRONIZABLE TEST SEQUENCES OF AN FSM

Definition. For a transition T in a transition sequence E of an FSM, its **port-predecessor transition**, denoted as **PPT**(T,E), is the closest predecessor of T in E that involves the input port of T. If such a predecessor of T in E does not exist, then PPT(T,E) is defined as null.

Let T be a transition in a test sequence E of an FSM M and let P:A be the input symbol of T. According to algorithm Port_Tester_Gen, the statement "send A to port P" appears in Tester(P,E). Consider the following two cases:

- PPT(T,E) is null. In this case, "send A to port P" is the first statement in Tester(P,E). During an execution of M and Tester(E), the statement "send A to port P" causes the port-based synchronization problem to occur if and only if P:A is a valid input for one of the states entered before the head state of T.

- PPT(T,E) is not null. In this case, "send A to port P" in Tester(P,E) appears immediately after the statements generated for PPT(T,E). Thus, T's input symbol arrives at M only after the completion of PPT(T,E). During an execution of M and Tester(E), the statement "send A to port P" causes the port-based synchronization problem to occur if and only if P:A is a valid input for one of the states entered after the head state of PPT(T,E) and before the head state of T.

Definition. For a transition T in a transition sequence E of an FSM, its **port-predecessor interval**, denoted as PPI(T,E), is defined as

the sequence of transitions in E after PPT(T,E) and before T,	if PPT(T,E) ≠ null, or
the sequence of transitions in E before T,	if PPT(T,E) = null.

Notes:

- If T is the first transition in E, then PPT(T,E) = null and PPI(T,E) = ε, where ε denotes the empty sequence.
- If PPT(T,E) is the immediate predecessor of T, then PPI(T,E) = ε.

Definition. Let T be a transition in a transition sequence E of an FSM M. T is said to be **port-synchronizable** for (E,M), if

(a) PPI(T,E) is ε, or

(b) PPI(T,E) is not ε and the input symbol of T is invalid for the head state of any transition in PPI(T,E).

Note: Assume that T is port-synchronizable for (E,M). The arrival of T's input symbol before head(T) becomes the current state of M does not create an additional eligible transition for the head state of any transition in PPI(T,E).

Definition. Let E be a transition sequence of an FSM M. E is said to be **port-synchronizable** for M if every transition in E is port-synchronizable for (E,M).

Consider the test sequence R2 = T1.T3.T4.T2.T3.T5 of M1. R2 is a postman tour of M1. Below we show the PPT and PPI of each transition in R2. Since T3 has two occurrences in R2, T3_1 and T3_2 denote the first and second occurrences of T3, respectively.

Transition T	T1	T3_1	T4	T2	T3_2	T5
PPT(T,R2)	null	null	T1	T4	T3_1	T2
PPI(T,R2)	ε	T1	T3_1	ε	T4.T2	T3_1

R2 is not pair-synchronizable, but it is port-synchronizable for M1. (Note that no pair-synchronizable postman tours of M1 exist.) By applying algorithm Port_Tester_Gen to R2, we obtain the following testers for ports P1, P2, and P3:

Tester(P1,R2)	Tester(P2,R2)	Tester(P3,R2)
(T1) send A to P1	(T4) receive Y from P2	(T3_1) send C to P3
(T1) receive X from P1	(T2) send B to P2	(T3_1) receive Z from P3
(T4) send A to P1	(T2) receive Y from P2	(T3_2) send C to P3
(T5) receive X from P1	(T5) send B to P2	(T3_2) receive Z from P3

Consider another test sequence R3 = T1.T3.T5.T2.T3.T4 of M1. R3 is also a postman tour of M1. Below we show the PPT and PPI of each transition in R3.

Transition T	T1	T3_1	T5	T2	T3_2	T4
PPT(T,R3)	null	null	null	T5	T3_1	T5
PPI(T,R3)	ε	T1	T1.T3_1	ε	T5.T2	T2.T3_2

R3 is not port-synchronizable for M1 since

- PPI(T5,R3) = T1.T3_1 and the input symbol of T5 is valid for head(T1).
- PPI(T4,R3) = T2.T3_2 and the input symbol of T4 is valid for head(T2).

Below we show some properties of pair- and port-synchronizable tests sequences of an FSM.

Theorem 6.1. Let E be a test sequence for an FSM M.

(a) E is pair-synchronizable if and only if for each transition T in E, PPI(T,E) is ε.

(b) If E is pair-synchronizable, then E is port-synchronizable for M. But the converse is not necessarily true.

(c) If M is completely specified, then E is pair-synchronizable if and only if E is port-synchronizable for M.

Proof. This theorem follows the definitions of pair- and port-synchronizable test sequences of an FSM. Q.E.D.

Theorem 6.2. Let E be a port-synchronizable test sequence of an FSM M and let Tester(E) be the set of testers generated by algorithm Port_Tester_Gen according to E.

(a) Any execution of M and Tester(E) is deterministic and successful. Thus, the port-based synchronization problem never occurs during any execution of M and Tester(E).

(b) At any time during an execution of M and Tester(E), two or more ports of M may contain input messages.

(c) If E is pair-synchronizable, then at any time during an execution of M and Tester(E), at most one port of M contains input messages.

(d) M does not contain a port-synchronizable test sequence F such that F≠E and Tester(F)=Tester(E).

(The proof is omitted.)

One interesting question is whether there exists a test sequence E of an FSM M such that E is not port-synchronizable for M, but any execution of M and Tester(E) is deterministic. The following theorem says such a test sequence of M does not exist. Based on this theorem, the definition of a port-synchronizable test sequence of M provides a necessary and sufficient condition under which a test sequence of M does not have the synchronization problem. To prove this theorem, we consider **the delayed execution** of M and Tester(E), which means that during an execution of M and Tester(E), the selection of a transition of the current state of M is delayed until (a) each tester in Tester(E) either has completed or is blocking on a receive, and (b) all messages sent by testers in Tester(E) have arrived at ports of M. Obviously, the delayed execution of M and Tester(E) is deterministic if and only if any execution of M and Tester(E) is deterministic.

Theorem 6.3. Let E be a test sequence of an FSM M and let Tester(E) be the set of testers generated by algorithm Port_Tester_Gen according to E. E is port-synchronizable for M if and only if any execution of M and Tester(E) is deterministic.
(The proof is omitted.)

7 GENERATION OF PORT-SYNCHRONIZABLE TEST SEQUENCES OF AN FSM

In this section, we briefly discuss possible approaches to extending a test sequence generation method for an FSM in order to generate port-synchronizable test sequences. One approach is to apply the concept of **backtracking**. Assume that during the application of a test sequence generation method to an FSM M, E is the partial test sequence generated so far and transition T is allowed to follow E. We need to determine whether T is port-synchronizable for E.T with respect to M. If the answer is yes, then we append T to the end of E and continue the original test sequence generation procedure. If the answer is no, then we try to find another transition, say T', that is allowed to follow E, and then we determine whether T' is port-synchronizable for E.T' with respect to M. We repeat this process until such a transition is found. If no such transition exists for E, then we replace the last transition in E with another transition and repeat the same process. Eventually either we find one port-synchronizable test sequence of M, or we conclude that no port-synchronizable test sequence of M exists, according to the original test sequence generation method. The concept of backtracking was also used in the generation of pair-synchronizable test sequences (Sarikaya and Bochmann, 1984) (Luo et al., 1993).

Some test sequence generation methods involve the generation of two types of transition sequences for a state S of an FSM M:

- transfer sequences for S, which start from the initial state of M and reach at S.
- state identification or validation sequences for S, which start from S.

Assume that for a transition T of M, a test sequence generation method has been extended to generate

- a set FS of port-synchronizable transfer sequences for head(T), and
- a set DS of port-synchronizable identification or validation sequences for tail(T).

We need to find one element of FS, say F, and one element of DS, say D, such that F.T.D is port-synchronizable for M. To search for such a port-synchronizable test sequence, we have the following observations:

(a) For each element D of DS, let Null(D) be the set of transitions in D with their PPT being null. To determine whether F.T.D is synchronizable for M, we only need to determine whether T and transitions in Null(D) are synchronizable for (F.T.D, M). Transitions in D that are not in Null(D) can be ignored, since they can never create a port-base synchronization problem for F.T.D.

(b) Let U be the shortest suffix of F that involves the input ports of T and transitions in Null(D) at least once. To determine whether F.T.D is port-synchronizable for M, we only need to determine whether T and transitions in Null(D) are port-synchronizable for (U.T.D, M). The reason is that T and transitions in Null(D) have their PPT in U.

Based on (b), we have developed algorithms for generating UIO-based port-synchronizable test sequences of an FSM.

8 RESULTS OF EMPIRICAL STUDIES

In this section, we describe the results of our empirical studies on pair- and port-synchronizable test sequences. We used a number of FSM-based protocols in our empirical studies. For each FSM-based protocol M, we applied the following steps:

(a) used **tsg**, a test sequence generation tool, to generate a set S of test sequences of M.
(b) for each test sequence E in S, determined whether E is pair-synchronizable, not pair-synchronizable, but port-synchronizable for M, or not port-synchronizable for M.

The tsg (*t*est *s*equence generation) tool was obtained from the Department of Computer Science at University of British Columbia. For a given FSM, tsg can generate test sequences according to the D-method (Kohavi, 1978), UIO-method (Sabnani and Dahbura, 1988), and W-method (Chow, 1978). For the input FSM, the tool inserts a reset transition for each state. If the input FSM is not completely specified, the tool inserts a self-loop transition for each unspecified input of a state. Each of the inserted transitions has no output.

We applied tsg to generate UIO-based test sequences for a number of FSM-based protocols. For each transition T in a given FSM, tsg generated one or more UIO-based test sequences, each starting with a reset transition, followed by u.T.v, where u is a transition sequence from the initial state to the head state of T and v is a UIO sequence for the tail state of T. After a set S of UIO-based test sequences for an FSM was generated by tsg, S was modified as follows:

(a) The test sequences in S for inserted transitions (i.e., reset transitions and the self-loop transitions for unspecified inputs) were deleted.
(b) For each of the remaining test sequences in S, the inserted transitions in this test sequence, if they existed, were deleted, and if the resulting test sequence contained zero or one transition, then this test sequence was deleted.
(c) After step (b), if two or more test sequences in S were identical, then only one of them was kept in S.

The following algorithm was used to determine whether a test sequence is pair-synchronizable, and if not, whether it is port-synchronizable for a given FSM M.

```
let E = E1.E2.....En be a test sequence of M;
pair_flag := true;      /* to indicate whether E is pair-synchronizable */
port_flag := true;      /* to indicate whether E is port-synchronizable for M */
for i = 2, 3, ..., n until port_flag = false do begin
    visit transition Ei;
    assume that the input symbol of Ei is P:A;
```

```
        if port P is not involved in E(i-1) then begin
            pair_flag := false;
            ppt_found := false;    /* to indicate whether PPT(Ei,E) has been found */
            for j = i-1, i-2, ..., 1 until port_flag = false or ppt_found = true do begin
                visit transition Ej;
                if port P is involved in Ej
                then   ppt_found := true;
                else   if P:A is a valid input for head(Ej) then port_flag := false;
    end; end; end;
```

After the completion of the above algorithm, if pair_flag is true, then E is pair-synchronizable. Otherwise, if port_flag is true, then E is not pair-synchronizable, but port-synchronizable for M. If port_flag is false, then E is not port-synchronizable for M.

The following table shows some of the results of our empirical studies.

protocol name	test_seq_#	pair_#	port_#	non_port_#
T-class-0	55	45 (82%)	52 (95%)	3 (5%)
T-class-4	111	85 (76%)	89 (80%)	22 (20%)
Q931	50	20 (40%)	48 (96%)	2 (4%)

where

- test_seq_# is the total number of UIO-based test sequences generated by tsg for the corresponding protocol.
- pair_# is the number of pair-synchronizable test sequences.
- port_# is the number of port-synchronizable test sequences.
- non_port_# is the number of non-port-synchronizable test sequences.
- T-class-0 refers to the transport class 0 protocol (Sarikaya and Bochmann, 1984) which has 4 states and 21 transitions.
- T-class-4 refers to the transport class 4 protocol (Sidhu and Leung, 1989), which has 15 states and 60 transitions.
- Q931 refers to the ISDN Q931 protocol (Zhu and Chanson, 1994), which has 8 states and 31 transitions.

Both T-class-4 and Q931 contain multiple types of timeout signals, which were treated as inputs from a unique port in our empirical studies.

The above table show that for the set of UIO-based test sequences generated by tsg for a protocol, the percentage of pair-synchronizable test sequences ranges from 40% to 82% and that the percentage of port-synchronizable test sequences ranges from 80% to 96%. The increase of percentage of synchronizable test sequences due to the definition of a port-synchronizable test sequence ranges from 4% to 56%. These results indicate that the use of port-synchronizable test sequences may significantly increase the number of synchronizable test sequences of an FSM.

9 CONCLUSIONS

In this paper, we have defined port-synchronizable test sequences of an FSM with multiple ports and shown that this definition is a necessary and sufficient condition under which a test sequence of an FSM does not have the synchronization problem. Our empirical studies show that an FSM may contain many port-synchronizable test sequences that are not pair-synchronizable. The use of port-synchronizable test sequences in protocol testing makes more test sequences to become synchronizable.

As mentioned in Section 4, group-based testing of an FSM M is to divide the ports of M into groups and allows the testers for ports in the same group to communicate with each other in order to synchronize the arrivals of inputs at M. For a given definition of groups of ports of M, we have defined a necessary and sufficient condition under which a test sequence of M either does not have the synchronization problem or can have the synchronization problem solved by allowing communication between testers for ports of M in the same group (Tai and Young 1995). For a test sequence of M that satisfies this condition, we have shown how to construct a set of testers with minimum communication between them. Also, we have studied the impact of timeout transitions on port- and group-synchronizable test sequences.

The use of the multi-port FSM model to specify communication protocols is increasing. We are currently investigating the generation of port- and group-synchronizable test sequences and the fault detection capability of port- and group-based testing.

10 ACKNOWLEDGMENTS

The authors wish to thank Dan Duvarney for his effort in carrying out empirical studies. Also, the authors are grateful to Pramod Koppol for his comments and to S. T. Chanson and J. Zhu for providing the tsg tool.

11 REFERENCES

Boyd, S. and Ural, H. (1991) The synchronization problem in protocol testing and its complexity, *Information Processing Letters*, Vol. 40, 131-6.

Carver, R. H. and Tai, K. C. (1991) Replay and testing for concurrent Programs, *IEEE Software*, Vol. 8, No. 2, 66-74.

Chen, W. H., Lu, C. S., Chen, L. and Tang, J. T. (1990) Synchronizable protocol test generation via the duplex technique, *Proc. IEEE INFOCOM,* 561-3.

Chen, W. H. and Ural, H. (1995) Synchronizable test sequences based on multiple UIO sequences, *IEEE/ACM Trans. Network*, Vol.3, No. 2, 152-7.

Chow, T. S. (1978) Testing software modeled by finite-state machines, *IEEE Trans. Software Eng.*, Vol. SE-4, No. 3, 178-87.

Guyot, S. and Ural, H. (1995) Synchronizable checking sequences based on UIO sequences, *Proc. Protocol Test Systems VIII.*

Kohavi, Z. (1978) Switching and Finite Automata Theory. 2nd edition, McGraw-Hill.

Luo, G., Dssouli, R., Bochmann, G. v., Venkataram, P. and Ghedamsi, A. (1993) Generating synchronizable test sequences based on finite state machine with distributed ports, *Proc. Protocol Test Systems VI*, 139-153.

Sarikaya, B. (1993) *Principles of Protocol Engineering and Conformance Testing*, Ellis Horwood Limited.

Sarikaya, B. and Bochmann, G. v. (1984) Synchronization and specification issues in protocol testing, *IEEE Trans. on Communications*, Vol. 32, No.4, 389-95.

Sabnani, K. and Dahbura, A. (1988) A protocol test generation procedure, *Computer Networks and ISDN Systems*, Vol. 15, no. 4, 285-97.

Sidhu, D. P. and Leung, T. K. (1989) Formal Methods for protocol testing: A detailed study, *IEEE Trans. on Software Engineering.*, Vol. 15, no. 4, 413-26.

Tai, K. C. and Ahuja, S. (1987) Reproducible testing of communication software, *Proc. IEEE Inter. Conf. on Computer Software and Applications (COMPSAC)*, 331-7.

Tai, K. C., Carver, R. H., and Obaid, E. E. (1991) Debugging concurrent Ada programs by deterministic execution, *IEEE Trans. Soft. Eng.*, Vol. 17, No. 1, 45-63.

Tai, K. C., and Carver, R. H. (1995) Testing of Distributed Programs, in *Handbook of Parallel and Distributed Computing* (ed. A. Zoyama), McGraw-Hill.

Tai, K. C., and Young, Y. C. (1995) Synchronizable test sequences of finite state machines, Technical Report, TR-95-11, Dept. of Computer Science, North Carolina State University.

Tarnay, K. (1991) *Protocol Specification and Testing*, Plenum Press.

Turner, K. J. (1993) *Using Formal Description Techniques: An Introduction to Estelle, Lotos, and SDL*, Wiley.

Ural, H. and Wang, Z. (1993) Synchronizable test sequence generation using UIO sequences, *Computer Communications*, Vol. 16, No. 10, 653-61.

Zhu, J. and Chanson, S. T. (1994) Toward evaluating fault coverage of protocol test sequences, *Proc. Protocol Specification, Testing, and Verification XIV*, 130-44.

12 BIOGRAPHY

Kuo-Chung Tai is a Professor in the Computer Science Department at North Carolina State University. He has published papers in the areas of software engineering, distributed systems, programming languages, and compiler construction. His current research interests include analysis, testing, and debugging of sequential and concurrent software. He received his Ph.D. degree in Computer Science from Cornell University in 1977. From 1989 to 1991, he served as the director of Software Engineering Program at the National Science Foundation. He is an associate editor of Journal of Computer Languages, Inter. Journal of Software Engineering and Knowledge Engineering, and Inter. Journal of Computer and Software Engineering.

Yu-Chiou Young is a Ph.D. student in the Computer Science Department at North Carolina State University. He is also a full-time employee of IBM in Raleigh, North Carolina.

25

Synchronizable Checking Sequences Based on UIO Sequences

S. Guyot and H. Ural
Department of Computer Science, University of Ottawa
150 Louis Pasteur, Ottawa, Ontario, K1N 6N5, Canada

Abstract

This study addresses the synchronization problem that arises during the application of a predetermined checking sequence in some protocol test architectures that utilize remote testers. The synchronization problem can usually be solved by adding an additional communication channel or an additional protocol for coordination between the remote testers. Such requirements can be eliminated by constructing a synchronizable checking sequence such that the corresponding sequence of transitions causes no synchronization problem. Based on UIO sequences, a synchronizable checking sequence construction method is proposed.

Keywords

Synchronization, test coordination, checking sequence construction

1 INTRODUCTION

Determining, under certain assumptions, whether any given "black box" implementation of a Finite State Machine (*FSM*) is functioning correctly is referred to as a *fault detection* (*checking*) *experiment*. Foundations of fault detection experiments can be found in sequential circuit testing literature (Gill, 1962), (Hennie, 1964). This experiment consists of applying an input sequence and comparing the resulting output sequence to the expected output sequence. The applied input sequence and the expected output sequence form a *checking sequence.*

Given a deterministic FSM M, the construction of a checking sequence from M with a designated initial state that determines whether any given FSM N is a correct implementation of M is sought in practice by considering some basic assumptions. The assumptions commonly made in the literature (Sabnani, 1988), (Chan, 1989), (Dahbura, 1990), (Aho, 1991) are that 1) M is *minimal, completely specified*, and represented by a *strongly connected* digraph, 2) N has the same set of inputs as M, and 3) faults in N do not increase the number of states in N. In addition, the construction of a checking sequence must deal with the "black box" nature of a given implementation N of M which allows only limited controllability and observability of N. The limited controllability refers to not being able to directly transfer N to a designated state and the limited observability refers to not being able to directly recognize the current state of N. In order to overcome the restrictions imposed by the limited controllability and observability, some special input sequences must be utilized in the construction of a checking sequence such that the output sequences produced by N in response to these input sequences provide sufficient information to deduce that every state transition of M is implemented correctly by N.

In order to verify the state transition from state *a* to *b* under input *x*, 1) before the application of *x*, *N* must be transferred to the state recognized as *a*, 2) the output produced by *N* in response to the application of *x* must be as specified in *M*, and 3) the state reached by *N* after the application of *x* must be recognized as *b*. Hence, a crucial part of testing the correct implementation of each transition is recognizing the starting and terminating states of the transition. The recognition of a state of an FSM *M* can be achieved by a distinguishing sequence (Hennie, 1964), a characterization set (Hennie, 1964) or a unique input-output (UIO) sequence (Sabnani, 1988). It is known that UIO sequences may not exist for every state of every minimal FSM (Sabnani, 1988) and determining the existence of a UIO sequence for a state of an FSM is PSPACE-complete (Lee, 1994).

Nevertheless, based on UIO sequences, various methods have been proposed in the literature to test FSMs (Sabnani, 1988), (Chan, 1989), (Dahbura, 1990), (Aho, 1991). In some of these methods, the UIO sequences obtained from an FSM *M* are used for verifying states of a given implementation *N* of *M* without confirming that they are also unique in *N*. Chan (1989) observed that when the uniqueness of UIO sequences does not hold in a faulty *N*, certain errors in *N* can not be detected. In order to overcome this problem, they proposed the *UIOv* method which is the UIO method (Sabnani, 1988) with the addition of a verification procedure to ensure that the UIO sequence for each state of *M* is also unique in *N*.

Fault detection experiments have been used to test the conformance of various protocol implementations to their specifications given as FSMs (Dahbura, 1990). It is observed that protocol testing can be carried out as a fault detection experiment in some specific test architectures. One such architecture is shown in Figure 1 where the lower interface and the upper interface of the protocol implementation *N* may be controlled and observed indirectly by the lower tester (*L*) and directly by the upper tester (*U*), respectively.

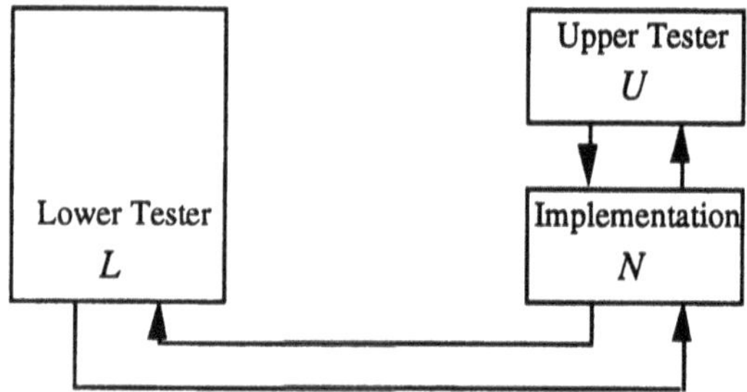

Figure 1 A Distributed Test Architecture.

During the application of a predetermined checking sequence, *U* and *L* are bound to synchronize with each other only through their interactions with *N*. However, this requirement may lead to a *synchronization problem* when *L* (or *U*) is expected to send an input to *N* after *N* responds to an input from *U* (or *L*) with an output to *U* (or *L*), but *L* (or *U*) is unable to determine whether *N* sent that output.

Synchronization between *U* and *L* can be achieved by any suitable test management protocol coordinating the actions of the testers (ISO TC97/SC21, 1987), by a ferry protocol with inter-process communication, or through manual coordination (by use of a telephone or terminal connection) (Rayner, 1987). However, these solutions either require an additional communication channel between *U* and *L* or an additional protocol to facilitate communication between *U* and *L*.

The necessity for such coordination is eliminated by constructing a synchronizable sequence of transitions that cause no synchronization problem (Ural, 1993), (Chen, 1995), (Tai, 1995).

In this paper, based on the work presented in (Chan, 1989), (Ural, 1993), (Perez, 1993), (Rezaki, 1994), a method for constructing synchronizable checking sequences is proposed. Related terminology is reviewed in section 2. In section 3, a method for synchronizable checking sequence construction using UIO sequences is presented and a proof for the resulting sequence to be a checking sequence is given. An illustrative example of the application of the proposed method is provided in section 4. In section 5, some minimization techniques are proposed. Concluding remarks are given in section 6.

2 PRELIMINARIES

2.1 FSM and its Graphical Representation

A *finite state machine*, *FSM*, is a quintuple $M = (S, X, Y, \delta, \lambda)$, where S is a finite set of states, X is a finite set of inputs, Y is a finite set of outputs, δ is a state transition function that maps $S \times X$ to S, and λ is an output function that maps $S \times X$ to Y. State $s_1 \in S$ is designated as the *initial state* of M, and $|S|$ is n. $- \in X$ and $- \in Y$ denote null input and null output, respectively. For a sequence of inputs $x_1x_2... x_k, x_i \in X$, $1 \leq i \leq k$ and a state $s \in S$, $\delta(s, x_1x_2... x_k) = \delta(\delta(s, x_1), x_2... x_k) = s_k$ and $\lambda(s, x_1x_2... x_k) = \lambda(s, x_1)\lambda(\delta(s, x_1), x_2... x_k) = y_1y_2... y_k$, where $y_i = \lambda(s_{i-1}, x_i)$, $s_i = \delta(s_{i-1}, x_i)$, $s_0 = s$, $y_i \in Y$, $s_i \in S$. In this case, it is said that $I = x_1x_2... x_k$ is an *input sequence*, $\lambda(s, I) = y_1y_2... y_k = O$ is an *output sequence*, and $I/O = (x_1x_2... x_k)/(y_1y_2... y_k) = (x_1/y_1)(x_2/y_2)... (x_k/y_k)$ is an *input/output sequence* (or IO-sequence, in short) starting at state s.

An FSM M is said to be *minimal* if $\forall (s_i, s_j) \in S^2, i \neq j, \exists I \in X^k$, such that $\lambda(s_i, I) \neq \lambda(s_j, I)$. An FSM M is said to be *completely specified* if $\forall x \in X, \forall s \in S$, $\exists (s, \delta(s, x); x/\lambda(s, x))$. An FSM M can be represented by a directed graph (digraph) $D = (V,E)$ where a set of vertices $V = \{1, 2,..., n\}$ represents the set S of states of M, and a set of directed edges $E = \{e_{ij} = (v_i, v_j; x/y); v_i, v_j \in V\}$ represents the set of transitions of M. Each $e_{ij} \in E$ represents a specified transition $t_{ij} = (s_i, s_j; x/y)$ of M from state s_i to state s_j with *label* x/y, input $x \in X$ and output $y \in Y$. Vertices v_i and v_j which represent respectively state s_i and s_j are called the *head* and the *tail* of e, denoted by *head*(e) and *tail*(e). The label of an edge e_{ij} (or transition t_{ij}) is denoted by *label*(e_{ij}) (or *label*(t_{ij})).

A *path* $P = (n_1, n_2; x_1/y_1)(n_2, n_3; x_2/y_2)... (n_{k-1}, n_k; x_{k-1}/y_{k-1})$, $k > 1$, in a digraph $D = (V, E)$ is a finite sequence of adjacent (not necessarily distinct) edges in D, where n_1 and n_k, are called the *head* and the *tail* of P, denoted by *head*(P) and *tail*(P), respectively, and $(x_1/y_1)(x_2/y_2)... (x_{k-1}/y_{k-1})$, is called the *label* of P, denoted by *label*(P). For convenience, a path $P = (n_1, n_2; x_1/y_1)(n_2, n_3; x_2/y_2)... (n_{k-1}, n_k; x_{k-1}/y_{k-1})$ will be represented by $(n_1, n_k; I/O)$ where I/O is the IO-sequence $(x_1/y_1)(x_2/y_2)... (x_{k-1}/y_{k-1})$, $I = x_1x_2... x_{k-1}$ is the *input portion* of I/O, $O = y_1y_2... y_{k-1}$ is the *output portion* of I/O, respectively. A sequence $(i_1i_2... i_k)$ is a *subsequence* of $(x_1x_2... x_m)$ if there exists a Δ, $0 \leq \Delta \leq m-k$, such that for all j, $1 \leq j \leq k$, $i_j = x_{j+\Delta}$. A sequence $(i_1i_2... i_k)$ is a *prefix* of $(x_1x_2... x_m)$ if for all j, $1 \leq j \leq k$, $i_j = x_j$.

A digraph $D = (V, E)$ is *strongly connected*, if for every pair of vertices v_j and v_k, there exists a path from v_j to v_k. An FSM M has the *reset feature* if there is an input r such that $\delta(s_i, r) = s_1$ (i.e. the initial state of M) and $\lambda(s_i, r) = -$, for every state s_i of M. A *transfer sequence* T of an FSM M from state s_i to state s_j is the input portion of the label of a path from s_i to s_j.

Let $\Phi(M)$ be the set of FSMs each of which has at most n states and the same input set as X of M. Let N be in $\Phi(M)$. M is *isomorphic* to N, denoted by $M \cong N$, if there is a one-to-one and onto function f on the state sets of M and N such that for any transition $(i, j; x/y)$ of M, $(f(i), f(j); x/y)$ is a transition of N. A *checking sequence* of M is an IO-sequence starting at a specific state of M that distinguishes M from any FSM of $\Phi(M)$ which is not isomorphic to M.

2.2 Synchronization Problem

Let each transition t_{ij} of an FSM M have one of the following labels, label$(t_{ij}) \in \{i^L/\text{-}, i^U/\text{-}, i^L/o^L, i^L/o^U, i^U/o^L, i^U/o^U, i^L/o^{U,L}, i^U/o^{U,L}\}$ where i^L (i^U) is an input from $L(U)$, $o^L(o^U)$ is an output to $L(U)$ and $o^{U,L}$ is an output to L and U. Then, considering two consecutive transitions of M, one of the testers, say L (or U), faces a *synchronization problem* if L (or U) did not take part in the first transition and if the second transition requires that it sends an input to M (Sarikaya, 1984). For example, a synchronization problem will occur if t_1 is followed by t_2 and label$(t_1) = i^L/o^L$ whereas label$(t_2) \in \{i^U/\text{-}, i^U/o^L, i^U/o^U, i^U/o^{U,L}\}$.

Two consecutive transitions t_{ij} and t_{jk} of M form a *synchronizable pair of transitions* if t_{jk} can follow t_{ij} without generating a synchronization problem. For example, a transition with label i^L/o^U forms a synchronizable pair of transitions when followed by any other transition of M. For a transition t_{ij} of an FSM, each transition t_{jk} that forms a synchronizable pair of transitions with t_{ij} is called an *eligible successor* of t_{ij}. A transition sequence of an FSM is *synchronizable* if for every two consecutive transitions in the sequence, the second transition is an eligible successor of the first one. A checking sequence of an FSM is synchronizable if it is the label of a synchronizable transition sequence of the FSM. A *synchronizable input sequence* for state s is the input portion of a synchronizable transition sequence that starts at s.

2.3 Order-specified Digraph and Synchronizable UIO Sequences

A digraph $D = (V, E)$ is called *order-specified* if for each edge $e_{ij} \in E$, a subset of the outgoing edges of vertex j is specified as eligible successors of e. A path in an order-specified digraph D is said to be *correctly-ordered* (CO) if for every pair of consecutive edges e_{ij} and e_{jk} in the path, e_{jk} has been specified as an eligible successor of e_{ij}. The label of a CO path is called a *correctly-ordered IO-sequence*.

A *Unique input-output* (UIO) sequence for a state of an FSM is a sequence of input/output symbols that are not exhibited by any other state of the FSM (Sabnani, 1988). A *synchronizable unique input/output* (SUIO) sequence for a state of an FSM is a correctly-ordered UIO sequence for that state. Although an SUIO sequence itself does not cause any synchronization problem, a synchronization problem can still arise if the transition corresponding to the first edge of the SUIO sequence is not an eligible successor of the transition which precedes the SUIO sequence in a given transition sequence. For example, when an SUIO sequence for a state s starts with an input sent by U, any incoming transition of state s with label i^L/o^L or $i^L/\text{-}$, will cause a synchronization problem if it precedes the SUIO sequence. To avoid a synchronization problem, for each state s, two SUIO sequences are needed, denoted by SUIO$^U(s)$ and SUIO$^L(s)$ where SUIO$^U(s)$ starts with an input sent by U and SUIO$^L(s)$ starts with an input sent by L.

3 CONSTRUCTION OF SYNCHRONIZABLE CHECKING SEQUENCES

Let $M = (S, X, Y, \delta, \lambda)$ hereafter stand for a minimal and completely specified FSM which is represented by a strongly connected and order-specified digraph $D = (V, E)$. Let $|S|$ be n and $s_1 \in S$ be the initial state of M. The construction of a synchronizable checking sequence of M is based on the construction of a correctly-ordered digraph $D' = (V', E')$ such that all edges of D are in one to one correspondence with edges in D', and all paths in D' are CO paths in D. Thus, finding a synchronizable checking sequence on D will be reduced to finding a checking sequence on D'.

3.1 Construction of a Correctly-ordered Digraph D'

The correctly-ordered digraph $D' = (V', E')$ was introduced by Ural (1993), and later modified by Perez (1993). Before $D' = (V', E')$ is constructed, the following sets of edges are formed :
for each vertex v in $D = (V, E)$,

$$Leave^U[v] = \{e \in E : head(e) = v \text{ and } label(e) \in \{i^U/\text{-}, i^U/o^U, i^U/o^L, i^U/o^{U,L}\}\}$$
$$Leave^L[v] = \{e \in E : head(e) = v \text{ and } label(e) \in \{i^L/\text{-}, i^L/o^L, i^L/o^U, i^L/o^{U,L}\}\}$$
$$Arrive^U[v] = \{e \in E : tail(e) = v \text{ and } label(e) \in \{i^U/\text{-}, i^U/o^U\}\}$$
$$Arrive^L[v] = \{e \in E : tail(e) = v \text{ and } label(e) \in \{i^L/\text{-}, i^L/o^L\}\}$$
$$Arrive^{U,L}[v] = \{e \in E : tail(e) = v \text{ and } label(e) \in \{i^L/o^U, i^U/o^L, i^L/o^{U,L}, i^U/o^{U,L}\}\}.$$

The edges in $Leave^U[v]$ are eligible successors of edges in $Arrive^U[v]$ and $Arrive^{U,L}[v]$, The edges in $Leave^L[v]$ are eligible successors of edges in $Arrive^L[v]$ and $Arrive^{U,L}[v]$. Furthermore, edges in $Leave^U[v]$ are the only eligible successors of edges in $Arrive^U[v]$, and edges in $Leave^L[v]$ are the only eligible successors of edges in $Arrive^L[v]$.

During the construction of $D' = (V', E')$ from $D = (V, E)$, for each vertex $v \in V$, 1, 2 or 3 vertices are created in D', depending on the labels of edges arriving and leaving v, so that if an edge arrives to a vertex in V', it can take any of the edges starting at this vertex. Thus, a path in D' becomes a CO path in D. Accordingly, the correctly-ordered digraph $D' = (V', E')$ is such that $V' = V^U \cup V^L \cup V^{U,L}$ and $E' = E_c \cup F$. The procedure of constructing the correctly-ordered digraph $D' = (V', E')$ from a strongly connected and order-specified digraph $D = (V, E)$ such that any edge in D has an eligible successor is as follows:

(1) For each vertex v in V,
If $Leave^U[v] \neq \emptyset$, then a vertex v^U is created in V^U.
If $Leave^L[v] \neq \emptyset$, then a vertex v^L is created in V^L.

(2) For each edge $(w, v; x/y) \in Arrive^U[v]$ (this implies that $(w, v; x/y) \in Leave^U[w]$), a directed edge from w^U to v^U is created in E_c. Similarly, for each edge $(w, v; x/y) \in Arrive^L[v]$ (this implies that $(w, v; x/y) \in Leave^L[w]$), a directed edge from w^L to v^L is created in E_c.

(3) For each edge $(w, v; x/y) \in Arrive^{U,L}[v]$, one of the following is performed:

a) In the case that vertex v^U exists but vertex v^L does not,
if $(w, v; x/y) \in Leave^U[w]$, then a directed edge from w^U to v^U is created in E_C,
else (i.e. $(w, v; x/y) \in Leave^L[w]$) a directed edge from w^L to v^U is created in E_C.

b) In the case that vertex v^U exists but vertex v^L does not,
if $(w, v; x/y) \in Leave^U[w]$, then a directed edge from w^U to v^L is created in E_C,
else (i.e. $(w, v; x/y) \in Leave^L[w]$) a directed edge from w^L to v^L is created in E_C.

c) In the case that both v^U and v^L exist, a vertex $v^{U,L}$ is created in $V^{U,L}$ (if it does not already exist) and two edges $(v^{U,L}, v^U)$ and $(v^{U,L}, v^L)$ with nil labels, denoted by -/- when needed, are constructed in F.
If $(w, v; x/y) \in Leave^U[w]$, then a directed edge $(w^U, v^{U,L}; x/y)$ is created in E_C,
else (i.e. $(w, v; x/y) \in Leave^L[w]$) a directed edge $(w^L, v^{U,L}; x/y)$ is created in E_C.

3.2 Proposed Method

The proposed method for the construction of a synchronizable checking sequence from M is based on the following assumptions:

1) The implementation N of M implements the reset feature of M correctly. A reset transition has label r/- and can be followed by any input from any tester without causing any synchronization problem.
2) For each edge e of D, there is a CO-path that starts at the initial vertex v_1 and contains e.
3) For each state s, represented by vertex v in D, if $\text{Arrive}^U[v]$ (resp. $\text{Arrive}^L[v]$) $\neq \varnothing$, then $\text{SUIO}^U(s)$ (resp. $\text{SUIO}^L(s)$) exists and if $\text{Arrive}^{U,L}[v] \neq \varnothing$, then at least one of $\text{SUIO}^U(s)$ or $\text{SUIO}^L(s)$ exists (which implies that each edge has an eligible successor).
4) There is at most one state s represented by vertex v in D, such that $\text{Arrive}^{U,L}[v] = \varnothing$ and both $\text{Leave}^U[v]$ and $\text{Leave}^L[v]$ are not empty.
5) The input portion of each SUIO sequence is a synchronizable input sequence for all states in M.

The proposed method utilizes $D' = (V', E')$ constructed from the given $D = (V, E)$ and proceeds as follows :

step1: Construct the sets of input sequences corresponding to input portions of SUIO sequences that will be applied to each vertex in D'.
I is defined as the set of input sequences corresponding to the input portions of the SUIO sequences of the states of M that have to be used for transition verification. Let I_{sU} and I_{sL} denote the input portions of $\text{SUIO}^U(s)$ and $\text{SUIO}^L(s)$, respectively. Note that $I_{s\theta}$ will stand for either I_{sU} or I_{sL}.

The construction of I is as follows: For each edge $(w, v; x/y)$ in D where v represents $s \in S$, consider the corresponding edge e in E_C:

if $\text{tail}(e) = v^U$, then $I = I \cup I_{sU}$ and define I_{vU}
if $\text{tail}(e) = v^L$, then $I = I \cup I_{sL}$ and define I_{vL}
if $\text{tail}(e) = v^{U,L}$, then mark e.

For each marked edge e, one of the following four cases will apply:

case 1: $I_{vU} \in I$, $I_{vL} \in I$, then $I_{vU,L}$ = shorter of $\{I_{vU}, I_{vL}\}$

case 2: $I_{vU} \in I$, $I_{vL} \notin I$, then $I_{vU,L} = I_{vU}$.

case 3: $I_{vU} \notin I$, $I_{vL} \in I$, then $I_{vU,L} = I_{vL}$.

case 4: $I_{vU} \notin I$, $I_{vL} \notin I$, then $I = I \cup$shorter of $\{I_{vU}, I_{vL}\}$ and $I_{vU,L}$ = shorter of $\{I_{vU}, I_{vL}\}$.

Let m denote $|I|$, $I = \{I_1, I_2, \ldots I_m\}$, and $I_i(k)$ denote the kth input of the input sequence I_i, $1 \le i \le m$. Since D' can be viewed as representing an FSM which is not completely specified, only a specific subset of I can be applied to each vertex v^U (resp. v^L). This subset of I is called R_{vU} (resp. R_{vL}) and defined as follows:

$\forall\ v^\theta \in V^U \cup V^L$, $\forall\ i$, $1 \le i \le m$, if $I_i(1)$ can be applied to v^θ, then $I_i \in R_{v\theta}$

Note that $|R_{v\theta}|$ is denoted by $m_{v\theta}$, $R_{vU} \cup R_{vL} = I$, $R_{vU} \cap R_{vL} = \varnothing$, $m_{vU} + m_{vL} = m$, the ith element (which is an input sequence) of $R_{v\theta}$ is denoted by $R_{v\theta}(i)$ and either R_{vL} or R_{vU} may be empty. When a reset input r is applied, the next vertex is $v_{1U,L}$.

step2: Construct an input sequence C, called *cover* of D', that contains:

a) a state recognition part which is the concatenation of
$rT_{iU}R_{iU}(1)rT_{iU}R_{iU}(2)r \ldots rT_{iU}R_{iU}(m_{iU})rT_{iL}R_{iL}(1)r \ldots rT_{iL}R_{iL}(m_{iL}), \forall i,\ 1 \le i \le n$.
where T_{iU} and T_{iL} are defined as follows:
- if vertex $v_{iU,L}$ exists, $T_{iU,L}$ is the shortest transfer sequence on D' from the initial vertex $v_{1U,L}$ to vertex $v_{iU,L}$ and $T_{iU} = T_{iL} = T_{iU,L}$
- If $v_{iU,L}$ does not exist, T_{iU} (resp. T_{iL}) is the shortest transfer sequence on D' from the initial vertex $v_{1U,L}$ to vertex v_{iU} (resp. v_{iL}).

b) a transition verification part which consists of the following *test segment* for each edge e in E_c corresponding to the edge $(w, v; x/y)$ in D: $rT_{head(e)}xI_{tail(e)}$ where $T_{head(e)}$ is the same transfer sequence on D' from vertex $v_{1U,L}$ to vertex $head(e)$ that was used in state recognition part.

Proposition: Cover C of D' exists.

Proof:
By assumption 2, for any edge e of E, there is a CO-path starting at the initial vertex and containing e. Thus, by the construction of D', for any $e' \in E_c$, there is a path on D' starting at the initial vertex $v_{1U,L}$ and containing e'. Every vertex in $V^U \cup V^L$ is the head of some edge in E_c and every vertex in $V^{U,L}$ is the tail of some edge in E_c. Therefore, for each vertex v in V', there is a transfer sequence on D' from vertex $v_{1U,L}$ to vertex v. Moreover, by assumption 3 and by construction of I, every element of I exists and, by assumption 5, every element of $R_{v\theta}$ is the input sequence of a path on D' starting at v^θ. Therefore, all the subsequences of C of the type rX are defined on D', i.e. cover C of D' exists. **EOP.**

Theorem: Let $D' = (V', E')$ be a correctly-ordered digraph constructed from an order-specified digraph $D = (V, E)$ representing an FSM M and let cover C of D' be the input portion of an IO-sequence Q which is the label of a path P' starting at any vertex v in D'. Then Q is a synchronizable checking sequence of M.

Proof:
This proof is composed of 3 parts.

First, it must be established that Q is a synchronizable IO-sequence of M. Clearly, this is the case since, by construction, cover C is an input sequence on D'. Therefore, Q is a CO-sequence as it is the label of a path in D'. Hence, Q is a synchronizable IO-sequence of M. Second, it must be shown that if Q is also an IO sequence for an implementation $N = (S', X, Y, \delta', \lambda')$ of $\Phi(M)$ then N has n states and I is also a set of SUIO sequences for N. In order to show that this is the case, note that, any state of M is in one of the following three disjoint subsets of S:

$S1 = \{s_i \in S \mid$ only one of v_i^U or $v_i^L \in V'\}$. Without loss of generality, say that $v_i^U \in V'$. Then, since D' is strongly connected, v_i^U is the tail of some edge in E_c and, by construction of I, I_{s_iU} is an element of I. Therefore, the state recognition for a state of $S1$ is achieved by $rT_{v_iU}I_1rT_{v_iU}I_2r...rT_{v_iU}I_m$

$S2 = \{s_i \in S \mid v_i^{U,L} \in V'\}$. Then, by construction of D', $v_i^{U,L}$ is the tail of some edge in E_c and, by construction of I, at least one of $\{I_{s_iU}, I_{s_iL}\}$ is an element of I. Therefore, the state recognition for a state of $S2$ is achieved by $rT_{v_iU,L}I_1rT_{v_iU,L}I_2r...rT_{v_iU,L}I_m$.

$S3 = \{s_i \in S \mid v_i^U \in V',\ v_i^L \in V'$ and $v_i^{U,L} \notin V'\}$. Note that by the 3rd assumption, $|S3| \leq 1$. Then, since D' is strongly connected, v_i^U and v_i^L are tails of some edges in E_c and, by construction of I, I_{s_iU} and I_{s_iL} are elements of I. Therefore, the state recognition part for a state of $S3$ is achieved by $rT_{v_iU}R_{v_iU}(1)rT_{v_iU}R_{v_iU}(2)r...\ rT_{v_iU}R_{v_iU}(m_{v_iU})rT_{v_iL}R_{v_iL}(1)r\ ...\ rT_{v_iL}R_{v_iL}(m_{v_iL})$, where T_{v_iU} and T_{v_iL} are different.

Since r is correctly implemented in N and N is deterministic, if T is some transfer sequence, N will always be in the same state after application of rT. Hence, one can rewrite the state recognition part of the input portion of the IO-sequence as in Table 1, where each column represents the state of N reached by the transfer sequence given as the label of the column, each row represents an element of I given as the label of the row and each cell contains the output produced by the state of N reached by the transfer sequence given as the label of its column after the application of the element of I given as the label of its row. Table 1 summarizes all realizable cases, with states s_i and s_j of M in $S1$, states s_k, s_p and s_q of M in $S2$ and state s_z in $S3$ and $I = \{I_{s_iU}, I_{s_jL}, I_{s_kL}, I_{s_pU}, I_{s_qU}, I_{s_qL}, I_{s_zL}, I_{s_zU}\}$.

For any state s of M in $S1 \cup S2$ of M, there is a column in Table 1 to which all m elements of I are applied, i.e. every cell of the column has a value. Let us take the $(|S1|+|S2|)\times m$ subtable containing only the states of M in $S1 \cup S2$. The value of a cell will be denoted o_{rc} with r the row number and c the column number. By construction, I contains at least an SUIO for each state of M. Thus, $\forall c_0, \exists r_0$ such that $\forall c, c \neq c_0, o_{r_0c} \neq o_{r_0c_0}$.

Therefore, N has $|S1|+|S2|$ distinct states. To each state s of M in $S1 \cup S2$, there corresponds a unique state s' of N, such that $\forall I_i \in I$, $\lambda'(s', I_i) = \lambda(s, I_i)$. As for the existence of a state in S3, there are two cases:

If $S3 = \emptyset$, then N has exactly n states and $\forall s' \in N, \forall I_i \in I, \lambda'(s', I_i) = \lambda(s, I_i)$

If $S3 \neq \emptyset$, say $S3 = \{s_z\}$, $|S1|+|S2| = n-1$, so N has already $n-1$ determined states.

The last two rows of Table 1 show that the state of N reached after the application of rT_{v_zU} is different from the $n-1$ states of N already determined and, identically, the state reached after the application of rT_{v_zL} is different from the $n-1$ states of N already determined. Since N is in $\Phi(M)$, N has at most n states, thus, after the application of rT_{v_zU} or rT_{v_zL}, N is in a unique state s'_z which is the same state as s_z. Hence N has exactly n states and the last two columns of Table 1 show that $\forall I_i \in I$, $\lambda'(s'_z, I_i) = \lambda(s_z, I_i)$.

Table 1 State Recognition Part

	$rT_{v_i}U$	$rT_{v_j}L$	$rT_{v_k}U,L$	$rT_{v_p}U,L$	$rT_{v_q}U,L$	$rT_{v_z}U$	$rT_{v_z}L$
$I_{s_i}U$	$\lambda(s_i,I_{s_i}U)$	$\lambda(s_j,I_{s_i}U)\neq\lambda(s_i,I_{s_i}U)$	$\lambda(s_k,I_{s_i}U)\neq\lambda(s_i,I_{s_i}U)$	$\lambda(s_p,I_{s_i}U)\neq\lambda(s_i,I_{s_i}U)$	$\lambda(s_q,I_{s_i}U)\neq\lambda(s_i,I_{s_i}U)$		$\lambda(s_z,I_{s_i}U)\neq\lambda(s_i,I_{s_i}U)$
$I_{s_j}L$	$\lambda(s_i,I_{s_j}L)\neq\lambda(s_j,I_{s_j}L)$	$\lambda(s_j,I_{s_j}L)$	$\lambda(s_k,I_{s_j}L)\neq\lambda(s_j,I_{s_j}L)$	$\lambda(s_p,I_{s_j}L)\neq\lambda(s_j,I_{s_j}L)$	$\lambda(s_q,I_{s_j}L)\neq\lambda(s_j,I_{s_j}L)$		$\lambda(s_z,I_{s_j}L)\neq\lambda(s_j,I_{s_j}L)$
$I_{s_k}L$	$\lambda(s_i,I_{s_k}L)\neq\lambda(s_k,I_{s_k}L)$	$\lambda(s_j,I_{s_k}L)\neq\lambda(s_k,I_{s_k}L)$	$\lambda(s_k,I_{s_k}L)$	$\lambda(s_p,I_{s_k}L)\neq\lambda(s_k,I_{s_k}L)$	$\lambda(s_q,I_{s_k}L)\neq\lambda(s_k,I_{s_k}L)$		$\lambda(s_z,I_{s_k}L)\neq\lambda(s_k,I_{s_k}L)$
$I_{s_p}U$	$\lambda(s_i,I_{s_p}U)\neq\lambda(s_p,I_{s_p}U)$	$\lambda(s_j,I_{s_p}U)\neq\lambda(s_p,I_{s_p}U)$	$\lambda(s_k,I_{s_p}U)\neq\lambda(s_p,I_{s_p}U)$	$\lambda(s_p,I_{s_p}U)$	$\lambda(s_q,I_{s_p}U)\neq\lambda(s_p,I_{s_p}U)$	$\lambda(s_z,I_{s_p}U)\neq\lambda(s_p,I_{s_p}U)$	
$I_{s_q}U$	$\lambda(s_i,I_{s_q}U)\neq\lambda(s_q,I_{s_q}U)$	$\lambda(s_j,I_{s_q}U)\neq\lambda(s_q,I_{s_q}U)$	$\lambda(s_k,I_{s_q}U)\neq\lambda(s_q,I_{s_q}U)$	$\lambda(s_p,I_{s_q}U)\neq\lambda(s_q,I_{s_q}U)$	$\lambda(s_q,I_{s_q}U)$	$\lambda(s_z,I_{s_q}U)\neq\lambda(s_q,I_{s_q}U)$	
$I_{s_q}L$	$\lambda(s_i,I_{s_q}L)\neq\lambda(s_q,I_{s_q}L)$	$\lambda(s_j,I_{s_q}L)\neq\lambda(s_q,I_{s_q}L)$	$\lambda(s_k,I_{s_q}L)\neq\lambda(s_q,I_{s_q}L)$	$\lambda(s_p,I_{s_q}L)\neq\lambda(s_q,I_{s_q}L)$	$\lambda(s_q,I_{s_q}L)$		$\lambda(s_z,I_{s_q}L)\neq\lambda(s_q,I_{s_q}L)$
$I_{s_z}L$	$\lambda(s_i,I_{s_z}L)\neq\lambda(s_z,I_{s_z}L)$	$\lambda(s_j,I_{s_z}L)\neq\lambda(s_z,I_{s_z}L)$	$\lambda(s_k,I_{s_z}L)\neq\lambda(s_z,I_{s_z}L)$	$\lambda(s_p,I_{s_z}L)\neq\lambda(s_z,I_{s_z}L)$	$\lambda(s_q,I_{s_z}L)\neq\lambda(s_z,I_{s_z}L)$		$\lambda(s_z,I_{s_z}L)$
$I_{s_z}U$	$\lambda(s_i,I_{s_z}U)\neq\lambda(s_z,I_{s_z}U)$	$\lambda(s_j,I_{s_z}U)\neq\lambda(s_z,I_{s_z}U)$	$\lambda(s_k,I_{s_z}U)\neq\lambda(s_z,I_{s_z}U)$	$\lambda(s_p,I_{s_z}U)\neq\lambda(s_z,I_{s_z}U)$	$\lambda(s_q,I_{s_z}U)\neq\lambda(s_z,I_{s_z}U)$	$\lambda(s_z,I_{s_z}U)$	

Therefore, the states of M are in one to one correspondence with the states of N. If an element of I is the input portion of an SUIO of a state s of M, then it is also the input portion of an SUIO of state $s' = f(s)$ of N.

In order to complete the proof, the following must also be shown : suppose Q is also an IO-sequence for an implementation N of $\Phi(M)$. Let $(a, b; x/y)$ be an edge of D, then $(f(a), f(b); x/y)$ is a transition of N, and reciprocally, if $(a', b'; x/y)$ is a transition of N, then $(f^{-1}(a'), f^{-1}(b'); x/y)$ is an edge of D.

In order to show that this is the case, let $(a, b; x/y)$ be an edge of D, then there is a corresponding edge e in E_C. The transition verification part is such that $(T_{head(e)}xI_{tail(e)})/(\lambda(v_1U,L, T_{head(e)})y\lambda(b, I_{tail(e)}))$ is a subsequence of Q and, by construction, I contains the input portion of the SUIO corresponding to the tail of each edge in E_C, thus $I_{tail(e)} \in I$. Then, since Q is an IO-sequence of N, $I_{tail(e)}$ is the input portion of an SUIO of state $b' = f(b)$ of N. Hence, the state reached by N after the application of $T_{head(e)}x$ is $b' = f(b)$.

Moreover, I contains the input portion of an SUIO for each state of M so $I_{a\theta} \in I$. Then, since Q is an IO-sequence of N, $I_{a\theta}$ is the input portion of an SUIO of state $a' = f(a)$ of N. The state recognition part is such that $(T_{head(e)}I_{a\theta}) / (\lambda(v_1U,L, T_{head(e)})\lambda(a, I_{a\theta}))$ is a subsequence of Q. Hence, the state reached by N after the application of $T_{head(e)}$ is $a' = f(a)$.

Therefore, $(a', b'; x/y)$ is a transition of N. This holds for every transition of M. Since M is completely specified, N is also completely specified, so for any transition $(a', b'; x/y)$ of N, there is a corresponding transition $(f^{-1}(a'), f^{-1}(b'); x/y)$ in M. i.e. N is isomorphic to M.

For any implementation N of $\Phi(M)$, if Q is an IO-sequence of N, then N is isomorphic to M. Hence, Q is a checking sequence of M. **EOP**

4 AN EXAMPLE

Consider the FSM M, represented by the digraph depicted in Figure 2, where input a is applied by upper tester U and input b is applied by lower tester L. The output 0 is sent to U and output 1 and 2 are sent to L.

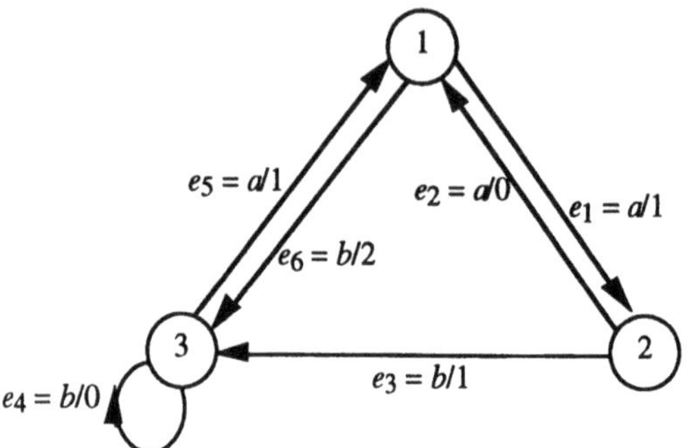

Figure 2 Digraph $D = (V, E)$.

Edge Types and Order Specifications:
$e_1 = (1, 2)$, label $(e_1) = i^U/o^L$, Eligible Successor: e_2, e_3
$e_2 = (2, 1)$, label $(e_2) = i^U/o^U$, Eligible Successor: e_1
$e_3 = (2, 3)$, label $(e_3) = i^L/o^L$, Eligible Successor: e_4
$e_4 = (3, 3)$, label $(e_4) = i^L/o^U$, Eligible Successor: e_4, e_5
$e_5 = (3, 1)$, label $(e_5) = i^U/o^L$, Eligible Successor: e_1, e_6
$e_6 = (1, 3)$, label $(e_6) = i^L/o^L$, Eligible Successor: e_4

Table 2 Arrive and Leave Sets for Each Vertex.

vertex	$Leave^U$	$Leave^L$	$Arrive^U$	$Arrive^L$	$Arrive^{U,L}$
1	e_1	e_6	e_2		e_5
2	e_2	e_3			e_1
3	e_5	e_4		e_3, e_6	e_4

The construction of the correctly-ordered digraph D' proceeds as follows:
Arrive and Leave sets for all vertices of M are determined as in Table 2.
For each vertex v in V, $Leave^U[v]$ and $Leave^L[v]$ are not empty so vertices $1^U, 1^L, 2^U, 2^L, 3^U, 3^L$ are created in V'. Then every edge of E is considered:

- $e_1 \in Leave^U[1]$ and $Arrive^{U,L}[2]$, then construct $2^{U,L}$ in $V^{U,L}$, edges $(2^{U,L}, 2^U; -/-)$ and $(2^{U,L}, 2^L; -/-)$ in F and edge $e'_1 = (1^U, 2^{U,L}; a/1)$ in E_c.
- $e_2 \in Leave^U[2]$ and $Arrive^U[1]$, then construct edge $e'_2 = (2^U, 1^U; a/0)$ in E_c.
- $e_3 \in Leave^L[2]$ and $Arrive^L[3]$, then construct edge $e'_3 = (2^L, 3^L; b/1)$ in E_c.
- $e_4 \in Leave^L[3]$ and $Arrive^{U,L}[3]$, then construct $3^{U,L}$ in $V^{U,L}$, edges $(3^{U,L}, 3^U; -/-)$ and $(3^{U,L}, 3^L; -/-)$ in F and edge $e'_4 = (3^L, 3^{U,L}; b/0)$ in E_c.
- $e_5 \in Leave^U[3]$ and $Arrive^{U,L}[1]$, then construct $1^{U,L}$ in $V^{U,L}$, edges $(1^{U,L}, 1^U; -/-)$ and $(1^{U,L}, 1^L; -/-)$ in F and edge $e'_5 = (3^U, 1^{U,L}; a/1)$ in E_c.
- $e_6 \in Leave^L[1]$ and $Arrive^L[3]$, then construct edge $e'_6 = (1^L, 3^L; b/2)$ in E_c.

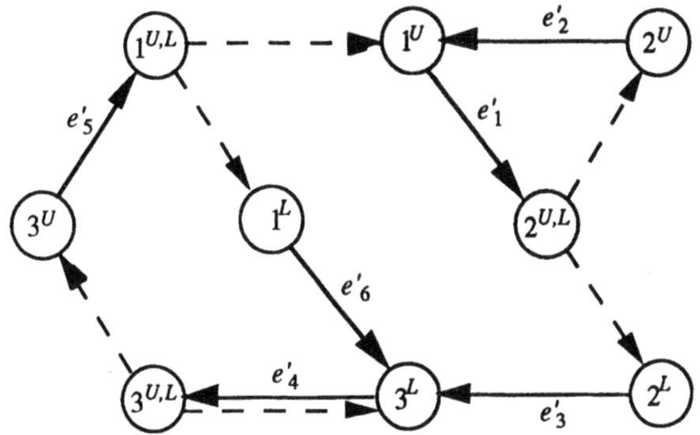

Figure 3 The Correctly-ordered Digraph D' of D in Figure 2.Edges in F are dashed.

SUIO sequences for each state are:

$SUIO^U(1) = a/1, a/0$	$SUIO^L(1) = b/2$
$SUIO^U(2) = a/0$	$SUIO^L(2) = b/1$
$SUIO^U(3) = a/1, a/1$	$SUIO^L(3) = b/0$

The construction of I proceeds as follows:

For each edge in E_c,

$tail(e'_1) \notin V^U \cup V^L$, mark e'_1

$tail(e'_2) = 1^U$ so, $I = I \cup I_{1U}$. Hence, $I = \{aa\}$

$tail(e'_3) = 3^L$ so, $I = I \cup I_{3L}$. Hence, $I = I \cup \{b\} = \{aa, b\}$

$tail(e'_4) \notin V^U \cup V^L$, mark e'_4

$tail(e'_5) \notin V^U \cup V^L$, mark e'_5

$tail(e'_6) = 3^L$, $I = I \cup I_{3L}$. Hence, $I = I \cup \{b\} = \{aa, b\}$.

For each marked edge in E_c,

$tail(e'_1) = 2^{U,L}$ and $I_{2U} = a \notin I$, $I_{2L} = b \in I$. Hence, $I_{2U,L} = b$.

$tail(e'_4) = 3^{U,L}$ and $I_{3U} = aa \in I$, $I_{3L} = b \in I$. Hence, $I_{3U,L} = b$.

$tail(e'_5) = 1^{U,L}$ and $I_{1U} = aa \in I$, $I_{1L} = b \in I$. Hence, $I_{1U,L} = b$.

Finally, $I = \{aa, b\}$, $R_{1U} = R_{2U} = R_{3U} = aa$ and $R_{1L} = R_{2L} = R_{3L} = b$

The construction of a cover C of D' proceeds as follows:

1) State Recognition Part:

$T_{1U} = T_{1L} = T_{1U,L} = -$

$T_{2U} = T_{2L} = T_{2U,L} = a$

$T_{3U} = T_{3L} = T_{3U,L} = bb$

$rT_{1U}aarT_{1L}brT_{2U}aarT_{2L}brT_{3U}aarT_{3L}b = raarbraaarabrbbaarbbb$

2) Transition Verification Part:

test segment($e'_1 = (1^U, 2^{U,L}; a/1)$) $= rT_{1U}\,a\,I_{2U,L} = rab$

which visits vertices $1^{U,L}1^U2^{U,L}2^L3^L$

test segment($e'_2 = (2^U, 1^U; a/0)$) $= rT_{2U}aI_{1U} = raaaa$
which visits vertices $1^{U,L}1^U2^{U,L}2^U1^U2^{U,L}2^U1^U$

test segment($e'_3 = (2^L, 3^L; b/1)$) $= rT_{2L}bI_{3L} = rabb$
which visits vertices $1^{U,L}1^U2^{U,L}2^L3^L3^{U,L}$

test segment($e'_4 = (3^L, 3^{U,L}; b/0)$) $= rT_{3L}bI_{3U,L} = rbbbb$
which visits vertices $1^{U,L}1^L3^L3^{U,L}3^L3^{U,L}3^L3^{U,L}$

test segment($e'_5 = (3^U, 1^{U,L}; a/1)$) $= rT_{3U}aI_{1U,L} = rbbab$
which visits vertices $1^{U,L}1^L3^L3^{U,L}3^U1^{U,L}1^L3^L$

test segment($e'_6 = (1^L, 3^L; b/2)$) $= rT_{1L}bI_{3L} = rbb$
which visits vertices $1^{U,L}1^L3^L3^{U,L}$

The input portion C of Q is *raarbraaarabrbbaarbbbrabraaaarabbrbbbbrbbabrbb*.
Q is a synchronizable checking sequence of length 46.

5 MINIMIZATION TECHNIQUES

To minimize the sequence Q produced by the proposed method, three complementary approaches can be used:

1) It is clear that the sequence derived in the example can substantially be reduced simply by eliminating the redundant subsequences. i.e. in cover C, eliminate rX if there is an rX' such that X is a prefix of X'. For example,
C = ***raa**rbr**aaa**rabrbbaarbbbrab**raaaa**rabbrbbbbrbbabrbb* where *raa* and *raaa* can be eliminated since they are prefixes of *raaaa*. This yields
C = ***rb**rabrbbaa**rbbb**rabraaaarabb**rbbbb**rbbab**rbb*** where *rb*, *rbb* and *rbbb* can be eliminated since they are prefixes of *rbbbb*. This yields
C = ***rab**rbbaa**rab**raaaa**rabb**rbbbbrbbab* where *rab* can be eliminated since it is a prefix of *rabb*.
Minimization using this approach yields C = *rbbaaraaaarabbrbbbbrbbab* of length 24.

2) In the example, the same transfer sequences are used in the state recognition and in the transition verification parts. By choosing different transfer sequences, one can increase the possibility of overlapping the test segments or of using shorter transfer sequences to a given state prior to verifying its outgoing edges. For example, since test segment(e'_6) = rbI_{3L} = *rbb*, one can take transfer sequence T_{3L} =*rb* to test e'_4 instead of *rbb*, yielding test segment(e'_4) = *rbbb*.

3) In the case that both children of a vertex $v^{U,L}$ are in I, the method requires that $I_{vU,L}$ should be the shorter of $\{I_{vU}, I_{vL}\}$, it would be better to choose the one that contributes to the greater reduction in the sense of (1). For example, for e'_4, there would be two possible test segments : *rbbbb* and *rbbbaa* and for e'_5 there also would be two possible test segments : *rbbab* or *rbbaaa*. Choosing test segment(e'_5) = *rbbaaa* implies that *rbbaa* can be deleted from C.

Combining these three approaches yields C = *raaaarabbrbbbrbbaaa* which is of length 19.

6 CONCLUSIONS

A method for constructing a synchronizable checking sequence of a given FSM M using UIOs has been proposed. The checking sequence constructed by this method has been shown to be easily reducible by eliminating redundancies and by making a wise choice of the transition sequences, and of the SUIO sequences to apply. The proposed method has been presented on the basis of some assumptions for the given FSM that are discussed hereafter. The first assumption states that an implementation of the given FSM possesses a correctly implemented reset feature which dramatically reduces the length of the state recognition part. Without this assumption, the resulting synchronizable checking sequence of the FSM will be restrictively long. In addition, the application of a reset breaks the connection in most real protocols, and thus can be utilized to form test cases from the resulting synchronizable checking sequence. The last four assumptions are necessary for any method that will attempt to construct a synchronizable checking sequence of a given FSM using UIOs. The second assumption assures that every transition of the FSM is part of a synchronizable transition sequence starting at the initial state. The third assumption assures that for each transition of the FSM, there is an SUIO synchronizable with the transition so that one can derive a test segment in the transition verification part. The fourth assumption requires that there should be at most one state that receives inputs from both U and L and is reached only by transitions with label $\in \{i^L/\text{-}, i^U/\text{-}, i^L/o^L, i^U/o^U\}$. The fifth assumption requires that any SUIO should be a synchronizable sequence for all states of the FSM. The last two assumptions assure the uniqueness of the SUIOs in an implementation of the given FSM.

Acknowledgments

This work is supported in part by the Natural Sciences and Engineering Research Council of Canada under grants STR0149338 and OGP0000976.

References

Aho, A.V., Dahbura, A.T., Lee, D. and Uyar, M.U. (1991) An optimization technique for protocol conformance test sequence generation based on UIO sequences and rural Chinese postman tours. *IEEE Trans. Comm.*, **39**, 1604-1615.

Chan, W.Y.L., Vuong, S.T. and Ito, M.R. (1989) An improved protocol test generation procedure based on UIOS. in: *Proc. SIGCOMM'89* , 283-294.

Chen, W.H. and Ural, H. (1995) Minimum-cost synchronizable test sequences based on multiple UIOs. *IEEE/ACM Trans. Networking*, **3**, 152-157.

Dahbura, A.T., Sabnani, K.K. and Uyar, M.U. (1990) Formal methods for generating protocol conformance test sequences. *Proc of IEEE*, **78**, 1317-1325.

Gill, A. (1962)*Introduction to the Theory of Finite-State Machines* . McGraw-Hill, New York, .

Hennie, F.C. (1964) Fault detecting experiments for sequential circuits. in: *Proc. Fifth Ann. Symp. Switching Circuit Theory and Logical Design* , 95-110.

ISO TC97/SC21, (1987) OSI conformance testing methodology and framework - Part 1-5, ISO 2nd and DP 9646-1 revised text (edited by D. Rayner), Vancouver, Canada

Lee, D. and Yannakakis, M. (1994) Testing finite state machines: state identification and verification. *IEEE Trans. Comput.*, **43**, 306-320.

Perez, M.E. (1993) Correctly-ordered Postman Tours and Synchronizable Test sequence generation for protocol conformance testing. *Master thesis*, Dept. of CSI, Univ. of Ottawa.

Rayner, D. (1987) OSI conformance testing. *Comput. Networks ISDN Systems.*, **14**, 79-98.

Rezaki, A., Ural, H. and White, G. (1994) Construction of checking sequences based on UIO sequences. *Proc. of ISCIS'94*, 315-326.

Sabnani, K.K. and Dahbura, A.T. (1988) A protocol test generation procedure. *Comput. Networks ISDN Systems.*, **15**, 285-297.

Sarikaya, B. and Bochmann, G.v. (1984) Synchronization and specification issues in protocol testing. *IEEE Trans. Comm.*, **32**, 389-395.

Tai, K.C. and Young, Y.C. (1995) Port-synchronizable test sequences for communications protocols. in this proceedings.

Ural, H. and Wang, Z. (1993) Synchronizable test sequence generation using UIO sequences. *Comput. Comm.*, **16**, 653-661.

26

Specification-driven Performance Monitoring of SDL/MSC-specified Protocols

P. Dauphin, W. Dulz, F. Lemmen

University of Erlangen-Nürnberg

IMMD VII, Martensstr. 3, 91058 Erlangen, Germany, Fax: +49-9131/85-7409,
email: {dauphin,dulz,lemmen}@informatik.uni-erlangen.de

Abstract

Protocol testing implies the functional analysis of a given implementation under test as well as its temporal performance evaluation. If protocols are formally specified sophisticated techniques and tools exist for analyzing functional properties, e.g. finding deadlocks or livelocks. Methods, however, for the temporal performance evaluation of formally specified systems are still in their infancy and tools for performance monitoring have to be developed.

In the telecommunications industry SDL and MSC specifications are common FDTs for developing reactive real-time systems. While SDL specifications describe the overall system architecture in detail, MSC scenarios are used to specify the load-dependent behavior of interacting system components.

For SDL/MSC specified systems, we present a method which allows a systematic quantitative temporal analysis of such systems. The basic idea is to apply MSCs in order to concentrate on the main parts of interacting SDL processes such as signal exchanges and timeouts. Thus, we can restrict the monitoring overhead by focussing on the most important points of interest in the interaction of communicating processes, and evaluating their behaviors by means of monitored event traces. Our approach allows us to integrate software and performance engineering techniques into a comprehensive methodology. A prototype tool called MISS was developed to proof the suitability and practicability of the method.

Keywords

Event-driven Monitoring, Message Sequence Chart (MSC), Performance Engineering, Specification Description Language (SDL), Specification-driven Monitoring

1 INTRODUCTION

Due to the ever increasing complexity of parallel and distributed systems integrated development and analysis tools which cover the whole life cycle of the system must be provided. The classic waterfall model for systems and protocol engineering defines all steps and phases which are necessary for managing the system's evolution, such as requirement analysis, design and specification, implementation, test and monitoring of the real system.

For reactive real-time systems like communication systems it is necessary to define models and to use paradigms which simplify the view of a complex system and which concentrate the attention on the development and analysis of the main parts of the system. Formal description techniques (FDTs) provide a unifying theoretical basis and permit

- to characterize the system behavior by abstractions such as creation and termination of concurrent parallel processes, sending and receiving of synchronizing messages between cooperating processes, setting and resetting of timeouts and transition conditions between different system states,
- to construct dedicated development tools such as graphical editors and user front-ends, functional and temporal validators, simulators, compilers, monitor and evaluation systems that support each step of the development process,
- to develop modular and hierarchical systems by reusing less complex subsystems with simpler functional or temporal properties,
- and last but not least to apply these techniques to a wide range of engineering problems.

An open issue concerning FDTs, however, is the development of methods and tools for studying the performance of systems which are specified and implemented by using an FDT approach. The performance evaluation starting from an abstract system model either can be achieved by mathematical analysis and simulation techniques in the design phase or by monitoring parts of the real system in the implementation and test phases. Because FDT environments support the automatic generation of executable code from a given formal system specification, monitoring tools can reuse the same specification as a formal monitoring model. One of the most promising ways of gaining insight into the temporal behavior of a system is *event-driven monitoring*, a technique which allows to construct event traces by monitoring the execution of real applications and selecting only those events which are related to interesting system states. The events are defined by inserting monitoring instructions into the program under investigation (*program instrumentation*). The monitoring instructions write event tokens to a hardware interface which is accessible for a hardware monitor (hybrid monitoring) or into a reserved memory area (software monitoring). Defining events with program instrumentation each monitored event token can be clearly assigned to a point in a program. This provides a problem-oriented reference. Thus, the evaluation of the event trace yields insight at the program level which is familiar to the program designer.

The main problems of event-driven monitoring, namely what to measure and where to find relevant instrumentation points inside the source program are answered implicitly using the technique of *model-driven monitoring* [Qui93, HQ93, HKM+94]. In model-driven monitoring the event, a common abstraction of event-driven monitoring and event-oriented models like graph models, Petri nets, queueing models, etc., is used to integrate modeling and monitoring. A so-called *monitoring model* is derived from the problem specification considering the aim of measurement. This monitoring model forms the basis for both the instrumentation of the program (*model-driven instrumentation*) and the evaluation of an event trace (*model-driven evaluation*) which is monitored during the execution of the instrumented program.

Formal monitoring models allow us to use monitoring procedures in a systematic and automatic way. The main disadvantage of model-driven monitoring, however, is the informal generation of the monitoring model, i.e. in general the model can not be derived automatically and may therefore be error-prone which means that the implementation and the monitoring model can be inconsistent [DKQ92].

Because model-driven monitoring is independent of the modeling technique used, FDT specifications may also serve for generating the monitoring model. Using the formal specification as monitoring model let us overcome the disadvantage of model-driven monitoring, that is the implementation is consistent with the monitoring model. We call this method *specification-driven monitoring*.

In this paper, we present the method of specification-driven monitoring by using the FDT technique SDL in conjunction with MSC. System design with SDL/MSC is reviewed in the next section. Section 3 describes the new method for MSC-based monitoring of SDL specifications. In section 4 the method is applied to the performance monitoring of the INRES protocol. Finally, in section 5 we review the key benefits of specification-driven monitoring.

2 SYSTEM DESIGN WITH SDL AND MSC

Among the most popular and widely accepted FDT techniques we can distinguish between the three standardized approaches: LOTOS [BB89], which is based on a process algebra with its origin in MILNERs CCS calculus; ESTELLE (ISO 9074) and SDL (ITU Z.100, [CCI92]), both of which describe the system behavior by communicating extended finite state machines. While process algebras are better suited in describing functional properties by embedded algebraic laws and logic inference techniques, finite state machines are more intuitive for decomposing a system into automata, state graphs and transition tables. The main advantage of the high-level specification language SDL compared to ESTELLE is the possibility to either use the user-friendly graphical SDL/GR syntax or the equivalent SDL/PR notation which is better suited for the integration of machine-oriented tasks and design tools. Besides, SDL is also a commonly accepted specification and description language in the telecommunications industry for the development of protocol architectures and communication systems.

While SDL allows the specification of architectural aspects by decomposing a system into a hierarchy of blocks and channels which consist of parallel processes and signal routes, the recently standardized MSCs (*M*essage *S*equence *C*harts, ITU Z.120) are well suited to describe the dynamic behavior of concurrent processes.

Each process is represented by a sequence of send and receive events which are logically ordered in an instance axis of the MSC diagram. It is also possible to define local and global process states which reflect major synchronization points between cooperating processes. For an example see Figure 4. The "*method for system development based on MSC, SDL and functional decomposition*" which is part of the "*SDL Methodology Guidelines*" in the appendices of Z.100 ([IT94]) describes the hierarchical top-down partitioning of a given system into lower-level subsystems until the process level is reached, thereby using SDL specifications and corresponding MSC scenarios.

Big efforts have been made recently to supply techniques and tools for SDL CAPE (*C*omputer-*A*ided *P*rotocol *E*ngineering) environments which provide automatic protocol validation, e.g. the formal proof that a given protocol specification is free of deadlocks or livelocks and does not contain unspecified message arrivals [Liu89], and directly produce implementations from given SDL specifications. A good overview of the state of the art can be found in [TOP94] which documents parts of the European TOPIC (Toolset for Protocol and Advanced Service Verification in IBC Environments) project.

Following [Liu89], the domain of protocol engineering is

- to provide a formal protocol specification by means of an appropriate FDT,
- to validate the functional correctness of the specification,
- to obtain some early indication of the expected efficiency,
- to compile major parts of the implementation directly from FDT specifications and finally
- to test the resulting implementation with respect to conformance, performance and reliability.

Most SDL CAPE environments like SDT [Tel93] or GEODE [GEO94] provide tool support for the functional validation, e.g. exploration of behavior trees in order to find system states which are involved in deadlock or livelock situations, but do not allow to explore temporal or performance aspects of SDL specified systems. Using SDL specifications as monitoring models allow us to integrate the method of model-driven monitoring in the SDL CAPE development cycle and thus, to remove the lack of performance evaluations tools for SDL specified systems.

3 MSC-BASED MONITORING OF SDL SPECIFICATIONS

Let us assume that the *system under study* (SuS) is totally specified in SDL and that various MSC scenarios are available which specify parts of the communication behavior within the SuS. One MSC together with the SDL system cover all the properties a monitoring model has to support for model-driven monitoring: the MSC indicates *what* to measure, whereas the SDL system shows *where* to instrument the SuS in order to get the desired insight into the dynamic behavior. The measurement preparation, during which the aim of the measurement is specified, consists of selecting a suitable MSC. Such an MSC which is consistent with the SDL specification of the SuS can be gained in numerous ways:

- The MSCs is derived in the requirement definition phase of the SDL/MSC software development cycle (see section 2).
- The MSC is constructed interactively by stepping through the description of the SDL system by means of an SDL editor.
- The MSC is generated while simulating the SDL system triggered by given signals, timeouts, and the values of internal state variables to enforce the desired behavior of the SDL system.
- The MSC is constructed from the reachability tree which was built up in the validation phase of the SDL system.
- The HASSEWithMSC tool (see section 4) was used to generate an MSC from a measured event trace.

In the following sections, we present a new technique which uses an SDL/MSC specification to carry out monitoring in a systematic way. The approach which applies the ideas of specification-driven monitoring is implemented in the tool MISS (*M*SC-based Mon*I*toring of *S*DL-*S*pecifications). MISS supports the monitoring process during the event selection, the program instrumentation, the measurement, and the analysis of a measured event trace.

3.1 MSC-based measurements of SDL systems

In order to prepare and execute the measurement of an SuS three tasks have to be carried out: *event selection*, *program instrumentation*, and the *measurement* itself. Figure 1 summarizes our approach for MSC-based measurements of SDL systems.

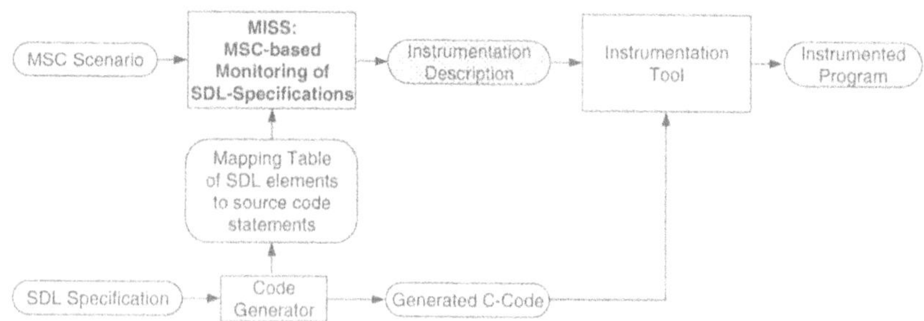

Figure 1 Generating an instrumented program from a given SDL/MSC specification.

Event selection:

The event selection is directly correlated with the amount of data to be collected during the measurement. To reduce the amount of data only the *points of interest* (PoI) should be selected for monitoring. Looking at an SDL/MSC-specified SuS PoIs can be found on three different levels of abstraction: With respect to the source code level each statement is a possible *program point of interest* (PPoI); concerning the SDL specification level each SDL element represents one possible *SDL point of interest* (SPoI); last but not least MSC elements in the MSC specification represent possible *MSC points of interest* (MPoI). Because an MSC specification focusses on the communication behavior which only reflects dynamic aspects of the complete SDL specification, and because each SDL element leads to numerous statements in the source code during program generation, the following relations between the different sets of PoIs hold:

$$\textit{number-of-PPoIs} \gg \textit{number-of-SPoIs} \gg \textit{number-of-MPoIs} \quad (1)$$

Eq. (1) shows a reduction of the complexity adhered to measurements: On the one hand, the number of points to be instrumented as well as the amount of data monitored will be limited by MSC-based monitoring. On the other hand, the MSC serves to focus on application-specific PoIs, and thus, supports the event trace analysis and the interpretation of evaluation results. In using an MSC for instrumentation, only the MPoIs indicating where to instrument the program are considered. The interdependences of MPoIs which are specified in the MSC are also used in MSC-supported event trace evaluation (see section 3.2). Therefore, an *MSC scenario* which is a MISS input structure cannot only be an ordinary set of MSCs, but also has to define relations between the MSCs of the MSC scenario in order to guarantee the desired communication behavior of the SuS. Well-defined relations are the *decomposition, repetition, concatenation, choice, and parallelization* of MSCs as described in the "method for system development based on MSC, SDL and functional decomposition" [IT94].

To use MSC-based event selection in the framework of specification-driven monitoring, a one-to-one mapping between the MPoIs and the PPoIs must be found. Because MSC and SDL specifications of an SDL/MSC-based monitoring model are consistent, a first straightforward one-to-one mapping between MPoIs and SPoIs is directly given. A second one-to-one mapping between SPoIs and PPoIs is generated by the source code generator which transforms the SDL specification automatically into compilable source code. This second mapping, however, is hidden by the transformation process.

In order to produce the second one-to-one mapping, the source code generator must reveal internals of the mapping of any given SDL elements to its corresponding source code statements (see Figure 1). Although an explicit assignment table would be the most convenient way to do this kind of mapping, SDL CAPE tools normally do not offer it. Instead, the following methods exist to define an implicit relation between SDL elements and corresponding source code statements:

- mapping of each SDL element to the *line number* inside the source code,
- mapping of each SDL element to the *identifier* of a specific procedure, and
- mapping of each SDL element to an unambiguous *annotation*, which is some kind of source code comment.

The selection of the method depends on the SDL/MSC CAPE tool used. Because we built the MISS prototype in conjunction with SDT (*S*DL *D*esign *T*ool) [Tel93] the third method was appropriate.

Instrumentation:

By using MPoIs MISS produces an *instrumentation description* (see Figure 1). For each instrumentation point, the instrumentation description contains a command which specifies

1. which annotation has to be instrumented, i.e. where to insert a monitoring statement, and
2. which monitoring statement must be inserted, that is a C-function which writes measurement events to a hardware interface or reserved memory areas.

We used the syntax and semantics of the instrumentation description language defined for the instrumentation tool AICOS (*A*utomatic *I*nstrumentation of *C* pr*O*gram*S*) [DHK+93].

Measurement:

The *instrumented program* must be compiled into an executable program. During the linking process the definitions of inserted monitoring statements (C-functions) have to be delivered. As MISS generates the instrumentation description, it exactly knows which bytes to send to the monitoring interface. Thus, in the case of software monitoring, the definition of monitoring statements can be generated by MISS automatically. In the case of hardware monitoring this function must be supplied by a linkable library.

During its execution each process of the instrumented SDL application produces a local event trace. If the local event traces are recorded by means of a common time basis*, they can be merged into a global event trace using the common time stamps in the local event traces for ordering. MISS generates a Unix shell script which carries out the merge.

3.2 MSC-based event trace analysis of SDL systems

For the analysis of measured event traces various tools are known, e.g. ParaGraph [HE91], TOPSYS [BR93] and TraceView [MHJ91]. We decided to use the universal monitor- and object-independent event trace evaluation environment SIMPLE (*S*ource-related *I*ntegrated *M*ultiprocessor and -computer *P*erformance evaluation, mode*L*ing, and visualization *E*nvironment) [Moh91]. The universality and independability of SIMPLE is achieved by

1. the *event trace description* language TDL [Moh92] which allows to describe the format of an arbitrary event trace in a problem-oriented way, and
2. *tool configuration files* which enables the adaption of each tool to the analysis of arbitrary object systems.

Both, the event trace description and the configuration files can automatically be generated by MISS given an MSC applied for the event selection (Figure 2). The problem-oriented TDL

*This can be achieved by synchronizing the distributed computers via the Network Time Protocol (NTP) [Mil94] or by carrying out the measurement using a distributed hardware monitor like ZM4 [HKM+94] which has a global clock.

event trace description is derivable due to the one-to-one mapping between MPoIs and PPoIs. The generation of tool configuration files depends on the aims of the measurement study.

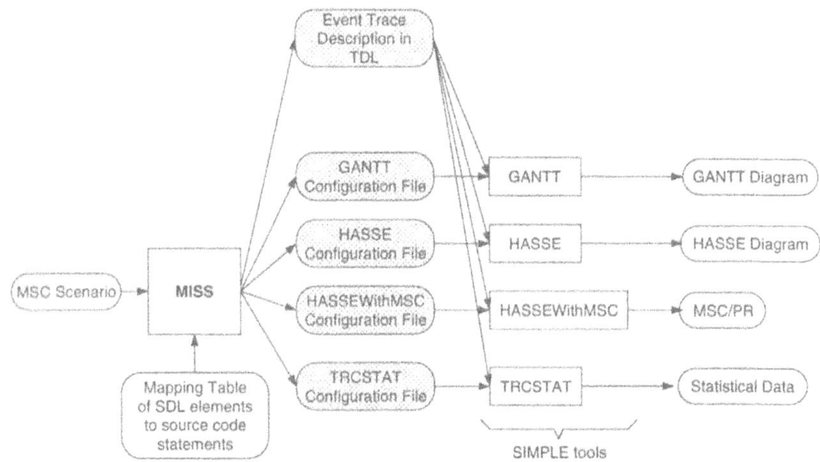

Figure 2 MSC-based trace evaluation with MISS and SIMPLE tools.

Event trace analysis is divided into *validation* where the consistency between the measured event trace and the corresponding SDL/MSC specification is checked, and *performance analysis* of the measured event trace which yields in various kinds of user-friendly diagrams (see Figure 5).

Validation:

From a given SDL specification an executable program only can be generated up to a certain degree - approximately 60% - by means of automatic procedures. The remaining part, called SDL environment, has to be provided by the user and therefore, is error-prone. Hence, an additional consistency check between the SDL/MSC specification which is the monitoring model and the measured event trace is necessary although all intermediate transformation steps are done automatically.

First, HASSEWithMSC generates an MSC from the measured event trace. The configuration file for HASSEWithMSC is yielded by MISS after interpreting causal dependences and state transitions provided in the SDL/MSC monitoring model. Then, the consistency between the derived MSC and the given SDL specification is validated using CAPE tools like SDT [Tel93] or GEODE [GEO94]. Note, that the validation should be performed before any other event trace analysis in order to be sure that the gained performance analysis results are meaningful.

Performance analysis:

For recognizing causal relationships HASSE is used. The PPoIs which are necessary for the generation of the HASSE configuration file are the setting, resetting, and expiring of timers, the sending and receiving of messages, and the creating and starting of process instances. All these PPoIs are directly related to MPoIs in the given MSC specification.

GANTT diagrams represent the state transitions of measured processes with respect to a common time. The application of GANTT for creating time-state-diagrams depends on the assignments of measured events to significant system states. Because states are defined in an MSC by using the keyword *condition* and events are defined by MISS while generating the instrumentation description, GANTT configuration files can be produced by MISS as well.

Besides behavior-oriented event trace analysis, MISS also supports statistics-oriented analysis using TRCSTAT which computes event frequencies and durations between events. To calculate message run times, events which represent the MPoIs for sending and receiving messages must be known; to compute timer expiration times MPoIs indicating the setting, resetting and expiring of the timer must be provided. The knowledge which events are necessary for calculating these values is contained in the MSC at corresponding MPoIs. Thus, TRCSTAT configuration files may be derived from an MSC too.

4 CASE STUDY: MONITORING THE INRES PROTOCOL

In this section we introduce the INRES protocol [Hog89] for discussing the main advantages of MSC-based monitoring for performance evaluation of SDL systems compared to conventional monitoring approaches. INRES offers a reliable, connection-oriented, and unsymmetrical service between two protocol entities called *Initiator* and *Responder* via its two service access points ISAPini and ISAPres. For data transmission INRES uses an unreliable, connectionless, and symmetric *Medium* service (see Figure 3). The protocol data units (PDUs) as well as the service primitives (SPs) of the INRES protocol are summarized in Table 1.

INRES is a three-phase protocol and supports a secure connection establishment between Initiator and Responder processes, an acknowledged data exchange and a final disconnection. During the connection establishment and the data exchange, the Initiator sets a timer directly after sending either a CR or a DT PDU. If the timer is adjusted too short a retransmission occurs and the communication will be delayed. In this situation, monitoring facilities are valuable in finding the reason of the delayed communication behavior.

Assuming a correct Medium service performance monitoring can be restricted to the INRES protocol instances exclusively. To prepare the specification-driven monitoring we choose an MSC specifying the communication behavior between the Initiator and Responder entities during connection establishment and data exchange, thereby neglecting all details concerning the Medium service. This MSC is shown in Figure 4 where the setting, resetting, and expiring behavior of the Initiator timer after sending each CR and DT PDU is of special interest.

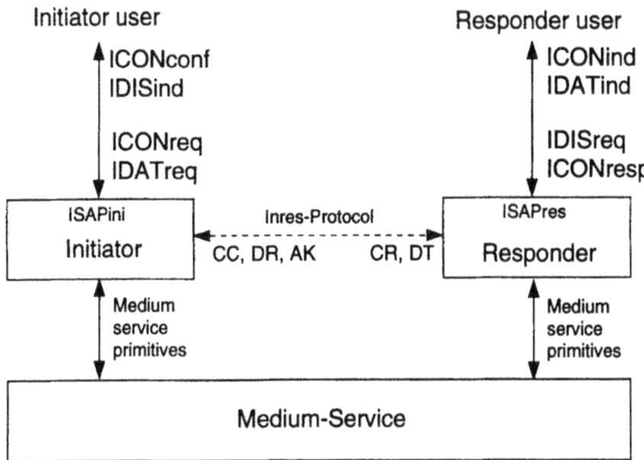

Figure 3 The INRES Protocol [Hog89].

PDU	Meaning
CR	connection request
CC	connection confirmation
DT†	data transfer
AK‡	acknowledgement
DR	disconnection request

SP	Meaning	PDU
ICONreq	connection request	CR
ICONind	connection indication	CR
ICONresp	connection response	CC
ICONconf	connection confirmation	CC
IDATreq	data request	DT
IDATind	data indication	DT
IDISreq	disconnection request	DR
IDISind	disconnection indication	DR

Table 1: INRES Protocol Data Units and Service Primitives

MISS is using this MSC for creating an appropriate instrumentation description (see Figure 1). For each specified state, i.e. `Disconnected`, `Waiting`, `Connected` and `Sending` as well as for each received and sent SP and PDU defined by the MSC, AICOS instrumentation statements are generated to carry out the instrumentation of the C code. The following representative extract of an instrumentation description describes the instrumentation for sending a CR PDU:

† Parameters of the DT PDU are a sequence number and the service data unit.

‡ The AK PDU has a single parameter, the sequence number.

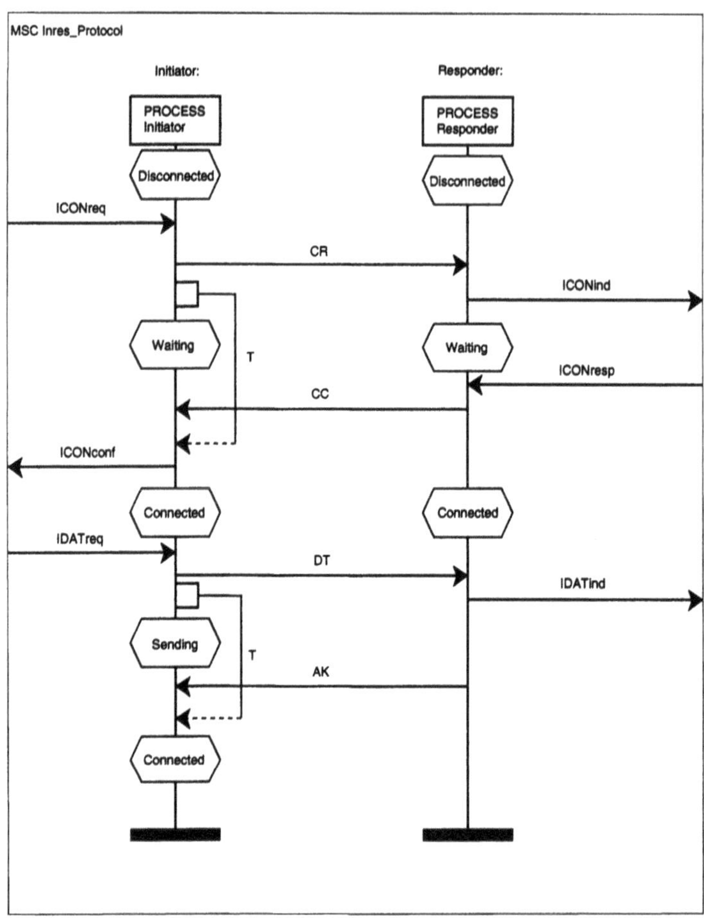

Figure 4 MSC used for monitoring the INRES protocol.

```
!OUT_cr
#BEFOREANNOTATEDCALL SDL_Output /* "6 ConnectionCreate 1 30 100" */
Initiator.c
trace_2(14, P->Self.GlobalNodeNr);
```

The first line indicates the event name (`OUT_cr`) which is automatically generated by MISS. The prefixes `OUT_` and `IN_` are used to denote outgoing and incoming events, respectively. In the second line the AICOS command `#BEFOREANNOTATEDCALL` indicates the insertion of the statement `trace_2(14, P->Self.GlobalNodeNr);` directly before

HASSE configuration file	GANTT configuration file

<pre>DEPENDENCE 'CR' IS
  OUT_CR {1,} ON PROCESS Initiator
     -> IN_CR ON PROCESS Responder
END
DEPENDENCE 'CC' IS
  OUT_CC ON PROCESS Responder
     -> IN_CC ON PROCESS Initiator
END
DEPENDENCE 'AK' IS
  OUT_AK ON PROCESS Responder
     -> IN_AK ON PROCESS Initiator
END
DEPENDENCE 'DT' IS
  OUT_DT {1,} ON PROCESS Initiator
     -> IN_DT ON PROCESS Responder
END
DEPENDENCE 'T' IS
  SET_T ON PROCESS Initiator
     -> { IN_T, RESET_T }
      ON PROCESS Initiator
END</pre>
	<pre>FOR PROCESS Initiator DO
	  GANTT DIAGRAM WITH
	    Disconnected: STATE_disconnected;
	    Waiting: STATE_waiting;
	    Connected: STATE_connected;
	    Sending: STATE_sending;
	  END
	END
	
	FOR PROCESS Responder DO
	  GANTT DIAGRAM WITH
	    Disconnected: STATE_disconnected;
	    Waiting: STATE_waiting;
	    Connected: STATE_connected;
	  END
	END</pre>

Table 2 HASSE and GANTT configuration file derived from the MSC§.

the first call of the function `SDL_Output` and after the occurrence of an annotation `/* "6 ConnectionCreate 1 30 100" */` in file `Initiator.c`. SDT annotations are unambiguous because they contain the X- and Y-coordinates of the corresponding SDL elements in the SDL/GR specification. Thus, there exists a one-to-one mapping between the SDL specification and its annotations. Instrumentation statements for other MPoIs given in the MSC of Figure 4 will be built in an analogous way.

After the automatic compilation by SDT the instrumented INRES protocol was executed on two SUN Sparc workstations interconnected via an IEEE802.3 LAN. The local clocks are synchronized using the network time protocol NTP [Mil94] which guarantees a common time base with a sufficient accuracy of 1 ms. This enables the merge of the local event traces into a single global one with events ordered according to increasing time stamps.

MISS also generates configuration files for the various SIMPLE event trace analysis tools. Table 2 shows the derived configuration files for GANTT and HASSE, words in upper-case letters are keywords of the configuration languages.

The HASSE configuration file contains a causal dependence for each PDU exchanged between the Initiator and Responder entities plus one dependence describing the causal relationship between the timer events. Each dependence consists of a causing event on the left hand side of the arrow (`->`) and a dependent event on its right hand side. The event name

§Events representing state entries or exits are prefixed with `STATE_`. For timers the prefixes `SET_` `RESET_`, and `IN_` for setting, resetting or expiring of the timer are used.

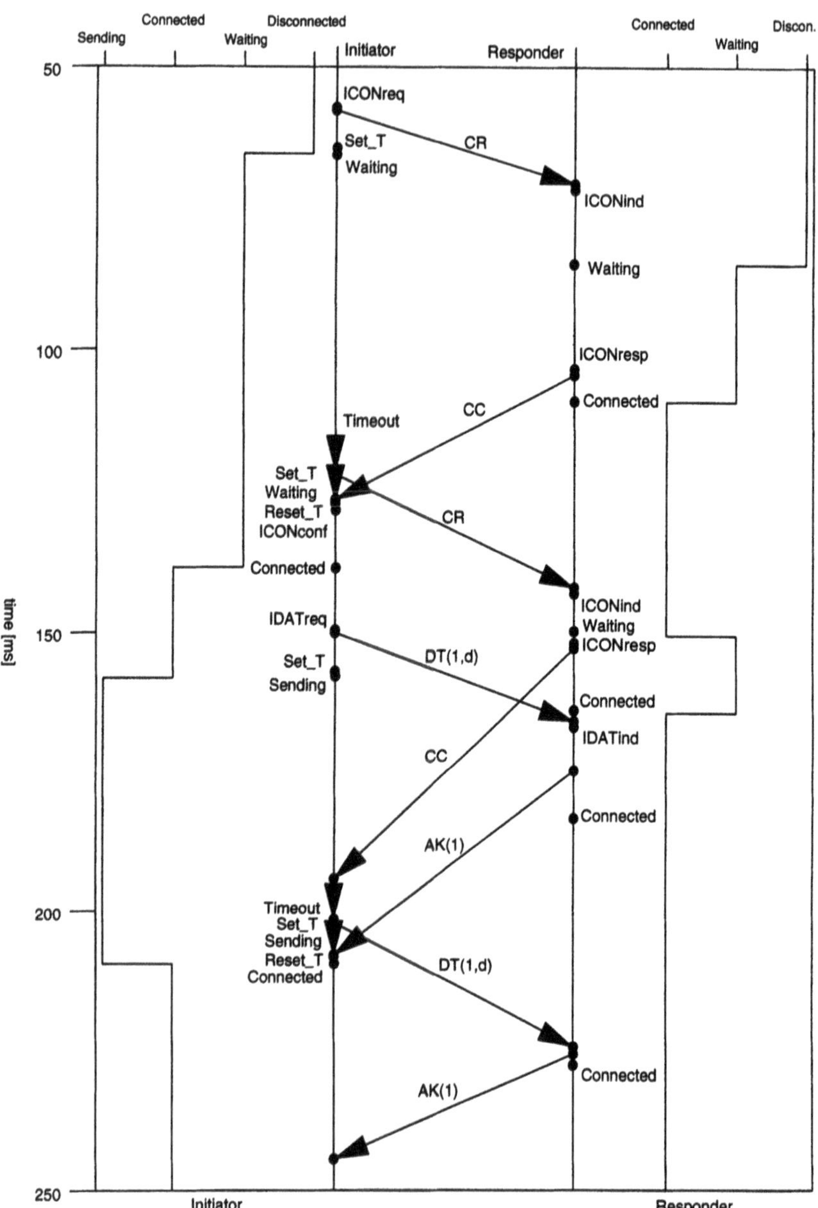

Figure 5 Combined HASSE and GANTT diagram.

together with the process where the event occurs are given. In the GANTT configuration file two time-state-diagrams are specified. The diagram for the Initiator consists of the four states `Disconnected`, `Waiting`, `Connected` and `Sending`, whereas in the Responder diagram the `Sending` state is missing. The events associated with the different states are given behind the state names separated by a colon.

Figure 5 shows the result of analyzing the measured event trace with GANTT, respectively HASSE, and using the configuration files of Table 2. Here, the output of both tools are visualized in one single diagram: the HASSE part is drawn in the middle while on the left and right GANTT time-state-diagrams of the Initiator and Responder are shown, respectively. The HASSE diagram reveals that two CR PDUs are sent by the Initiator before the connection is established successfully. Shortly, after the first CR PDU is sent the timer T is set to an initial value of 60 ms and expires after this time period. This causes a second CR PDU to be sent by the Initiator because the expected Responder CC PDU arrives too late. *Obviously, the timer's expiration time is adjusted too short.* Despite of the late arrival of the first CC PDU, the Initiator consumes it, notifies a successful connection establishment to the user (ICONconf event), and changes into the `Connected` state. While connected a user IDATreq request causes the Initiator to send a DT PDU to the Responder, to set a timer, and to turn into the `Sending` state while waiting for a Responder AK PDU. Again, the timer expires before the Responder AK PDU arrives which causes the Initiator to repeat sending the first DT PDU. In contrast to the receipt of the first DT PDU where the Responder indicates the exchange of user data with an IDATind, the Responder ignores doubled data¶ by sending a repeated AK PDU to the Initiator.

Comparing the given MSC in Figure 4 with the event trace evaluation results presented in Figure 5 explains the delayed communication behavior of the INRES protocol. The obvious conclusion from Figure 5 is to *enlarge the timer value* in order to save Medium bandwidth and to speedup the communication between Initiator and Responder entities. In general, the presented method of MSC-based monitoring of SDL systems gives us a tool to adjust time-sensitive parameters of complex parallel and distributed systems.

5 CONCLUDING REMARKS

In this paper we have introduced a new method which is suitable for monitoring complex real-world applications specified with SDL: MSC-based monitoring. The main advantages of the method can be summarized as follows:

- Event definition, the hardest job when dealing with monitoring, is supported by the MSC scenario in a systematic way.
- MSC-based event definitions enables automatic source code instrumentation.
- MSC scenarios focus on application-specific PoIs for performance evaluation and neglect surplus details.
- Interdependences provided in MSC scenarios support automatic generation of meaningful and application-specific monitoring experiments.

¶The Responder recognizes doubled data by means of the sequence number transmitted in each DT and AK PDU.

- Only fully automatic specification-driven monitoring allows the user-friendly integration of monitoring tools in the SDL/MSC software engineering cycle.
- MSC-based techniques can be applied for analytical and simulative performance studies of SDL specified systems which is outlined in [Dul94] and is the topic of further research.

Note, that the method described so far may also be applied in a cyclic and hierarchical way. Suppose the performance results of an MSC-based event trace analysis indicate a strange behavior in parts of the SuS. The interpretation of the performance results will require a hierarchical refinement of the MSC-based monitoring to get deeper insight in the identified parts of the SDL system. It will lead to new measurements and detailed performance results concerning the observation of the refined parts of the SDL system. The hierarchical refinement is supported by both, the hierarchical structure of the MSCs which may consist of sub-MSCs, and by utilizing detailed knowledge of the SDL system itself. This cyclic and hierarchical approach serves for iteratively improving an SDL specification, e.g. to eliminate a bottleneck found.

The proposed method of specification-driven monitoring is also suitable to examine whether the automatic generation of C code from a given SDL specification produces an implementiation which meets a required efficiency.

Besides, the statistical values calculated with MSC-based monitoring can be applied to attribute a given MSC with realistic parameters in order to carry out performance simulations for performance prediction of alternative solutions of the system under study.

REFERENCES

[BB89] T. Bolognesi and E. Brinksma. Introduction to the ISO Specification Language LOTOS. In P.H.J. van Eijk, C.A. Vissers, and M. Diaz, editors, *The Formal Description Technique LOTOS*, pages 23–73, Amsterdam, 1989. North-Holland.

[BR93] T. Bemmerl and B. Ries. Programming Tools for Distributed Multiprocessor Computing Environments. *International Journal of High Speed Computing*, 5(4):595–615, December 1993.

[CCI92] CCITT. *Recommendation Z.100: Specification and Description Language SDL, Blue Book.* ITU General Secreteriat — Sales Section, Place des Nations, CH-1211 Geneva 20, 1992.

[DHK+93] P. Dauphin, F. Hartleb, M. Kienow, V. Mertsiotakis, and A. Quick. PEPP: Performance Evaluation of Parallel Programs — User's Guide – Version 3.3. Technical Report 17/93, Universität Erlangen–Nürnberg, IMMD VII, September 1993.

[DKQ92] P. Dauphin, M. Kienow, and A. Quick. Model-driven Validation of Parallel Programs Based on Event Traces. In Topham, Ibbett, and Bemmerl, editors, *Proceedings of the "Working Conference on Programming Environments for Parallel Computing", Edinburgh 6–8 April*, pages 107–125, 1992.

[Dul94] W. Dulz. Leistungsbewertung von SDL/MSC-spezifizierten Protokollen . In *GI/ITG-Arbeitsgespräch "Modellierungs- und Bewertungstools 94", Uni Dortmund*, 1994.

[GEO94] GEODE. *Geode - Technical Presentation* . VERILOG SA, France, 1994.

[HE91] M.T. Heath and J. A. Etheridge. ParaGraph: A Tool for Visualizing Performance of Parallel Programs. Technical report, Oak Ridge National Laboratory, Tennessee, November 1991.

[HKM+94] R. Hofmann, R. Klar, B. Mohr, A. Quick, and M. Siegle. Distributed Performance Monitoring: Methods, Tools, and Applications. *IEEE Transactions on Parallel and Distributed Systems*, 5(6):585–598, June 1994.

[Hog89] Dieter Hogrefe. *Estelle, LOTOS und SDL*. Springer, Berlin, 1989.

[HQ93] F. Hartleb and A. Quick. Performance Evaluation of Parallel Programms – Modeling and Monitoring with the Tool PEPP. In B. Walke and O. Spaniol, editors, *Proceedings der 7. GI–ITG Fachtagung "Messung, Modellierung und Bewertung von Rechen- und Kommunikationssystemen", Aachen, 21.–23. Spetember 1993*, pages 51–63. Informatik Aktuell, Springer, 1993.

[IT94] ITU-T. *Recommendation Z.100: Appendices I and II*. International Telecommunication Union - Telecommunication Standardization Sector, Place des Nations, CH-1211 Geneva 20, 1994.

[Liu89] Ming T. Liu. Protocol engineering. In Marshall C. Yovits, editor, *Advances in Computers*, volume 29, pages 79–195. Academic Press Inc., Boston, 1989.

[MHJ91] A.D. Malony, D.H. Hammerslag, and D.J. Jablonowski. Traceview: A Trace Visualization Tool. *IEEE Software*, September 1991.

[Mil94] D.L. Mills. Improved algorithms for synchronizing computer network clocks. *Computer Communication Review*, 24(4):317–327, October 1994.

[Moh91] B. Mohr. SIMPLE: a Performance Evaluation Tool Environment for Parallel and Distributed Systems. In A. Bode, editor, *Distributed Memory Computing, 2nd European Conference, EDMCC2*, pages 80–89, Munich, Germany, April 1991. Springer, Berlin, LNCS 487.

[Moh92] B. Mohr. SIMPLE — User's Guide Version 5.3.. Technical Report 3/92, Universität Erlangen–Nürnberg, IMMD VII, März 1992.

[Qui93] A. Quick. *The M^2–Cycle: Model-driven Monitoring for the Evaluation of Parallel Programs (in German)*. PhD thesis, University of Erlangen–Nürnberg, 1993. to be published.

[Tel93] TeleLOGIC. *SDT Version 2.3 Reference Manual, Volume 1 and 2, SDT Users Guide, SDT Technical Presentation*. TeleLOGIC, Malmö, 1993.

[TOP94] Project TOPIC. Integration of the Toolset Prototypes (V1) . Technical report, CEC Deliverable Number R2088/GMD/SEM/DS/L/013/b5, March 26, 1994, 1994.

PART TEN

Test Management

Test Management and TTCN based Test Sequencing

Jining Tian and Jianping Wu
Department of Computer Science
Tsinghua University, Beijing, 100084, P. R. China
wjp@tsinghua.edu.cn

Abstract

This paper presents a test management system named TMS which is capable of providing necessary management and organization of conformance testing according to the definition of conformance assessment process in the OSI Conformance Testing Methodology and Framework [1]. TMS is also able to dynamically sequence the TTCN test cases during test execution. The test sequencing is based on the ordering relation among test cases, which can solely derived from the TTCN test suite. An operational semantics of TTCN is defined here using the Labeled Transition System (LTS) to model the derivation of this relation. Design of the TMS also provided. Moreover, the TMS is an integrated part of our Protocol Conformance Testing System (PCTS). This paper also demonstrates how certain features of TMS, when combined with the characteristics of PCTS, can promote the overall system performance.

Keywords

Conformance Testing, Test System, Test Management, Test Sequencing

1 INTRODUCTION

The ISO has investigated considerable efforts in the elaboration of a common framework for protocol conformance testing, namely the OSI Conformance Testing Methodology and Framework [1]. According to this document, the Conformance Assessment Process (CAP) is the key activity in protocol conformance testing and the Means of Test (MOT) is the principal facility to carry out the process. Test Management System plays an important role in the protocol conformance testing process. Its target is to manage and organize the conformance assessment process carried out by the test laboratory for a specific test client. It covers the complete test preparation phase and the test operation phase as defined in [1].

The Protocol Conformance Testing System (PCTS) developed by Tsinghua University is an integrated platform for protocol conformance testing. PCTS has three major subsystems: the Test Management System, the Test Execution System, and the Test Result Report and

Analysis System. Three main functions of TMS are test organization, test suite management, and test document management.

The test management system we presented here has some unique characteristics. First, we claim a strict accordance to ISO standards. All the concepts, methods and procedures adopted in PCTS should strictly follow their definitions in the conformance test standard. Second, we have a TTCN based test suite management. TTCN is adopted as the basic notation in TMS. It is the only tangible form of test suite that exists in our TMS and PCTS. Third, we have a TTCN based test sequencing. All the ordering relations are derived solely from the TTCN test suite. By doing so, the TMS has become protocol independent and has gained great flexibility. Moreover, we adopted formal methods in the specification of our ordering relations. we here put forward a formal model of the operational semantics of TTCN in terms of the Labeled Transition System (LTS). Finally, we provide a user friendly interface and a convenient document management system.

The importance of test management can be seen from the following several aspects. First, it is an important step towards test automation. The test organizer will guide you step by step through the conformance assessment process without human interference. Second, the test sequencing module within TMS will systematically execute the test cases in the most efficient order. Besides, the TMS is also an important way to ensure the accordance with the standard and to enhance the comparability and the acceptability of the test results. Finally, the TMS provides a bunch of tools to promote test efficiency. The TMS, together with our test case generation tool TUGEN [4], the TTCN editing tool and the test execution system, have covered the whole life cycle of a test case. These tools thus formed an integrated platform for developing protocol conformance test suites.

The outline of this paper is as follows. Section 2 presents the basic functions of test management. Section 3 gives out a formal approach towards TTCN based test sequencing. Section 4 describes a design of the TMS according to the models presented above and discusses some technical issues. Section 5 shows an example of test sequencing. Section 6 concludes the paper.

2 TEST MANAGEMENT

2.1 Overview

There exists many useful tools for protocol conformance testing, each of which can more or less perform the management functions defined above. The XRTLE is a general purpose OSI tester from Alcatel TITN Inc. It can perform the SUT management function and can provide some test case selection criterion such as selection by PICS/PIXIT, by historical information or by human interference [18]. The IDACOM PT500 series testers are proprietary. They enjoy a high reputation in the testing of OSI lower layer protocols. They have a terminal based management system and also provide PICS/PIXIT selection [19]. The Test Case Management System (TCMS) from UBC can manage multiple relations among the test cases. Its static selection module supports PICS/PIXIT selection and its dynamic selection module supports test sequencing [6]. But none of them can perform the whole range of functions provided by TMS and they seldom can directly manage the TTCN test suites. The TMS distinguishes itself by the following characteristics.

Test organization has a full coverage of all the steps that must be carried out in the test preparation and the test operation phase. It is a powerful tool which can provide useful guidance through out these two phases.

Test suite management is based on TTCN, which is now the only standard test notation in CTMF and is widely accepted as a test specification language. We gained remarkable advantages by adopting it as the basic format of test suite in TMS.

- The TMS is protocol independent and protocol test method independent;
- The test case can be created dynamically with our embedded TTCN editor;
- Confusion, disintegration and ambiguity caused by intermediate notations can be avoided;
- The exchange of test suite with other test laboratories will be easy;
- Interpretation based test execution strategy [3];
- Powerful test suite management and selection functions.

Convenient document management and user friendly interface have been invested to ease the cross-reference and information retrieval.

The structure of TMS is shown in Figure 1. Following are the detailed discussions of the major roles of TMS.

2.2 Test Organization

The test organization is responsible for the organization of the conformance assessment process. It is used here to ensure that all the procedures involved in PCTS are carried out in exactly the same order and manner as they are defined in CTMF. It contains the following sub-modules.

Static Conformance Review
This step is used to check whether the mandatory options in the conformance requirements has been claimed in the PICS. The capability test is the actual test of these options.

PICS/PIXIT Based Test Case Selection
The TMS must maintain a relation between the test cases and the PICS/PIXIT items, then select the test cases accordingly. Other selection methods are also available: selection by test group, by PDU/ASP used, by variables involved, by test steps it invoked, or by any attribute associated with test cases. Manual adjustment of the selection is also possible.

Test Sequencing
Test selection can be extended into the test campaign by the test sequencing. This action is beyond the scope of CTMF. It is done solely for the purpose of promoting test efficiency.

Test Parameter Preparation
Since we adopt the interpretation based test execution, no tangible Executable Test Suite (ETS) exists in the system. Only necessary parameter tables are obtained from the client information module and are provided to the TE.

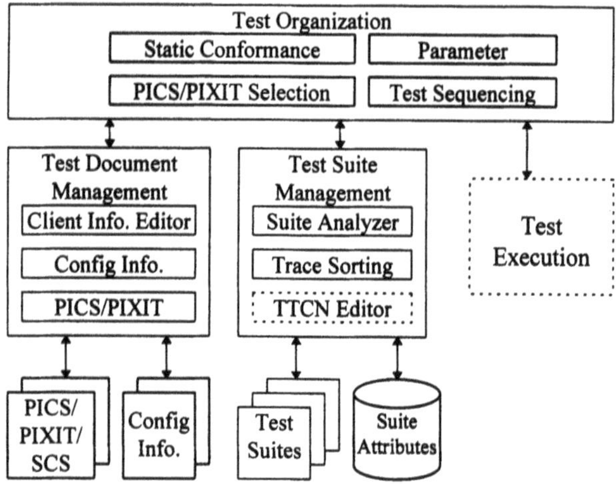

Figure 1 Structure of Test Management System

2.3 Test Suite Management

Test suite is the basis of the whole protocol conformance testing. It is involved in nearly every part of the conformance assessment process. This function of the TMS is to prepare the test suite for the conformance testing, especially for the PICS/PIXIT selection and test sequencing. It can be further divided into suite analysis and trace sorting.

The suite analyzer extracts various attributes from the TTCN test suite and stores them in a relational database. These attributes include test suite structure, selection expression, and other attributes of each test case such as its ID, variables used, PDU/ASP involved, etc. The analyzer also calculates the traces of each test case according to our definition of the TTCN machine which is introduced in section 3. These traces are then used by the trace sorting module to obtain the ordering relations which are defined in section 3. These relations are used by the sequencing module to determine the execution sequence of the test cases.

The TTCN editor can also be seen as an integrated part of TMS although it can also be invoked as an independent utility. This X-windows based editor can perform the basic maintenance role for TTCN. It can be invoked from any point within PCTS or TMS to support on-the-spot modifications of the test suite. It can translate test cases between GR form and MP form. The embedded TTCN editor together with our interpretation based TTCN execution gives PCTS the greatest flexibility. We can write a new test case on the spot and then execute it immediately.

2.4 Test Document Management

Several X-window based tools are provided here to facilitate user information registration and document generation.

- Check list generator: It indicates the information provided and needed by the test laboratory. The exchange of checklists are helpful in the choice of Abstract Test Method (ATM) and Abstract Test Suite (ATS).
- Client Information Editor: This editor defines the SUT and the IUT based on the user information, then generates the SCS describing the relation among SUT, IUT and relevant protocol standard. Administrative information is stored in database to ease future information retrieval.
- Universal PICS/PIXIT editor: It can read the PICS/PIXIT proforma written in a special format, display it in a window after the user has filled it in, then generate the PICS and PIXIT document.
- Configuration Information Editor: According to the PICS/PIXIT document, especially based on the ATM and ATS selected, the configuration of the tester can be determined. This includes the configuration of underlying service provider, auxiliary channels, and all the other entities in the test environment. Entities in PCTS are connected by a global message delivery system. Virtual channels must be established between related entities.

3 TTCN BASED TEST SEQUENCING

3.1 Basic Concepts

The test cases in TTCN are organized into a hierarchy of test groups. This grouping can provide some information about the relations among test cases, because test cases within the same group usually have related test purposes. However this do not imply any execution order among test cases.

The CTMF distinguishes four types of testing according to the extent to which they provide indication of conformance [1]: basic interconnection test (BIT), capability test, behavior test, and conformance resolution test (CRT). A list may be included in the test suite indicating which test cases are appropriate for BIT, and capability tests may be grouped in a separate group. These are just some rough indication of the test execution order, e.g., usually the BIT are execute first, then capability test, last the behavior test. They can not provide any finer information regarding to testing order within a certain test type, especially within the behavior tests which is the major part of an ATS. The above distinction is still an insufficient indication of execution order.

According to CTMF, each test case can be executed independently. But there do exists some relation among the results of test cases [5]. When a test case fails, some other test cases may be very likely to fail. We use "very likely" here because we even can not guarantee the failure of the same test case if it is executed again. In order to determine the test case execution order, we can define a "ordering relation" among test cases.

3.2 Obtain ordering relation from protocol specification

The traditional way of deriving the ordering relation is to start from the protocol specification. The Test Case Management System developed at University of British

Columbia [5] adopted this method. They called it "dynamic selection". According to their method, a relational model named Abstract Test Case Relation Model (ATCRM) can be constructed from the state machine of the protocol specification. Then, the test cases in a test suite can be mapped onto this relational model according to their IO paths to establish the ordering relations.

This method has several disadvantages. First, if we want to derive a relation on a test suite automatically, we must first obtain the formal descriptions of the related protocol specification written in a certain FDT. Second, the relation model may not match well with the test suite, in terms of the granularity of states. Third, the test case and the specification must be written in the same FDT or at least, has the same semantic model. Otherwise the mapping from test cases to the relational model will be very difficult.

3.3 Obtain ordering relations from test suite

In order to get rid of the above restrictions, we adopted a new approach to derive the ordering relation solely from the test suite. Here we can not obtain the state information from the protocol specification, but this information is equally unavailable during the test campaign. The MOT can understand the IUT only through interactions at PCO. We thus can only model the IUT together with the underlying service provider as a black box. In another word, we are going to give the MOT some intelligence to let the MOT learn gradually from the IUT, predict its future based on its past, and adjust the tester's behavior accordingly.

This approach has the following advantages. Since it is derived solely from the test suite, there will be no matching or mapping problems. Moreover, we make the sequencing method protocol independent because we do not assume the existence of protocol specification. Last, we can enjoy great flexibility introduced by this method due to its minimum requirements on outer conditions. Once we obtained a test suite in TTCN, we can right away carry out the sequencing against it.

3.4 A formal approach towards test sequencing

Although TTCN has already become an international standard, a standard concerning its formal semantics has yet to be developed. An operational semantics of TTCN based on the Labeled Transition System (LTS) is discussed in [8,9] which emphasizes its compositional structure.

In the following discussion, we are going to establish a basic semantic model for TTCN using the LTS. Our model focuses on the traces of a test case and it intends to provide an unambiguous definition of our ordering relations and also to achieve a sound mathematical foundation for test sequencing. This enables us to reason about the properties of these relations so that further optimization of test execution is possible. This formal model is also of great importance to our PCTS system because it has a TTCN based test case interpreter which bares a great similarity to the model. The properties of this model may also provide some useful guidance to the further development of the TTCN interpreter.

We first give out the definition of the TTCN machine and a simplified basic TTCN machine in terms of the LTS. Then, according to this model, the traces of the test cases and the prefixing relations among them are presented. We further put forward some auxiliary notations based on the previous definitions. Next, the ordering relations are discussed. Unless explicitly explained, we always assume that the test results are repeatable. Of course this may not always be the case. At last, the non-determinism is discussed.

3.5 Labeled Transition System and TTCN machine

Def. 1 A LTS is a 4-tuple $<S, L, T, s_0>$, where S is a set of states, L is a set of observable actions. Let τ denotes the unobservable action, and δ denotes the successful termination. $T \subseteq S \times \{L \cup \{\tau\} \cup \{\delta\}\} \times S$ is the set of transitions. It can also be written as $s \xrightarrow{u} s'$ where $s, s' \in S$, $\mu \in L \cup \{\tau\}$, $(s,\mu,s') \in T$.

Def. 2 The syntax of a Test Behavior Expression (TBE) is defined as follows:
$B =_{def}$ stop | exit | ?a;B | !a;B | B[]B | B>>B | [q];B | [I:=v];B

The operator precedence is ';' > '[]' > '>>'. q is the qualifier which is a Boolean expression. Its syntax and semantics are ignored here just for simplicity. This is an algebraic representation of the Abstract Evaluation Tree defined in [1]. The TBE can express the most important syntactical components in the dynamic part of TTCN, but it is much simpler. So that it can be handled by our basic TTCN machine defined below. This test behavior expression is a formal representation of the Abstract Evaluation Tree defined in [1].

We can first translate TTCN into TBE. Let *Texp* denotes the set of all TBEs. Let *Texpr*(t) denotes the function that does this translation. *Texpr*(t) = B_t . $B_t \in$ *Texp*.

Ex 1: Sequential composition is mapped onto action prefix.

```
!a
  ?b        ⇒     !a (?b; stop [] ?c; stop )
  ?c
```

Ex 2. Attached trees are viewed as a separate process:

```
!a            B:    ?d
  +B                  !e   ⇒          !a ; B >> ?c
      ?c            ?f(Inconc)   or   !a ; ( ?d ; !e ; exit [] ?f ; stop ) >> ?c
```

Ex 3. Loops are translated into recursions:

```
?a            L1
  !b                ⇒     L1 = ?a ; !b ; L1
      → L1
```

Def. 3 A TTCN Machine, is a virtual machine that can perform the evaluation of the TTCN test case. The state of this TTCN machine can be represented by a 4-tuple: (*stack*, *env*, *ctrl*, *sto*). It has four parts:

1) Internal stack: *stack* is the place to store some intermediate results;
2) Environment: *env* preserves the information about variable type and scope;
3) Control Part: *ctrl* contains the TTCN test case to be evaluated;
4) Storage Part: *sto*=(m, i, o) $\in$ *Mem*×*Input*×*Output*, where m $\subseteq$ *Mem*=*Ident*×*Value*, is a set of pairs represents the identifier and its

corresponding value. *i* and *o* are the input queue and output queue respectively. Elements in these queues are called messages and each representing an event. Each element has a PCOid and an event type.

Since our aim is to specify the ordering relations, we can omit some of the details in this module, such as the stack for the calculation of expressions and the analysis of complex sentence constructions, and the environment.

Def. 4 A Basic TTCN Machine is a simplified TTCN machine. Its state has only two parts: the control part and the storage part:

1) Control part: $ctrl \in Texp$, is the TBE to be evaluated. In order to avoid getting entangled into the complicated TTCN syntax, we use TBE instead of TTCN. Thus we can achieve a simple and elegant model which is nevertheless sufficient to describe our ordering relations.
2) The storage part is the same as that of a TTCN machine.

The behavior of a Basic TTCN machine can be modeled by a Labeled Transition System:

Def. 5 A Test Behavior Expression B_t is defined as in Def 2, its semantics is defined by a Labeled Transition System $Lts(B_t)$: $Lts(B_t) =_{def} <S, L', Tran, s_0>$ where
$S=\{(ctrl, sto)\} \subseteq (Texp \times Store)$ denotes all the possible states of the TTCN machine;
$L'=\{L \cup \{\tau\} \cup \{\delta\}\}$ is the set of all possible actions. ?x denotes a receive event and !x denotes a send event;
$s_0 = (B_t, sto_0)$ where $sto_0 = (m_0, i_0, o_0)$ is the initial state;
$Tran \subseteq S \times L' \times S$ are the transitions satisfying the following rules:

$(exit, sto) \xrightarrow{\delta} (stop, sto)$;
$(?a;B, (m, a \cdot i, o)) \xrightarrow{?a} (B, (m, i, o))$;
$(!a;B, (m, i, o)) \xrightarrow{!a} (B, (m, i, a \cdot o))$;
$(B_1[]B_2, sto) \xrightarrow{u} (B_1', sto')$, if $(B_1, sto) \xrightarrow{u} (B_1', sto')$;
$(B_1[]B_2, sto) \xrightarrow{u} (B_2', sto')$, if $(B_2, sto) \xrightarrow{u} (B_2', sto')$;
$(B_1 >> B_2, sto) \xrightarrow{u} (B_1' >> B_2, sto')$, if $(B_1, sto) \xrightarrow{u} (B_1', sto')$;
$(B_1 >> B_2, sto) \xrightarrow{\tau} (B_2, sto')$, if $(B_1, sto) \xrightarrow{\delta} (B_1', sto')$;
$([q];B, sto) \xrightarrow{\tau} (B, sto)$, if q=Ture;
$([q];B, sto) \xrightarrow{\tau}$ stop, if q=False;
$([I:=v];B, (m, i, o)) \xrightarrow{\tau} (B, (m[v/I], i, o))$ where $m[v/I] = (m-\{(I, x)\}) \cup \{(I, v)\}$

3.6 Traces and Prefix

Def. 6 Traces of a LTS
Let $<S, L, T, s_0>$ be a LTS, $s, s', s_1, s_2, \ldots s_n, s_{n+1}, s_1', s_2', \ldots s_n' \in S$, $\mu_1, \ldots \mu_n \in L'$.
Let $\sigma = \mu_1 \cdot \mu_2 \ldots \mu_n$ be a sequence of labels from L', then σ is said to be a trace over L'.

The set of all possible traces over L′ is noted as L′*. We further have the following notions:

1) if $s = s_1 \xrightarrow{\mu_1} s_2 \xrightarrow{\mu_2} \ldots \xrightarrow{\mu_n} s_{n+1} = s'$, then $s \xrightarrow{\sigma} s'$.
2) if $s = s_1 \xrightarrow{\tau^*} s_1' \xrightarrow{\mu_1} s_2 \xrightarrow{\tau^*} \ldots \xrightarrow{\tau^*} s_n' \xrightarrow{\mu_n} s_{n+1} = s'$, then $s \overset{\sigma}{\Rightarrow} s'$.
3) if $\exists s'$, $s \xrightarrow{\sigma} s'$, then $s \xrightarrow{\sigma}$.
4) if $\exists s'$, $s \overset{\sigma}{\Rightarrow} s'$, then $s \overset{\sigma}{\Rightarrow}$.

Def . 7 B_t is the test behavior expression of test case t and $Lts(B_t) = <S, L, T, s_0>$ is the LTS for B_t where $s_0 = <B_t, sto_0>$. Let T_c denotes the set of all test cases in a test suite. t is a test case.

$Trace(t) =_{def} \{ \sigma | <B_t, sto_0> \overset{\sigma}{\Rightarrow} \}$
$AllTrace(t) =_{def} \{ \sigma | <B_t, sto_0> \xrightarrow{\sigma} \}$

Trace(t) denotes all the possible observable traces of test case t, and *AllTrace*(t) denotes all possible observable and unobservable traces of test case t.

Def . 8 Prefix is a relation among traces: $\prec \subseteq L^* \times L^*$. Let '.' denotes the concatenation of traces. $\sigma,\sigma' \in L^*$, if $\exists \sigma'' \in L^* \wedge \sigma = \sigma' \cdot \sigma''$, then σ is a *prefix* of σ, noted as $\sigma' \prec \sigma$, and $\sigma' = \sigma \backslash \sigma''$.

3.7 Auxiliary Definitions

Def . 9 Auxiliary Definitions

1) *Log*: $T_c \mapsto L^*$. $Log(t) =_{def} \sigma$, where $\sigma \in Trace(t)$.
   Log emulates the function of the MOT: it executes the test case and gives out a log. σ is the trace observed during the test campaign.
2) *Verdict*: $L^* \times T_c \mapsto$ {Pass, Inconc, Fail}
   $Verdict(\sigma,t) =_{def} v$, where $\sigma = Log(t)$, $v \in$ {Pass, Inconc, Fail}.
   v is the verdict assigned in test case t after trace σ is observed.
3) $Purpose(t) =_{def} \{ \sigma_p \in Trace(t) \mid Verdict(\sigma_p,t) = Pass \}$
   These are the purpose traces, ie. traces that have "Pass" verdicts.
4) $Expect(\sigma,t) =_{def} \{ x \mid \exists \sigma' \in Purpose(t)$ satisfy $\sigma \cdot x \leq \sigma' \}$
   Expect(σ,t) denotes the proper actions expected by the test case after trace σ has been observed.
5) Let σ_F denotes a trace, $t \in Tc$, $(Log(t) = \sigma_F) \wedge (Verdict(\sigma_F, t) = Fail \mid Inconc)$, then $Correct(\sigma_F, t) =_{def} \sigma_c$ where $\sigma_F = \sigma_c \cdot f \wedge f \notin expect(\sigma_c)$
   Correct(σF, t) denotes the correct trace that had been observed, just before the even leading to a fail or "Inconc" verdict happened.
6) $Proper(\sigma, t) =_{def} \{ \sigma_p \mid \sigma_p = \sigma \cdot x, x \in Expect(\sigma, t)\}$
   Proper(σ, t) denotes the proper traces expected by the test case t after trace σ has been observed.
7) $Der(Tr) =_{def} \{ \sigma \mid \sigma' \prec \sigma \wedge \sigma' \in Tr\}$, where $Tr \subseteq L^*$.
   Der(Tr) denotes all the possible extensions of the traces in set Tr.

3.8 Ordering Relations

Def. 10 Ordering relation 0:

If $\forall\ \sigma_2 \in Purpose(t_2)$, $\exists\ \sigma_1 \in Purpose(t_1)$ satisfy $\sigma_1 \prec \sigma_2$, then $t_1 \leq_0 t_2$.

The meaning is if the test purpose traces of t_2, ie. those have "Pass" verdicts, are prefixed by the test purpose traces of t_1, then t_1, t_2 has ordering relation $\leq_0$.

Def. 11 Ordering relation 1

If $\forall \sigma \in Purpose(t)$, $\exists t_i \in T_c$ satisfy $\exists \sigma_i \in Purpose(t_i) \wedge \sigma_i \prec \sigma$, then $\{ t_i \} \leq_1 t$.

This relation takes into account the situation that several test cases together determines the result of another test case. We can further refine this relation by using a more detailed test report , we can refer to the log to obtain the trace of the test case that has just been observed. Since we broke the test cases into traces, then instead of obtaining a relation among test cases, we get a relation between a set of traces and a test case.

Def. 12 Ordering relation 2

$Tr \subseteq L^*$, $t \in Tc$. If $Purpose(t) \subseteq Der(Tr)$, then $Tr \leq_2 t$.

We can obtain more relations according to this definition because usually σ_F is shorter than *Purpose*(t), so it has a higher possibility of becoming a prefix of another test case.

3.9 Properties of Ordering Relations

Here we demonstrate how to use the ordering relations to carry out test sequencing. Using the following theorems, we can determine the result of a test case t without executing it.

Theorem 1. t', t are test cases. If

1) $Verdict(Log(t'), t)$ = Fail | Inconc,
2) ttt $t' \leq_0 t$,
3) test results are repeatable, then $Verdict(Log(t), t)$ = Fail | Inconc.

Theorem 2. T_c is a set of test cases, t is a test case. If

1) $\forall\ t_i \in Tc$, $Verdict(Log(t_i), t_i)$ = Fail | Inconc,
2) $T_c \leq_1 t$,
3) test results are repeatable, then $Verdict(Log(t), t)$ = Fail | Inconc.

Theorem 3. T_c is a set of test cases, t is a test case. If

1) $\forall\ t_i \in T_c$, $Log(t_i) = \sigma_{Fi}$, $Verdict(\sigma_{Fi}, t_i)$ = Fail | Inconc,
2) $\exists\ Tr_{unreach} = \bigcup_{t_i} Proper(Correct(\sigma_{Fi}, t_i)) \wedge Tr_{unreach} \leq_2 t$,
3) test results are repeatable, then $Verdict(Log(t), t)$ = Fail | Inconc.

The proofs are straight forward except the last theorem which needs some explanation. When test case t_i fails (or get the "Inconc" verdict) with trace σ_{Fi} observed, we can mark *Proper*(*Correct*(σ_{Fi}, t_i)) as an unreachable set of traces provided that the test results are repeatable. The size of the set of unreachable traces grows as the test continues. This set can be written as $Tr_{unreach} = \bigcup_{t_i} Proper(Correct(\sigma_{Fi}, t_i))$ where t_i is the test case failed(or get the "Inconc" verdict). Whenever the purpose traces of a test case t are covered by *Der*($Tr_{unreach}$), or, in another word $Tr_{unreach} \leq_2 t$, then t is doomed to fail or (to get the "Inconc" verdict), and can thus be skipped. In short, the sequencing can be done according to the above relations. Efficient algorithms are needed to calculate the $Tr_{unreach}$ and to determine the '$\leq$' relations.

3.10 Non-determinism

Actually the test results may not be repeatable. Even the re-execution of the same test case against the same IUT may give out different results. The IUT's choices at the same forking point may not be the same. But there may exist some statistical probabilities among the IUT's choices. Those test cases whose purposes are within *Der*($Tr_{unreach}$) merely have higher probability to fail. We say a test case is doomed to fail under the assumption that the test results are repeatable. In reality, some adjustment is needed: we mark a trace unreachable only when it still fails after a certain number of retries. This number is proportional to the number of valid IUT responses at this branching point.

4 DESIGN OF THE TEST MANAGEMENT SYSTEM

4.1 PCTS and its TMS

The general design guidelines of PCTS are as follows.

- Strict accordance with CTMF
- Use of Formal Methods
- TTCN based Test Execution
- X-window based test presentation
- Object-Oriented software environment

Because of the unique functions carried out by it, the TMS also has its own highlights when applying these general guidelines. Since it has a broad coverage of the Conformance Assessment Process, the procedures defined in CTMF can be best presented here. TTCN based test notation is also a unique characteristic of TMS. All the selection and sequencing are done based on TTCN. Since there are numerous human-machine interactions in the test preparation phase, a convenient and friendly interface is of great importance here. Great efforts have been made in order to ease cross-reference, information retrieval and document generation.

4.2 Modules in TMS

TMS has six major modules as they are shown in Figure 2.

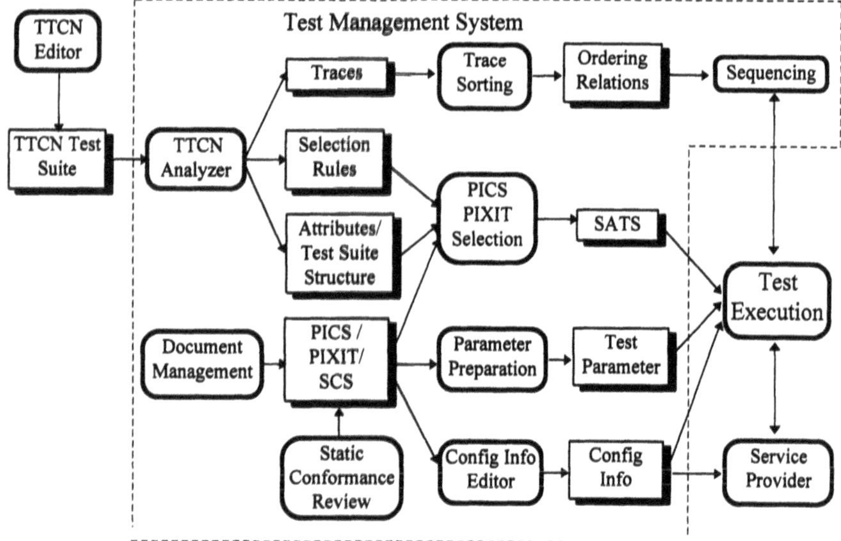

Figure 2 Modules of TMS in Conformance Assessment Process

Document management

It obtains necessary information from the client, this includes PICS/PIXIT information required by PICS/PIXIT selection and parameterization. It will also get other administrative information from the test client. It will provide some necessary check lists to facilitate the negotiation between test laboratory and test client on the selection of test method and test suites. SCS, PICS and PIXIT documents are generated by this module.

TTCN analyzer

This module carries out the syntax analysis of test suites written in TTCN.MP. Because the BNF definition for TTCN has more than 300 production rules, so we use Lex and Yacc to facilitate the analysis. This module will generate three parts of information.

- Test suit structure and test case attributes. The structure is an index of the test suite and the attributes include the test case ID, parameters and PDU/ASP used, etc. They are used for PICS/PIXIT selection.
- PICS/PIXIT selection rules. This information is extracted from the selection expressions part in the test suite, or from a separate selection table that comes along with the test suite. It sets up a selection criterion for each test case by using a Boolean expression of PICS/PIXIT variables.
- Purpose traces of each test case. We first derive the test behavior expression of each test case, then calculate its purpose traces, i. e. the traces with "Pass" verdicts. To avoid infinite traces, loops are unfolded just once.

Thus for each test case, we can create a behavior tree by merging the common prefixes of its traces. Each path between the root and the leaf is a possible trace.

PICS/PIXIT selection

For each test case, its selection expression is calculated, the result will explicitly indicate whether it will be selected or not. Then a new index is generated, only the selected test cases are included. This is the Selected ATS (SATS). User interference and selection according to other criterion can be introduced at any time during the selection.

Trace sorting

This module is the central part of test sequencing. First we construct a global behavior tree by merging the purpose traces of each test case together. This is done also by merging the common prefix of the traces. We then mark the ending node of each purpose trace in the global behavior tree with the identifier of its corresponding test case. Second, in the global behavior tree, at each node that corresponding to a purpose trace, we are going to include a reference to that test case. The reference count of a test case is the number of its purpose traces. With the global behavior tree, it's much easier for us to determine prefix relations among traces: traces corresponding to a parent node in the tree are always the prefix of those traces corresponding to its children.

Sequencing

The execution sequence of the test case will be determined by the following. We traverse the global behavior tree in a width first strategy and execute the test cases corresponding to the node we reached. If the test case fails or gets an "Inconc" verdict with a logged trace σ_F, then the fail count of all the traces within *Purpose*(*Correct*(σ_F,t)) are increased by 1. If the fail count of any trace exceeds its upper limit, it will be marked unreachable and will be added to $Tr_{unreach}$. For each purpose trace within *Der*(*Purpose*(*Correct*(σ_F,t))), the reference count of its corresponding test case will be decreased by 1. If the reference count of any test case reaches 0, then it will not be executed.

Since we are executing the test case in parallel with the interpreting of another [3], the sequencing module is providing the Test Execution module with both the current and the most possible next test cases. When traversing the behavior tree, we adopted the width first strategy, so that the ordering relations between the current and the next test cases will be reduced to the minimum. Thus, when the current test case fails, the next test case which has already been pre-analyzed by the TE is unlikely to become doomed. In this way, we improved overall performance.

Static Conformance Review, Parameter Preparation and Test Configuration

The static conformance review module check the PICS to see if all the mandatory requirements have be fulfilled. The parameter preparation module collects various test parameters from test client and feed them to the test execution. The test configuration module set up the correct environment for testing.

5 EXAMPLE OF TEST SEQUENCING

Suppose we have three test cases Tc1 ={t1, t2, t3}. Their TTCN.GR representations are shown in Table 1. Their test behavior expressions are:

Bt1 = (!a;((?b;exit)[](?c;exit)[](?other;stop)))

Bt2 = (!a;(?b;!g;(?e;exit[]?other;stop)[]?other;stop))
Bt3 = (!a;(?b;!g;(?f;exit[]?other;stop)[]?other;stop))

t_1: TTCN	Verdict
!a	
?b	Pass
?c	Pass
?otherwise	Fail

t_2: TTCN	Verdict
!a	
?b	
!g	
?e	Pass
?otherwise	Fail
?otherwise	Inconc

t_3: TTCN	Verdict
!a	
?b	
!g	
?e	Pass
?otherwise	Fail
?otherwise	Inconc

Table 1 TTCN.GR of test cases

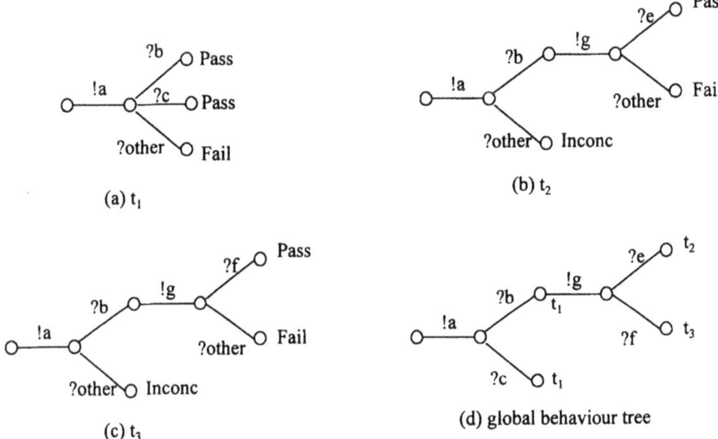

Figure 3 Behaviour tree of test cases

Their behavior trees and the global behavior tree of the three test cases are shown in Figure 3. The purpose traces calculated by the basic TTCN machine are:

Purpose(t_1) = { !a?b, !a?c }, *Purpose*(t_2) = { !a?b!g?e }, *Purpose*(t_3) = { !a?b!g?f }.

According to Def 10, because (!a?b) $\prec$ (!a?b!g?e), (!a?b) $\prec$ (!a?b!g?f), so we have $t_1 \leq_0 t_2$ and $t_1 \leq_0 t_3$. The execution sequence will be (t1,t2,t3) or (t1,t3,t2). According to Theorem 1, if t_1 fails then t_2 and t_3 are skipped.

But consider another test suite $T_{c2}=\{t_2,t_3\}$. This time we still can find some relations between t_2 and t_3. If t_2 fails with *Log*(t_2)=!a?x, x≠b. Then we have *Proper*(*Correct*(!a?x))={!a?b,!a?c}. Then the fail count of !a?b and !a?c increases by one. Here we set the upper limit of all the fail counts to one. Then {!a?b,!a?c} will be marked unreachable. $Tr_{unreach}$={!a?b, !a?c}. Now *Der*({!a?b})={!a?b, !a?b!g?e, !a?b!g?f , ...}.

According to Def 12, we have $\{!a?b,!a?c\} \leq_2 t_2$, and $\{!a?b,!a?c\} \leq_2 t_3$. So we have $Tr_{unreach} \leq_2 t_3$. According to Theorem 3, t_3 is skipped.

In reality, both the reference count of t_1(!a?b!g?e) and t_2(!a?b!g?e) are originally set to one. After t_2 fails, we decrease the reference count of t_3 and it becomes zero. So it is dropped.

6 CONCLUSIONS

The TMS presented above has been used in the testing of many real OSI protocols such as X.25 LAPB, X.25 PLP, TP and X.400, etc. Sequencing has been applied to the X.25 LAPB and X.25 PLP tests. Standard test suites in ISO 8882 are adopted[2]. In X.25 LAPB and PLP test suites, although ordering relations among test cases are few, the common traces between test cases are numerous. Usually the test cases within a same test group has the same preambles. So our sequencing strategy works especially well under the following situations: when a preamble gets "Inconc", and after several retries the whole group will be discarded.

For example, ISO 8882-2 contains 288 test cases for X.25 LAPB and 287 cases are counted for ordering (except DL4_113). In the global behavior tree, totally there are 25888 purpose traces with average length to be 11.57 (test events). The number of common prefixes can be seen from the following Common Prefix Ratio (CPR):

$$CPR =_{def} 1 - \frac{\text{(number of test events in the global behaviour tree)}}{\text{(total number of test events in all purpose traces)}} = 1 - \frac{67062}{299465} = 0.78$$

This means that in the test suite about 80% of test trace length is included in another trace.

The choice of LTS in the establishment of a formal model of TTCN seems to be intuitive. Currently it appears to be the only attempt to formalize TTCN [8,10,23,24]. We believe this is largely due to the tree nature of TTCN test cases and the succinctness of expressing behavior trees in LTS.

In general, the sequencing method we introduced here can greatly increase the testing efficiency especially during the testing of immature systems.

The interpretation based test execution strategy has been proofed to be a great success in the testing of OSI lower level protocols such as X.25 LAPB, X.25 PLP and TP. Now we are focusing on the testing of higher level protocols. Problem aroused when ASN.1 was introduced into TTCN due to the difficulty in handling complex data types which is a common and in-born weakness of all interpreters (e.g. BASIC, Tcl, etc.). This is a new challenge for the TMS to manage TTCN test suite with ASN.1 extensions. However we deployed an interpreting-compiling combined method to overcome this type handling problem. All the ASN.1 definitions embedded in TTCN and the definitions by reference are collected together by the TMS and are then fed to an ASN.1 compiler to create a ASN.1 library which includes encoding, decoding, assignment and comparison functions that can be accessed during test execution. In this way we enjoys the efficiency brought by the compilation while at the same time maintains the flexibility introduced by the interpretation. This method is feasible but is subject to further verification.

Current research works in this area include the development of more efficient algorithms for sequencing and the refinement of the TTCN semantic model .

6 REFERENCES

[1] ISO/IEC IS 9646 1-7 1991(E), OSI-Conformance Testing Methodology and Framework.

[2] ISO/IEC 8882 1-3, ISO/IEC 8882:1992(E), X.25 DTE Conformance Testing.

[3] Yamin Wang and Jianping Wu, An approach to TTCN-based test execution, IWPTS VII, p299-306, IFIP 1994.

[4] Ruibing Hao, Jianping Wu and Samual T. Chanson, Design and implementation of an automatic test suite generator, Chinese Journal of Advanced Software Research, vol. 1, No.2, p152-165 1994.

[5] Samual T. Chanson and Qin Li, On Static and Dynamic Test Case Selections in Protocol Conformance Testing, IWPTS IV, 1991.

[6] Samuel T. Chanson, Sijian Zhang and Qin Li, A Test Case Management System, IWPTS, 1992.

[7] La Briere, Testing in Practice OSI Test Center, IFIP Trans. C vol:C-11, p19-29, 1993.

[8] Walter and B. Plattner, An operational semantics for concurrent TTCN, IWPTS V, p131-143. IFIP, NorthHolland, 1992.

[9] Finn Kristoffersen and Thomas Walter, TTCN test case correctness validation, IWPTS VII, p 37-53, IFIP 1994.

[10] G. J. Tretmans, A Formal Approach to Conformance Testing, PhD Thesis, 1992.

[11] R. Milner, A Calculus of Communicating Systems. Lecture Notes in Computer Science 92. Spring-Verlag, 1980.

[12] Hoare, Communicating Sequential Processes. Prentice-Hall, 1985.

[13] T. Bolognesi and E. Brinksma, Introduction to the ISO specification language LOTOS, Brinksma, et al, A Formal Approach Computer Networks and ISDN Systems, IFIP Transactions. North Holland, 1992.

[14] CCITT. Specification and Description Language (SDL), CCITT Recommendation Z.100, CCITT/ITU, 1992.

[15] ISO/IEC, OSI-Specification of Abstract Syntax Notation 1 (ASN.1), IS 8824, ISO/IEC, 1987.

[16] K. Inan and P. Varaiya, Finitely recursive process models for Discrete Event Systems, IEEE Trans. on Automations. Vol. 33, No. 7, July 1988.

[17] de Meer and V.Heymer,etc, An approach to a Conformance Testing Methodology and the COAST Test System, IFIP, 1991.

[18] Alcatel TITN Inc., XRTLE User Guide, 1992.

[19] UBC/IDACOM, OSI PT Environment, 1990.

[20] H. Ural and Z. Wang, Synchronizable test sequence generation using UIO sequences, Computer Communications, Vol:16, p653-63,1993.

[21] M. E. Koblentz, Issues in testing fast packet services over the broadband (ISDN), IFIP Transaction C vol:c-11, p3-18, 1993.

[22] A. D. Varvitsiotis and G. I. Stassinopoulos, Extending ASN.1 into a full-fledged constraint language in the context of OSI protocol Conformance Testing, Computer Networks and ISDN Systems, p1243-63, 1993.

[23] E.Brinksma, et al, A Formal Approach to Conformance Testing, IWPTS IV, Netherlands, Oct. 1991.

[24] Hogrefe, Status Report on the FMCT Project, IWPTS VII, p.165-180, IFIP, 1994.

28

Towards a "Practical Formal Method" for Test Derivation

R. L. Probert and L. Wei
TSERG, Department of Computer Science, University of Ottawa
150 Louis Pasteur, Ottawa, Canada K1N 9B4
Tel. (613) 562-5800 Ext. 6709 Fax (613) 562-5185 bob@csi.uottawa.ca

Abstract

We present a new semantic approach, called Nondeterministic Ripple Set (NRS), for modelling process behaviors in a distributed system. Then, we describe its application and contributions to the theory and practice of system testing. This approach considers both *environment control* (*test purpose*) and *process nondeterminism* as complementary factors determining system behaviours, and captures system behaviours in terms of the mutual influences of these two factors. We then illustrate systematic test derivation by applying this approach to the INRES service as a case study. This semantic approach brings a new dimension, *environment control,* into a formal semantics for testing in distributed systems, and shows a means of capturing the practical notion of "test purpose" within an algebraic context. It thus contributes to bridging the gap between theory and practice in testing communications systems.

Keywords

semantics of distributed processes, communications systems, testing, use cases, scenarios.

1 INTRODUCTION

Computer communications systems are notable for their strategic importance to many sectors of society, including government, industry and commerce, the military, and the educational infrastructures. At the same time, interoperability of these systems remains a major concern, even when two communicating systems both claim conformance the same an international standard (specification). To promote interoperability, ISO and ITU (formerly CCITT) have developed and published standardized tests which must be passed by systems which claim conformance to the service or protocol. The process of developing and maintaining these standardized tests generally helps to improve the correctness and completeness of manufacturers' understanding of the international standards, and the quality of the standards themselves.

Product test engineers are aware that products must satisfy customer requirements as well as international standards. Accordingly, product tests are designed in such a way that the purpose of each test is aligned with one or more of these customer requirements. A complete computer communications system consists of the interactions between a distributed service provider (called the process) and a user (called the environment). In this paper, we sketch the main parts of a formal theory of modelling the behaviours of such systems. We show how this new theory of Nondeterministic Ripple Sets (NRS) captures the interactions between environment (which tries to satisfy requirements by making choices in a controlled way) and the service process (which can make choices which appear externally to be nondeterministic). We then show how to apply this theory to the INRES service for test design. The final sections of the paper describe the general contributions of the approach to conformance testing, and conclude with a brief discussion of benefits of our approach.

2 BACKGROUND

In this paper, we use a standard representation of processes, namely a labelled transition system (Bolognesi, 1989; Brinksma, 1989; Milner, 1988). Labelled transition systems (LTS) are widely used to represent distributed systems. Figure 2.1 depicts (a) a given distributed communicating system, (b) its abstraction into an environment interacting with a non-deterministic process, and (c) an example LTS which represents the interactions between the process and its environment.

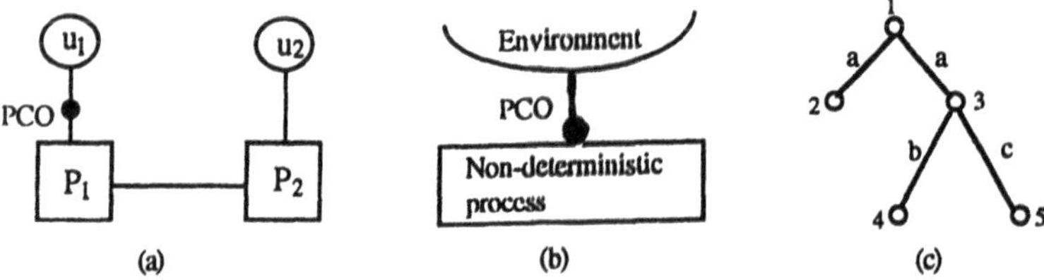

Figure 2.1 Three related representations of a distributed communications system.

In Figure 2.1(a), user entities u_1 and u_2 communicate via a link between devices P_1 and P_2. In Figure 2.1(b), if the PCO (Point of Control and Observation) corresponds to the • between u_1 and P_1 in (a), then u_1 constitutes the environment and the combined interplay of P_1, P_2, and u_2 constitute a nondeterministic process (from u_1's viewpoint). In Figure 2.1(c), the LTS shows that when u_1 offers a message "*a*", the nondeterministic process (P_1, P_2, u_2) can go from state 1 to either state 2 or state 3. Labelled transition systems are a generalization of finite state machines/automata (Hopcroft, 1979), and provide a natural operational model for formal languages such as CCS, CSP, and Lotos (Milner, 1980; Hoare, 1985; Brinksma, 1987). As well, LTS provide a useful starting point for the NRS approach presented here.

In verification or testing of a nondeterministic process (or a distributed system), three practical aspects, among others, should be addressed:

(1) Environment Constraint:

Each time an observer or tester(environment) interacts with a process, he has a *specific purpose* in his mind, usually called a *verification goal* or *test purpose*. From the environment point of view, the purpose of the tester(or test case) is to interact with the process at the PCO and try to *constrain* the process to follow a path which corresponds to the implementation of a customer requirement. Thus, at every state of the process encountered during a test execution, the tester knows which action he should offer according to the test purpose. His test strategy is called *environment constraint.*

(2) Non-deterministic choice of process:
Nondeterminism in the process can interfere with the purpose of the tester. For example, the tester tries to guide the process to follow one specific path determined by his purpose; while the nondeterministic transition at a state may steer the process into another path which fails the test purpose. A practical scenario is that a tester tries to establish a connection between two communicating entities to test a transfer of a data file. But, one process disconnects itself. In this case, he is not successful in achieving his test purpose.

(3)Termination of a test and test management:
When a path satisfying the test purpose cannot be followed because of the process's nondeterminism, test execution can usually not be stopped right away. The process has to be guided further until it reaches a proper stopping point. For example, the operation of a communication system can be continuous, however, a reasonable *session* could be a complete data transfer from connection request to disconnection confirmation, or a complete duration from login to logout. Stopping at a recognizable state is practical for managing test executions.

According to these aspects of testing, we know that the behavior of a process cannot be decided by either the process or its environment alone. Given a *test purpose* of an environment, the environment constraints can be derived in advance from the *test purpose*. Then, in test execution, *a set of traces* has the potential to be followed due to nondeterminism inside the process. This forms a semantic denotation to characterize behaviors of the process (Wei, 1994).

Test cases in practical test suites such as for X.25 (ISO/IEC 8882-3) reflect this observation. For example, each test case in the standard X.25 test suite described in TTCN (Probert, 1992) has two essential parts, (1) test purpose, and (2) dynamic behavioral description. Each dynamic behavioural description is derived based on the test purpose and has a tree structure as follows:

```
!Act a          (send "a")
  ?Act b        (receive "b")
      !Act c

          ...

  ?Act d        (or receive "d")
      !Act e
```

where *!Act a* is an action (send message *a* to u_2) performed by entity u_1 in Figure 2.1, and *?Act b* and *?Act d* are potential responses from u_2 to u_1's action. Because u_1 does not know which response will come from u_2, u_1 must be prepared to deal with either of them in the test case. Such a behavioral tree can be unwound into a set of traces. The test cases derived from other approaches such as Canonical Tester (Brinksma, 1989; Langerak, 1989; Wezeman, 1989) also have this characteristic. Test generation in their approach is not directly based on customer requirements.

Note that to limit the length of the paper, we intend here to illustrate our approach rather than present a complete mathematical theory of NRS (Wei, 1994). For our purpose, NRS should be understood to be a set of potential traces of execution of the nondeterministic process by the environment exercising environmental constraint to achieve a specific test purpose. We formalize this in the next section.

3 NRS: A NEW BEHAVIORAL VIEW OF PROCESSES

In this section, we formalize the semantic view and nondeterministic ripple sets based on representing processes as labelled transition systems (Bolognesi, 1989; Cleaveland, 1993). We begin with reviewing basic terminology and discussing some operational characteristics in labeled transition systems.

Definition 3.1 A *labeled transition system*(LTS) is a triple $(P, Act, \longrightarrow)$, where

i) P is a set of states(processes)
ii) Act is a set of actions
iii) $\longrightarrow$ is a transition relation in $P \times Act \times P$

Intuitively, $\longrightarrow$ is the set of transitions which may result when an action is executed in a state. For simplicity of presentation, we assume that Act does not include invisible actions such as τ (This is not necessarily a restriction (ISO/IEC JTC1/SC21/P.54)). We write $p \text{—}a \rightarrow p'$ in place of $(p, a, p') \in \longrightarrow$ for convenience. Furthermore, the relations $\text{—}a \rightarrow$ are extended to the sequence of actions(trace) $p \text{—}s \rightarrow p'$, for every s in Act^*, as follows,

i) $p \text{—}\varepsilon \rightarrow p'$ if p' is p
ii) $p \text{—}as \rightarrow p'$ if $p \text{—}a \rightarrow p''$ for some p'' such that $p'' \text{—}s \rightarrow p'$.

This means that $p \text{—}s \rightarrow p''$ if p can evolve to p' by performing the sequence of actions s. We also use $p \text{—}s \rightarrow$ to mean that there exists a p' such that $p \text{—}s \rightarrow p'$. For technical convenience, we shall only consider LTSs with *finite* branching at their states, that is, for each p in P, the set $\{p' \mid \exists\ a, p \text{—}a \rightarrow p'\}$ is finite. For a state r in P, we will represent a process with initial state r as

$$LTS_r = (P_r, Act, \longrightarrow_r) \text{ derived from } LTS = (P, Act, \longrightarrow),$$

where P_r is the set of all states reachable from r by some finite sequence of transitions, and $\longrightarrow_r$ is the restriction of $\longrightarrow$ to P_r. Therefore, a state in P can be used to define a process. We will use the graphical notation for labelled transition systems denoted *synchronization tree* or *behavioral tree* (Guillemot, 1989; Milner, 1980; Velthuys, 1992) in examples for simplicity.

Now, we discuss some operational characteristics of testing based on labelled transition systems, that is, *environment constraint and process choice.*

This can be shown with a state transition as in Figure 3.1. There are three groups of transitions at state p_1 labelled with a, b and c, respectively. In process execution, at this state, which group of transitions will be executed depends first on the environment choice (say "c"). We call this the *environment constraint* of the system behavior. However, after c is offered, all the transitions labelled with c have the potential to be executed. Which transition will be chosen depends on the process choice. Thus, both environment and process exert influences on each other's behavior (transition choice).

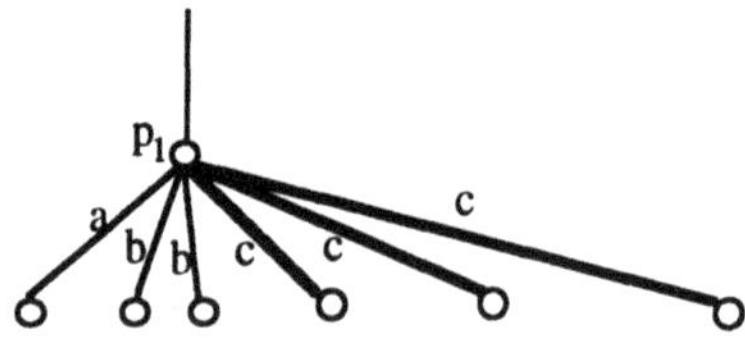

Figure 3.1 Environment constraint and process choice (environment chooses "c", process has 3 choices of transitions).

Thus, a path traversed during execution is decided by two factors as a point in a plane is decided by the point's two coordinates. This path cannot be determined ***in advance***, as it depends on the dynamic choices of both sides.

We now consider a process behavior from its environment (testing) side. Testing treats the implementation as a black box. However, the tester can work with the process *specification*, and *imagine* its potential execution paths based on its operational semantics (LTS) (this corresponds to a test plan). Thus, it is not necessary to look inside the process. The tester's strategy (plan) is driven by a test purpose.

Due to nondeterministic choices of the process, the tester has to cooperate with the process by choosing proper actions at different potential states to keep the execution aligned to his test purpose. Thus, test purpose implies a specific, biased set of paths has the potential to be traversed. The set of traces labeling these paths is a test purpose oriented behavioral characterization of the process.

We now formally define the ideas discussed above. First, we formalize the process choice.

Definition 3.2 (*Ripples or Derivatives at a state for process choice*) For an LTS = $(P, Act, r, \longrightarrow)$ and $p, p' \in P$, $a \in Act$, we define

$D(p, a) = \{p' \mid p \xrightarrow{a} p'\}$, $D(p, a)$ is called the *Ripples*, or *Derivatives* at p with respect to a.

At any state p of an LTS, let $D(p, a) \neq \varnothing$. Then, during a transition with action a at p, the process has the potential to reach any p' in $D(p, a)$ after performing action a. We call this ***nondeterministic rippling effect of the process with respect to action a at state p***, or simply *nondeterministic rippling. Nondeterministic rippling effect* intuitively denotes the spreading of possible execution paths from a state, and formalizes the effect of **process choice** on system behavior.

Definition 3.3 For an LTS = $(P, Act, r, \longrightarrow)$ and $p, p' \in P$, $a \in Act$, $s \in Act^*$, respectively,

i) $D(p, s) = \{p' \mid p \xrightarrow{s} p'\}$, the s *-Ripples*, or *-Derivatives* of p.

ii) $S(p) = \{a \mid p \xrightarrow{a}\}$, the *Successors* of p.

Now, we define the environment constraint. All choices made by the environment at all potential states encountered during an execution can be represented as a set C which is a subset of the set of pairs: $\{(p, a) \mid p \in P \text{ and } a \in S(p)\}$, i.e.,$C \subseteq \{(p, a) \mid p \in P \text{ and } a \in S(p)\}$.

Each (p, a) in C stands for the action a chosen by the environment at state p. We take PS to represent the set of all states which may be potentially encountered during an execution. Then, the environment constraint can be described by the choice pattern as defined below.

Definition 3.4 (*Choice pattern for environment constraint*). C is called a *choice pattern*, if C and PS satisfy the following conditions.

1) root condition: $r \in PS$,

2) control condition: If $p \in PS$ and $S(p) \neq \varnothing$, then for exactly one $a \in S(p)$, $(p, a) \in C$,

3) rippling condition. If $(p, a) \in C$, then for all $q \in D(p, a)$, $q \in PS$.

Note that in definition 3.4, the auxiliary set of states, PS, is defined with the choice pattern C in parallel. in a choice pattern, exactly one action can be selected at each state. We represent the set of all choice patterns of a process P as $CS(P)$. In practice, the choice pattern is derived from the test purpose.

Definition 3.5 (*Nondeterministic ripple set(nrs)*)

A ***nondeterministic ripple set*** of a process is a non-empty set of maximal traces followed potentially during an execution which is under the constraint of the nondeterministic rippling effect of the process and action selections specified by a choice pattern C at all states encountered during the execution.

Operationally, given a process P, and a choice pattern C, a nondeterministic ripple set of P, denoted nrs, can be calculated inductively as follows, where nrs_n is an auxiliary variable denoting a set of traces:

$nrs \leftarrow \{\varepsilon\}$.

repeat $nrs_n \leftarrow \emptyset$.

for each $s \in nrs$, and $p \in D(r, s) \cap PS$

if there exists a $(p, a) \in C$,

then $nrs_n \leftarrow nrs_n \cup \{s.a\}$, $C \leftarrow C - \{(p, a)\}$

else $nrs_n \leftarrow nrs_n \cup \{s\}$.

$nrs \leftarrow nrs_n$.

until $C = \emptyset$.

The name *nondeterministic ripple set* is used because it is the result of nondeterministic rippling effect of the process choices on its traces. For an empty choice pattern C, the corresponding *nrs* is $\{\varepsilon\}$, while for a process with infinite or recursive behaviors, the computation in this definition may not terminate unless some pre-selected exit point in the process is specified as in real testing activities.

Each nondeterministic ripple set characterizes a specific possible nondeterministic behavior of a process during an execution. All possible nondeterministic ripple sets generated by all different choice patterns during all possible executions characterize all possible nondeterministic rippling behaviors of the process. Thus, we can define NRS to characterize all behaviors of the process.

Definition 3.6 (Characterization set NRS of a process) For a process P, the ***characterization set of the behaviors*** **of** P, represented as ***NRS(P)***, is the set of all nondeterministic ripple sets of P, i.e.
NRS(P) =$\{nrs \mid nrs$ is a nondeterministic ripple set for some choice pattern of$P\}$.

Thus, the *NRS(P)* is the set of all *nrs* sets derived from all choice patterns of *P*. We also call *NRS(P)* the ***characterization*** of P, playing the same role as failure sets or acceptance trees (Brookes; Hennessy, 1985; Hennessy, 1988). Examples of characterization given in Figure 3.2.

The sets of choice patterns for these processes are CS(P1) = $\{\{(r_1, a), (1, b)\}, \{(r_1, a), (1, c)\}, \{(r_1, a), (1, b)\}\}$, CS(P2) = $\{\{(r_2, a), (1, b), (2, d)\}, \{(r_2, a), (1, c), (2, d)\}\}$, and CS(P3) = $\{\{(r_3, a), (1, b), (2, c), (3, d)\}\}$, respectively. Their corresponding characterization sets are shown in (b). Intuitively, that any of the three *nrs* 's in NRS(P1) contains only one trace implies that a tester can control P1 to follow any of its three traces *ab*, *ac*, and *ad* in an execution depending on his purpose. Note that P1 is deterministic. For process P2, no matter which trace the tester wants to follow, there is another trace which has the potential to be followed due to the nondeterministic rippling at the root node. This is why each *nrs* in NRS(P2) contains two traces. P3 is fully nondeterministic, no matter which trace in it is intended to be followed, the other two traces has the potential to be traversed. Thus, P3 has only one *nrs* including all its three traces *ab*, *ac*, and *ad* .

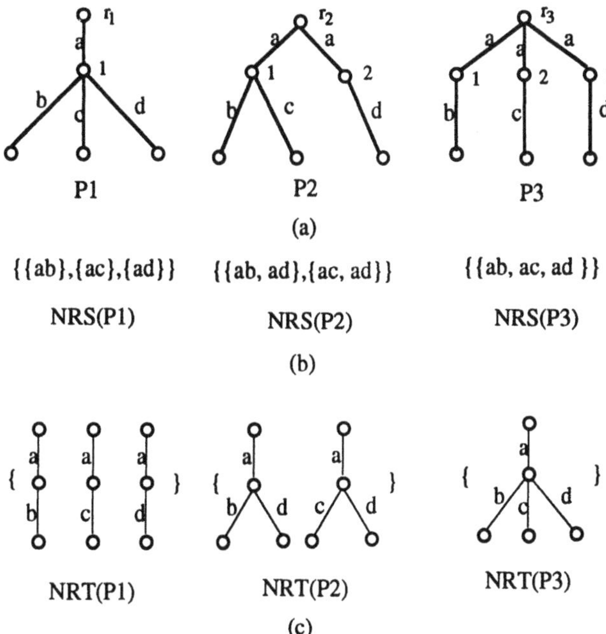

Figure 3.2 Characterization sets of processes.

Finally in this section, we prove an alternative representation for NRS(P) to help apply this theory to test generation and test specification. We transform the trace set representation of NRS(P) as in definition 3.6, into a set of deterministic tree representation. To do this, we first attach a distinguished symbol σ ($\notin Act$) to the end of each finite trace in nrs $\in$ NRS(P). This is to prevent a short trace from being covered by a longer one in the transformation if the short one is a prefix of the longer one. Next, given a set of traces, it is straightforward to arrange all the traces in the set to share all their common prefixes among them to form a deterministic tree. Formally,

Let $nr\sigma = \{s\sigma \mid s \in nrs\}$, and

$\text{NR}\sigma(P) = \{nr\sigma \mid nr\sigma \text{ is computed for each } nrs \in \text{NRS}(P)\}$.

Let *nrt* represent the deterministic tree obtained from a $nr\sigma$ by arranging all the traces in the $nr\sigma$ to share all their longest common prefixes, and

$\text{NRT}(P) = \{ nrt \mid nrt \text{ is the deterministic tree obtained from each } nr\sigma \in \text{NR}\sigma(P)\}$.

We have the following proposition.

Proposition 3.1: Given a process *P* and its NRS(*P*), a one to one correspondence exists between the elements of NRS(*P*), NRσ(*P*) and NRT(*P*).

Proof: The following is a one to one correspondence between the elements of NRS(*P*), NRσ(*P*) and NRT(*P*),

$nrs \Leftrightarrow nr\sigma \Leftrightarrow nrt$

where $nrs \in$ NRS(P), $nr\sigma \in$ NRσ(P), and $nrt \in$ NRT(P); $\Leftrightarrow$ means "corresponds to". The correspondence relations are

$nrs \Leftrightarrow nr\sigma$ iff $nrs = \{s \mid s\sigma \in nr\sigma\}$, and $nr\sigma = \{s\sigma \mid s \in nrs\}$

$nr\sigma \Leftrightarrow nrt$ iff $nr\sigma = \{s\sigma \mid s\sigma$ is a trace of $nrt\}$, and nrt is the deterministic tree obtained from $nr\sigma$

.♦

Theorem 3.1 A process P can be characterized by a set of deterministic trees, NRT(P). Each tree in the NRT(P) describes a behavior of the P resulting from the mutual influences between P's nondeterministic rippling and the environment/test purpose control.
Proof: We prove by construction.
1. Compute the characterization set NRS(P) of P by Definition 3.6.
2. Transform the NRS(P) into NRT(P) according to Proposition 3.1.♦

The deterministic tree representations NRT(P)'s of the NRS(P)'s for processes in Figure 3.2(a) are shown in Figure 3.2(c). Deterministic trees are popular specification method for test cases. They have been practically used in the test suite specification written in TTCN (Probert, 1992) and the tests derived from Canonical tester (Brinksma, 1989), for example. As similar testing mechanisms defined in (Brinksma, 1989) or (Hennessy, 1988), each tree in a NRT(P) can be seen as a test case for the process concerned, if we add some related verdicts such as *pass* or *fail* to each end of the trace in the tree and synchronize the tree with the process. From this point of view, the characterization set of a process can actually serve a test suite for test selection, and the procedure for generating it be a test generation process. This show that this semantic approach is motivated from testing, and applying it back to the practical testing environment seems promising. In next part of this paper, we show an application of this theory for test generation and test specification.

4 A CASE STUDY OF APPLICATION OF NRS TO TEST GENERATION FOR INRES SERVICE

To illustrate the practicability of this semantic approach, in this section, we apply it to a small but practical data transfer service protocol INRES for test generation. INRES is a simplified version of the Abracadabra service introduced in (Hogrefe; Velthuys, 1992) and has been selected by the international standards organizations ISO/IEC JTC1/SC21/P.54 and CCITT SG X/Q.10 as a reference protocol for studying testing techniques in the context of formal methods(FMCT) (ISO/IEC JTC1/SC21/P.54). The detailed structure and Lotos specification of the service are included in the Appendix of this paper. We shall focus on the testing issues related to this application. We first review some testing terminology and then, turn to the discussion for test generation for INRES by applying the NRS theory of the last section. Some basic OSI (Open Systems Interconnection) concepts are used in this application and we refer the reader to the ISO document: conformance testing framework and methodology (ISO/IS9646) for more details.

The INRES (INitiator-RESponder) data transfer service is a connection oriented service. It consists of a Service Provider, an Initiator and a Responder. The service can be accessed from two SAPs(Service Access Point). On the one SAP(ini), the initiator must initiate a connection request before sending data. On the other SAP(res), the responder can accept the connection or reject it. After a connection is established, the responder can receive data from the initiator until a disconnection request is sent by the responder.

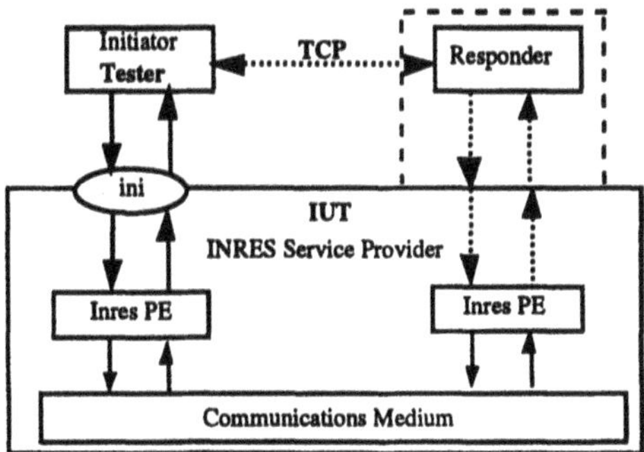

Figure 4.1 Test configuration for INRES Service.

Suppose that we use the **remote testing method** (ISO/IS9646) to test this service. Then, the testing configuration looks like the one shown in Figure 4.1. This testing method assumes that there is no test interface at the top of the IUT(Implementation Under Test) and no formal Test Coordination Procedures(TCP) between LT(Lower tester) and UT(Upper Tester, not appear in Figure 4.1) exist (only manual coordination, if any). The method relies solely on the protocol to be tested for synchronization between LT and IUT. The state of the IUT is assumed to be known from actions specified for the LT. Verdicts are formulated based on stimulus provided by the LT and responses of the IUT observed by the LT. The remote testing method is widely used for testing implementations of X.25 protocol, where the IUT is the Data Terminal Equipment(DTE) and the LT emulates the Data Communications Equipment(DCE). In our case, the IUT is the entire service provider.

Now, we transform the remote testing configuration and the INRES service specification into the structures which satisfy the conditions for applying our NRS approach.

According to the remote testing method, the whole system behaviors have to be controlled and observed at the SAP: ini. Thus, it is sufficient to generate test cases from the specified behaviors at this point. We can transform the configuration in Figure 4.1 into the similar interactive structure as shown in Figure 4.2 which is modelled by our NRS approach where Tester represents the environment, and both INRES Service Provider and the Responder user (called system under test) represents the process which interacts with the tester.

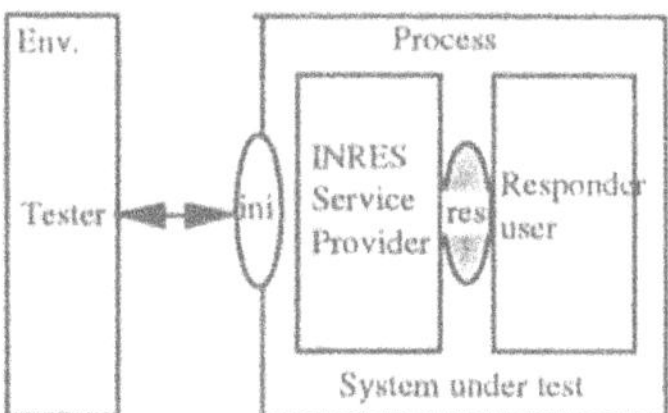

Figure 4.2 The process to be tested and its environment.

Next, we need the system behavioral specification at the point: ini, i.e., between the process and its environment. This is obtained by projecting the whole system specification into the SAP: ini, and is shown as the behavioral tree in Figure 4.3 which consists of all interactive actions controllable and observable at ini and time orders between them. The black and grey states represent the states in which the system has the same behaviors.

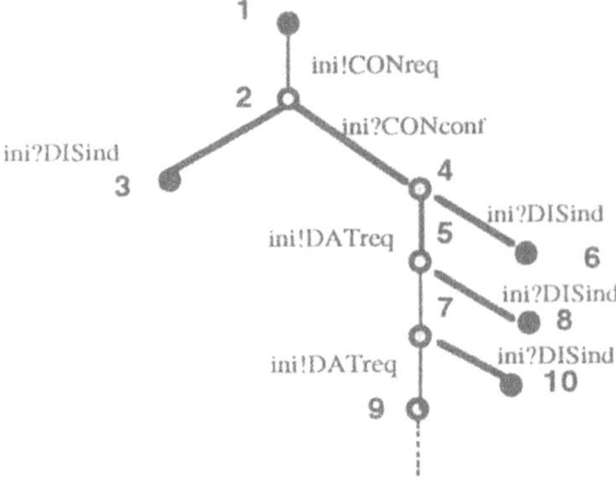

Figure 4.3 Behavioral tree at the SAP: ini.

Furthermore and importantly, we need to determine the nondeterministic rippling effect in the behavioral tree of Figure 4.3 because what is described in the behavioral tree still reflect the system overall behaviors at the interaction point, and the nondeterministic factors in the process to its environment is not specified. This usually depends on the specific configuration(i.e. testing method used) and many practical conditions which may not be described in a formal theory. Note that this is also the normal case when a formal theory is applied to practical situations, many practical concerns must be handled, and often, the detailed realization can only be informally dealt with. For example, in this case, the behavioral tree derived is not like the abstract ones where the initiators of actions are not distinguished. Here, for an action, we now understand that it is the combination of both its initiator and its action symbol. As such, !.a and ?.b are examples of actions.

We decide the nondeterminism in the process according to the usual sense in the context of interactions between two communicating entities: two branches emitting at a state are nondeterministic if the actions(prefixed with ?) labelling these branches are the potential answers to the action(prefixed with !) labelling the branch leading to this state. A similar method is used in (Phalippou). According to this definition and the conditions in our testing configuration, three cases in a behavioral tree should be considered as shown in Figure 4.4.

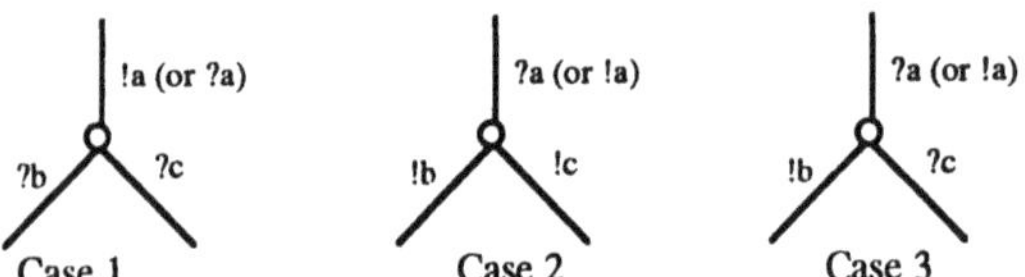

Figure 4.4 Nondeterministic factors in a behavioral tree.

In Case 1 of Figure 4.4, the two branches labeled with ?b and ?c are nondeterministic because ?b and ?c are responses from the responder entity to action !a. Note that the action ?a is possible before ?b and ?c and it results in two consecutive received actions. This situation(also two consecutive sending actions prefixed with ! such as in case 2) is called *unsynchronized interaction* in protocol design and is undesirable (Sarikaya, 1984). The two branches in Case 2 which are labelled by the two actions initiated from the initiator are deterministic because which one of them can be sent is under the control of the initiator(i.e., environment). Finally, Case 3 describes a situation where after receiving action a, the initiator wants to send action b, but unexpectedly, action c may arrive. In this situation, whether to accept c or send b first is under the control of the initiator, if the medium uses two channels for output and input messages, which is our assumption in this configuration. The two branches labeled by !b and ?c can be deterministic. However, if action c contains a critical message such as disconnection indication as in the behavioral tree of Figure 4.3. Then, whenever in such a state, the initiator(environment) has to check for its arrival and accept it first before doing anything else. Here, the action ?c should not happen is a pre-condition of !b. This argument concerns selection of a *choice pattern* in a test case derivation.

For the behavioral tree in Figure 4.3, by the method discussed above, we can decide that the relation between branches at state 2 labeled with ?DISind and ?CONconf are nondeterministic, and the relation between branches at state 4, 5, and 8 labeled with !DATreq and ?DISind is deterministic. *Note that this behavioral tree has infinite behaviours, that is, action ?DATreq can repeatedly appear, and after action !Disind, the tree would repeat the same behaviour as at root state r. In this case, as discussed in introduction section, some proper exit point should be made in test generation.* We can decide that the nondeterministic rippling at each state is,

(*) R(r, !CONreq) = {2}, R(2, ?CONconf) = {3, 4}, R(2, ?DISind) = {3,4 },
R(4, !DATreq*) = {5}, R(4, ?DISind) = {6}, R(5, !DATreq*) = {7},
R(5, ?DISind) = {8}, R(7, ?DISind) = {10}, ...

Note that we replace action !DATreq with action !DATreq* which is similar to !DATreq, but with an additional check for action ?DISind's arrival before doing !DATreq. Formally, DATreq* is understood as an atomic action: "if not (?DISind) then !DATreq".

Now, we are in a position to generate the tests for the INRES service under the remote testing configuration. We first determine test purposes. As usual, test purposes include:

1) check for connection establishment.
2) check for data transfer of one data unit.
3) check for connection release

Next, choice patterns determined by these test purposes are, (*note that an exit point is made at the completion of each test purpose*)

c1 = {(r, !CONreq), (2, ?CONconf), (4, ?DISind)},
c2 = {(r, !CONreq), (2, ?CONconf), (4, !DATreq*)}
c3 = {(r, !CONreq*), (2, ?CONconf), (4, !DATreq*), (5, ?DISind)}

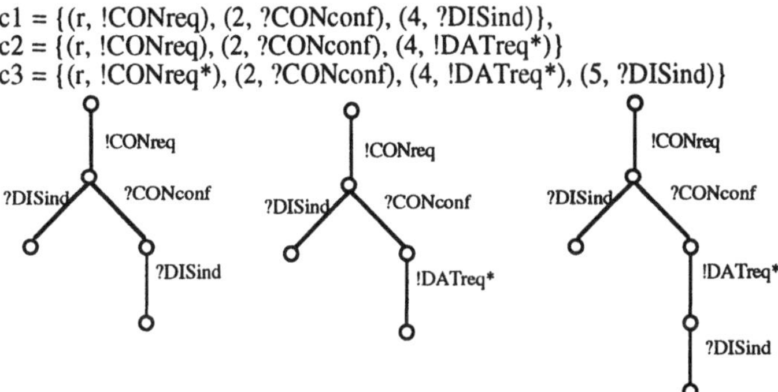

Figure 4.5 Three finite elements in NRT(ini) determined by choice patterns c1- c3.

Now, we can generate the finite elements in NRT(ini) based on the choice patterns c1 - c3. *Note that nondeterministic rippling in process ini is based on ones in (*)* (not the ones as in the ideal case that branches are labelled by identical actions). The elements in Figure 4.5 are finite deterministic tree representations of nondeterministic ripple sets for process *ini*. If used as test cases, they can be readily specified by formal description techniques such as Lotos. The deterministic tree representations of NRS can *also* be easily specified in practical situations by TTCN, the international standard test specification notation (Probert, 1992). In TTCN-like notation, the three elements in Figure 4.5 can be described as test cases with verdicts added as in tables 1 - 3:

Test case 1. Test purpose: check for connection establishment.		
Dynamic behavior tree:	!CONreq.	
	?CONconf	Pass
	?DISind	Inconclusive
	?DISind	

Table 1 Test case 1 for test purpose 1

Here, for assigning a verdict to a test, we only consider the path which completes the function of the test purpose as Pass (success), if it is executed. For other paths, any verdicts such as Pass, Fail or Inconclusive may be assigned depending on the practical considerations.

Test case 2. Test purpose: check for data transfer of one data unit.		
Dynamic behavior tree	!CONreq.	
	?DISind	Inconclusive
	?CONconf	
	!DATreq	Pass

Table 2 Test case 2 for test purpose 2

Test case 3: Test purpose: check for connection release		
Dynamic behavior tree	!CONreq.	
	?DISind	Inconclusive
	?CONconf	
	!DATreq*	
	?DISind	Pass

Table 3 Test case 3 for test purpose 3

These three tests are the main representatives in the test suite for INRES service. Other tests actually consist of one of these three tests with some repetitions of its subpart such as sending more data units or disconnection after sending more data units.

In the above discussion, we have outlined a method of applying our test-purpose directed semantics to test generation in conformance testing for a formal Lotos specification of INRES service protocol. We may remark that the method is also applicable to FSM-based formal specification techniques such as Estelle and SDL. We can also combine the D-method under the same framework of this test generation methodology (Chan, 1989; Sidhu, 1989). D-method is a commonly-used test generation method for protocols specified by FSM-based specification techniques. This can be illustrated by using the same example above, if we directly start with the behavioral tree of the service shown in Figure 4.3. By combining the states of the same behaviors together in Figure 4.3, a finite state machine can be formed as the FSM-based behavioural specification of the INRES service at SAP *ini* (similar Estelle and SDL specifications of INRES service can be found in (Hogrefe)). Then, the same procedure as described above can be followed for test generation. If the *distinguishing sequence* for each state is available, it can be used to verify the state following the transition to be tested as practised in D-method (Sarikaya, 1984). Accordingly, the dynamic behavioral tree for test purpose 1 is as in Table 4:

Preamble !CONreq.		
Test-body		
	?DISind	
	?CONconf	(Pass)
Post-amble	+Check(state r)	
	+Reset	Pass

Table 4 The test case with preamble and postamble for test purpose 1.

The test above has a similar format to a real test such as in the test suite for X.25 protocol (ISO/IEC 8882-3; Probert, 1992).

In summary, this section outlined a methodology for test generation in conformance testing by applying our NRS semantic approach to the INRES service. The result obtained is close to the ones obtained by informal and manual methods, and shows support for the use of formal semantics as a disciplined means of test generation and test specification in practical situations.

5 CONTRIBUTIONS OF NRS APPROACH TO INTEGRATING THEORY AND PRACTICE OF TESTING

As mentioned in the introduction section, This NRS semantic approach comes from our experience in observing process behaviours in testing telecommunications systems. The approach considers both contributing factors to a process behaviour, nondeterminism inside a process and the environment control determined by a test purpose, and thinks that a process behaviour is the result of the mutual influences of these two factors. Different from other semantic modelling methods, by considering the role of the environment control in a process behaviour, the NRS semantic approach brings a new dimension into the formal modelling techniques of process behaviours in a distributed system. This makes possible to handle the practical notion of "*test purpose*" within a formal algebraic context. Furthermore, by describing a process behaviour to be the result of the mutual influences of the two factors, the NRS semantic view appears to be natural for modelling interactive behaviours of processes in testing. These considerations may give rise to some *significant* effect in applying a formal theory to conformance testing of telecommunication systems.

In the following, we discuss a major contribution of the NRS approach to the field of conformance testing of telecommunications systems. We first briefly discuss the current state of conformance testing, and point out a gap between theoretical studies and practical applications in this field. Then, we discuss the special contributions of NRS semantics to bridging this gap.

•*A gap between theory and practice of conformance testing*

In recent years, theory and practice on conformance testing have been actively carried out in both academic and industrial worlds. In theoretical aspect, the representative testing models: bisimulation testing (Abramsky, 1988; Milner, 1980), failure/testing equivalence (Brookes; Hennessy, 1988; De Nicola, TCS) have been developed. Many other theories also exist but they mainly follow the formats of these models. The main advantage of these models is that they are precise, mathematically-based theories and possess well-defined theorems for discussing process properties. But, in considering applications of these theories to practical testing, some unfavorable features may be observed as.

1. Tests in these models are terms of some algebra. The set of tests appears identical for every process without discrimination, and thus, even for trivial processes, the tests considered can be infinite. This seems less practical.

2. There are no direct relations between tests and requirements-directed test purposes. This point is extremely relevant to practical testing. Thus, the present formal testing models still need to be improved to be more practical for their applications in realistic situations.

In the practical world of testing, the situation is different. At present, aspects 1. and 2. listed above are practised in realistic testing as follows:

(i) Tests are derived from the specific process(specification) whose implementation is under test.

(ii) Each test is derived directly from a test purpose.

In general, the practical testing techniques are ad hoc and informal. Test generation and selection are mainly manual which is tedious and error-prone. Furthermore, most practical techniques deal with only deterministic processes.

Most systems are specified in natural languages or informal notations. However, formal specification and testing techniques are gradually becoming of interest the industry world. Current testing techniques are not adequate for test generation and verification. Therefore, applying formal methods to testing is important.

•NRS contributions to bridging the gap in theory and practice of testing

The above discussion points out a gap between theoretical studies and practical applications in conformance testing, that is, the existing testing theory is formal but less practical, and practical testing techniques are practical but informal. Two *crucial* points of testing are handled differently in the two sides. For a formal testing theory to be practically applicable, its processing of these two points has to be matched with realistic testing techniques, that is, the gap between theory and practice of conformance testing has to be bridged.

The NRS approach is considered having made some contributions to bridging this gap. How the theoretical and practical differences of conformance testing are harmonized in the NRS semantics is indicated as follows:

(1-i) NRS approach contributes to harmonizing the differences in test generation between theory and practice. That is, the testing framework derived from NRS approach generates and selects tests from a specification if an implementation of the specification is under test. Furthermore, the three important aspects in testing as mentioned in the introduction section have been precisely modelled in the approach. This renders the number of tests generated from this approach to be in the comparable level required in realistic situation for test selection. This is illustrated as in Figure 3.2 and the case study in section 4.

(2-ii) NRS semantics contributes to solving the problem that a test in a testing theory should be related to a test purpose. In NRS semantics, a nondeterministic ripple set is derived from a choice pattern which is directly determined by a test purpose. This allows that in a formal context, each semantic object has a corresponding test purpose attached. As a whole, the NRS(P) for a process P characterizes all test purposes for testing the process P in the sense that each test purpose is to test a transition or a path of the process.

In addition, NRS approach shows a direct application of a semantic theory to testing practice. The tests derived in NRS semantics can be directly mapped to TTCN (Probert, 1992) or specified by Lotos (Brinksma, 1987; Brinksma, 1989), since each deterministic tree representation of the elements in NRS set can be actually seen as a dynamic behavioral tree of the implementation under test. This is illustrated by the example in section 3.

6 CONCLUSION

We have defined a new behavioral view of processes based on mutual influences between nondeterminism inside a process and the environment control determined by a test purpose. Studies of this semantic view is in the similar range to other well-known semantic views such as bisimulation (Milner, 1980) and failure/testing semantics (Brookes; De Nicola, TCS; Olderog, 1983). Compared with other semantic views, the special contribution of our semantic view (and also the main difference from others) is its consideration of both contributing factors to a process behavior. Other semantic views seem focusing more on the nondeterminsitic side of a process.

By considering the role of the environment control in a process behaviour, this method introduces a new dimention into the semantic modelling techniques. As such, it appears that our semantic approach is closer to a practical model of process behaviors in a testing situation. This is why some attributes of our NRS approach do not appear in other semantic theories.

As discussed in sections 3 and 4, application of this semantic method to practical testing is promising: (1) the number of semantic items (i.e., deterministic trees in NRT(P) produced from each process P is similar to those generated in practice; (2) each deterministic tree is already in a format similar to the graphical representation of Lotos specifications or the standard test specification in TTCN, and supports direct transformation into a real test case described by these formal description techniques; (3) each tree has a test purpose attached. These are our observations based on our experence, and proof of practicality can only be done in practical applications of this approach to realistic examples of telecommunication system specification and testing.

More theoretical studies of this semantic approach have been done. As usual, the semantic object NRS(P) can induce an equivalence between processes. This equivalence has been shown to imply failure equivalence. In addition, an effective algorithm has been derived to generate *nrs* sets by calculus from CCS-like process notations without enumerating all choice patterns. Furthermore, this semantic view also helps performing trace analysis, and formalizing process relations in system design. Interested readers are referred to (Probert, 1995; Wei) for studies of these aspects.

Finally, we hope that this approach supports and improves on application of formal methods to specification and testing of telecommunication/distributed systems by addressing the practical issue of test purpose. However, compared with other well-studied semantics, much remains to be studied for this semantic approach. Presently, we are extending its theoretical aspects and formulating a testing theory in a realistic environment.

ACKNOWLEDGEMENTS

The authors are very grateful for useful discussions with Professor. L. Logrippo and other colleagues of the University of Ottawa. As well, the authors are grateful to Professor M. Hennessy for many comments and suggestions. This research was supported by a grant from the Telecommunications Research Institute of Ontario.

REFERENCES

Abramsky S. (1988) Observational equivalence as a testing equivalence, *Theoretical. Computer Science* 53, 225-241.

Bolognesi T, Caneve M. (1989) Equivalence verification: theory, algorithms and a tool, *The Formal Description Techniques Lotos*, Elsevier Science Publishers B. V.

Brinksma, H. (1987) An introduction to Lotos, in *Protocol Specification, Testing, and Verification VII*, North-Holland.

Brinksma H. (1989) A theory for derivation of tests, *The Formal Description Techniques Lotos*, Elsevier Science Publishers B. V.

Brookes S., Hoare C. and Roscoe A. *A theory of communicating sequential processes*, JACM, 31, No.7, 560-599.

ISO/IS 9646, *Conformance Testing Methodology and Framework*, Part 1-Part 6.

Chan W.M., Vuong S.T., Ito M.R. (1989) The UIOv-method for protocol test sequence generation, *Proc. 2nd Intl. Workshop on Protocol Test Systems*, Berlin, Germany.

Cleaveland R. and Hennessy M. (1993) Testing equivalence as a bisimulation equivalence, *Formal Aspects of Computing* 5, 1-20.

ISO/IEC JTC1/SC21/P.54 and CCITT SG X/Q.10 (1993) *Formal methods in conformance testing*, (working Draft for review and comments.

Luo, G. Bochmann, G., Das, A. and Wu, C. (1992) Failure-equivalent transformation of

transition systems to avoid internal actions, *Information processing letters* 44, 333-343.
Guillemot, R. and Logrippo, L. (1989) Derivation of useful execution trees from Lotos specifications by using an interpreter, *FORTE'89*, 311-320.
Hennessy M. (1985) Acceptance trees, *JACM* 32, 896-928.
Hennessy M. (1988) *Algebraic theory of processes* ,The MIT Press.
Hoare C.A.R. (1985) *Communicating Sequential Processes*, Prentice-Hall.
Hogrefe, D. *OSI formal specification case study: the Inres protocol and service, revised*, Institut fur Informatik, Universitat Bern.
Hopcroft J.E., Ullman J.D. (1979) *Introduction to automata theory, languages and computation*, Addison-Wesley.
ISO/IEC 8882-3 *X.25 DTE Conformance testing-Part 3 Packet Level confoemante test suite.*
Langerak R. (1989) A testing theory for Lotos using deadlock detection, 9th IFIP WG 6.1 Intel Symposium on *Protocol Specification,Testing, and Verification*, The Netherlands.
Milner R. (1980) *A Calculus of Communicating System*, LNCS, Vol 92, Springer Verlag.
Milner R. (1988) *Operational and algebraic semantics of concurrent processes*, ECS-LFCS-88-46, Department of Computer Science University of Edinburgh.
De Nicola R. and Hennessy M. *Testing equivalences for processes*, TCS.
Olderog E. and Hoare C. (1983) Specification-oriented semantics for communicating processes, *Acta Informatica*, Vol.23, 9-66, 1986.34, 83-133.
Phalippou M. Test suite generation for the Inres protocol with TVEDA tool, *CNET/LAA/SLC/EVP*, BP. 40 F-22301 Lannion CEDEX France.
Probert, R., Monkewich O. (1992) TTCN - The international notation for specifying tests of communication systems, *Computer Networks and ISDN Systems*, Vol. 23, No. 5, 417-438.
Probert R., and Wei L. (1995) Testing Behaviours of Processes Based On Nondeterministic Rippling and Environment Control, The 7th International Conference on computing and information, ICCI'95, Trent University Peterborough, Ontario, Canada.
Sarikaya B. and Bochmann G. (1984) Synchronization and specification issues in protocol testing, *IEEE Transactions on Comm.unications*, Vol. 32, No. 4.
Sidhu, D. and Leung T. K. (1989) Formal methods for protocol testing: a detailed study, *IEEE Transactions of Software Engineering*, Vol 15, 413-426.
Velthuys R. J., Schneider J.M. and Zorntlein G. (1992) A test derivation method based on exploiting structure information, Proc. 12th IFIP *Protocol specification, testing, and verification.*
Wezeman C.D. (1989) The CO-OP method for Compositional derivation of conformance testers, Proc. 9th IFIP on *Protocol Specification, Testing, and Verification.*
Wei L., *Semantics of specification and testing of distributed systems*, Ph.D. Thesis, Department of Computer Science University of Ottawa, Canada, December, 1994.

BIOGRAPHY

Robert L. Probert is Professor of Computer Science and Coordinator of the Telecommunications Software Engineering Research Group at the University of Ottawa. He was the founding Chair of the ACM Principles of Distributed Computing Conformance, co-chair of the 10th IFIP Symposium on Protocols Specification and Testing, and Scientific Leader for the team that developed the TTCN Workbench, a comprehensive environment for designing test suites for interactive systems. Probert has worked with industry in the area of Software Quality Engineering. His current interests include grey-box analysis, a use-case driven technique for enhancing the quality of software designs.
Linsheng Wei received his doctorate (Computer Science) from the University of Ottawa in the Spring 1995. He is currently employed at Bell-Northern Research Ltd. in Ottawa.

INDEX OF CONTRIBUTORS

KEYWORD INDEX

GPSR Compliance
The European Union's (EU) General Product Safety Regulation (GPSR) is a set of rules that requires consumer products to be safe and our obligations to ensure this.

If you have any concerns about our products, you can contact us on

ProductSafety@springernature.com

In case Publisher is established outside the EU, the EU authorized representative is:

Springer Nature Customer Service Center GmbH
Europaplatz 3
69115 Heidelberg, Germany

www.ingramcontent.com/pod-product-compliance
Ingram Content Group UK Ltd.
Pitfield, Milton Keynes, MK11 3LW, UK
UKHW022322190726
13856UKWH00001B/159

* 9 7 8 1 4 7 5 7 6 3 1 1 9 *